Metal Toxicity in Agriculture Crops:
Emerging Trends

Dr. Prasann Kumar, Assistant Professor cum Research Intensive Faculty in the Department of Agronomy, School of Agricultural (ICAR-Accredited), Lovely Professional University, Punjab, specializes in stress physiology, plant tissue culture, physiology of plant-microbe/ pathogen interaction, and biodiversity management. Dr. Kumar has completed the International Agricultural Rural Development Programme from Cornell University, Ithaca, New York, USA. A doctorate from Banaras Hindu University, Varanasi. He has teaching and research experience of over 03 years and 03 years, respectively. He has to his credit around 300 publications including research papers in national and international journals, book chapters, abstracts, etc. He also has authored 40 books published by national and international publishers, which find a place in national as well as various Overseas Institutions' libraries. He has guided 19 M.Sc. students for their thesis. Currently, 17 masters and 02 Ph. D. (Full Time) students taking guidance under the supervision of Dr. Kumar. He has completed research projects (LPU/REG/EC/190320/002/01/07) funded by Punjab Biodiversity Board, Punjab, India. Besides, Dr. Kumar has received the seed money (Rs.100000) LPU, to conduct the initial trail of research on different aspects of heavy metal research.

Ms. Shipa Rani Dey is a Research Scholar in the Department of Agronomy, School of Agricultural, Lovely Professional University, Punjab, where she is working on heavy metal toxicity research under the guidance of Dr. Prasann Kumar. She has over fifty publications to her name, including a research paper, review articles, book chapters, popular articles, and abstracts of International and National Reputation. She has attended and presented her research papers at national and international conferences/seminars/symposiums.

Metal Toxicity in Agriculture Crops: Emerging Trends

Dr. Prasann Kumar
Ms. Shipa Rani Dey
Department of Agronomy
School of Agriculture
Lovely Professional University,
Punjab-144411, India

Published by
SCIENTIFIC PUBLISHERS (INDIA)
5 A, New Pali Road, P.O. Box 91
Jodhpur 342 001 (INDIA)
E-mail: info@scientificpub.com
Website: https://www.scientificpubonline.com

ISBN: 978-93-89832-97-6
ISBN: 978-93-89832-99-0 (EBook)

Printed in India

Division of Research & Development

Foreword

I am happy to write foreword for this book titled "Emerging Trends in Heavy Metals Toxicity in Agricultural Crops" by Dr. Prasann Kumar and Ms. Shipa Rani Dey. I have gone through Dr. Prasann Kumar's research works and writings in the past two years. He is a passionate writer and always can be seen among students mentoring and handholding them.

This edited book is the need of the hour as the world is very much concerned on food quality and safety. Agricultural soils in many parts of the world are contaminated by heavy metal toxicity such as cadmium, copper, zinc, cobalt, nickel, chromium, lead, arsenic etc. There may be several reasons for the same. But it is a global concern that it affects the health of humans and animals. Great people across the world have come together to pen down their thoughts and recent advancements on heavy metal toxicity in agricultural crops.

I encourage not only the students and academicians, but most importantly the policy makers and industrial stakeholders to read and take appropriate action to save the planet and to leave behind a healthy generation in the days to come.

I appreciate all the authors of this book for sparing their time for a global cause. I also wish the editors of this book to continue doing great work.

Dr. G. Geetha
Professor and Additional Dean
Head
Division of Research and Development
Lovely Professional University
Phagwara-Pubjab 144411
India

List of Contributors

1. **Metals Homeostasis and Chelators**

Poonam Kumari, Department of Agronomy, School of Agriculture, Lovely Professional University, Jalandhar, Punjab, 144411

Email: poonamnaik8988@gmail.com

2. **Vacuole, Trichome and Hydropotes Sequestration**

Monika Sharma, Department of Agronomy, School of Agriculture, Lovely Professional University, Jalandhar, Punjab, India, 144411

Email: ms9375033@gmail.com

3. **Role of Growth Hormone in Mitigation of Heavy Metal Stress**

Monika Sharma, Department of Agronomy, School of Agriculture, Lovely Professional University, Jalandhar, Punjab, India, 144411

Email: ms9375033@gmail.com

4. **Use of Algae in Wastewater Treatments**

E Lokesh Goud, Department of Agronomy, School of Agriculture, Lovely Professional University, Jalandhar, Punjab, India,144411

Email: lokeshgoud1832@gmail.com

5. **Significance of Microbial Volatiles in Agro-Ecological Health**

E Lokesh Goud, Department of Agronomy, School of Agriculture, Lovely Professional University, Jalandhar, Punjab, India,144411

Email: lokeshgoud1832@gmail.com

6. **Avenue of Phytoremediation Strategies for Fluorine Toxicity**

Lalit Saini, Department of Agronomy, School of Agriculture, Lovely Professional University, Jalandhar, Punjab, 144411, India

Email: sainilalit994@gmail.com

7. **Phytohormones: Master Mediators in Abiotic Stress Tolerance in the Crop Plants**

Priyanka Devi, Department of Agronomy, School of Agriculture, Lovely Professional University, Jalandhar, Punjab, 144411, India

Email : parichib17@gmail.com

8. **Foliar Heavy Metal Uptake, Toxicity and Detoxification In Plants**

Priyanka Devi, Department of Agronomy, School of Agriculture, Lovely Professional University, Jalandhar, Punjab, 144411, India

Email : parichib17@gmail.com

Contents

Metals Homeostasis and Chelators

Poonam Kumari

Department of Agronomy, School of Agriculture
Lovely Professional University, Jalandhar, Punjab, 144411
Email: poonamnaik8988@gmail.com

Abstract

The plant faces many problems due to the accumulation of heavy metals in large quantities, and this will disturb the homeostasis of the ions. There should be detoxification of metals to reduce the effect of these metals or to maintain homeostasis between plants. Heavy metal chelation should be there to maintain homeostasis. Many transition elements such as Copper (Cu), Iron (Fe), Zinc (Zn), Manganese (Mn) and Nickel (Ni) etc. are important for plant physiological processes. However, at a higher dose, these metals are even toxic to plants. The homeostasis process minimizes damage to the different cell compartments and protects against such non-essential ions. Chaperons are proteins that absorb the heavy shock produced by metals and help in the unfolding and folding of macromolecule structures. Hyperaccumulators or Metallophores plants are mostly used for the accumulation of soil metals. The hyperaccumulator plant protects against many diseases, fungal attacks, etc. due to the presence of metals in it. In homeostasis, many transporters are used to help transport cells at the cellular level and to maintain the balance of cells internally.

Introduction

Various metals like Iron, Zinc, Copper, Manganese and Nickel etc. are used for the proper development of plants. These elements are required to the plants in some specific amount. Even the toxicity and deficiency of these elements causes the improper growth of plants. It causes many diseases due to the improper presence of these elements. So, in case of deficiency of these transition metals like Iron, Manganese etc. causes lack of chlorophyll even it causes inter venial chlorosis in plants. Also, the toxicity of these elements causes many disorders, which completely lead to a reduction in crop yield. Therefore, to get rid of these deficiencies, metal chelates which are mainly organic in nature and are used to fulfil the requirement of these micronutrients (Pilon et al., 2014). Chelates are the organic form of metals, which are provided to the plants for their good growth and development. They contain the positive charge, which is easily absorbed by the roots of the plants, and their translocation is present in a systematic way. Chelates are the form of micronutrients, which are adsorbed in the soil surface that is why it is not leached during the excess of irrigation and rainfall. They are organic in nature so is eco-friendly and does not pollute the environment. Homeostasis is a process in which during the environmental change, the internal structure or stability of plants is maintained. Homeostasis reduces the loss of water while in the opening and closing of stomata. It maintains the regulation of plant system. This helps in the maintenance of all physiological processes occurs in the plants (Mai et al., 2016). Different plants have different habitats and they acquire and adapt different habitats according to the changing environmental conditions and their locality. During homeostasis, the internal stability of the plants is maintained by the removal of unwanted substances from the plant system by respiration, photosynthesis etc. Inside the cell, the metals like zinc are present in the form of Zn^{2+} in its stable state while the other metals like Manganese (Mn), Iron (Fe) and Copper (Cu) are present in its redox state. These metals can also participate in the transfer of electrons during various reactions occurs inside the body of the cell. Many elements can help in the discard of the formation of radicles during catalysis process. The transition elements like Zinc and Copper can easily be displaced by the activity of Iron and Manganese that is why the deficiencies of these metals are present in a large amount. Various transporters are used in the process of homeostasis. All the transporters used in the process of homeostasis belong to the family of proteins. The plant species *Arabidopsis thaliana* is mostly used in the homeostasis process and the Fer like iron deficiency-induced transcription factor that is also known as the FIT is helpful in the uptake of iron through the roots of the plant species *Arabidopsis thaliana.* During the transportation of transition elements, they require energy and that energy is provided through the hydrolysis of ATP molecules, from which they get energy for transportation and that type of transportation, is known as active transport. However, there is another case through which transportation of trace metals is through antiport and symport mechanism, that mechanism of transportations is known as passive transport. The ATP molecules or Adenosine triphosphate is also known as energy molecules are hydrolysed with the help of protein motive force. In this review the main aim is to shortlisted the new findings in the topic homeostasis of various metal like Iron, Zinc, Manganese, Copper and Nickel ions as well as to study the new concepts of the mitochondrial, Golgi, Ribosomal and Vacuolar type of homeostasis occurs in the cell. Along with this, the process of chelation is also discussed in this chapter.

Metal Accumulators

Metal accumulators plants are also known as hyperaccumulator plants. The plants, which can absorb toxic metals or heavy metals from soil in some high concentrations, are known as hyperaccumulator plants. These plants accumulate heavy metals without causing any damage to that plant. The plant absorbs metals in various forms like with the uptake of water from the soil (Bashir et al., 2016). Mostly serpentine soils are used to cultivate these plant species. These soils contain very less amount of macronutrients and micronutrients. The hyperaccumulator plant species provide resistance from the fungal infections, diseases, pests etc. due to the presence of metals in the plant species. This type of plant species is also known as Metallophytes. Many plant families are used as hyperaccumulator species. Families like Fabaceae, Crassulaceae, Brassicaceae, Asteraceae, Caryophyllaceae etc. are used. Many plant species are used for a hyperaccumulation of metals. Among these *Arabidopsis thaliana, Brassica napus, Helianthus annus, Berkeya coddii, Alyssum murale, Leptoplas emarginata, Sebertia accuminata* etc. any many more are most common. (Boominathan et al., 2004).

Homeostasis

The process that maintains the internal stability of the plant during the changing environment is known as homeostasis. Homeostasis helps to control the opening and closing of the stomata from which there is a minimum loss of water. It maintains an adequate supply of water and nutrients from the soil to the leaves (Banaker et al., 2019). Homeostasis can be performed in plants using two processes, Osmoregulation and Excretion. In the Osmoregulation process, the appropriate amount of water and solutes is maintained in the body fluid for homeostasis. Plants are adapted to different habitats according to their purpose and adaptation. Habitats such as Hydrophytes, Mesophytes and Xerophytes decide the amount of water the plant needs. In the process of excretion, unwanted substances or materials are removed from the plant body by falling leaves, photosynthesis and respiration processes (Clemens, 2001).

Cellular Metal Homeostasis

Vacuoles act as a reservoir for making a balance in the cell. Metals are present in the soil from where plant accumulates and transfer it to various cell organelles like cytoplasm, mitochondria, chloroplast etc. Various membranous proteins are used to transfer metals into the cell (Andreson et al., 2018). These metals are somehow toxic for the plant. So, for this many chelating agents are in work like Ascorbate, Phenols, Nicotianamine, Citrate and Deoxymugineic acids. These agents protect the cell from oxidative damage and help to solubilise the metals. Many transporter-like Vacuolar Iron transporter families (VIT), Yellow Stripe Like family (YSL), cation diffusion facilitator family (CDF), Cation Exchange Family (CAX), Zinc regulated transporter family (ZRT), Natural resistance association macrophage protein family (NRAMP), Iron regulated transporter protein family (IRT), Zinc Iron regulated protein family (ZIP), Heavy metal ATPase family (HMA) plays the main role in metal homeostasis at the cellular level. These transporter families help in maintaining the internal balance of the cells (Bashir et al., 2016).

Chloroplast Metal Homeostasis

For the synthesis of chlorophyll, the occurrence of the process of photosynthesis, its stabilization, etc., helps in all these processes as well as in the maintenance of homeostasis in chloroplasts. Chloroplast helps in the storage of Iron (Fe) or acts as a storage house for Iron. As a result, ferritin protein is present in the chloroplast (Cui et al., 2017). Elements such as Copper (Cu), Iron (Fe) and Manganese (Mn) help in the process of photosynthesis and are mainly present in the early stages of plant growth. Hyperaccumulator species such as Arabidopsis ferric chelate reduce ferric ions to ferrous ions, which is important for Iron (Fe) homeostasis (Zhang et al., 2019). The synthesis of chlorophyll, the occurrence of photosynthesis, its stabilization, etc. helps in all these processes as well as in the maintenance of homeostasis in chloroplasts. C1 (Permease in chloroplast 1) plays an important role in the development of chloroplasts and the transport of iron (Fe) to chloroplasts. The reduction and increase in permease in chloroplast 1 directly affect the toxicity and deficiency of Iron. AtMfL1 (Arabidopsis Mitoferrin Like 1) also helps to transport Iron to the chloroplast. During the release of Chloroplast Fe-NA, the Arabidopsis YSL-4 and YSL-6 present in chloroplast help during the process of embryogenesis (Wang et al., 2019). The transportation of Copper (Cu) in the chloroplast is mainly due to the HMA6 which is also known as PAA1 or PIB- type ATPase. It also helps in photosynthesis. Besides this PAA2 or HMA8, transport copper to the thylakoid membrane of the chloroplast. Manganese (Mn) like Iron and Copper also acts in photosynthesis. HMA1 maintains the homeostasis of Zn in the chloroplast that's why its transport is somehow controversial (Pilon et al., 2009). The transporter of chloroplast helps in to regulate the metal homeostasis, which leads to the improvement of crops. This will improve the photosynthetic rate, provide better adaptability to the crop stand, metabolism of cell also improves and provide resistance from various stress conditions in the altering environmental conditions (Banaker et al., 2016).

Vacuole Metal Homeostasis

Most metal homeostasis occurs in vacuoles. Most of the metals are present in a very large quantity and are stored in vacuoles. Many inflows and efflux transporters are connected to the vacuole membrane. Cytoplasmic Iron is stored in vacuoles with Arabidopsis thaliana FPN2, also known as IREG2 (Clemens, 2001). Vacuole transporters, such as VITI, help with the deposition of Iron and Manganese. Along with VITI, many other Fe transporters are also involved in the transport of cytoplasmic iron to vacuoles that help in the homeostasis of Mn, Zn and Fe. These proteins are VTL1, VTL2, and VTL5. Some vacuoles transporters, such as VIT1 and VIT2 in the rice plant, contribute to the homeostasis of Zinc, Manganese and Iron. These transporters are rich in rice plants and their alteration causes a change in flower colour due to the accumulation of metals in other plants (Andreson et al., 2018). Like the rice, the plant absorbs more manganese than the root portion in the leaves. The NRAMP3, NRAMP4 and ZIPI proteins help to remobilize vacuoles manganese in the Arabidopsis species. The COPT or Copper Transporter / Ctr family, such as COPT1, COPT2, COPT3 and COPT5, are helping to transport copper. COPT5 helps in the photosynthesis and transport of electrons in the chloroplast, thereby pumping vacuolar copper into the cytoplasm. Some proteins control gene behaviour to remove excess metals from Iron deficiency (Wang et al., 2019; Zhang et al., 2019).

Mitochondrial Metal Homeostasis

Mitochondria develop through the process of endosymbiosis, which has various functions, such as the production of lipoic and heme acid cofactors, together with the ISC (Fe-S) clusters. This ISC machine is important for other cells because Iron-sulfur proteins are mainly dependent on Sulfur compounds present in Mitochondria (Banaker et al., 2019). From mitochondria to cytoplasm STA-1 or Arabidopsis ATM3, the trisulphide compounds of Glutathione are exported and cause transport restrictions that affect Molybdenum co-factors and Iron sulphur co-factor assembly and lead to plant chlorosis. The additional importance of the Iron-Sulfur Cluster Machinery is that the metal is readily available to the plant through Glutathione. It also tolerates iron deficiency in Arabidopsis (Boominathan et al., 2004). Mitochondria have different functions in plants, rather than producing energy for the cells, and it also regulates certain signals that are useful in controlling the Iron deficiency. Since mitochondria are known as the cell powerhouse, changes in cells due to mitochondrial activity also cause changes in other cell organelles, such as changes in gene expression or photosynthesis, etc. Mitochondria is the central part of the redox regulation of cells and provides a metabolic pathway for energy conservation as well. Only a very small amount of protein is used to transport metals to plant mitochondria (Clemens 2001). Some mitochondrial proteins, such as Arabidopsis FRO3 and FRO8, can actively participate in the reduction of ferric acid and can help in the transport of mitochondrial iron. Mitochondrial iron transporters are present in rice, which are important for the proper growth and development of the plant (Cui et al., 2017). These proteins (MITs) transport the Iron from the cytoplasm to the mitochondria. Some mutants like (mit-1) or MIT knockout mutants in rice are lethal and the other mutants (mit-2) reduce the chlorophyll content of the plant, which leads to poor plant growth. Due to the presence of this mutant, there is less Iron accumulation in mitochondria, which causes Iron deficiency in plants (Pilon et al., 2009).

Golgi and Vesicles Cellular Metal Homeostasis

The distribution of metals in vesicles and the Golgi complex is also important for the proper functioning of the cell, and many proteins are attached to these cells. An appropriate amount of iron is present in the nucleus. In the nucleus, the movement of metals is easy enough because of the large pore size of the nucleus, so many metals like Iron, Copper, Manganese and Zinc etc. can move easily in the nucleus (Zhang et al., 2019). Due to the high presence of toxic elements, the nucleus is toxic. In the hyperaccumulator plant species Arabidopsis, the IRT2 protein makes a pathway between the vesicles and the root epidermal cells and causes Iron deficiency. Arabidopsis MTP11 helps in the transport of manganese (Mn) and provides resistance or tolerance from manganese to the internal vesicular organelles and the Golgi complex (Wang et al., 2019). For transport of Mugineic acid and Iron, the YSL5 barley is attached to the membrane of the vesicles. In rice, OsYSL6 acts as a detoxifying agent and removes the excess amount of manganese, while YSL6 in rice is sensitive to the high concentration of manganese. MTP8.1 and MTP8.2 are the two Manganese efflux which transports the Golgi localized barley elements. In MTP8.1, manganese deficiency and toxicity increase and the expression of MTP8.2 decrease for manganese deficiency. The expression of both leaves decreases if there is an excess of Manganese (Bashir et al., 2016; Mai et al., 2016).

METAL HOMEOSTASIS

Iron Homeostasis

Iron is one of the main micronutrients needed for proper plant growth and development. Iron helps to maintain various cellular processes in plants. Iron acts as a co-factor for the synthesis of many enzymes and helps in the respiration of mitochondria, maintains protein structure and helps to provide proper photosynthesis, etc. Iron plays an important role in the homeostasis of metals (Connorton et al., 2017). Iron is present in very large quantities in soil, but the plant does not absorb enough iron through its roots mainly in Calcareous soils due to the lower solubility of iron, which is why the plant shows symptoms of deficiency due to low iron solubility. This deficiency leads to a reduction in yields. Crops or plants deficient in Iron because various disorders, such as anaemia, in humans when consumed. As a result, by improving the physiological and molecular mechanism of the plant, it will be able to adopt an Iron deficiency (Forester et al., 2017). Soil pH, aeration, microbial activity, presence of organic matter and redox potential are the same factors, which affect the availability of iron in the soil. Iron is the micronutrient, which takes place in many processes for maintaining the solubility and stability of proteins, chemical reactions going in the plant body, assimilating the level of nitrogen and the regulation of hormones. The deficiency of iron leads to the chlorosis between the young leaves of plants due to which the rate of chlorophyll reduces. In this, the veins remain green and the whole leaf turns pale yellow in colour and ultimately reduces the yield of crops (Cui et al., 2017). Under the conditions of iron deficiency, iron immobilization is one of the important aspects required for proper growth and development of plants. During the transportation through the cellular membrane, the ferric form is directly reduced to the ferrous form. The stored iron in vacuoles is converted into ferritin, which is necessary for various metabolic reactions going in the plant body. Vacuole stores the high amount of iron that is 50% of the total iron present in Arabidopsis seeds (Baker et al., 2016). The iron transporters in mitochondria are the proteins of Mitochondria carrier family. These proteins are present in the inner membrane of mitochondria from which they transport many acids and solutes as well. The gene AtMfl1 or Mitoferrin like 1 present in *Arabidopsis thaliana* helps in the transport of iron (Fe) in the cell mitochondria. Along with this, the gene Mitoferrin like 1 also helps in the trafficking of iron in the chloroplast of the cell. The mitochondrial ion transporter gene or MIT gene reduces the chlorophyll content, which ultimately reduces the growth rate of the plant. This gene also causes many alterations in the location of iron within the cells (Mai et al., 2016). In vacuole, the important place for iron storage is Globoids. In the pea plant, embryo stores 92% of total iron ferritin amount. Therefore, there are different ways and different places of Iron storage. In aerated soils, Iron is not available properly and not easily, the plants because in these types of soils iron is present in trivalent form, due to which it do not absorb uptake by the plants. A little bit of iron is only absorbed by the plants. Trivalent form of iron easily precipitates and forms oxides and hydroxides of iron, which are poorly soluble by the plants. Iron can be accessed by two strategies, which are physiological, by nature (Robert et al., 2015). These strategies depend on reduction and chelation of Iron that is Strategy I and Strategy II. The Strategy I is the condition used for the reduction of Fe (III) and the Strategy II Fe (III) is the condition on chelation. In the Strategy I (Fe III) reduction, there is P-type ATPase, which helps in acidification of rhizosphere, and Apoplast of roots. In acidification, solubility of 1000 in a factor is increased but the increase in the level of pH at one time. Phenolic compounds are used in soil solution for the chelation of ferrous ions. These phenolic compounds are mainly produced by the root portion of the plants and by the catechol type Coumarins. Coumarins are used to reduce the amount of apoplast (Andreson et al., 2018). After the proper reduction of these coumarins, the Fe^{2+} ions are transferred through the plasma membrane of the cells with the help of proteins present in the ZIP family, which act as ion transporters. During the development of the seeds, the iron should be dissolved from the mother tissue of the symplasm area and then again absorbed into the embryo of the plant. There is some space between the tissues that are filled with malate and citrate. Some transporters, such as the NRAMP, NRAMP3 and NRAMP4 families, are important for the establishment of seedlings in soils, which are iron deficient and calcareous (Forester et al., 2017). Various plant expressions are related to iron-associated genes in iron deficiency. There are many complications in the signal network for the adoption of iron deficiency. Many messenger molecules, plant hormones, are used in this process. In French bean, the hormone Auxin increases the reactivity of FCR (Ferric Chelate Reductase) in the roots. Abscisic acid increases the iron-deficient response, while cytokinin prevents this type of response in plant species of Arabidopsis (Connorton et al., 2017; Robert et al., 2015). The hormone Gibberellin may also increase the response of iron deficiency, and the response of Jasmonic acid may decrease. Abscisic acid regulates the iron efflux in the vacuole through AtNRAMP3 and increases the secretion of phenolic compounds. As well as abscisic acid, the translocation of Iron from the bottom parts of the plants to the upper parts of the plants is also improved. Cell wall contains negative charges that are useful for metal binding (Kramer et al., 2005). From the site of the apoplast, pectin begins secrets to the site of the apoplast. Demethylation of pectin occurs with pectin methylation, which also increases the negative charge of iron for the cell wall and the carboxylic group in the cell. As iron is an important element for the performance and maintenance of physiological processes in the cell. Many genes are used at transcriptional levels to reduce the iron deficiency and some post-transcriptional factors to adjust the iron condition deficiency (Andreson et al., 2018). First, in tomatoes, the FER or Iron Efficiency reactions were identified and encoded with the transcriptional factor bHLH. This reaction causes the plant to adapt to the iron deficiency or to make the plant resistant to iron deficiency, as well as to control the morphology and physiology of the roots. Transcriptional factors are the basic helix loop helix-like induced FIT transcriptional factor that regulates the expression FRO2 and IRT1 through acquisition, iron sensing and response in the species of Arabidopsis thaliana. EIN3 and EIN3-like 1 that is EIC1-the transcriptional ethylene factor that interacts with the FIT and activates the expression of the Iron transcriptional factor. The activity of the iron transcription factor is only started when it interacts with proteins (Forester et al., 2017).

Zinc Homeostasis

After Iron, zinc is the most common transition element found in the earth crust. It is used as a cofactor and has six classes of enzymes. Zinc is mostly present in 2+ biologically relevant oxidation state. Zinc is not much active but at a certain high level of zinc is toxic to the plant because the transition metal zinc changes the position of other metals in the cell. For the survival of cells even at a very small amount of zinc is essential (Pilon, 2009). Therefore, the availability of zinc is necessary for the cell even in a small amount. In the cellular level, the homeostasis of zinc should be maintained with the help of transporters mainly in the cell membrane and the membranes of vacuoles. In some plant species like *Arabidopsis thaliana,* the transporters of ZIP family helps in the uptake of Zinc (Clemens, 2001). Some transporters, such as AtMP1 and AtMP3, also allow zinc to move into vacuoles, which are cation diffusion family transporters. AtZIFI, a protein present in tonoplast, transports organic zinc ligands to vacuoles (Sinclair et al., 2012). Although Zn is not redox dynamic much higher levels of Zn are harmful because Zn can dislodge different metals in the cell. Dissimilar to other metals such as Fe, Cu and possibly Mn, there are no specific frameworks for such known Zn cofactors. Along these lines, proteins seem to get their Zn cofactors together without restraint. This should be allowed as a bioaccessible Zn pool to be kept in cells (Andreson et al., 2018). Strikingly, lower basal cell Zn levels may also be required for cell endurance as Zn levels have been found to intercede with apoptosis. Zn present in cellular homeostasis is maintained by all accounts in cellular and vacuolar film transporter exercises. In Arabidopsis thaliana and various plants, the ZIP family transporters allow the cell to be taken up. The CDF transporters AtMTP1 (root and shoot) and AtMTP3 (root) move Zn to the vacuoles (Haert, 1963). AtZIF1 is a protein present in the tonoplast, which may export Zn to the natural ligand into the vacuoles. AtMTP1, AtMTP3 and AtZIF1 all can help with the buffering of Zn levels in cells. AtHMA2 what is more, AtHMA4 transporters intervene by sending Zn out of the cell. AtHMA4, which is dynamic between the root vasculature, is particularly important in controlling the transport of the root to the shoot by allowing Zn to be sent out with the help of xylem in the root symplast. Zn is limited in nicotinamide and YSL family transporters can contribute through Zn and re-assignment to vegetative tissue (Sinclair et al., 2012). Some studies are going on AtHMA2, which showed the regulation of AtHMA2 and other transporters like P1B- type ATPase. There are two terminals present in zinc specific ATP dependent transporters of HMA that are the zinc-binding terminals which are also known as C terminal and another one is known as N terminal. If the N terminal binding site of metal is not present in the transporters still then the metal binds in the sight of translocation. Some hypothesis states that the N terminal present in transporters inhibits the ATPase and the activity of ATPase in the absence of N terminal, which is a relative metal-binding site (Andreson et al., 2018). A cytoplasmic loop is rich in histidine, which is present in the transporter AtMTP1. This loop is important for the binding of zinc ions together and the transportation of Zn^{2+} ions. This loop also helps to regulate the free zinc level of cytosols and to control the buffer, but this loop does not transport the Zn^{2+} ions to the vacuoles. Zinc homeostasis helps with transcription control. In Arabidopsis thaliana plant species, ZIP family transporters lead to an up-regulation of various elements, such as zinc and iron, which ultimately causes iron and zinc deficiency in Arabidopsis thaliana plant species (Kramer et al., 2005). While IRT1 and IRT2 members of the ZIP cause iron deficiency family, resulting in excess zinc uptake, while ZIF1 is used in excess zinc uptake in vacuolar transport (Heart et al., 1963).

Chelated zinc is present in the divalent form. Ferric ion has much stability constant than the metal chelates. When they are applied in the soil, the efficiency of metals is retraced in plants, which are growing in field conditions. That is why these chelates are applied mostly as a foliar application. The results formed from the foliar application are not much good as compared to the use of inorganic forms of metals. If there is an increase of metals in the tissue portion of the plant that can help in the remediation of the deficiency of metal. (Clemens, 2001). Mostly the plant shows deficiency symptoms of zinc at shoot portion of the plant when the concentration of zinc is below 15 to 20 mg Zn per kg of the dry mass of the plant. The chlorosis occurs by the deficiency of metals like Iron, which can be corrected, with the supply of chelates of Iron to the plants. Due to the application of metal chelates like Iron, there will be some change in the iron content of the plant, mainly in stem and leaf portion of the plant system. The ionic form of zinc is easily fixed in most of the soil but the anionic form of zinc is more mobile and is also chelated. The chelated zinc is more efficient, is mobile in nature, and helps the zinc to interact with the deep roots of the plants. The deep shallow-rooted plants having zinc deficiency is corrected with the application of chelated zinc (Pilon et al., 2009). Zinc is an element, which can easily bind with soil, forms precipitates, mainly in the form of hydroxides, phosphates, and carbonates. The solubilisation of Zinc is through acidification and the secretion of organic chelators having low molecular weight present in the rhizosphere. Among these, Iron (Fe III) is less soluble. The plants through the semi-permeable membrane of roots mainly absorb zinc. This process mostly occurs in grasses. The homeostasis of Iron and Zinc interacts with each other (Sinclair et al., 2012).

Nickel Homeostasis

Nickel is present in the soil through human intervention or naturally. Through natural activities, nickel can enter the soil through the weathering of parent material and human activities, primarily through anthropogenic activities such as industrial waste, coal mining, phytomining, fossil fuel combustion and vehicle intervention, municipal waste through the use of fertilizers and the smelting process (Bani et al., 2015). In the environment the nickel is added through some metallurgic processes, batteries etc. and this is the great source from where the nickel is added to the environment. Nickel is present in the concentration of 3.0 to

1000 mg kg^{-1}, but as in well-polluted soil, the concentration of nickel reaches up to 200 to 26000 mg Kg-1. Nickel is also found in many forms like Nickel hydroxide that is Ni $(OH)_2$, Nickel oxide that is NO and the Nickel trioxide that is Ni2O3 (Sheoran et al.,2009; Rosenkranz et al.,2019). All these are water-soluble. Due to these anthropogenic activities, there is an increase in the concentration of Nickel, which causes the toxicity of this element. Toxicity plant may suffer from many problems like Chlorosis of leaves mainly inter venial chlorosis, decrease in the production of biomass causes a change in the balance of nutrients and inhibits the formation of lateral roots with the increase in the amount of nickel there is a decrease in the amount of copper, zinc etc (Kramer et al.,2005). However, along with the number of iron increases with the increase for Nickel. Like other transition elements, Nickel is also one of the important nutrients for the proper growth of plant even in a small quantity. A very small concentration of Nickel is required for the normal growth of the plant. Along with this enzyme, Urease is needed for the development of the plant. Urease enzyme is required for the detoxification process used in the formation of urea due to the degradation of proteins. Nickel is also an important element, which is used for the following mechanisms of tolerance, phytoremediation and hyperaccumulation in plants. Many plant species like *Berkeya coddii, Alyssum murale, Alyssum bertolonni, Thlaspi caerulescens, Leptoplax emarginata, Sebertia accuminata, Alyssum markgraftii* and *Bornmuellera tymphaea* etc. are used for the accumulation of Nickel from these hyperaccumulator plant species (Tognacchini et al.,2020). Various processes are used for the extraction of these metals from plants like Phyto volatilization, Phytoextraction; Rhizo filtration, Compartmentalisation, Phyto stabilization and Phyto degradation are used. For the detoxification of Nickel, there is the induction of growth regulators like Ethylene, Jasmonic acids, Proteins like Histidine and proline, amino acids, phenolic compounds, acids like malic acids, magnetic acid, oxalic acid, citric acid etc. (Xin-zhang et al., 2014). The chelation process occurs in the cytosol portion of the cell and this portion is important for the tolerance and the detoxification of the metals. Various Phytochelatins, Metallothioneins etc. are used as metal chelators. They have very low molecular weight and metal-binding capacity as well as they have peptides which are rich in cysteine (Ravet et al., 2013). The phytochelatins are used in the detoxification and are synthesized from Glutathione. After that through the recycling process of urea, the nitrogen is utilised in the growth of plants. Along with this area is also formed from the deterioration of some acids like Allantoic and Ureides, which helps in the exportation of nitrogen from the nodules of legumes. Urease is the most common and first-ever enzyme that was crystallized (Sachan et al., 2019). On the active site of this enzyme, the Nickel element is bounded with the terminal water and on another end on the active site of the enzyme, two molecules of histidines proteins are attached with Nickel. First, the site, which binds with terminal water, is replaced with carboxyl oxygen. As the transporter of copper is found in Eukaryotes, likewise the transporter of nickel is found in Prokaryotes like Ni-SOD present in blue-green algae (Keeling et al., 2010). Ni performs an extraordinary, extra job in hyperaccumulator plants. The organic pertinence of hyperaccumulation contrasted and rejection based metal resilience is to pick up advantage over non-gatherer family members by providing resistance against herbivores, which was explored in different metal–plant-pathogen/ herbivore mixes (Rosenkranz et al., 2019). The accessibility of chelating specialists in the plant cell might be significant for metal particles maintenance in the cell-like the transport over the tonoplast and capacity present in vacuole the exchange to nearer cells is through plasmodesmata and the discharge into the apoplast for the stacking of xylem or phloem. The receptive associations between the metal particles and some ions like Sulphur, Nitrogen and Oxygen prepares the natural acids and amino acids which are filled in as potential ligands for the chelation of metals (Sheoran et al., 2009). The huge coordination of Nickel with the proteins histidine was uncovered in all-encompassing X-beam ingestion of the fine structure in the investigation of the *Alyssum lesbiacum* xylem sap. Plant species *Arabidopsis montanum* when provided with equimolar centralizations of exogenous analysis of the free histidine also known as His. Its role in xylem stacking in plant species *Arabidopsis lesbiacum* and *Brassica juncea* and saw that this protein His moved more Ni2+ to the top portion of shoots because of upgraded Ni^{2+} transition into the xylem and demonstrated resilience to applied Ni fixations (Tognacchini et al., 2020). The information is steady with a model where Ni^{2+} take-up is autonomous to His protein take-up and the upgraded arrival of Ni into the xylem is related to the simultaneous arrival of Histidine protein from an expanded root free His pool. The useful job of high levels of histidine proteins has been appeared in transgenic hyperaccumulator plant species Arabidopsis thaliana communicating a *Salmonella typhimurium* in ATP phosphoribosyl transferase protein that is StHisG (Bani et al., 2015). The bacterial catalyst is heartless toward criticism hindrance by the histidine protein. Transgenic plants gathered around two overlays of histidine protein in higher level in their leaf tissues than the plants which are growing in wild habitats and indicated more than ten times more biomass creation within the sight of harmful Ni^{2+} in the development which provides a demonstration in between the histamine protein and the Ni^{2+} ions tolerance (Xin Zhang et al., 2014).

Manganese Homeostasis

After iron, manganese is the second most important metal present in the earth crust. Manganese is present in Mn^{2+} form in the soil solution and is divalent in nature. The concentration of Manganese present in soil solution is about 0.1 to 800 u M. If there is an increase in the one unit of pH then it causes a decrease in the concentration of Manganese. Mainly, the deficiency of Manganese is founded in the alkaline soils whereas in acid soils there is the toxicity of Manganese (Andreson et al., 2018). The decrease in the concentration of redox potential there is an increase in the concentration of manganese (Mn^{2+}) ions. This is the only reason, which causes the toxicity of Manganese in most of the soils, which are poorly drained. Plants require manganese for its proper growth and development because its deficiency reduces yield production. Various plants have acquired the strategies to cope from these deficiencies (Shao et al., 2017). In leaf area of plant the concentration of manganese present in less than 10-20 mg /kg. The deficiency of manganese mainly shown in the plants having symptoms in young leaves, that is of inter venial chlorosis. The brown spots on the leaves of the plants show the toxicity symptoms of manganese. Like other metals, the availability of Manganese is also identified with the pH of the soil and its redox potential. This metal is mostly absorbed by the epidermal cells of the cortex cells present in the roots in the form of Mn^{2+} with the help of many proteins (Vigani et al., 2018). This transportation mechanism through epidermal cells of roots is also available in metals like Calcium. Manganese is present in a very less concentration that is in nanomolar to micromolar. Manganese influx can be controlled through IRT1 protein that has a very low specificity towards the substrate and is responsible for the uptake of ions in the plants through the help of Strategy I. Likewise, the Manganese is related to Iron, and so the control of IRT1 protein in iron is same as that of Manganese (Andreson et al., 2018). The protein IRT1 is in the control of the transcription factor FIT which helps in the control of the amount of Iron and Manganese present in the plant. However, during the deficiency of these elements like Manganese and Iron, this affects the transportation of each other because they have a link between them. During the deficiency of Iron, there is a need for detoxification of Manganese because the uptake of Manganese through IRT1 protein reaches very low (Shao et al., 2017). Likewise, if there is the toxicity of Manganese, then it induces the deficiency of Iron as well as it increases the mechanism of IRT1 protein. The influx of Manganese is high due to IRT1 protein but the influx is not low due to IRT1 protein as per Strategy I. In the reduced conditions of Manganese, the transporter NRAMP1 is important for the uptake of Manganese in *Arabidopsis thaliana* plant species. The transporter NRAMP1 helps in protecting the irt1 mutant phenotype due to the deficiency of Iron. As well as this transporter is poorly selective (Kramer et al., 2005). The expression of the protein IRT1 and transporter NRAMP1 is regulated by the deficiency of Manganese. Instead of all these, the Strategy II plants works on a large extent even at low availability of Manganese. It is noticed in Barley that the tolerance towards the deficiency of Manganese is related to the protein IRT1. In the barley crop, the protein IRT1 helps in the transport of Iron and Manganese also, as in *Arabidopsis thaliana* plant species. In Strategy II the uptake of Iron with the help of chelation based Strategy II (Sachan et al., 2019). The influx pathway is through the transporter NRAMP1 for Manganese in rice crop, but in *Arabidopsis thaliana,* the transporter NRAMP1 is present in all the cell layers of roots while the transporter OsNRAMP5 is present mainly in the endodermal cells and exoderminal cells of the roots. In the proximal sides of the cells, the protein MTP9 (Metal tolerance proteins) is present which is also a member of the family CDP. The main work of these proteins is the transportation of various metals to the cell organelles or the apoplast of cytosol. This mutant MTP9 helps in the translocation of transition element Manganese towards the shoot portion of the plant (Keeling et al., 2010). This will leads to the reduced accumulation of Manganese which reduces the growth and development of plants due to which there is a reduction of grain yield. When there is less supply of Manganese the transporter, NRAMP3 changes the supply of Manganese to the xylem, which provides the Manganese to the young leaves, and young panicles of the plant. However, when there is a high supply of Manganese, this protein is destroyed and the supply of Manganese is towards the older leaves of the plants. Other proteins like YSL, it also helps in the transportation of Manganese

(Shao et al., 2017). The transporter OsYSL2 helps in the translocation of Manganese within the phloem and the transporter OsYSL6 increases the tolerance towards the high amount of Manganese and helps in keeping the level of Manganese in apoplastic pathway less than the critical level of threshold. Plant by its own detoxifies the levels of Manganese for its proper growth. Some cell organelles like Golgi bodies, vacuoles, endoplasmic reticulum etc. helps in the process of chelation which tolerates the high level of Manganese in plant system (Kramer et al., 2005). In rice plant, for the detoxification of Manganese, the transporters like OsYSL6 and the transporter OsMTP8.1 helps in the transportation of manganese ions, along with this the transporter OsMTP8.1 is a member of the CDF family that is commonly known as cation diffusion facilitator and this is mostly found in the tonoplast area. Between these two transporters, if any of them is not available then it causes an increase in the concentration of manganese (Andreson et al., 2018). As well as the transporter, OsYSL6 increases the concentrations of Manganese in the apoplast of leaf and causes a decrease in the concentration of manganese in the symplast of the leaf. The mutant OsMTP8.1 is important in vacuoles for the sequestration of manganese whereas the mutant OsYSL6 is important for the transport in the apoplast area to the symplastic area for manganese nicotianamine (Vigani et al., 2018).

Copper Homeostasis

The plant needs various elements for its proper growth and development. The amount of nutrient consumed by the plants depends upon the growth of the plants. The metal nutrient is sessile in nature, due to which they face changes in the availability of these nutrients (Robert et al., 2015). Copper is the mineral nutrient which is needed to the plant at a concentration of 5 to 20 ug/gram of biomass of dry tissues but whereas in soil, it is present in ionic concentrations. Due to the variability in the doses of copper various symptoms like toxicity, deficiency symptoms appear on the plants. In soil, about 20% of copper is present free (Keeling et al., 2016). Most of the copper is present in the form of salt like CuSO4 and CuCl2 etc. Copper is generally immobile so the deficiency symptoms appear firstly in the newly formed leaves and the emerging reproductive parts of the plants. The deficiency of Copper reduces the respiration, growth and photosynthesis also (Jeong et al., 2009). Deficiency of such elements can be solved by providing chemical fertilizers or organic chelates to the crop and the toxicity of such elements may be caused due to mining, human interventions and excessive use of chemical fertilizers and irrigation through saline water, which even causes soil pollution. Copper is believed to be fused in two stages, where Cu (II) is initially to be decreased to Cu (I) through the activity of Cu (II) that is the chelate reductase action and Cu (I) is caught by plant establishments in a transporter-intervened process (Shao et al., 2017). This process is employable for iron take-up in dicots and convincing information recommend that a similar technique could be used too for copper take-up. Successfully copper insufficiency has been appeared to instigate Fe (III) that is chelate reductase action from which there is a decrease in both Fe (III) and Cu (II), in underlying foundations of pea seedlings

(Chang et al., 2009). Additionally, the mutants of *Arabidopsis thaliana* contain some recessive genes encoding with Fe (III) chelate reductase, which loses the capacity to decrease Cu (II) chelates. The transporters of copper are mostly found in eukaryotes. It has been proved that the Copper II is easily absorbed by the plant's roots but the Copper I is not easily absorbed by the plant's roots (Vigani et al., 2018). In *Arabidopsis thaliana,* the transporter of copper of C DNA helps in providing resistance from the mutants of yeast, which does not have much affinity uptake system of Copper. The transporter here used is COPT1. Many studies revealed that there are many transporters like COPT1 that are COPT-2, COPT-3, COPT-4 and COPT-5, etc. The transporters COPT-1 is made up of a chain of polypeptides containing 170 amino acids. COPT-1 is full of histidine and methionine residues, which are same as binders of copper in bacteria. These properties only have 44 amino acid terminals of COPT-1 and are present in the extracytoplasmic region of the cell (Andreson et al., 2018). The terminal of COPT-1 is important for the uptake of copper at a high-affinity level in the yeast. Arabidopsis has five family members in CTR genes, which are named as COPT-1, COPT-2, COPT-3, COPT-4 and COPT-5. The proteins contain trans-membrane-spanning domains (TMSD) which are three in number and are hydrophobic in nature derived from CTR family. The Trans membrane-spanning domains (TMSD) are acquired with COPT family (Musselman et al., 2004). There are variable size and length of amino-terminal proteins and carboxyl proteins. COPT-2 is 72% identical to COPT-1 whereas COPT-3, COPT-4 and COPT-5 are not much identical to COPT-1 and this does not have any copper binders whereas in COPT-2 and COPT-1 relation there is methionine rich amino-terminal and are also rich in histidine protein (Pilon et al., 2014). From many research, it was found that COPT-1 mRNA is not present in roots whereas the COPT-2 is present in roots and protects the high-affinity copper uptake in yeast. In Copper, proteins are bounded in a specific control manner with Metallochaperones. The concentration of ions of copper present in the yeast cell is always less than 10^{-18}M that is lower than one atom present in the cell (Hasegawa et al., 2009). Copper ions have less molecular weight and the proteins, which help in the delivery of these ions, are known as Metallochaperones. These Metallochaperones helps in the action of copper ions to the copper proteins on its active binding site (Mira et al., Andreson et al., 2018). In *Arabidopsis thaliana,* there are seven copper Metallochaperones shown in (figure 1) that are Antioxidant I (ATX1), HCCI, Scol (SCO) chaperones, CCS, Cytochrome coxidase 17 (Cox 17), Copper chaperone and the COX11 (Cytochrome C oxidase 11). Based on functions and their interaction with each other, these copper Metallochaperones has further three classes. The ions of Copper at the cellular level are attached mostly with Metallochaperones and Metallothioneins as well. Whenever the Copper ions cross the membrane then the chaperones allow the ions to cross the membrane to the transporters that will facilitate the pathway of copper actively and passively (Shao et al., 2017).

Fig. 1: Types of Copper Chaperones

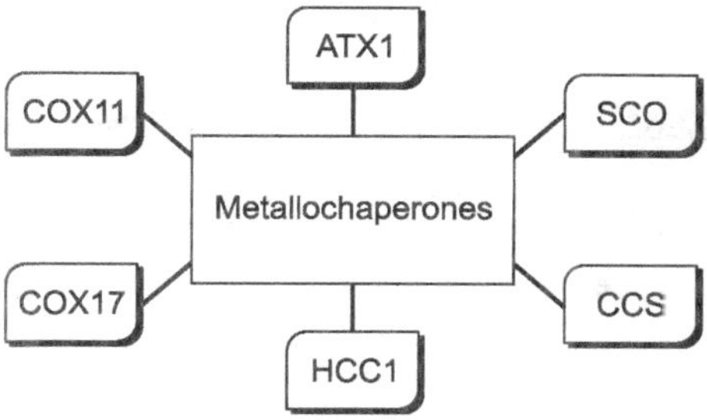

ATX1
COX11
SCO
Metallochaperones
COX17
CCS
HCC1

In reducing conditions, copper ions are available in redox state because there is zero solubility of copper salts mainly when they are in redox state. When other habitats become oxidising at that time the iron oxides are not available, if available then in a very small amount because of the hydroxides and oxides of iron are not soluble properly. The oxidised enzymes of copper are used as a defence against the attack of pathogens etc. (Kramer et al., 2005). There are three classes of proteins. The first class of protein has a very strong absorption band range up to 600 which is a blue colour portion and mononuclear in nature. This absorption band is produced by the charge caused during the transfer of ligand metal to the cysteine residue of sulphur and copper. The second class of copper protein is non-blue and is mononuclear in nature. The third class of copper protein is binuclear in nature and has a coupled centre (Robert et al., 2015). The best copper protein is a blue colour protein which is also known as type-I protein. In this type of protein, there is a transfer of electrons to the PSI reaction centre from cytochrome cf complex to the cytochrome b 6 f complex. The copper protein binds at least two copper ions with the cysteine residues of sulfhydryl group, that exhibits the distortion of Copper II to Copper I and there is a loss of blue colour. Copper proteins have three-dimensional structures, distorted geometry and in this one methionine, cysteine and two histidines are present (Shao et al., 2005). This structure is present in the enatic state, due to which there is a fast transfer of electrons. The family of copper proteins are also known as Phyto cyanins, which have further some subgroups like Ulclacyanins, Plantacyanins and Stellacyanins also. These all proteins are soluble and have a blue colour and are Type-I proteins. In this group, there is another group of proteins which is also associated with this, that is Early Nodulin like protein, but the important point of this group is that it doesn't bind with Copper (Kramer et al., 2005, Andreson et al., 2018).

CHELATORS

Chelates are the natural complex substances to which the atoms are held together with a piece of metal so that it cannot be in contact with other substances and forms insoluble structures (Anjum et al., 2015). The chelating operators are the natural particles, which can absorb the metals and utilizes the particles of Calcium, Magnesium, Iron, Cobalt, Zinc and Manganese etc. Chelates are polydentate structures because their molecules can form many bonds with the single metal atom. They are formed from the bonds of polydentate ligands with the central atom. It is a process in which two or more than two different coordinates bonds are formed in between the central atom and the ligands (Nguyen et al., 2017). The ligands here are some sequestering agents etc. The metals bounded in the chelate rings lost their cationic activity. The structure of chelates is not precipitated easily that is why the use of chelates is good in farming. Chelates are very important because they are used for the softening of hard water as well as for the separation of lanthanides and actinides. Ethylenediaminetetraacetic acid (EDTA) is regarded as the best chelating agent because EDTA can forms about 4 to 6 bonds with the metal ions of transition metals and other metals also. They can also form bonds with the complex forms of soaps and the detergents. EDTA binds the Magnesium and Calcium ions, request them and prevent them from interference. The use of chelates in micronutrients increases the efficiency of the micronutrients (Viehweger K; 2014). Various chelating agents help in the natural disintegration, for example, natural acids, amino acids, lignin sulfonates, lignin polycarboxylates, sugar acids and subordinates, phenols, poly flavonoids, siderophores and Phyto siderophores. Many chelating operators have been grown artificially. The two classes of chelating operators or complexes increment micronutrient solvency (Yadav K.S; 2010). Chelating agents have some important characteristics, which are utilized in the general dependability of different metal chelates (Toppi et al., 2002). Dependability of the micronutrient applied to the plant determines the accessibility of the applied chelate (Mohan et al., 2007). Some synthetic micronutrients are also produced through the chelating agents like Citric acid, Ethylene di amine di glycine, Nitrilo tri acetic acid, Ethylene di amine di acetic acid, Glucoheptonic acid, Di ethylene tri amine Penta acetic acid etc (Cobbett C.S; 2000).

Advantages of Chelators

Chelates have various advantages and are useful for the plants:-

- Chelates are present in the organic form so they are easily absorbed with the help of plant roots.
- The process of chelation easily removes the positive charge of micronutrients.
- Easy assembling of chelates is there in a plant system.
- The physiological disorder like Scorching in plants is less due to the use of chelates because they are present in organic form and does not cause any harm.
- The action of chelates are systematic that is why they are translocated easily in the plant (Cobbett, 2000).
- Chelates are easily adsorbed at the surface of the soil, through which they are not leached from the soil particles.
- Chelated Copper (Cu), Iron (Fe), Manganese (Mn) and Zinc (Zn) are provided as a micronutrient to crop in alkaline conditions.
- Chelated micronutrient is easily and properly assimilated by the plants and required in very less quantity as compared to the inorganic substances or fertilizers.

- Chelates do not cause any reaction with other components and are easily compatible with different fertilizers and pesticides.
- Chelates can also be usually mixed with liquid fertilizers and dry mixtures (Anjum et al., 2015).

Conclusion

The homeostasis of metals in plant cells helps in the control of trafficking of various transitional elements. Many processes like detoxification, chelation, transportation etc. maintain the balance of various elements within the cell. These metals are present naturally or through many anthropogenic activities in the soil due to which, the high amount of metals causes toxicity of that element and even causes many deficiency symptoms and disorders to the plant. Therefore, for maintaining the balance of metals in the cell, these processes are important. Chelation is one of the processes, which helps in the homeostasis. Chelation can tolerate the effect of accumulated metals with the help of the phytoremediation process. Many transporters are used in this process. All the transporters used in the process of homeostasis belong to the family of proteins. The plant species *Arabidopsis thaliana* is mostly used in the homeostasis process and the Fer like iron deficiency-induced transcription factor that is also known as the FIT is helpful in the uptake of iron through the roots of the plant species *Arabidopsis thaliana.* With the help of many papers and reports, we have presented this chapter as the name of Metal homeostasis and chelators. Various organic acids like amino acids, malate, citrate etc. contributed to the chelation process. As well as the protein, histidine also works as a derivative. Hyper accumulator plants are best suited for the tolerance of such reactions than non-hyper accumulated plants.

References

Andresen, E., Peiter, E., & Küpper, H. (2018). Trace metal metabolism in plants. Journal of Experimental Botany, 69(5), 909-954.

Anjum, N. A., Hasanuzzaman, M., Hossain, M. A., Thangavel, P., Roychoudhury, A., Gill, S. S., ... & Ahmad, I. (2015). Jacks of metal/metalloid chelation trade in plants—an overview. Frontiers in Plant Science, 6, 192.

Baker, A. J. M. (2000). Metal hyperaccumulator plants: a review of the ecology and physiology of a biological resource for phytoremediation of metal-polluted soils. Phytoremediation of contaminated soil and water.

Banakar, R., Fernandez, A. A., Zhu, C., Abadia, J., Capell, T., & Christou, P. (2019). The ratio of phytosiderophores nicotianamine to deoxymugenic acid controls metal homeostasis in rice. Planta, 250(4), 1339-1354.

Bani, A., Echevarria, G., Sulçe, S., & Morel, J. L. (2015). Improving the agronomy of Alyssum murale for extensive phytomining: a five-year field study. International Journal of Phytoremediation, 17(2), 117-127.

Bashir, K., Rasheed, S., Kobayashi, T., Seki, M., & Nishizawa, N. K. (2016). Regulating subcellular metal homeostasis: the key to crop improvement. Frontiers in Plant Science, 7, 1192.

Boominathan, R., Saha-Chaudhury, N. M., Sahajwalla, V., & Doran, P. M. (2004). Production of nickel bio-ore from hyperaccumulator plant biomass: Applications in phytomining. Biotechnology and bioengineering, 86(3), 243-250.

Chang, T. C., You, S. J., Yu, B. S., Chen, C. M., & Chiu, Y. C. (2009). Treating high-mercury-containing lamps using full-scale thermal desorption technology. Journal of Hazardous Materials, 162(2-3), 967-972.

Clemens, S. (2001). Molecular mechanisms of plant metal tolerance and homeostasis. Planta, 212(4), 475-486.

Cobbett, C. S. (2000). Phytochelatins and their roles in heavy metal detoxification. Plant physiology, 123(3), 825-832.

Cui, W., Cao, H., Yao, P., Pan, J., Gu, Q., Xu, S., ... & Shen, W. (2017). Methane enhances aluminum resistance in alfalfa seedlings by reducing aluminum accumulation and reestablishing redox homeostasis. Biometals, 30(5), 719-732.

Connorton.JM; Balk.J; Celma.JR (2017): Iron homeostasis in plants – a brief overview: Mini Review: Royal Society of Chemistry.

Di Toppi, L. S., Prasad, M. N. V., & Ottonello, S. (2002). Metal chelating peptides and proteins in plants. In Physiology and Biochemistry of Metal Toxicity and Tolerance in Plants (pp. 59-93). Springer, Dordrecht.

Forester, N. T., Lane, G. A., Steringa, M., Lamont, I. L., & Johnson, L. J. (2018). Contrasting roles of fungal siderophores in maintaining iron homeostasis in Epichloë festucae. Fungal Genetics and Biology, 111, 60-72.

Haertl, E. J. (1963). Chelation in nutrition, metal chelates in plant nutrition. Journal of Agricultural and Food Chemistry, 11(2), 108-111.

Hanikenne, M., Merchant, S. S., & Hamel, P. (2009). Transition metal nutrition: a balance between deficiency and toxicity. In The Chlamydomonas Sourcebook (pp. 333-399). Academic Press.

Hasegawa, H., Rahman, M. A., Matsuda, T., Kitahara, T., Maki, T., & Ueda, K. (2009). Effect of eutrophication on the distribution of arsenic species in eutrophic and mesotrophic lakes. Science of the Total Environment, 407(4), 1418-1425.

J. F. Musselman and QEP., (2004) Sources of Mercury in Wastewater, Pre-treatment corner.

Jeong, J., & Guerinot, M. L. (2009). Homing in on iron homeostasis in plants. Trends in plant science, 14(5), 280-285.

Keeling, S. M., Stewart, R. B., Anderson, C. W. N., & Robinson, B. H. (2003). Nickel and cobalt phytoextraction by the hyperaccumulator Berkheya coddii: implications for polymetallic phytomining and phytoremediation. International Journal of Phytoremediation, 5(3), 235-244.

Krämer, U., & Clemens, S. (2005). Functions and homeostasis of zinc, copper, and nickel in plants. In Molecular biology of metal homeostasis and detoxification (pp. 215-271). Springer, Berlin, Heidelberg.

Mai, H. J., Pateyron, S., & Bauer, P. (2016). Iron homeostasis in Arabidopsis thaliana: transcriptomic analyses reveal novel FIT-regulated genes, iron deficiency marker genes and functional gene networks. BMC Plant Biology, 16(1), 1-22.

Mohan, D., & Pittman Jr, C. U. (2007). Arsenic removal from water/wastewater using adsorbents—a critical review. Journal of hazardous materials, 142(1-2), 1-53.

Nguyen, M., Meunier, B., & Robert, A. (2017). Catechol-Based Ligands as Potential Metal Chelators Inhibiting Redox Activity in Alzheimer's Disease. European Journal of Inorganic Chemistry, 2017(25), 3198-3204.

Palmer, C. M., & Guerinot, M. L. (2009). Facing the challenges of Cu, Fe and Zn homeostasis in plants. Nature chemical biology, 5(5), 333.

Pilon, M., & Tapken, W. (2011). Copper homeostasis: regulation in plants. Encyclopedia of Inorganic and Bioinorganic Chemistry, 1-11.

Pilon, M., Cohu, C. M., Ravet, K., Abdel-Ghany, S. E., & Gaymard, F. (2009). Essential transition metal homeostasis in plants. Current opinion in plant biology, 12(3), 347-357.

Ravet, K., & Pilon, M. (2013). Copper and iron homeostasis in plants: the challenges of oxidative stress. Antioxidants & redox signaling, 19(9), 919-932.

Robert, A., Liu, Y., Nguyen, M., & Meunier, B. (2015). Regulation of copper and iron homeostasis by metal chelators: A possible chemotherapy for Alzheimer's disease. Accounts of chemical research, 48(5), 1332-1339.

Rosenkranz, T., Hipfinger, C., Ridard, C., & Puschenreiter, M. (2019). A nickel phytomining field trial using Odontarrhena chalcidica and Noccaea goesingensis on an Austrian serpentine soil. Journal of environmental management, 242, 522-528.

Sachan.P; Lal.N (2019): Molecular Mechanisms of Nickel (Ni2+) Homeostasis in Plants Uptake, Tolerance and Hyper accumulation: Chapter; Environmental Toxicity and Bioremediation; Pages: 178-196

Shao, J. F., Yamaji, N., Liu, X. W., Yokosho, K., Shen, R. F., & Ma, J. F. (2018). Preferential distribution of boron to developing tissues is mediated by the intrinsic protein OsNIP3. Plant physiology, 176(2), 1739-1750.

Sheoran, V., Sheoran, A. S., & Poonia, P. (2009). Phytomining: a review. Minerals Engineering, 22(12), 1007-1019.

Sinclair, S. A., & Krämer, U. (2012). The zinc homeostasis network of land plants. Biochimica et Biophysica Acta (BBA)-Molecular Cell Research, 1823(9), 1553-1567.

Tognacchini, A., Rosenkranz, T., van der Ent, A., Machinet, G. E., Echevarria, G., & Puschenreiter, M. (2020). Nickel phytomining from industrial wastes: Growing nickel hyperaccumulator plants on galvanic sludges. Journal of environmental management, 254, 109798.

Viehweger, K. (2014). How plants cope with heavy metals. Botanical Studies, 55(1), 1-12.

Vigani, G., & Hanikenne, M. (2018). Metal homeostasis in plant mitochondria. Annual Plant Reviews online, 111-142.

Wang, Z., Ma, T., Huang, Y., Wang, J., Chen, Y., Kistler, H. C., ... & Yin, Y. (2019). A fungal ABC transporter FgAtm1 regulates iron homeostasis via the transcription factor cascade FgAreA-HapX. PLoS pathogens, 15(9), e1007791.

Yadav, S. K. (2010). Heavy metals toxicity in plants: an overview on the role of glutathione and phytochelatins in heavy metal stress tolerance of plants. South African journal of botany, 76(2), 167-179.

Zhang, X., Houzelot, V., Bani, A., Morel, J. L., Echevarria, G., & Simonnot, M. O. (2014). Selection and combustion of Ni-hyperaccumulators for the phytomining process. International Journal of Phytoremediation, 16(10), 1058-1072.

Zhang, X., Houzelot, V., Bani, A., Morel, J. L., Echevarria, G., & Simonnot, M. O. (2014). Selection and combustion of Ni-hyperaccumulators for the phytomining process. International Journal of Phytoremediation, 16(10), 1058-1072.

Zhang, X., Zhang, D., Sun, W., & Wang, T. (2019). The adaptive mechanism of plants to iron deficiency via iron uptake, transport, and homeostasis. International journal of molecular sciences, 20(10), 2424.

Vacuole,Trichome and Hydropotes Sequestration

Monika Sharma

Department of Agronomy, School of Agriculture

Lovely Professional University, Jalandhar, Punjab, India, 144411

Email: ms9375033@gmail.com

Abstract

Worldwide soil contamination with heavy metals has become a great problem in the field of agriculture now days. Various anthropogenic activities and industrial exposure lead to the release of many toxic gases, which creates a problem in the soil and environment and affects the quality of both air and soil. With this, there is a decrease in the quality and quantity of agricultural produce, which ultimately affects the health of the population. Plants take up various metals from their surroundings by different parts or different transporter membranes for adequate growth and development of the plant. Inside the plant, many physiological processes take place through which metals travel a longer distance and cause various hazard to the plant system. Therefore, sequestration of heavy metals from plants is necessary to achieve good production and a quality yield. Plants have different mechanisms to bear any type of stresses so that the plant can achieve good growth and high yield. However, the high toxicity of heavy metal ions greatly affects the growth and functioning of the plant system. In this chapter, I will discuss the mechanism of sequestration of heavy metals from different parts of the plant system. Sequestration is the process of removal of heavy metals or toxic materials by the plant system so that they can grow significantly. Vacuole sequestration proves very important to protect the plants from the toxicity of heavy metal ions. Trichomes and hydropotes also serve as a shield to protect the plants from any toxic metabolites. In this chapter, we will discuss vacuole sequestration and removal of heavy metals by trichomes and hydropotes from the plant system in detail and this review focuses on the current scenario to protect the plants from heavy metal toxicity.

Introduction

Naturally, the metals are present in various forms in the crust of the earth. They are released in the environment in different ways, both naturally and through human activities. Most of these metals are released into the environment by human activities such as burning fossil fuels, mining ores, use of fertilizers and pesticides, disposal of municipal waste, etc. Metals at higher concentrations are toxic to plants and animals, as well as to humans and the environment. The level of toxicity of metals depends on their availability in the soil and the environment. Higher concentrations become toxic to microflora and fauna in the soil. For plants at lower concentrations, different metals are essential for proper growth and development of the plant. However, its higher concentration also poses a high risk to the plant and its development. Metals such as Zn, Mn, Cu, B are essential for the plant and its growth and development, but their higher concentration also poses a major risk to the plant. Some metals, such as cadmium, lead, etc., are toxic even at low concentrations and are not essential for the plant system (Todeschini et al., 2011). Heavy metals are released into the environment due to various anthropogenic factors such as the use of sewage or sludge water for irrigation purposes, the use of various insecticides, excess pesticides and fertilizers, a mixture of industrial and mining gaseous smoke, etc. Any metal that is not required by plants present at a high concentration harms the plant. Plants at very low concentrations generally require metals. Some heavy metals require a concentration of less than 1 mg/kg. The high concentration of metals in the soil degrades the quality of the soil and reduces the yield of crops and the quality of the product, which ultimately affects the health of both the population and the animals (Blaylock and Huang, 2000). Sequestration is the process by which the plant removes harmful materials, i.e. heavy metals from the soil and water, which are toxic even at lower concentrations. Sequestration is also called phytoextraction in which plants, through any external or internal organ or part, remove the toxic effects of heavy metals so that they can function well or grow and develop. Plants have different mechanisms for coping with the type of stress. They have their organs developed that help them to tolerate stressful conditions. Sequestration of heavy metals can be carried out by vacuoles, trichomes and hydropotes which are internal and external organs of the plant system and which help to remove heavy metals from the plant system. These systems help the plant to get rid of heavy metal stress and make it suitable for good growth and maximum productivity. Plant vacuoles are the main organelles of the plant cell, which temporarily store the necessary and unnecessary materials, minerals and nutrients and other elements. Vacuole sequestration controls and concludes the transport of heavy metals in the plant system. Vacuole sequestration capacity also involves the regulation of chelators and other ions, which subsequently contribute to the long-distance transport of heavy metals within the plant system. Similarly, trichomes are structures or small hairs which are externally present on the plant body and which help to eliminate the effect of heavy metals on the plant system.

Heavy Metals Sources

Heavy metals are present in the soil due to various natural and anthropogenic factors. Plants, along with water and minerals, absorb heavy metals through the soil. The main sources of heavy metal accumulation in the soil are industrial exposure or disposal of waste, i.e. sewage and sludge. They 're mixed up with water and transported from one place to another. They pollute rivers and lakes and infect our groundwater. Different mining and smelting industries are responsible for the high release of heavy metals to the environment. They are deposited through the water in the plant system and affect the normal functioning of the plant. Volatilization of various fertilizers, insecticides, pesticides and herbicides also results in the release of heavy metals to soil and the environment. In urban areas, the sewage and sludge are disposed of near the water and, along with the water; the soil is polluted and transported from one place to another. This polluted water is sometimes used for irrigation purposes by farmers and is introduced into the plant system. As a result, different heavy metals are introduced into the plant system and affect the quality of the product and ultimately the health of the population. These metals enter plants using root exudates and various transport membranes.

Fig. 1. Sources of heavy metal toxicity

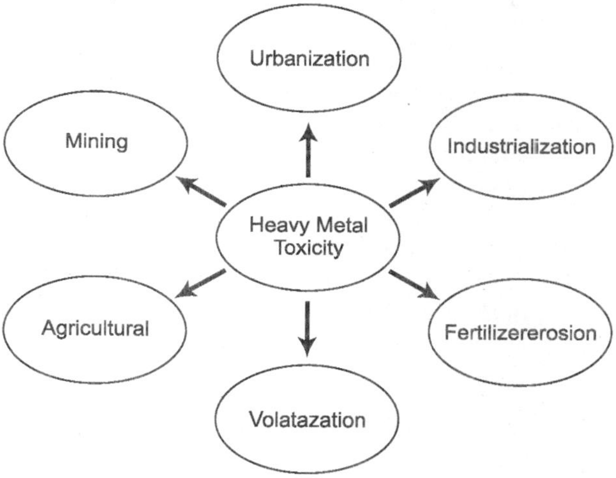

Uptake of Heavy Metals by Plants

The cells of the root hairs accumulate the heavy metal ions in the soil and absorb them along with water and nutrients through the plasma membrane of root cells. The metal ions are bonded in the membranes of the plant cells and then from there are transported from roots to all other parts of the plant with the transport mechanism of the plant. From soil, these metal ions are taken up by the membrane cells. The uptake of the metal ions generally takes place through secondary metabolites, which further translocate these metals in the other parts of the plant. The potential of the membrane to bind the metal ions is possible by the action of secondary transporters (Hirsch et al., 1998). Zinc (Zn) and Cadmium (Cd) ions are transported by some proteins, which make the availability of these metals ions and transport it to the shoot through plant transport mechanism of the plant. Cadmium can also enter in the plant system through the action of various cations (Thomine et al., 2000). Passive absorption of lead is required to enter or transport in the plant (Tung and Temple, 1996). For the entry of lead in the plants, some proteins and secondary transporters also involved. They facilitate the binding of the lead at a place and then the entry in the plant system. The uptake of heavy metal ions in tobacco plant can only possible through a kind of proteins.

Physical and chemical properties of soil also affect the metal ion absorption in the plant. The soil properties include its particle size, cation exchange capacity, soil pH, organic matter content relates to the accumulation of the heavy metals in the soil. Excess accumulation of heavy metals can leads to their transport at high rate. When the plant takes up heavy metals then most of the metal ions remains accumulated in the roots itself. The plant transport mechanism i.e. xylem involves in the transport of these metal ions from roots to shoot part. After the accumulation of heavy metals inside all the parts of the plant, it activates its detoxification mechanism. The plant has various mechanisms to detoxify the effect of heavy metals so that the [plant can survive under stressful conditions.

Heavy metals are also absorbed or enter in the plants by foliar or aerial parts. The factors responsible for the entry of these heavy metal ions through the aerial parts are the layout of the foliage, the structure of the epidermis and capacity of the epidermal cells, the composition of the cuticle etc. mostly the entry of heavy metals in the plants takes place through the root portion. After entering from these parts the metal ions are further transported to the other parts through phloem or plant transport mechanism. If the metal ions are absorbed through aerial parts then the only 1% of the heavy metal ions transported to the root part. From the foliar part, the metal ions transported accordingly within the plant system.

Mechanism Involved in Heavy Metal Tolerance

The toxic effect of heavy metals has a significant impact on plant growth, development and results in a decrease in quantity and quality of production. Heavy metal tolerance is the ability of the plant to withstand the stress caused by heavy metals. The atmosphere and soil are two of the most important natural factors in the heavy metal tolerance mechanism (Macnair et al., 2000). Various gene types are also involved in the tolerance of heavy metals in the plant system. When a plant begins to grow in the soil, it meets many useful or harmful metabolites. The presence of heavy metals in the soil can have an impact on its growth and development. Along with water and nutrients, it absorbs many harmful metabolites or metal ions that ultimately affect its growth habits. Various stresses arise in the plant system due to the accumulation of heavy metals within the plant. The heavy metal affected plant has low soil nutrient uptake and a nutrient deficiency, much oxidative stress in seed germination, and various other metabolic or genetic stress is occurring (Macnair et al., 2000; Khanand and Khan, 2010). Extracellular metabolites and membranes help the plant to withstand heavy metal stress. The plasma membrane of the plant cell also has an important part of the intolerance mechanism. It accumulates heavy metal ions to the vacuole where its long-distance transport and sequestration takes place. Different plant cells, i.e. mesophyll cells or protoplasts, and various heat shock proteins (HSPs) accumulate heavy metal ions and their long-distance transport from root to shoot. Different proteins, amino acids, plant chelators and mutants help in the vacuoles and trichome sequestration of heavy metals. The sequestration mechanism using vacuoles, trichomes and hydropotes helps in a great way to tolerate heavy metal stress in the plant and to make the plant actively growing under heavy metal stress conditions. Copper and zinc toxicity is affected by mycorrhizae in the soil. The adverse effect of Arsenic is accumulated by decreased or decreased plasma membrane uptake. Cadmium toxicity is controlled by the action of various plant chelators to protect the activity of the plant system. Metallothioneins and other heat shock proteins or mutants remove chromium and other ions. Lead is removed by sequestration or compartmentalization of these metal ions in the vacuole or trichomes. Vacuole sequestration also involves the accumulation of metal ions in the shooting part and in the vacuoles of the trichome cells, which enables them to cause any toxic effect on the plant body. Trichomes and hydropotes are the outgrowth of epidermal cells and have the function of accumulating heavy metal ions in their vacuoles.

Sequestration

Sequestration is the process by which the plant removes harmful material, i.e. heavy metals or any other toxic material from the soil and water that is toxic even at lower concentrations. Sequestration is also called phytoextraction in which plants, through any external or internal organ or part, remove the toxic effects of heavy metals so that they can function well, grow, and develop properly. Plants have a variety of mechanisms to cope with metal stress. They have different organs or organelles, which serve as tolerance against these types of stress in which vacuole sequestration is very common.

Vacuole Sequestration

A vacuole is a very crucial organelle of the plant cell, which plays a vital role in the storage of water, mineral and nutrients in the plants. It can also store many toxicants in most of the higher plants. Vacuoles functions in the long-distance transport of the metals through its sequestration capacity. The cytosol metal chelators and other transporters control vacuole sequestration capacity. Vacuole sequestration is very important to know the long-distance transport of the minerals and other materials in the plant system. To know the root to shoot transport of the heavy metals vacuole sequestration plays a good role. Vacuole sequestration lets the essential materials to function in the plant system and similarly check or phytoremidiate the toxic metals or materials from the plant body.

In plants, the sequestration of the metal ions (Cd) with plant chelators are taken place in the vacuole. This is the process, which performs the action of metal detoxification in the plant cells. The action of HMT1 mutants shows a clear-cut effect of accumulation and transportation of Cd in the plant. Gel filtration chromatography shows a visual diagram of the transport or mobility of these metal ions. This HMT1 mutant does not form HMW complexes and transport proteins, which are located in the membranes. This gene (HMT1) expresses or encodes the members of ATP-binding cassette (ABC) membrane protein which transport the metal ion is present in the vacuolar membrane of the cell (Ortiz et al., 1992). For the transport or long-distance translocation of the cadmium ions, these genes (HMT1) and ATP is required for the transportation of these metals or other metabolites to the membrane cells of vacuoles (Ortiz et al., 1995). Some plants in which plant chelator synthesis does not takes place, the members of ABC family transport the molecules or ions of the metal (Cd) to the vacuole (Li et al., 1997). Different genes involve in the transport of heavy metals in some plants and affect the ABC transporter proteins. These transporters are involved in the transport of the metabolites and sequestration of the metal ion complex in some plants. The sequestration of these metabolites to the vacuole is derived by the action of metal ions transporter proteins. The sequestration of metals from source to sink is taken to the vacuoles of the plant top cell (Vogeli-Lange et al., 1990). In a tobacco plant, the accumulation of plant chelators and metal ions is transported to the vacuoles of the mesophyll cells. A proton of ATP dependent activity is also contributing in the transport of plant chelators and metal ions (Cd) complex from root to shoot part of the plant (Salt et al., 1995). Vacuole sequestration of heavy metals takes part in the bioavailability of the mineral and nutrients metabolites and protects the plant from the toxic effects of the heavy metal stress. Different genes and plant chelator express their functions in the transport of heavy metal ion from the source to vacuole of the sink part (shoot portion) (Fig. 2).

Fig. 2. Vacuole sequestrations of heavy metals in plant cell

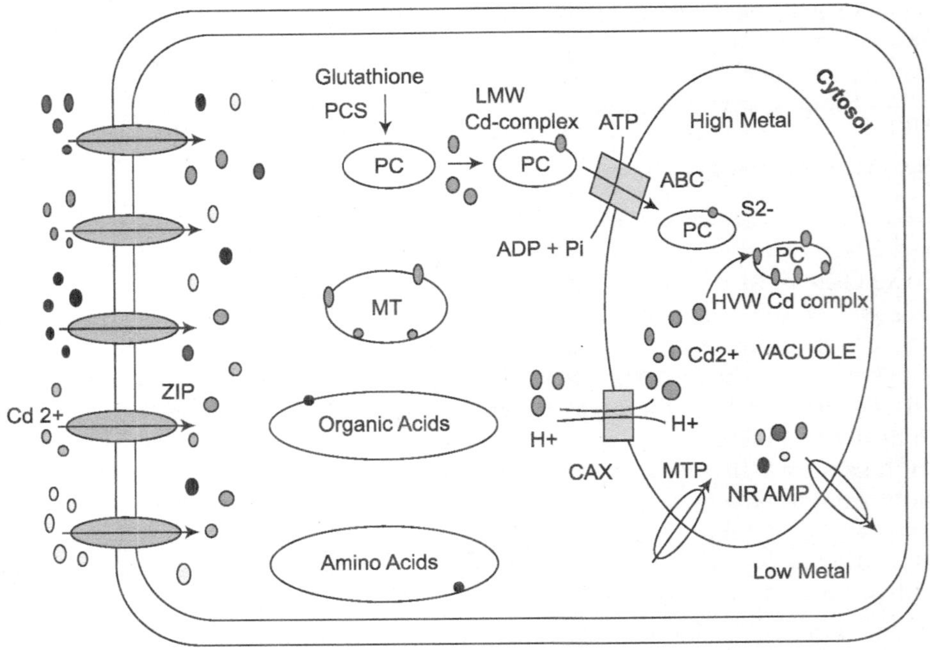

Vacuole Sequestration Capacity (Vsc) Regulation

Regulation by Transporters

The majority of metal transporters located in tonoplast (Martinoia et al., 2012) regulates vacuole sequestration capacity. NRAMP 3 and NRAMP4 genes are involved in the mobilization of vacuoles to cytosols to control nutrient deficiency in the plant system by reducing vacuole sequestration capacity (Thomine et al., 2000, 2003). Various other genes are also involved in the regulation of metal sequestration in vacuoles such as Metal Transporter 3 (MTP 3), Zinc Induced Facilitator 1 (ZIF 1), Vacuolar Iron Transporter 2 (VIT2), Cation Exchanger 4 (CAX4) (Mei et al., 2009). In the case of vacuole sequestration of heavy metals, vacuoles are regulated when metals are found to be in excess in the environment and with excess heavy metals, there is a significant increase in the activity of vacuole sequestration capacity and the maximum accumulation of heavy metals in plant cell vacuoles. With the status of heavy metals present in the environment, some genes are not affected and remain unaffected by the lower and higher concentrations of heavy metal present. These genes are VIT1, MTP1, and Copper Transporters 5 (COPT 5) (Garcia Molina et al., 2011; Klaumann et al., 2011). The functional properties of these genes change with the accumulation of heavy metals at the cell, subcellular or tissue level. The fine expressions of these genes do not permit the proper accumulation of heavy metals inside the plant cell (Table 1).

Table 1: Impact of Tonoplast-located transporter on the transport of toxic metals in the plant

Transporters	Mineral	Reference
MTP1	Zn	Ricachenevsky et al., (2013)
MTP3	Zn	Arrivault et al., (2006)
ZIF1	Zn	Haydon and Cobbett (2007), Haydon et al., (2012)
NRAMP3/NRAMP4	Fe	Thomine et al., (2003), Lanquar et al., (2005)
VIT1/VIT2	Fe, Zn	Kim et al., (2006), Zhang et al., (2012)
COPT5	Cu	Klaumann et al., (2011)
CAX2/CAX4	Cd	Koren kov et al., (2007)
HMA3	Cd	Ueno et al., (2010, 2011)

The vacuole is a storehouse of various minerals, nutrients, ions and other materials within the body of the plant. Vacuole sequestration plays an important role in reducing the impact of heavy metals on the plant. Vacuole sequestration can be performed on any heavy metals such as cadmium, nickel, arsenic; zinc, etc. Vacuole sequestration of heavy metals alters the path of heavy metal ions from the essential or functionally active parts and reduces the adverse effects of heavy metal ions. These active compartments are cytosols, chloroplasts and mitochondria. In the plant system, various transporters are present internally in the membranes that regulate the storage of metal ions in the plant organelles. The accumulation of metal ions is determined by the presence of the electrochemical potential of the positive ions present in the plant system by the enzyme vacuolar H+-ATPase (Kakinuma et al., 1993). Various heavy metals are transported to tonoplasts present in the plant system using a proton anti-port mechanism (Carrier et al., 2003). Vacuoles and extra vacuolar localization of metal ion Ni are produced by a pH gradient that is further dependent on the type of yeast. In-plant Thalspi goes intense some proteins are involved in the transport of metal ions in the shooting part of the vacuoles. These proteins are referred to as metal tolerance proteins

(TgMTPs) (Persans et al., 2001). The 'Zn-malate shuttle' carried out the transport of Zn ion to the vacuoles. Malate is released during the exchange of oxalate and citrate and is then transported back to the cytoplasm (Ernst et al., 1992). Zinc transporter genes are widely identified in plant sp. Arabidopsis affecting the accumulation of zinc in roots under high Zn treatment (Van der Zaal et al., 1999).

Regulation Vacuole Sequestration by Plant Chelators

Plant chelators are important molecules in the plant that detoxify the toxic metabolites inside the plant. Plant chelators are molecules or agents that form a stable metal PC complex and are significantly sequestered into the vacuoles (Cobbett and Goldsbrough, 2002). Plants have a variety of chelator molecules to protect the plant. In Arabidopsis, long-distance transport of metal ion (Cd) is delayed due to overexpression of the SpHMT1 fission yeast gene encoding the PC-Cd transporter (Huang et al., 2012). This delay in the transport of metal ion is due to an increase in the sequestration capacity of the vacuoles in the root zone of the plant. If the SpHMT1 gene is inserted in the root zone, the accumulation of heavy metals in the seed part is significantly reduced. There is a reduction of up to 70% in the accumulation of heavy metal in the seed part. The sequestration capacity of the vacuole in the plant is therefore regulated by PCs and the accumulation of metal ion is decreased. These chelators play a vital role in the detoxification of toxic compounds and metabolites from the plant system and protect the plant from adverse conditions (Table 2).

Table 2: Transporter genes in heavy metal uptake

PLANT	GENES	ELEMENTS	REFERENCE
Rice	OsNRAMP1, OsNRAMP2	Mn	Belouchi et al., 1997
Arabidopsis	Cpx-type ATPase	Cu, Zn. Cd, Pb	Tabata et al., 1997
Arabidopsis rice	NRAMP	Cd	Thomine et al., 2000
Arabidopsis	ZAT1, ZAT2, ZAT3	Cd, Zn, Mn	Wang et al., 2000

Another type of plant chelator is found in higher plants called Nicotianamine. Various reviews indicate that plant chelator Nicotianamine regulates the vacuole sequestration capacity (VSC) of heavy metals Zn and Fe and its long-distance transport from root to plant. Some plants are heavy metal hyperaccumulators that accumulate a large amount of heavy metals. In Arabidopsis halleri, a large amount of nicotianamine (NA) is present in the root tissues of higher plants, which contributes to the long-distance transport of Zn metal ion to shoots compared to the reduced amount of nicotianamine (NA) present in plants that have to translocate less Zn in the xylene and shoot regions (Deinlein et al., 2012). Several other studies also show that overexpression of the AtZIF1 gene increases the sequestration of nicotianamine (NA) vacuoles in the plant cell. This increase in NA sequestration leads to an accumulation of Fe in the xylene sap and tissues of the shoot region (Haydon et al., 2012). The study shows that the increase in sequestration of nicotianamine (NA) reduces cytosolic nicotianamine (NA) and affects the transport of Fe-NA ions within the plant cell (Haydon et al., 2012).

Regulation of VSC and Metal Transport in Plants

Transport of Mineral Nutrients and VSC

Vacuole sequestration function in photosynthesis and most of the cell Fe is found in the chloroplast of the plant cell. A sufficient amount of mineral nutrient is stored in the plant's leaf vacuoles. In a study of two tonoplast-localized metal transporters, i.e. the genes OsVIT1 and OsVIT2 is located in the rice plant (Zang et al., 2012). Overexpression of these genes increases the accumulation of Fe and Zn metal ions across the tonoplast of the yeast. The genes OsVIT1 and OsVIT2 are most commonly seen in the upper or flag leaf of the rice plant, which plays a major role in photosynthesis. The activity of this gene reduces the accumulation of Fe and Zn ion in the last leaf if there is any functional damage to it. It increases the accumulation of rice grain and phloem from the uppermost leaf of the flag leaf and the last node of the plant shoot. Generally, there is no alteration in the uptake or transport of metal ions from root to shoot between the species and other wild species of the plant. Thus, the study found that the genes OsVIT1 and OsVIT2 are very important and play a key role in the long-distance transport of heavy metals between the seed (lowest part) and the uppermost leaf (flag leaf). These control the sequestration capacity of the vacuole for metal ions, i.e. Fe and Zn are either in the sink or in the flag leaf. In Arabidopsis, under conditions of Fe deficiency, the tonoplast localized Fe transporter AtVIT1 (Kim et al., 2006), AtNRAMP3 (Natural Resistance Macrophage Protein 3) and AtNRAMP4 (Lanquar et al., 2005) were affected by functional damage and a major blockage of plant growth and development was observed. Vacuole Sequestration Capacity (VSC) further regulates the movement of metal ions and promotes the growth of the plant system (Fig. 3).

Fig. 3. Transport of mineral nutrients and heavy metals ions vacuole sequestration

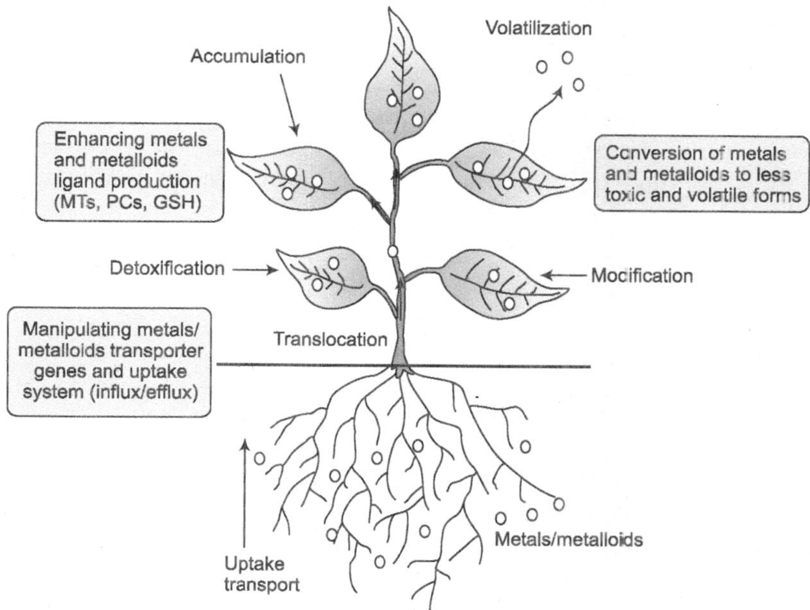

Vacuole Sequestration Capacity and Transport of Toxic Metals

Heavy metals at high concentrations are toxic to the plant and adversely affect the growth and development of the plant. Some non-essential metal ions, such as Cd and As metal ions, harm the plant. They inhibit the necessary metabolites of minerals and essential nutrients and include oxidative or plant-degrading injury or competition. In that case, the portion of the sink must be protected to facilitate the process of photosynthesis. The plant has developed various mechanisms and machines to regulate the long-distance transport of metals from root to shoot and to help protect the above-mentioned soil part of the plant for the process of photosynthesis and other essential metabolic processes. Vacuole sequestration plays a vital role as a whole in the process of metal detoxification. In a study, Cd ion accumulation in rice plant shoots is observed by the activity of the OsHMA3 gene in the rice plant (Ueno et al., 2010). OsHMA3 facilitates the accumulation of Cadmium ion in the rice plant and regulates its long-distance transport. The overaccumulation of these metal ions also occurs because the functional activity of these genes and alleles is damaged or disrupted (Ueno et al., 2010). These genes also possess the transport of these metals to the tonoplast in the roots of the rice plant and then regulate the transport to the vacuole part of the plant cell. Limited action of this type of gene facilitates the accumulation and transport of metal ions, and over-expression of these specifically reduces accumulation in the shoot region of the plant system. This reduces the accumulation of metal ions and there is no visible effect on the accumulation of other important nutrients and mineral compounds. In the rice plant, therefore, these genes, i.e. OsHMA3 regulates the vacuum sequestration capacity (VSC) of heavy metals and mediates the long-distance transport of the Cadmium metal ion from source to sink (aerial part). The accumulation of heavy metals can also be accomplished by the action of many hyperaccumulators. Hyperaccumulators are compounds that can store or accumulate a large concentration or toxic effect of heavy metals in the upper part of the plant compared to other or non-hyperaccumulators (Kramer, 2010). These hyperaccumulators do not show any effect on the plant that can be seen visually. Vacuole sequestration involves the regulation of the transport of toxic metals and results in a high tolerance of heavy metals. Vacuole sequestration capacity (VSC) for Zn ion by the roots of hyperaccumulator plants decreases compared to non-hyperaccumulators that regulate the more translocation of zinc metal ions from the source to the sink region. Transport of Zn is facilitated and regulated by an increase in the accumulation of Zn in xylem (Lasat et al., 1998).

Trichomes and Hydropotes Sequestration

Trichomes are a fine growth of epidermal hair on plants, algae, etc. They are special structures in the plant system and perform a great function. These are the covering of the hair on the plant or on any part of the plant, called indumentum, i.e. the covering of the hair on the plant and the surface on which these epidermal hairs lie, called pubescent. Trichomes and hydropotes are the outgrowth of the epidermal hair and perform different functions in the development of the plant system. These structures vary from plant to plant, and from species to species.

They may also vary depending on the species and organs of the individual plant system. The structural features of the trichomes are that they are unicellular or multicellular, straight upright and have no branches, spiral or curved, glandular, present on the epidermis of the plants on the upper and lower surfaces of the leaf and any other organ (Payne & Willard, 1978). Trichomes protect the plant from harmful molecules and insects such as pest, UV light, transpiration, etc. Trichomes and hydropotes are hairy outgrowths present in the epidermis of the plant or plant parts. These are tiny pinches and the outgrowth of the hair on the surface of the leaf. These are non-trachomatous epidermal cells and allow these surface hairs to emerge so that they can perform various tasks to protect the plant from harmful metabolites or toxic metals. Trachoma is a hair that can be unicellular or multicellular, either branched or unbranched. Multicellular hair has many layers of cells and may have a specific function in the protection of the plant. Branched hairs appear to be shaped like a tree or a star in some plants and have the ability to access many tasks at a time. Another type of trichome structure is the type of scale or shield where many cells are joined together to form a trichome. They are directly attached to the surface of the plant. Examples of this type of trichome structure are pineapple, rhododendron and sea buckthorn, which prevent the long-distance transport of toxic metal and protect plant life. Some of the trichome structures are glandular, which secrete the essential oils to protect the plant against various metabolites. A perfect example of this type of trichome structure is the mint or Lamiaceae family (Levin & Donald, 1973). The surface appearance of the organs of the plant system is described by the presence on the surface of the plant of different types of trichomes. These trichomes and hydropotes take part in the sequestration and detoxification of toxic compounds from plants and help them to grow better and produce high-quality yields. Sequestration through trichomes and hydropotes leads to the accumulation of these metal ions in the vacuoles of the trichome cells and enables the plant to survive under stress conditions. Trichomes help the plant to tolerate the toxic effects of heavy metals and protect plants against other toxic metabolites. Plants have different mechanisms for the sequestration of heavy metals. They have other morphological features for sequestration or detoxification of heavy metals. The epidermal structures of trichomes and hydropotes are involved in the process of chelation, sequestration or detoxification of metal ions from plants (Figure 4).

Fig. 4. Structure of Trichomes

(a) Trichomes of *Copsicum pubescens* plant

(b) Trichomes on *Cannobis*, rich in cannabinoids

Source: /sequestration/Trichome%20-%20Wikipediahtml

Trichome and Hydropote Sequestration of Different Heavy Metals

Trichomes are tiny epidermal hairs on the surface of the leaves that can protect the plant from any kind of damage or stress. Trichomes play a crucial role in protecting the plant against excessive weather conditions such as sunlight and wind, removing various harmful metabolites from plants and storing metal ions, etc. In tobacco seedlings exposed to cadmium toxicity, there is a significant decrease in plant growth and a double increase in the number of trichomes in the experiment (Choi et al., 2001). Halleri metal Zn remains tetrahedral and complex to carboxyl and hydroxyl groups of some species in the trichomes of Arabidopsis (Sarret et al., 2002). Specific overexpression of the metallothionein coding gene (MT2) that accumulates heavy metal ions and detoxifies toxic metals from the plant system is present in trichomes (Garcia-Hernandez et al., 1998). These functional features of the trichomes show that plant trichomes and hydropotes are of great importance in the plant system to detoxify the effect of metal ions. Trichomes are modified extracellular epidermis cells that can be classified in different types (Fahn, 1990). These are specialized structures present on the surface of the leaves and have a different composition of the elements compared to other cells in the epidermal region. These trichomes are rich in calcium elements in most plant species. Several studies have ensured that the surface of trichomes of many species of the plant contains crystalline rich in calcium (Broadhurst et al., 2004a, Kupper et al., 2001, Psaras et al., 2000). With a high concentration of Ca-rich trichomes on the surface of the leaves and the stem region, there is an increase for element in the shoots of many plants (Broadley et al., 2003, Ghaderian et al., 2007). The study shows that there is less concentration of K ions compared to Ca ions in the plant shoot.

The ion K is located on the cytoplast and is the main site for the accumulation of potassium ions in the trichomes of many plants (Marschner, 1995). Because the concentration of calcium ion is maximum in trichomes, sometimes K ions are replaced by calcium ions and their cations. The concentration of the Ni element accumulated in the growing region is less than the amount accumulated by the Ni hyperaccumulators in soil (Broadhurst et al., 2004a). Several case studies have observed the sequestration of Ni ions by trichomes of different plant species. The experiment was conducted on Plant A. Inflammation in which different concentrations of metal ion are used to know the growth and production of biomass at different concentrations. This study concludes that the concentration of Ni above 350 M increases the accumulation of Ni in the shoot region of a plant that can become toxic to the plant and cause damage to the plant. Besides, there is a significant decrease in biomass production (Ghaderian et al., 2007). From this study, we can conclude that this plant does not detoxify the higher concentration of Ni as compared to the other hyperaccumulators of Nickel (Ni). Accumulation of Ni in plants more than its detoxification capacity may contribute to the toxicity of Nickel (Ni) in the shooting part and other parts show Ni toxicity (Kramer and Clemens, 2006; Seregin and Kozhevnikova, 2006). The sequestration of the Ni ion is mediated by the long-distance transport of the plant chelator to the trichomes. The concentration of the Ni in trichome is naturally present, which is the same as the accumulation of the Ni in the shoot portion. The same concentration of Ni present either naturally in the trichomes or accumulated as a result of the shoot that, naturally, trichomes are not responsible for the accumulation of Ni at a higher level. The concentrations of the different metal ions are accumulated by trichome and hydropote structures present on the surface of the leaves and stem. Low concentrations of heavy metals such as Cd, Zn, Mn, etc. lead to lower accumulation and transport of metal ions in the shoot portion. Glandular and non-glandular trichomes and hydropotes are the essential components of the plant system for the accumulation of various heavy metals in the vacuoles and the shoot portion. The concentration of nickel (Ni) is already present in the trichome cells of the leaf and stem regions. Plant chelators (PCs) and proteins are therefore active to reduce the effect of these heavy metals on the plant. Naturally, the concentration of calcium (Ca) is present in hydropote cells and the higher concentration of heavy metal present may replace the concentration of calcium (Ca) ion in trichomes and hydropotes. Different heavy metals show different signs of toxicity in plants. They affect plant growth and cause a nutrient deficiency in the plant. Plant cells are shrivelled due to the high toxicity of these heavy metals. Cadmium (Cd) and Arsenic (As) are major toxic heavy metals that have a significant impact on the plant and lead to deterioration of the plant. It completely damages the plant and reduces the yield potential. Products derived from heavy metal toxicity are of poor quality and may affect the health of the population and the animals that feed on them. Various plant system structures, i.e. vacuoles, trichomes and hydropotes, help the plant to cope with this type of stress.

Fig. 5. Accumulation of zinc and cadmium and localization of zinc in Picris divaricata Vant.

(A) (B) (C)

Source: https://www.ars.usda.gov/ARSUserFiles/333/Bauchan/

The accumulation of Ni by trichomes is not the same, whether the trichomes are of the stem region or the leaf region. Trichome structures of different parts contribute to the different concentration and quantity of Ni ions in the plant itself. The concentration of Ni in trichome is linked to the concentration of Ni in the shoot portion of the plant. The accumulation of this metal ion can be done by the trichomes of the stem part at the higher concentration of the Ni present in the soil. Under natural conditions, the low concentration of Ni in the shoot can lead to a decrease in the concentration of Ni in the stem and leaves trichomes. From this, we can conclude that the Ni element hyperaccumulator plant cannot accumulate the Ni ion at a higher concentration. Hyperaccumulation of the Ni ion may be possible through other hyperaccumulator plants or the expression of genes related to the sequestration of heavy metals in the plant system. The concentration of the metal ion in the trichomes of the leaves increases with the action of the epidermal mesophyll cells because the epidermal cells have a high concentration of nickel (Ni) metal ions in the leaves of the plants (Asemaneh et al., 2007). Accumulation of Ni under natural conditions may also occur in different regions, such as apoplast, epidermal cell vacuoles, which are an important area for the accumulation of metal ions in it. Either glandular trichomes are present in the stem or in the leaf, region and can play a major role in the accumulation of metal ions in plants. Nonglandular trichomes are physiologically active in the early stages of leaf development. They remain active in the early stages and accumulate the concentration of heavy metals in the early stages of leaf growth. Nonglandular trichomes remain active in the earlier stages of plant growth and later become inactivated and block its connections between other cells (Fahn, 1986;

Uphof, 1962). The low concentration of Ni in the soil results in less accumulation of metal ion in the trichomes of the leaves. The concentration of the Ni ion is maximum compared to the hyperaccumulators used for the accumulation of the Ni element (Baker and Brooks, 1989). In low concentrations of Ni, other parts of the plant system, such as plant cell walls, vacuoles, etc., become more important in the accumulation of Ni than in trichomes. In heavy metal hyperaccumulators, the accumulation of the metal ion is in the apoplastic of the leaf (Kramer et al., 2000). It is a major site in the heavy metal accumulation plant system. As a result, the high apoplastic capacity to bind or accumulate heavy metals develops a high tolerance to heavy metals in plants. If there is less concentration of Ni in the leaves, there is less bonding of Ni in the trichomes. This can be derived from the ability of the apoplastic to bind these heavy metals. By increasing, the concentration of heavy metals in the leaves, the activity of apoplast and trichomes increases and may accumulate a higher concentration of heavy metals in the plant. The maximum accumulation of heavy metals takes place in the shooting part of the plants and then in the vacuoles and trichomes. Hydropotes are the glandular structure of the epidermis cells on the surface of the plant. They are like trichomes and can perform the accumulation activity of heavy metals to protect the plants or to make them tolerate heavy metals. These are specifically different types of epidermal cells that can bind heavy metals to the vacuoles. In heavy metal hyperaccumulators, the sequestration of the metal ion in the vacuole or the binding of the metal ion in the apoplastic can provide better tolerance for heavy metal stress in the plant (Hall, 2002). Heavy metal cannot, therefore, come in connection with other sites or locations and remain protected.

Fig. 6. Glandular trichome regulates stress in plants

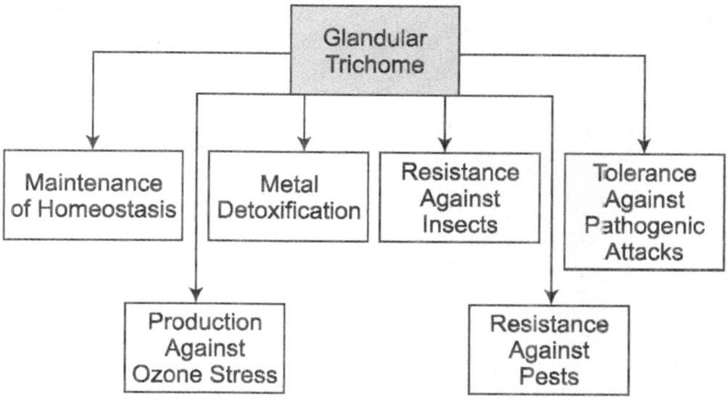

The study shows that the accumulation of metal ions occurs mainly in the vacuoles from the other components of the epidermal cells. This shows that there is less accumulation of heavy metals in the cytoplasmic area of the plant. Preferential accumulation of heavy metals occurs in the epidermis cell region (Kupper et al., 2001). These accumulations may result in the tolerance of heavy metals to toxic metals. The binding of metal ions to epidermal cell vacuoles can be found in the hyperaccumulators of these heavy metals (Bidwell et al., 2004). It is clear from various studies that trichomes are the basis for the accumulation of heavy metals such as Ni, Zn, Mn, etc. and the sequestration of these heavy metals into vacuoles can also be done based on trichomes (Broadhurst et al., 2004a, 2004b, 2009; de la Fuente et al., 2007). Trichomes also take part in the vacuole sequestration of metal ions so that the plant can achieve its proper growth and development. The binding of these metals is facilitated by trichomes and hydropots. The concentration of heavy metals indicates sequestration capacity. If there is less concentration of heavy metals in the soil or surrounding area, there is less binding of metal ions in the shoot and therefore there is less transport of metals from the root to the shoot portion. The activity of trichomes is different as the very high or toxic concentration of heavy metals is present in the shoot region of the plants. The ultra-peripheral cell wall of the trichomes is very rich in calcium elements. Excessive concentration of heavy metals may be translocated or moved to the outer cell wall of the trichomes. The concentration of calcium ions is then reduced and occupied by the other heavy metals. These trichomes are covered by heavy metal bindings and the concentration of their calcium ion is less or less negligible. The heavy metal transport of the root to the shoot leads to the accumulation of these heavy metals at a higher concentration in the shooting part and allows the plant to be tolerated against the toxicity of these metals.

Table 3: Heavy metal accumulation in different compartments

TISSUE	ELEMENT	PLANT	REFERENCE
Trichome	Zn	*Arabidopsis halleri*	Kupper et al., 1999
	Cd	*Brassica juncea*	Salt et al., 1995
	Ni	*Alyssum lesbiacum*	Kramer et al., 1997
Epidermal	Zn	*T. caerulescens*	Kupper et al., 1999
	Zn	*T. caerulescens*	Vazquez et al., 1994
	Ni	*Alyssum*	Kramer et al., 1997
Mesophyll	Zn	*Arabidopsis halleri*	Kupper et al., 1999
	Cd	*Sedum alfredii* H.	Xiong et al., 2004
Cell wall	Ni	*T. goesingense*	Kramer et al., 2000
	Cu	*Elsholtzia splendens*	Yang, 2002
	Zn	*Sedum alfredii* H.	Kramer, et al., 2000
	Pb	*Sedum alfredii* H.	He, et al., 2003
Vacuole	Zn	*T. caerulescens*	Kupper et al., 1999
	Cd	*Sedum alfredii* H.	Xiong et al., 2004
	Zn	*Sedum alfredii* H.	Kramer et al., 2000

Staining Material for Trichome and Hydropote

Trichomes and hydropotes are the main sites for the accumulation and transport of heavy metals from the root to the shoot or trichome region. Of course, the heavy metal ion present in the trichome is not the physiological response of the plant. The higher concentration of these heavy metals is therefore toxic to plants. The toxicity of these metals varies between species. When the concentration of metal ion exceeds the trichomes, the metal ion moves or moves to the other compartments of the plant. Therefore, in the hyperaccumulator plant of heavy metal accumulation, trichomes are the main source of accumulation of heavy metal ion in the metal ion concentration of the leaves. The concentration of heavy metal in the leaves is affected by the binding of the metal ion in the trichomes. The metal concentration in the soil medium interacts with the other elements of the plant and has a toxic effect on the plant compartments. Trichomes, therefore, play a role in binding these heavy metals to the shoot portion or within the vacuoles of the trichome cells so as not to affect the growth and development of the plant. The accumulation of these heavy metals can be mediated by a variety of plant chelators and mutants. The accumulation capacity in the plant system changes with the time and structure of the trichomes. In earlier stages, when the leaves grow, non-glandular trichomes become active and, in later stages, inactivate their function and activation of glandular trichomes serve as a binding function for heavy metals in the vacuoles. To visually see the accumulation of these metals in the trichomes, the staining of the leaf is done so that a clear image can be seen under a scanning electron microscope. DMG is not generally used for the staining of leaves to determine nickel (Ni) ions, as this crystal formation can occur in cells (Bhatia et al., 2004). Besides, the DMG staining compound can not penetrate the wall of the plant cells or the membranes. It cannot be fused in non-water-soluble compounds of the plant cell, i.e. suberin, wax, etc. (Smart et al., 2007). To solve this problem, the semi-quantitative DMG method is used to determine the accumulation of heavy metals inside the trichomes and hydropots of heavy metal hyperaccumulators (Northeastern Naturalist, 2008).

Conclusion

Various anthropogenic activities and industrial exposure lead to the release of many toxic gases, creating problems in the soil and the environment and affecting the quality of both air and soil. This reduces the quality and quantity of agricultural products, which ultimately affects the health of the population. The inoculation of heavy metals in soil and water pollutes the soil and the environment. This polluted water, when absorbed by plants, affects its growing conditions and developmental characteristics. As a result, there is a significant decrease in the quality and quantity of the product. Heavy metal toxicity significantly affects the quality of the product and causes health hazards when the population of animals ingests the product. It is therefore important to have some mechanisms or processes in place to deal with this type of stress. The plant itself has some mechanisms to cope with all kinds of stress. Sequestration of heavy metals may play a better role in the tolerance of metal ions in plants. Vacuole sequestration accumulates heavy metal ions and their long-distance transport from the roots to the shoot portion. The growth of epidermal cells, i.e. trichomes and hydropotes, also involves the sequestration process. They accumulate or bind the metal ion without the vacuole or other parts of the trichome cells. Nowadays, the use of hyperaccumulator plants is used to detoxify heavy metals, which is a very good concept. Phytoremediation of the polluted environment should continue to increase so that healthy products can be achieved. The alteration of genes and mutants, plant chelators, and hyperaccumulators can contribute to a significant improvement in the production of quality yields. The sequestration of heavy metals using vacuoles, trichomes or hydropotes plays a major role in detoxifying the effects of heavy metals and in protecting the plant from any other toxic metabolite.

Reference

Asemaneh, T., Ghaderian, S. M., Crawford, S. A., Marshall, A. T., & Baker, A. J. M. (2006). Cellular and subcellular compartmentation of Ni in the Eurasian serpentine plants Alyssum bracteatum, Alyssum murale (Brassicaceae) and Cleome heratensis (Capparaceae). *Planta, 225*(1), 193-202.

Baker, A. J., & Brooks, R. (1989). Terrestrial higher plants which hyperaccumulate metallic elements. A review of their distribution, ecology and phytochemistry. *Biorecovery., 1*(2), 81-126.

Belouchi, A., Kwan, T., & Gros, P. (1997). Cloning and characterization of the OsNramp family from Oryza sativa, a new family of membrane proteins possibly implicated in the transport of metal ions. *Plant molecular biology, 33*(6), 1085-1092.

Blaylock, M. J. (2000). Phytoextraction of metals. *Phytoremediation of toxic metals: Using plants to clean up the environment,* 53-70.

Broadhurst, C. L., Chaney, R. L., Angle, J. S., Maugel, T. K., Erbe, E. F., & Murphy, C. A. (2004). Simultaneous hyperaccumulation of nickel, manganese, and calcium in Alyssum leaf trichomes. *Environmental Science & Technology, 38*(21), 5797-5802.

Broadley, M. R., Bowen, H. C., Cotterill, H. L., Hammond, J. P., Meacham, M. C., Mead, A., & White, P. J. (2003). Variation in the shoot calcium content of angiosperms. *Journal of Experimental Botany, 54*(386), 1431-1446.

Chen, Y., Graziano, J. H., Parvez, F., Hussain, I., Momotaj, H., Van Geen, A., ... & Ahsan, H. (2006). Modification of risk of arsenic-induced skin lesions by sunlight exposure, smoking, and occupational exposures in Bangladesh. *Epidemiology,* 459-467.

Chibuike, G. U., & Obiora, S. C. (2014). Heavy metal polluted soils: effect on plants and bioremediation methods. *Applied and environmental soil science, 2014.*

Cobbett, C. S. (2000). Phytochelatins and their roles in heavy metal detoxification. *Plant physiology, 123*(3), 825-832.

Cobbett, C., & Goldsbrough, P. (2002). Phytochelatins and metallothioneins: roles in heavy metal detoxification and homeostasis. *Annual review of plant biology, 53*(1), 159-182.

Cummings, D. E., Caccavo, F., Fendorf, S., & Rosenzweig, R. F. (1999). Arsenic mobilization by the dissimilatory Fe (III)-reducing bacterium Shewanella alga BrY. *Environmental Science & Technology, 33*(5), 723-729.

Deinlein, U., Weber, M., Schmidt, H., Rensch, S., Trampczynska, A., Hansen, T. H., ... & Clemens, S. (2012). Elevated nicotianamine levels in Arabidopsis halleri roots play a key role in zinc hyperaccumulation. *The Plant Cell, 24*(2), 708-723.

Fahn, A. (1986). Structural and functional properties of trichomes of xeromorphic leaves. *Annals of Botany, 57*(5), 631-637.

Fahn, A. 1990. Plant Anatomy, 4th Edition. Pergamon Press, Oxford, UK. 588 pp.

Garcia-Molina, A., Andrés-Colás, N., Perea-García, A., del Valle-Tascón, S., Peñarrubia, L., & Puig, S. (2011). The intracellular Arabidopsis COPT5 transport protein is required for photosynthetic electron transport under severe copper deficiency. *The Plant Journal, 65*(6), 848-860.

Ghaderian, S. M., Mohtadi, A., Rahiminejad, R., Reeves, R. D., & Baker, A. J. M. (2007). Hyperaccumulation of nickel by two Alyssum species from the serpentine soils of Iran. *Plant and Soil, 293*(1), 91-97.

Ghasemi, R., Ghaderian, S. M., & Krämer, U. (2009). Accumulation of nickel in trichomes of a nickel hyperaccumulator plant, Alyssum inflatum. *Northeastern Naturalist, 16*(sp5), 81-92.

Gramlich, A. (2008). *Development of a semi-quantitative method to determine the distribution of Ni in hyperaccumulator plants* (Master's thesis, ETH, Swiss Federal Institute of Technology Zurich, ITES-Institute of Terrestrial Ecosystems).

Hall, J. Á. (2002). Cellular mechanisms for heavy metal detoxification and tolerance. *Journal of experimental botany, 53*(366), 1-11.

Haydon, M. J., Kawachi, M., Wirtz, M., Hillmer, S., Hell, R., & Krämer, U. (2012). Vacuolar nicotianamine has critical and distinct roles under iron deficiency and for zinc sequestration in Arabidopsis. *The Plant Cell, 24*(2), 724-737.

He, B., Yang, X. E., Ni, W. Z., Wei, Y. Z., & Ye, H. B. (2003). Pb uptake, accumulation, subcellular distribution in a Pb-accumulating ecotype of Sedum alfredii (Hance). *Journal of Zhejiang University-SCIENCE A, 4*(4), 474-479.

Huang, J., Zhang, Y., Peng, J. S., Zhong, C., Yi, H. Y., Ow, D. W., & Gong, J. M. (2012). Fission yeast HMT1 lowers seed cadmium through phytochelatin-dependent vacuolar sequestration in Arabidopsis. *Plant physiology, 158*(4), 1779-1788.

Jin-Yuan, L. I. U., Guang-Rong, Z. H. A. O., & Ji, L. I. (2000). Classification and nomenclature of plant metallothionein-like proteins based on their cysteine arrangement patterns. *Journal of Integrative Plant Biology, 42*(6), 649.

Khan, M. R., & Khan, M. M. (2010). Effect of varying concentration of nickel and cobalt on the plant growth and yield of chickpea. *Australian Journal of Basic and Applied Sciences, 4*(6), 1036-1046.

Kim, S. A., Punshon, T., Lanzirotti, A., Li, L., Alonso, J. M., Ecker, J. R., ... & Guerinot, M. L. (2006). Localization of iron in Arabidopsis seed requires the vacuolar membrane transporter VIT1. *Science, 314*(5803), 1295-1298.

Klaumann, S., Nickolaus, S. D., Fürst, S. H., Starck, S., Schneider, S., Ekkehard Neuhaus, H., & Trentmann, O. (2011). The tonoplast copper transporter COPT5 acts as an exporter and is required for interorgan allocation of copper in Arabidopsis thaliana. *New Phytologist, 192*(2), 393-404.

Koncz, C., & Schell, J. (1986). The promoter of TL-DNA gene 5 controls the tissue-specific expression of chimaeric genes carried by a novel type of

Agrobacterium binary vector. *Molecular and General Genetics MGG, 204*(3), 383-396.

Krämer, U., & Clemens, S. (2005). Functions and homeostasis of zinc, copper, and nickel in plants. In *Molecular biology of metal homeostasis and detoxification* (pp. 215-271). Springer, Berlin, Heidelberg.

Krämer, U., Pickering, I. J., Prince, R. C., Raskin, I., & Salt, D. E. (2000). Subcellular localization and speciation of nickel in hyperaccumulator and non-accumulator Thlaspispecies. *Plant physiology, 122*(4), 1343-1354.

Krämer, U., Pickering, I. J., Prince, R. C., Raskin, I., & Salt, D. E. (2000). Subcellular localization and speciation of nickel in hyperaccumulator and non-accumulator Thlaspispecies. *Plant physiology, 122*(4), 1343-1354.

Kramer, U., Smith, R. D., Wenzel, W. W., Raskin, I., & Salt, D. E. (1997). The role of metal transport and tolerance in nickel hyperaccumulation by Thlaspi goesingense Halacsy. *Plant physiology, 115*(4), 1641-1650.

Küpper, H., E. Lombi, F-J. Zhao, G. Wieshammer, and S. P. McGrath. 2001. Cellular compartmentation of nickel in the hyper accumulators Alyssum lesbiacum, Alyssum bertolonii, and Thlaspi goesingense. Journal of Experimental Botany 52:2291–2300.

Küpper, H., Zhao, F. J., & McGrath, S. P. (1999). Cellular compartmentation of zinc in leaves of the hyperaccumulator Thlaspi caerulescens. *Plant physiology, 119*(1), 305-312.

Lanquar, V., Lelièvre, F., Bolte, S., Hamès, C., Alcon, C., Neumann, D., ... & Thomine, S. (2005). Mobilization of vacuolar iron by AtNRAMP3 and AtNRAMP4 is essential for seed germination on low iron. *The EMBO journal, 24*(23), 4041-4051.

Lasat, M. M., Baker, A. J., & Kochian, L. V. (1998). Altered Zn compartmentation in the root symplasm and stimulated Zn absorption into the leaf as mechanisms involved in Zn hyperaccumulation in Thlaspi caerulescens. *Plant physiology, 118*(3), 875-883.

Li, Z. S., Lu, Y. P., Zhen, R. G., Szczypka, M., Thiele, D. J., & Rea, P. A. (1997). A new pathway for vacuolar cadmium sequestration in Saccharomyces cerevisiae: YCF1-catalyzed transport of bis (glutathionato) cadmium. *Proceedings of the National Academy of Sciences, 94*(1), 42-47.

Macnair, M. R., Tilstone, G. H., & Smith, S. E. (2020). 13 The Genetics of Metal Tolerance and Accumulation in Higher Plants. *Phytoremediation of contaminated soil and water*, 235.

Marschner, H. 1995. Mineral Nutrition of Higher Plants. 2nd Edition. Academic Press, New York, NY, USA. 889 pp.

Mei, H., Cheng, N. H., Zhao, J., Park, S., Escareno, R. A., Pittman, J. K., & Hirschi, K. D. (2009). Root development under metal stress in Arabidopsis thaliana requires the H+/cation antiporter CAX4. *New Phytologist, 183*(1), 95-105.

Ortiz, D. F., Kreppel, L., Speiser, D. M., Scheel, G., McDonald, G., & Ow, D. W. (1992). Heavy metal tolerance in the fission yeast requires an ATP-binding cassette-type vacuolar membrane transporter. *The EMBO Journal, 11*(10), 3491-3499.

Ortiz, D. F., Ruscitti, T., McCue, K. F., & Ow, D. W. (1995). Transport of metal-binding peptides by HMT1, a fission yeast ABC-type vacuolar membrane protein. *Journal of Biological Chemistry, 270*(9), 4721-4728.

Pal, R., & Rai, J. P. N. (2010). Phytochelatins: peptides involved in heavy metal detoxification. *Applied biochemistry and biotechnology, 160*(3), 945-963.

Payne, W. W. (1978). A glossary of plant hair terminology. *Brittonia, 30*(2), 239-255.

Persans, M. W., Nieman, K., & Salt, D. E. (2001). Functional activity and role of cation-efflux family members in Ni hyperaccumulation in Thlaspi goesingense. *Proceedings of the National Academy of Sciences, 98*(17), 9995-10000.

Psaras, G. K., Constantinidis, T. H., Cotsopoulos, B., & Manetas, Y. (2000). Relative abundance of nickel in the leaf epidermis of eight hyperaccumulators: evidence that the metal is excluded from both guard cells and trichomes. *Annals of Botany, 86*(1), 73-78.

Reise, S. P., & Waller, N. G. (2009). Item response theory and clinical measurement. *Annual review of clinical psychology, 5*, 27-48.

Salt, D. E., & Rauser, W. E. (1995). MgATP-dependent transport of phytochelatins across the tonoplast of oat roots. *Plant Physiology, 107*(4), 1293-1301.

Salt, D. E., Blaylock, M., Kumar, N. P., Dushenkov, V., Ensley, B. D., Chet, I., & Raskin, I. (1995). Phytoremediation: a novel strategy for the removal of toxic metals from the environment using plants. *Bio/technology, 13*(5), 468-474.

Seregin, I., & Kozhevnikova, A. D. (2006). Physiological role of nickel and its toxic effects on higher plants. *Russian Journal of Plant Physiology, 53*(2), 257-277.

Smart, K. E., Kilburn, M. R., Salter, C. J., Smith, J. A. C., & Grovenor, C. R. M. (2007). NanoSIMS and EPMA analysis of nickel localisation in leaves of the hyperaccumulator plant Alyssum lesbiacum. *International Journal of Mass Spectrometry, 260*(2-3), 107-114.

Tabata, K., Kashiwagi, S., Mori, H., Ueguchi, C., & Mizuno, T. (1997). Cloning of a cDNA encoding a putative metal-transporting P-type ATPase from Arabidopsis thaliana. *Biochimica et Biophysica Acta (BBA)-Biomembranes, 1326*(1), 1-6.

Thomine, S., Lelièvre, F., Debarbieux, E., Schroeder, J. I., & Barbier-Brygoo, H. (2003). AtNRAMP3, a multispecific vacuolar metal transporter involved in plant responses to iron deficiency. *The Plant Journal, 34*(5), 685-695.

Thomine, S., Wang, R., Ward, J. M., Crawford, N. M., & Schroeder, J. I. (2000). Cadmium and iron transport by members of a plant metal transporter family in Arabidopsis with homology to Nramp genes. *Proceedings of the National Academy of Sciences, 97*(9), 4991-4996.

Thomine, S., Wang, R., Ward, J. M., Crawford, N. M., & Schroeder, J. I. (2000). Cadmium and iron transport by members of a plant metal transporter family in Arabidopsis with homology to Nramp genes. *Proceedings of the National Academy of Sciences, 97*(9), 4991-4996.

Todeschini, V., Lingua, G., D'agostino, G., Carniato, F., Roccotiello, E., & Berta, G. (2011). Effects of high zinc concentration on poplar leaves: a morphological and biochemical study. *Environmental and Experimental Botany, 71*(1), 50-56.

Ueno, D., Yamaji, N., Kono, I., Huang, C. F., Ando, T., Yano, M., & Ma, J. F. (2010). Gene limiting cadmium accumulation in rice. *Proceedings of the National Academy of Sciences, 107*(38), 16500-16505.

Vázquez, M. D., Poschenrieder, C., Barceló, J., Baker, A. J. M., Hatton, P., & Cope, G. H. (1994). Compartmentation of zinc in roots and leaves of the zinc hyperaccumulator Thlaspi caerulescens J & C Presl. *Botanica Acta, 107*(4), 243-250.

Vögeli-Lange, R., & Wagner, G. J. (1990). Subcellular localization of cadmium and cadmium-binding peptides in tobacco leaves: implication of a transport function for cadmium-binding peptides. *Plant physiology, 92*(4), 1086-1093.

Xiong, Y. H., Yang, X. E., Ye, Z. Q., & He, Z. L. (2004). Characteristics of cadmium uptake and accumulation by two contrasting ecotypes of Sedum alfredii Hance. *Journal of Environmental Science and Health, Part A, 39*(11-12), 2925-2940.

Yang, M. J. (2002). *Copper hyperaccumulation in Elsholtzia splendens and its mechanisms* (Doctoral dissertation, PhD dissertation, Zhejiang University).

Zhang, Y., Xu, Y. H., Yi, H. Y., & Gong, J. M. (2012). Vacuolar membrane transporters OsVIT1 and OsVIT2 modulate iron translocation between flag leaves and seeds in rice. *The Plant Journal, 72*(3), 400-410.

3

Role of Growth Hormone in Mitigation of Heavy Metal Stress

Monika Sharma

Department of Agronomy, School of Agriculture
Lovely Professional University, Jalandhar, Punjab, India, 144411
Email: ms93750033@gmail.com

Abstract

The increasing population today is responsible for industrialization; urbanization and other man include activities which involves the use of heavy metals in a large amount. Heavy metal-containing products during processing release heavy metals to the atmosphere and pollute air and water. The processing and disposal of these products lead to the pollution of soil and water by the exposure of heavy metal ions in the atmosphere which can further create the toxicity among plants. The agriculture sector mostly depends upon environmental factors i.e. soil, water, air, light etc. Heavy metals contaminated soil is deficient in nutrients and its higher concentration leads to many health hazards. Heavy metal toxicity in the environment contributes lower yield in terms of both quality and quantity which ultimately affect the health of human and animals. The entry of toxic metal ions in the plant system may affect various morphological, biochemical and physiological processes. From various studies, it has been confirmed that the growth hormones play a pivotal role in regulating the growth of the plant under heavy metal stress conditions. In this review, we will discuss the role of plant growth regulators in mitigating heavy metal toxicity and influencing the survival of plant under metal stress conditions.

Introduction

Heavy metals are found in earth's crust in different forms. The exposure of the toxic metal ions in the natural environment takes place by the different natural and fabricated factors. These are released in the soil, water and environment by different sources and affect the population in a great way. Heavy metals are released in the environment by the process of industrialization, mining, smelting and other agricultural activities. Due to invasion of human activities, the heavy metal ions release to the environment and pollute the soil and water. The quality of crops has been decreasing day by day due to the presence of a higher concentration of heavy metals in the soil and source of irrigation. It is estimated that about 70% of the heavy metal ions are introduced in the human body due to means of food (Jaishankar et al., 2014). The concentration of heavy metals affects all the processes of plant life and decreases its quality, which leads to the various health hazards to the animals and the human population. Pollution of soil and water by human activities has become a great concerned issue to the scientists. In agricultural activities, the use of an excessive amount of fertilizers, pesticides and insecticides leads to the release of heavy metals in the soil and creates a big problem of formation of barren land. Heavy metal affected soil is not suitable for growing of the crops. Among pollutants, heavy metals are placed in the second category in terms of causing pollution of the soil, water and environment. The heavy metal ions enter the food chain and cause symptoms of toxicity in the animal and human system. There is a need to enhance the properties of the food products for the wellbeing of the population. Plants have various mechanisms fighting against the stress of heavy metals. Among these, plant growth hormones are very important to regulate the heavy metal stress in the plant system. Plant growth regulators play a very important role in protecting the plant system and its various processes from the toxic effect of heavy metals. Plant growth regulators are very important for the growth and development of the plant. They play an important role in the plant system for the proper growth and to influence all the properties of the plant. Plant growth hormones are very essential compounds of the plant system, which performs the various functions in the cellular compartments. In stress conditions, plant growth hormones activate various plant processes to enhance plant growth and actively perform their functions under stresses conditions. Abscisic acid (ABA), salicylic acid (SA), and Jasmonic acid (JA) are the hormones involved in the stress tolerance in the plants. Growth hormones function in the physiological mechanism of the plant and protect the plant against the toxic effect of heavy metal stress. Therefore, the main purpose of this review is to understand the role of plant growth hormones in tolerating or regulating the stress caused by heavy metals. Plant growth hormones involve Indole-3Acetic Acid (IAA), Cytokinins (CK), Ethylene (ET), Abscisic Acid (ABA), Jasmonic Acid (JA), Salicylic Acid (SA), and Brassinosteroids (BR) to play a vital role in regulating the heavy metal stress and are very important for the mitigation of heavy metal stress in the plant. Plant growth hormones protect plants from different types of biotic and abiotic stresses and regulate their functions (Bucker-Neto et al., 2017). Plant growth hormones are applied externally on the plants to achieve good results and it is a very safe method (Masood et al., 2016). Cellular absorption of these hormones helps in regulating the toxicity of heavy metal ions and enhances the survival of plants under heavy metal stress conditions. Different plant growth hormone has a significant role in tolerating the toxicity of heavy metals. They perform various functions in the plant system to avoid the toxic effect of heavy metals and influence all the processes of the plant for its development and higher yield potential. Phytohormones priming is a recently developed concept, which plays a major role in regulating metal toxicity in the plants.

Importance of Plant Growth Hormones in Plant

Plant growth hormones are chemical compounds, which are important in every stage of the plant life for the proper growth and development of the plants. The different hormone plays a different role in plant system for the growth and development of the plant. A single hormone can perform multiple tasks and sometimes all hormones combined to perform a single task in the plant system (Gray, 2004). These are an essential part of the plant system and are very important for the betterment of the plant processes. These are also important in tolerating various biotic and abiotic stresses of the plant system. Strigolactone (SL) is a plant growth hormone, which is identified recently and plays an important role in protecting the plants from different types of stresses and help to influence their growth and development. Plant growth hormones involve Indole-3Acetic Acid (IAA), Cytokinin (CK), Ethylene (ET), Abscisic Acid (ABA), Jasmonic Acid (JA), Salicylic Acid (SA), and Brassinosteroid (BR), which plays a major role in stimulating various processes of the plant life. Phytohormones protect plants against various biotic and abiotic stresses (Xu et al., 2016). These regulate the toxicity of heavy metals and help agronomical crops to produce higher yield (Piotrowska-Niczyporuk et al. 2012). External application of Phytohormones regulates the stress caused by heavy metals and influences further its development. It is a very significant and widely used method of application of Phytohormones, which is easily absorbed by the plants, and it is a very safe method. Different Phytohormones has a various role in signalling, biochemical pathways, defence mechanism and tolerating the stress of heavy metals. Heavy metal stress is a topic of interest now days and is there is a need to increase the quality of the product and reducing the toxicity of heavy metals from them. Therefore, the application of plant growth hormones in regulating heavy metal stress is a topic of great interest among scientists to keep the level of toxicity of heavy metals among agronomical crops (Fig. 1).

Fig. 1. Significant role of Phytohormones in plants

ROLE OF PHYTOHORMONES IN PLANTS

Brassinosteroids
- Enhance antioxidant
- Synthesis of IAA
- Stimulate growth of cell
- Regulate seed germination and stomatal differentiation

Auxins
- Stimulate root & shoot growth
- Increase metal phytoextraction
- Inhibit lateral buds growth

Gibberellins
- Stimulate cell elongation
- Regulate N assimilation
- Regulate metal transport

Salicylic Acid
- Help in photosynthesis
- Increase electrolyte leakage
- Decrease ROS
- Improve ion uptake

Cytokins
- Stimulate cell division
- Increase transpiration rate
- Increase antioxidant capacity

Strigolactones
- Promote parasitic plant germination
- Inhibit branch formation
- Recognize plant by symbiotic fungi

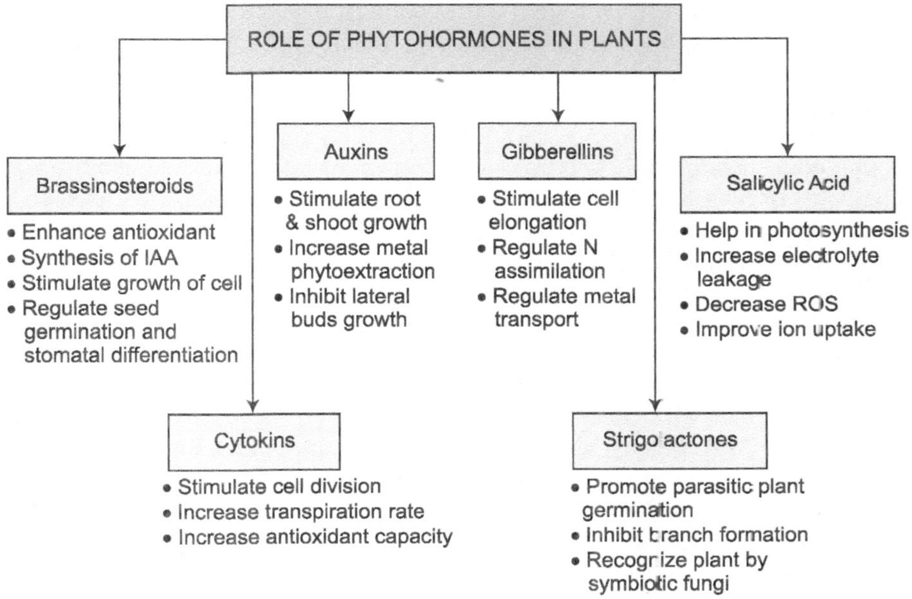

Phytohormones have various functions to perform in the plant system and to enhance the physiological and biochemical processes of the plant. They influence the cell division, stimulate the growth of root, and shoot. Phytohormones regulate the seed germination and stomatal and vascular differentiation to achieve good growth and development of the plant. Phytohormones help to increase the rate of photosynthesis, transpiration and improve the uptake of ions and nutrients from the soil. They promote plant system to cope with any type of biotic and abiotic stresses and functions to survive under stressed conditions. Plant growth hormones function in the transport mechanism of the plant and help in the transportation of minerals, nutrients and ions. They enhance the rate of cellular mechanism and assimilation of nitrogen in the plants. Plant growth hormones help the plant to achieve enhanced growth and higher yield potential so that they can meet the demand of the population.

Causes of Metals Toxicity on Plants

Heavy metals are toxic compounds, which affect the growth of the plant and population. The accumulation of heavy metals in the plants affects the plant system in a great way. The heavy metals show toxicity in all parts of the plant when it is transported from one place to another. It shows adverse effects as it is transported from root to shoot part. Heavy metals mostly affect the cellular activities of the plant system so that there is a decrease in all the processes of plant life including - various physiological and biochemical processes. Toxicity of heavy metals affects the growth and development of the plant and decreases the quality yield potential of the plant system. Various heavy metal cause toxicity in the cellular mechanism and affect the physiological activities of the plant, which may decrease the development and yield of the plant under heavy metal stress (Singh and Prasad, 2011). Various studies show the toxicity of Copper (Cu) that heavy metal toxicity causes a great reduction in the quality yield of the crops. Toxicity of heavy metal ions causes a great reduction in the root and shoots growth of the different crops. The Copper (Cu) Cadmium (Cd) and Chromium (Cr) present in higher concentration in the crop plants and damage the internal activities and mechanisms of the plants. The toxicity of chromium (Cr) causes a decrease in the process of photosynthesis in the plants and declines the growth of the plant. The inhibition of the photosynthetic activity leads to a decrease in the rate of enzymatic processes, exchange of gases, chlorophyll content and assimilation of N and CO_2 in the plant system. The toxicity of heavy metal decreases all the processes of plant life, which are important for the development of the plant and damage the plant cells and other organs. Heavy metal stress leads to the existence of other abiotic stresses to the plant also. The toxicity of chromium (Cr) causes a decrease in the accumulation of biomass in the leaves of the barley plant and further decrease in its grain and fodder yield. Chromium (Cr) toxicity leads to the formation of a large structure on the leaves of the barley plant and leads to the decline in quality yield by decreasing the growth and development. These structures, which are formed on the leaves, include thickening and swelling of the chloroplast and membrane cells disruption. This toxicity of Chromium (Cr) inhibits the photosynthetic rate, chlorophyll content, transpiration rate, CO_2 accumulation and regulation of the stomatal conductance (Ali et al., 2013).

Arsenic (As) is a very toxic element and can adversely affect plant growth and quality yield production. It can cause serious health hazards to humans and animals. The toxicity of Arsenic (As) causes a great decline in all the physiological and biochemical activities of the plant system. Its toxicity depends upon its concentration, time and stage of the plant (Dho et al., 2010). The heavy metal Arsenic (As) interferes with the cellular components of the plant and causes a great reduction in the mechanism and action of different cells and tissues. Arsenic (As) replace phosphate from the ATP molecule and decrease the amount of energy in the plant. The different forms of Arsenic (As) inhibit the different functions of the plant system. The stress of Arsenic (As) decreases the relative growth rate of the plant and inhibits physiological activities of the plant which leads to diminishing development and low yield (Gusman et al. 2013). The toxicity of Arsenic decreases the activities of various enzymes, which are very helpful for the growth, and development of the plant. Various enzymes like nitrogenase, nitrate reductase, nitrite reductase etc. inhibit their function under the toxicity of heavy metals (Porter and Sheridan, 1981; Singh et al., 2009). The toxicity of Cadmium (Cd) also affects the growth of the plant and causes a great decline in the yield potential of the plant (Wu et al., 2007). The Cadmium (Cd) toxicity also causes a great reduction in the growth and damage the plant by causing chlorosis and necrosis in the leaves. Heavy metal lead (Pb) also causes great damage to the plant system by affecting the morphological and physiological functions of the plant system. Lead (Pb) toxicity affects the biosynthesis of chlorophyll and rate of photosynthesis that cause a great decrease in the growth and development of the plant. It decreases the biomass of the plant and inhibits the functioning of secondary metabolites. Heavy metal causes toxic effects to the plants by decreasing all the cellular and molecular activities of the plant, which help the plant for their proper development, and production of the higher yield. The higher amount of heavy metal Lead (Pb) in the plant's cause difference in the permeability of the cell membrane disturbs plant water relation and inhibits the enzymatic activities of the plant (Sharma and Dubey, 2005).

Plant Growth Regulators under Metal Toxicity

Plant growth regulators are chemical compounds, which play an important role in proper growth and development of the plant and protect the plant from different biotic and abiotic stresses. Phytohormones play a major role in protecting the plant under heavy metal stress conditions. Plant growth hormones influence the physiological and molecular processes of the plant and play a major role in tolerating the heavy metal stress (Fahad et al., 2015). Plant growth hormone influences the permeability of cell membrane, enzyme activities and growth of secondary metabolites at very low concentration (Wani et al., 2016). They regulate the growth and reproduction process in the plants. Growth regulators perform the functions in the cellular compartments to regulate the physiological functions of the plant under heavy metal stress. Heavy metal stress in plants causes a decrease in all the structural and functional activities of the plants. The toxic ions of heavy metals combined in the plant root, shoot portion and decrease the cellular mechanism of the plants. Plant growth hormones also help in vacuole and trichome sequestration of heavy metals and help the plant to grow under heavy metal stress conditions. Plant growth hormones regulate root and shoot biomass, which is decreases due to heavy metal stress. It increases the cellular activities of the underground and aboveground region and enhances the biomass of the plants, which help them to survive under metal stress conditions. Many heavy metals are very toxic and affect the germination of seed in the soil. These heavy metals of high concentration include - Arsenic (As), Cadmium (Cd), Copper (Cu), Lead (Pb), Chromium (Cr), Nickel (Ni) etc. These heavy metals in higher concentration greatly affect the plant system (Sethy and Ghosh 2013). They affect the germination of the seed or delay the seed germination and unable them to grow further. Heavy metals at higher concentration affect all the activities of the plant and decrease the growth and yield potential of the plants. Heavy metal Nickel (Ni) regulates the enzymatic activities of the plant and influence the functioning of various enzymes like amylase, ribonuclease, protease etc. Plant growth hormones enhance these enzymatic activities of the plant and regulate its growth and development (Ahmad and Ashraf, 2011). Under lead (Pb) toxicity, plant growth hormones influence the activity of enzyme peroxidase, polyphenol oxidase, phosphatic acids, carbohydrate metabolism, α and β-amylases, and another metabolism of the plant (Singh et al., 2011). Growth hormones perform its functions in the plant cells and within the tissues of the plant and stimulate the functioning of various enzymes. Toxicity of Copper (Cu) also inhibits the catalase activity of the plants and cause oxidative damage to the plant system. Cellular activities of the plants are influenced by the plant growth hormones, which lead to the survival of plants under heavy metal stress conditions. Under Copper (Cu) stress, the toxicity of the Copper (Cu) decreases the protein oxidative damage (Pena et al., 2011). Under this, plant growth hormones stimulate plant system to cope with this stress and increase all the activities of the plant system. Toxicity of Cadmium (Cd) regulates some other enzymes of the plants like glutathione peroxidase and reductase activity, which causes a great effect to the plant. Increase and decrease in the enzymatic activities of the plant system depend upon the type and concentration of heavy metal present in the plant system. The model plants are used to detect the effect of plant hormone on the toxicity of the heavy metal ions on the plants (Bulak et al., 2014). The higher and lower concentrations of the heavy metals are responsible for the harmful effect caused to the plant system. Lower concentration of heavy metal does not cause much damage to the plant as compared to their higher concentrations (fig. 2).

Fig. 2. Plant growth hormones under heavy metal stress

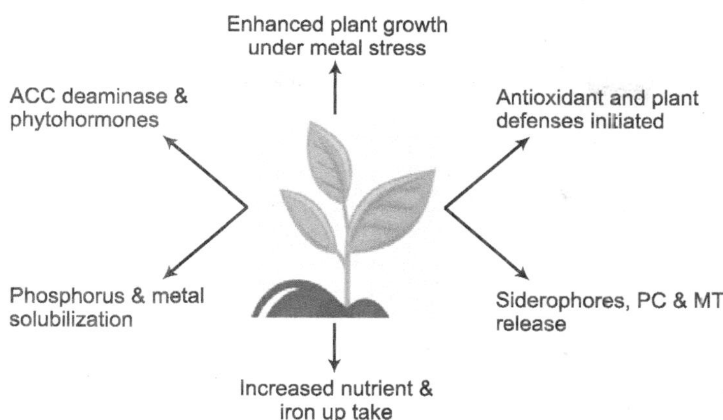

Heavy Metals Toxicity and Role of Phytohormones in Plants

Plant growth hormones play a vital role in protecting the plants under various biotic and abiotic stresses. Plant growth hormones are also called as Phytohormones. These are organic substances, which regulate the growth and development of the plants. Plants have various mechanisms to survive under heavy metal stress conditions. Plants contribute to different plant growth hormones to protect them from the toxic effect of heavy metals. Plant growth hormones include Auxins, Cytokinin (CK), Salicylic Acid (SA), Ethylene (ET), Jasmonates (JA) etc. Plant growth regulators are the fabricated compounds, which are applied externally to the plants to protect them from heavy metal toxicity. Plant growth hormones are very important for protecting the plant under various types of stress conditions and help them to survive during metal stress and increase its yield potential.

Abscisic Acid (ABA)

Abscisic acid (ABA) is a very important plant growth hormone that is needed for plant growth and development and plant protection under heavy metal stress conditions. It helps the survival of plants under different biotic and abiotic stresses and regulates the stress caused by heavy metals. Heavy metal stress affects the level of abscisic acid (ABA) that regulates the other components of the plant system for access to plant growth and development. With the increase in the level of endogenic abscisic acid (ABA), the activation of signaling pathways and genetic expressions is taking place which influences the overall development of the plant and regulates the structural and physiological functions of the plant (Brien and Benkova, 2013). Under the toxicity of metal ions, plant growth hormones are used in the cellular mechanism of the plant system to stimulate the growth and development of plants. Abscisic acid (ABA) regulates the encoding of the gene in the plant and helps the survival of the plant under heavy metal stress. Abscisic acid (ABA) has specific functions in the plant's cellular mechanism. The toxicity of heavy metal ions causes the synthesis of different growth hormones in the plant. Due to the toxicity of Cadmium (Cd) in plants, there is a malfunction in the physiological and biochemical activities of the plant. Higher concentrations of heavy metals maximize the level of abscisic acid (ABA) that regulates physiological functions such as stomata opening and closing, plant water relation and signaling pathways in plants (Sharma and Kumar, 2002). Abscisic acid (ABA) regulates the genetic activity of the plant and stimulates the growth of plants under heavy metal stress. Abscisic acid (ABA) regulates heavy metal stress and balances the survival and growth of the plant system under heavy metal stress conditions. The role of growth hormones in tolerating the toxicity of heavy metals is very specific, which helps stimulate plant endogenous activity and shows promising results in tolerating heavy metal stress.

Auxins

Auxins play an important role in the regulation of plant metabolites. Under stress conditions, there is an increase in the level of auxins, which regulates the root and shoots growth of the plant. There is an increase in the protection of the cellular compartments of the plants and regulation of the metabolic processes. The increase in the level of auxins in the plants decreases the toxicity of the metals inside the plant and helps them to survive under heavy metal stress. The role of auxin is clear to enhance the development of the plant and increase its production. In heavy metal stress conditions, the exogenous application of the growth hormone auxin increases the activity of Indole Acetic Acid (IAA) and other hormones to protect the plants from metal stress. A known concentration of auxin is applied externally on the surface of the plant to protect them from the toxicity of heavy metal ions. It has been observed from a study that the exogenous application of auxin promotes the growth, yield potential of the Brassica juncea, and help it to survive under metal toxicity (Srivastava et al., 2013). Application of Indole Acetic Acid (IAA) also plays an important role in tolerating heavy metal stress to the plant. Auxins regulate the toxicity of the heavy metal ions and stimulate the growth and development of the plant system, which cause a significant increase in the quality yield production of the plants. The growth hormone auxin also indicates the sign of stress and leads to the influence of the synthesis of other plant growth hormones to protect the plants under heavy metal stress. Auxin is a major plant growth hormone, which plays a significant role in different processes of plant life. It is very important for regulating the stress caused by heavy metals in the plants. Auxin stimulates the growth regulators and helps in the phytoextraction of the lead (Pb) from the plants. It increases the level of growth hormones (GHs) in the plants and helps them to survive under heavy metal stress conditions. Auxin plays a promising role in protecting the plants against heavy metal toxicity and helps them to perform their physiological and biochemical functions for the proper growth and development of the plants. An increase in the level of ROS enhances the response of the auxin under heavy metal stress conditions. Thus, auxin is a major plant growth hormone, which performs significant functions in the plant system to secure them from the various biotic and abiotic stresses.

Gibberellic Acid

Gibberellic acid (GA) is a very important plant growth hormone, which is required for stimulating the growth and development of the plant. This hormone is required during different processes of the plant life at a different stage of plant growth from seed germination to maturity. Gibberellins (GA) influence cell division, cell elongation and cell development in every stage of the plant life. The Gibberellins (GA) are plant growth hormones that belong to a large family of tetracyclic diterpenoid. It is joined with various processes of the plant growth like seed germination, stem elongation and helps the plant to achieve good yield potential. All the functions of leaf expansion, flower development, grains production and development are regulated by gibberellins (GA) (Matsuoka, 2003). Gibberellin is the plant growth hormone, which influences the plant system to mitigate the heavy metal stress and protect the plant from heavy metal stress. This hormone functions to survive the plant even in metal stressed conditions and help to achieve great growth and development. Gibberellins (GH) hormone influences the rate of cell division and development, stimulate seed and bud dormancy, growth of root cells and tissues and stimulate plant growth even at very low temperature. Gibberellin hormone (GA) present in young fruits, germinating seeds and tissues and cells of root and shoot which help the plant for their growth and development. Gibberellic acid (GA) influences the loading of phloem and synthesis of fructose-1, 6 bisphosphatases and sucrose phosphatase which helps the plants in regulating physiological activities of the plant system ((Iqbal et al., 2011). It performs a great role in protecting the cells from heavy, metal stress also. The role of Gibberellic acid (GA) in protecting the plant under heavy metal stress maximizes the growth and yield of the plant. Gibberellic acid functions in the formation of source and sink and is a hormone, which plays a significant role in protecting the plants under heavy metal stress. Various studies declare that Gibberellic acid (GA) regulates the heavy metal stress in the plant and leads to the phytoextraction of the heavy metals from the plants. It reduces the toxic effect of the heavy metals and stimulates the functioning of the plant for its growth and development. Gibberellic acid (GA) decreases the toxic effect of cadmium (Cd) by reducing the uptake by the roots and lipid peroxidation (Zhu et al., 2012). Gibberellic acid (GA) also functions in the cellular mechanism and helps in vacuole sequestration of the heavy metals. It regulates the functions of tissues and cells in the plants and influences the phytoextraction of cadmium and other ions of heavy metals from the plant system. Nickel (Ni) is reported in a wheat plant to cause damage to the growth, chlorophyll content and other enzymatic activities by increases the oxidative stress in the plant (Siddiqui et al., 2011). Gibberellic acid (GA) decreases the ROS activity of the plants and cause a reduction in the oxidative stress of the wheat plant. Gibberellic acid (GA) also reduces the adverse effect of Chromium (Cr) and regulates the antioxidant properties of the plant system. These growth hormones succeed in the process of protection of plants from the toxic effect of heavy metals and increase the growth and yield production. Gibberellic acid (GA) plays a pivotal role in promoting the plant biomass and physiological processes of the plant system, which leads to an increase in the yield, and production of the plant's system. Gibberellic acid (GA) is a specific and significant hormone, which involves the phytoextraction of the heavy metals from the plants (Table 1).

Table 1: Plant growth hormone and plant response under heavy metal stress

Plant growth hormone	Plants	Plant parts	Heavy metals	Plant reaction	References
	Mustard	7 Days old seedlings	Copper (Cu)	Shoot emergence improved, block the absorption of Cu metal	Sharma and Bhardwaj (2007)
	Radish	at the harvesting stage	Copper (Cu)	reduce the toxicity of Cu	Choudhary et al. (2010)
Brassinosteroids	Brassica sp.	Aerial part 40DAS	Nickel (Ni)	Decrease Ni toxicity, stimulate photosynthetic rate, chlorophyll content and enzymatic activities	Alarm et al. (2007)
	Rape-seed	14 days old cotyledons	Cadmium (Cd)	Decreases Cadmium toxicity, cause damage to photochemical reactions	Janeczko et al. (2015)
	Radish	Leaves	Cadmium (Cd)	Stimulate the process of photosynthesis	Anuradha and Rao (2009)
	French bean	Biomass and roots of 30DAS plant	Cadmium	Increase all the enzymatic activities of the plant	Rady (2011)
	Chick-pea	60 DAS leaves	Cadmium (Cd)	Influence the level of the antioxidant system, protects the plant growth	Hasan et al. (2008)
Gibberellin	Pea	Roots and shoots of seedlings 11DAS	Chromium (Cr)	Increase N assimilation and activities anti-oxidative of enzymes	Gangwar et al. (2011)
	Arabidopsis thaliana	Roots of the plant	Cadmium	Reduce the accumulation of metal	Zh et al. (2012)
	Parthenium hysterophorus	Roots, stem and leaves of 60 DAS	Cadmium (Cd)	plant growth and biomass Increases	Hadi et al. (2014)

	rice	Above ground and underground portion of seedlings	Iron (Fe)	Transport of metal ions	Wang et al. (2017)
Salicylic acid	Cucumber	Aerial parts of the plant 33DAS	Manganese	Reduction of the level of ROS	Shri and Zhu (2008)
	Linum usitatissimum L.	Stem and aerial parts of seedlings	Cadmium	enhance total lipid content in the plants	Belkhadi et al. (2010)
	Mustard	Stem and leaves of 30 DAS	Nickel	Stimulate antioxidant enzymes activities	Kazemi et al. (2010)
	Peanut	Stem and leaves of 30 DAS	Cadmium	Improve plant growth and decreases metal toxicity, improve photosynthesis and chlorophyll content	Xu et al. (2015)
	Malting Barley	Three leaf stage leaves	Zinc (Zn), Cu, Mn, Cd, Hg, Pb	Inhibit the activities enzymes at higher amounts	Song et al. (2014)
	Mustard	Aerial and underground parts of a 15-day-old plant	Lead	Increases the heavy metal tolerance and plant-water relation in the plant.	Kohli et al. (2018b)
	Mustard	The above and underground part of 28DAS of plant	Lead	Regulated the gene expression and enzymatic activity of the plant. Increase the changes in defence mechanism.	Kohli et al. (2018c)

Cytokinin	Sunflower	50DAS of plants	Lead, Zinc	Mitigation of metal ions, Increase the biomass production, stimulates cell growth, Increases the rate of transpiration, chlorophyll and photosynthesis.	Tassi et al. (2008)
	Alyssum murale	Whole plants	Nickel	Increases the biomass of the plant	Cassina et al. (2011)
	Chlorella	Cells	Cadmium, copper, lead	Regulates heavy metal accumulation, Increases primary metabolites, enhance algal growth and defence system.	Piotrowska-Niczyporuk et al. (2012)

Cytokinins

Cytokinin is a plant hormone, which is very important for plant development. It is required for the regulation of growth and development and utilized by the plant as stress tolerance. A cytokinin is an important group of plant growth regulators, which is involved in cellular maintenance and prevents premature leaf senescence (Schmulling, et al., 2002). Overexpression of phytohormone cytokinin increases the ability of the plants to tolerate drought stress and helps them to survive under drought stress conditions. Cytokinin is translocated in one cell to another through the vascular bundles and then promotes the enzymes to protect the plant from drought stress. Plant growth hormone Cytokinin (CK) synthesized in the plants itself under the increased concentration of ROS activities. These are produced in roots, shoot, fruits and seeds of the plant. cytokinin are transported from root to shoot portion by the transport mechanism of the plant. Xylem helps in the transport of growth hormone cytokinin (CK) to all parts of the plant system. Continuous supply of cytokinin (CK) helps the plant to grow faster. If a plant is cut from the shoot portion or from where the supply of cytokinin is going on, the ageing of the plant occurs as compared to the plant, which is attached to the roots or where a continuous supply of cytokinin (CK) is in progress. Cytokinin (CK) plays an important role in the functioning of the plant system under heavy metal stress conditions. It stimulates the development, and response in adverse conditions of the plant growth. Cytokinin (CK) plays an important role in acclimatization of the plant under the stressed condition as like to the function of auxin and gibberellins. Cytokinin (CK) interacts with other phytohormones also and helps in regulating the various types of biotic and abiotic stresses

(Ha et al., 2012). Cytokinin (CK) also performs the role along with other plant growth hormones and stimulates the physiological and biochemical functions of the plant along with them. Heavy metals stress leads to a decrease in the level and functioning of cytokinin (CK) hormone. Various biotic and abiotic stresses like light, temperature, water deficit, salt and heavy metals cause a reduction in the level of the cytokinin (CK) in the roots, shoot and other parts of the plants and leads to a decrease in the growth and development of the plant, which ultimately decreases the quantity of quality produce. Cytokinin (CK) in the plants leads to an increase in the internal functions of the plant and make them suitable to achieve a good yield potential which is becoming a great issue now a days. The external application of cytokinin (CK) can help increases the level of this hormone in the growing region of the plant. An increase in the level of cytokinin (CK) can help in promoting the capacity of stress tolerance in the plants. It helps the plants to tolerate the different type of environmental stresses. It has been found that there is a need to increase in the cytokinin (CK) level to promote the growth of the plant under toxicity of heavy metal ions. It is important to influence the agronomical production. Cytokinin (CK) extracts the heavy metals from the plant system and performs the function of acclimatization of the plants under stress conditions and deficiency of nutrients. Cytokinin (CK) promotes the uptake of minerals and nutrients from the soil and help in the transportation of beneficial metabolites in the plant system. From a study, it has been shown that in the tobacco plant, a gene synthesises the production of cytokinin (CK) in the plants, which cause changes in the oxidative metabolism of the plant and regulate the growth rate of the plant. It causes an increase in the biomass of the plant and increases the accumulation of water and other nutrients from the soil (Rubio-Wilhelmi et al., 2011). Maximization for cytokinin (CK) improves the nitrogen use efficiency in the plants. In Arabidopsis, plant growth hormone Cytokinin (CK) plays a major role in regulating the deficiency of potassium ions (Nam et al., (2012). Various studies show the use of cytokinin – deficient mutants and cytokinin- receptors to decrease the level of potassium deficiency in plants. Cytokinin leads to a decrease in ROS accumulation and stimulates the growth of the hairs of the root zone. Thus, cytokinin (CK) proves to be an essential hormone in promoting the growth and development of the plant and to tolerate the heavy metal stress at any stage of the plant life.

Plant Response to Heavy Metal Stress with External Application of Plant Growth Hormones

The application of plant growth hormones externally on the surface of the plant has great potential to tolerate the toxicity caused by heavy metals. It enhances the tolerance capacity of the plants and promotes them to survive under heavy metal stress. Various studies show that plant growth hormones play an important role in response to heavy metal stress. The application of plant growth hormone decreases the toxic effect of heavy metals and enhances the growth and development of the plant. The exogenous application of plant growth hormones regulate the physiological properties of the plant and make them able to survive in the toxicity of heavy metal. The function of the Cytokinin (CK) is observed by the exogenous application of this hormone to the plant surface, which helps in the biosynthesis of proteins and other essential metabolites (Pospisilova, 2003). Cytokinin (CK) hormone also regulates the function of secondary metabolites in the plant system, which performs the functions of plant growth under heavy metal stress. In the wheat plant, the application of growth hormones externally on the surface of the plant can stimulate soluble compounds like proline and phenols in the plants (Ergun and Oncel, 2012). The exogenous application of abscisic acid (ABA) and Gibberellin (GA) also increases the phenolics and protein content of the plant. The toxicity of heavy metal Cadmium (Cd) is regulated by exogenous application of plant growth hormones (Ahmad et al., 2015). Different growth hormone performs different functions in regulating heavy metal stress in the plant system. Plant growth hormones perform their functions in alone and in combination with other growth hormones. The functioning of plant growth hormones depends upon plant type, the growth stage of the plant, the concentration of heavy metal inside the plant, type of heavy metal and plant growth hormone. The placement of Gibberellin (GA) on the surface of the plant at adequate concentration leads to an increase in the level of other growth hormones in the plant and they joined to function in the heavy metal stress conditions. The application of Gibberellin (GA) regulates the toxicity of lead (Pb) in the plants (Nuray and Isil, (2012). The Cadmium (Cd) toxicity is regulated by the exogenous application of Indole Acetic Acid (IAA) in the plants (Bashri and Prasad, 2015). The concentration of Indole Acetic Acid (IAA) regulates the toxic effect of heavy metals in the plants. Plant growth hormone regulates the extractions of ions of heavy metals Cadmium, Lead and Zinc from the plants and increases the growth of the plant (Fassler et al and Hadi et al., 2010). Plant growth hormones stimulate the root and shoot growth of the plants. The application of growth hormones externally on the plant surface is a very efficient method by which the concentration of growth hormones is easily absorbed by the plants. The application of Indole Acetic Acid (IAA), Gibberellin (GA) and Cytokinin (CK) extract the heavy metal ions from the plant and stimulate an increase in the biomass of the plant (Aderholt et al., 2017). External application of plant growth hormones regulates the N assimilation and cause a reduction in the absorption of metal ions in the plants. The toxicity of heavy metal ions is mitigated by the exogenous application of the plant growth regulators on the plants. The rapid absorption of the plant growth hormones takes place as it is applied externally on the surface of the plant. The activities of N assimilation, increase in photosynthetic and chlorophyll content and further enhances the growth and development of the plant (Zhu et al., 2012). The application of Gibberellic Acid (GA) increases the physiological activities of the plant and extracts the toxicity of Iron (Fe) from the plants. Gibberellic Acid (GA) regulates the translocation of iron (Fe) in the plant system through the transport mechanism of the plant (Wang et al., 2017). Cytokinin (CK) stimulates the cell division, cell elongation and cell development and cause addition in the aerial parts of the plant. The insertion of Cytokinin (CK) enhances the root and shoots development by regulating the growth and development of the plant cells (Tassi et al., 2008). The external application of Cytokinin (CK) influence the rate of photosynthesis and regulates the rate of transpiration, which leads to an increase in the plant biomass and quality yield production (Piotrowska - Niczyporuk et al., 2012) (fig. 3).

Fig. 3. Mechanism of Phytohormones under HM Toxicity

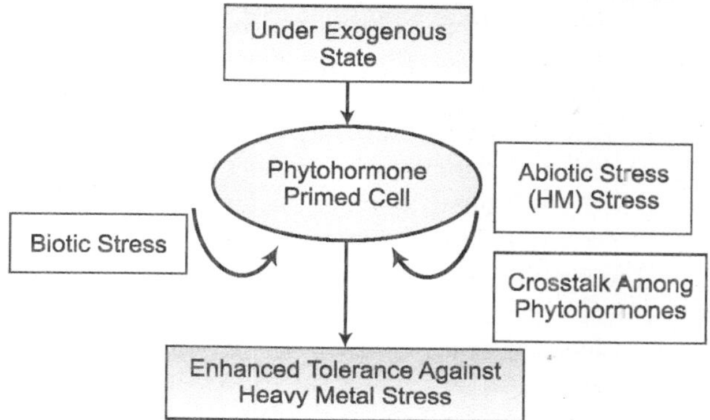

External application of Phytohormones on the aerial parts of the plant can enhance the rapid absorption of growth hormones in the plants and causes an increase in the functioning of plant growth hormones to tolerate heavy metal toxicity. Arabidopsis thaliana is a model plant, which is used to check the concentration of growth hormone to be used to detoxify the effect of heavy metals. During heavy metal toxicity, it is used to find the interaction of signalling pathways with growth hormones. The concentration and function of brassinosteroid are used interaction with other secondary metabolites and signalling pathways to enhance the growth and functioning of root and shoot of the plants. The external application of Brassinosteroids (BR) suppresses the association of microbes and activates the metabolites which response to heavy metal stress in the plants. The application of Brassinosteroids (BR) also increases the rate of wound triggered the response of the plants under the toxicity of heavy metal ions (Bjornson et al., 2016). Brassinosteroid (BR) is an important plant growth hormone, which plays a significant role in tolerating the toxicity of heavy metal in the plant. It increases the antioxidant mechanism of the plant and enhances the growth and functioning of various enzymes. The activities of enzymes catalase, peroxidase, superoxide dismutase etc. are influenced by the exogenous application of the Phytohormones on the plants. The application of Brassinosteroids helps in decreasing the Relative Oxygen Species (ROS) activity and level of carbonyls in the plant (Rady, 2011; Ramakrishna and Rao, 2012). Application of Brassinosteroids (BR) increases the synthesis of other plant growth hormones like auxins, Indole Acetic Acid (IIA), and Gibberellins (GA) that further functions in the growth and development of the plants (Choudhary et al., 2011). Plant growth hormone Salicylic acid (SA) also helps in promoting the growth of the plants. The application of Salicylic acid (SA) also play a crucial role in decreasing the Relative Oxygen Species (ROS) and influence the activation of enzymes lipid peroxidase, electrolyte leakage, chlorophyll and photosynthesis rate to increase the plant biomass and its yield potential (Kazemi et al., 2010). Salicylic acid increases the photosynthesis and chlorophyll content in the plants. It also increases the mineral and nutrient uptake from the soil

(Xu et al., 2015a, b). Plant growth hormones tolerate the heavy metal stress and influence the metabolic and biochemical processes of the plants. Many studies have been observed that the application of plant growth hormones like Gibberellins (GA), Cytokinins (CK), Auxins, Indole Acetic Acid (IAA) cause reduction in the absorption of heavy metal ions in the plants and helps them to grow under metal stress conditions (Bulak et al., 2014). Plant growth hormones regulate the growth and development of the plants and reduce the toxicity of heavy metals from the plant system. Plants growth hormones prove a lead role in tolerating the plants physiological and biochemical functions to protect them from heavy metal stress conditions and to make them able to survive under metal stress. Different growth hormones have a different mode of action to protect the plants and crops from damage caused by the toxic effect of heavy metals. The exogenous application of plant growth hormones proves reliable outcomes in tolerating the toxic effect of heavy metals. Application of Jasmonic acid (JA) protects the plant from the toxicity of Cadmium (Cd) and regulates the functioning of many enzymes in the rice plant. This hormone is easily absorbed by the cell membrane as applied externally on the plants and stimulates the activities of CAT and SOD (Per et al., 2018). The application of an accurate concentration of Jasmonic acid (JA) on the surface of the plant can improve the physiological properties of the plant and can increase its growth and quality yield. The application of Jasmonic acid (JA) on the plants can reduce the toxicity of Cadmium (Cd) and regulates the photosynthetic and stomatal activities. It also regulates the rate of transpiration and reduces the damage caused by heavy metal toxicity (Chen et al., 2014; Per et al., 2018). Jasmonic acid (JA) in plants regulates various morphological, physiological and biochemical activities and promote them to grow the quality yield. Along with the tolerance against heavy metals, Jasmonic Acid (JA) also regulates the action of genes for the stability of the plant under heavy metal toxicity. The external application of growth hormones promotes a significant growth of the plant and regulates various internal functions of the plant concerning climate alteration (Song et al., 2014; Piotrowska-Niczyporuk and Bajguz, 2014). Use of plant growth hormone Auxin for tolerating heavy metal stress proves to perform a specific function in the development of plant and protection of the plant from heavy metal stress. The growth and development of the plant is depending upon the rate of physiological activities of the plant system. The quality yield generally depends on the rate of photosynthesis, chlorophyll content, rate of transpiration and uptake of different micro and macronutrients (Xu et al., 2015a, b). The rate of photosynthesis in the plant regulates the function and quality yield of the plant. These physiological activities are stimulated by the action of plant growth hormones for the production of the quality yield of the agricultural produce. Indole Acetic Acid (IIA) enhances the rate of PSII and reduces the toxicity of heavy metals to the plant (Ouzounidou and Ilias, 2005). The chlorophyll content and stomatal activities are regulated by Brassinosteroids (BR) and improve the development and physiological processes of the plants. The surface application of Brassinosteroids (BR) enhances the physiological activities of the plant and enhances the functioning of PSII for the proper development of the biomass and quality yield. Thus, it is clear that the surface application of plant hormones proves promising results in tolerating the stress caused by heavy metals in the plants. The application of plant growth hormones plays a lead role in reducing the toxicity of heavy metals in the plants and stimulates the growth and production of quality yield.

Plant growth hormones are the important components of the plant system which help the plant in every adverse condition. They help the plant to function under various kinds of biotic and abiotic stresses and influence the growth and development of the plant. Plant growth hormones protect agronomical as well as horticultural crops from the toxicity of heavy metals. The application of Ethylene on the horticultural crops influences the process of ripening of fruits (Agarwal et al., 2012). Plant growth hormone Ethylene (ET) also influences the rate of photosynthesis and transpiration. It regulates the chlorophyll content and stimulates the activities of other growth hormones in the plant system. The application of Ethylene (ET) increases the activities of amino acid and various enzymes which help in to protect from the toxic effect of the heavy metals. Now days, various new hormones come into existence which helps in the growth and development of the plant and survival of the plant under heavy metal stress conditions. Strigolactone (SL) is the most recently developed hormone which functions in the protection of the plant from the toxicity of the heavy metals. These hormones are the derivatives of carotenoids and possess the functioning of growth promoter and help in the mitigation of heavy metal stress in the plant system. Strigolactone is the plant growth hormone which can protect the plants from heavy metal stressed conditions and are found in the plants which accumulate higher concentration of heavy metals in them e.g. Arabidopsis thaliana (Xie et al., 2010). Strigolactone (SL) plant hormone regulates the growth and development of the plant and its interaction with the environment (Gomez-Roldan et al., 2008; Leyser 2009). It is a recently discovered hormone and has great importance for the good growth and protection from the heavy metal stress. This hormone is widely used for the proper functioning of the plant even in the conditions of heavy metal toxicity by the plants. The exogenous application of the Strigolactone (SL) mitigates the toxicity of the Cadmium (Cd) from the leaves and other parts of the plant system. It has been studied in the various reviews that the recently developed hormone Strigolactone (SL) helps in the mitigation of the Cadmium (Cd) toxicity from the plants. Plants have various other mechanisms to tolerate the toxicity of heavy metals. These include vacuole sequestration; trichome and hydropotes sequestration etc. but among these, the use of plant growth hormones for regulating the stress of the heavy metal is the most important and promising. The performance of growth regulators in mitigating the toxicity of heavy metal ions from the plant system proves a believable result and enhances the quantity and quality of the product which is the basic requirement of the population. Along with the quantity, the quality of the agricultural produce should be best. The plant growth hormone Strigolactone (SL) also regulates the expressions of some mutants or genes which functions in the process of stress tolerance. The exogenous application of the Strigolactone (SL) helps in enhancing the synthesis of various proteins and metabolites which promotes the physiological functions of the plant. The application of Phytohormones externally on the surface of the plant activates the enzymatic and physiological mechanisms of the plant system and increases their synthesis and functioning to mitigate the stress of the heavy metal from the plants and enhancing the growth and development of the plants. Plant growth hormones are the main component of the plant system for promoting the development of the plant and mitigate the heavy metal toxicity from the plants. The level of the heavy metals in the plant depends upon the concentration of the heavy metals in the soil. Lower concentration of plant growth hormones are used to mitigate the lower level of the heavy metals and a little increased concentration is used to mitigate the higher amount of the heavy metal ions in the plant system.

Conclusion

Various biotic and abiotic factors affect plant growth and development. Among them, heavy metal stress is one of the major stresses to the plant species which decrease the plant biomass and leads in lower yield potential of the plants. The toxicity caused by heavy metal ions affects the quality of the agricultural produce which is an emerging issue related to the health of the population. There is a need to improve the quality of our agricultural products for improving the health of the people. Various mechanisms in the plant system help them to grow under the stress of heavy metals. Plant growth hormones play a significant role in protecting the plant from the toxic effect of heavy metal ions. They influence the physiological, morphological and biochemical activities of the plants to attain the proper growth and high production even under heavy metal stress. From this review, we conclude that the application of plant growth hormones proves promising results in mitigation of heavy metal stress from the plants. Various studied shows the importance of plant growth hormones in regulating the molecular mechanisms under heavy metal stress. Plant growth hormones are important at every stage of plant development. They regulate the various biotic and abiotic stresses to the plant and enhance the cellular mechanism of the plant to achieve a higher growth and quality yield. Different plant hormones perform the function of improving the cellular and molecular mechanism of the plant system to influence the yield. Thus, plant growth hormones are very important for the growth and functioning of the plant as well as tolerating the toxic effect of heavy metal stress. Toxicity of heavy metal stress causes a great loss of plant biomass and quality yield production. It decreases the quality of the product and affects the health of the population. Application of plant growth hormones helps to survive the plant in the toxicity of heavy metal and increase the quality of the produce by regulating the heavy metal stress.

Reference

Aderholt, M., Vogelien, D. L., Koether, M., & Greipsson, S. (2017). Phytoextraction of contaminated urban soils by Panicum virgatum L. enhanced with application of a plant growth regulator (BAP) and citric acid. *Chemosphere, 175*, 85-96.

Agarwal, G., Choudhary, D., Singh, V. P., & Arora, A. (2012). Role of ethylene receptors during senescence and ripening in horticultural crops. *Plant signaling & behavior, 7*(7), 827-846.

Ahmad, M. S. A., & Ashraf, M. (2012). Essential roles and hazardous effects of nickel in plants. *Reviews of environmental contamination and toxicology*, 125-167.

Ali, S., Farooq, M. A., Yasmeen, T., Hussain, S., Arif, M. S., Abbas, F., ... & Zhang, G. (2013). The influence of silicon on barley growth, photosynthesis and ultra-structure under chromium stress. *Ecotoxicology and environmental safety, 89*, 66-72.

Bashri, G., & Prasad, S. M. (2015). Indole acetic acid modulates changes in growth, chlorophyll a fluorescence and antioxidant potential of Trigonella foenum-graecum L. grown under cadmium stress. *Acta physiologiae plantarum, 37*(3), 49.

Belkhadi, A., Hediji, H., Abbes, Z., Nouairi, I., Barhoumi, Z., Zarrouk, M., & Djebali, W. (2010). Effects of exogenous salicylic acid pre-treatment on cadmium toxicity and leaf lipid content in Linum usitatissimum L. *Ecotoxicology and environmental safety, 73*(5), 1004-1011.

Bjornson, M., Dandekar, A. M., Chory, J., & Dehesh, K. (2016). Brassinosteroid's multi-modular interaction with the general stress network customizes stimulus-specific responses in Arabidopsis. *Plant Science, 250*, 165-177.

Bücker-Neto, L., Paiva, A. L. S., Machado, R. D., Arenhart, R. A., & Margis-Pinheiro, M. (2017). Interactions between plant hormones and heavy metals responses. *Genetics and molecular biology, 40*(1), 373-386.

Bulak, P., Walkiewicz, A., & Brzezińska, M. (2014). Plant growth regulators-assisted phytoextraction. *Biologia plantarum, 58*(1), 1-8.

Cassina, L., Tassi, E., Morelli, E., Giorgetti, L., Remorini, D., Chaney, R. L., & Barbafieri, M. (2011). Exogenous cytokinin treatments of a Ni hyper-accumulator, Alyssum murale, grown in a serpentine soil: Implications for phytoextraction. *International journal of phytoremediation, 13*(sup1), 90-101.

Choudhary, S. P., Kanwar, M., Bhardwaj, R., Gupta, B. D., & Gupta, R. K. (2011). Epibrassinolide ameliorates Cr (VI) stress via influencing the levels of indole-3-acetic acid, abscisic acid, polyamines and antioxidant system of radish seedlings. *Chemosphere, 84*(5), 592-600.

Ergün, N., & Öncel, I. (2012). Effects of some heavy metals and heavy metal hormone interactions on wheat (Triticum aestivum L. cv. Gun 91) seedlings. *Afr J Agric Res, 7*(10), 1518-1523.

Fahad, S., Hussain, S., Bano, A., Saud, S., Hassan, S., Shan, D., ... & Tabassum, M. A. (2015). Potential role of phytohormones and plant growth-promoting rhizobacteria in abiotic stresses: consequences for changing environment. *Environmental Science and Pollution Research, 22*(7), 4907-4921.

Fässler, E., Evangelou, M. W., Robinson, B. H., & Schulin, R. (2010). Effects of indole-3-acetic acid (IAA) on sunflower growth and heavy metal uptake in combination with ethylene diamine disuccinic acid (EDDS). *Chemosphere, 80*(8), 901-907.

Gangwar, S., Singh, V. P., Srivastava, P. K., & Maurya, J. N. (2011). Modification of chromium (VI) phytotoxicity by exogenous gibberellic acid application in Pisum sativum (L.) seedlings. *Acta Physiologiae Plantarum, 33*(4), 1385-1397.

Gray, W. M. (2004). Hormonal regulation of plant growth and development. *PLoS Biol*, *2*(9), e311.

Ha, S., Vankova, R., Yamaguchi-Shinozaki, K., Shinozaki, K., & Tran, L. S. P. (2012). Cytokinins: metabolism and function in plant adaptation to environmental stresses. *Trends in plant science*, *17*(3), 172-179.

Iqbal, N., Nazar, R., Khan, M. I. R., Masood, A., & Khan, N. A. (2011). Role of gibberellins in the regulation of source–sink relations under optimal and limiting environmental conditions. *Current Science*, 998-1007.

Jaishankar, M., Tseten, T., Anbalagan, N., Mathew, B. B., & Beeregowda, K. N. (2014). Toxicity, mechanism and health effects of some heavy metals. *Interdisciplinary toxicology*, *7*(2), 60.

Kazemi, N., Khavari-Nejad, R. A., Fahimi, H., Saadatmand, S., & Nejad-Sattari, T. (2010). Effects of exogenous salicylic acid and nitric oxide on lipid peroxidation and antioxidant enzyme activities in leaves of Brassica napus L. under nickel stress. *Scientia Horticulturae*, *126*(3), 402-407.

Khan, M. I. R., & Khan, N. A. (2014). Ethylene reverses photosynthetic inhibition by nickel and zinc in mustard through changes in PS II activity, photosynthetic nitrogen use efficiency, and antioxidant metabolism. *Protoplasma*, *251*(5), 1007-1019.

Kohli, S. K., Handa, N Khan, M. I. R., & Khan, N. A. (2014). Ethylene reverses photosynthetic inhibition by nickel and zinc in mustard through changes in PS II activity, photosynthetic nitrogen use efficiency, and antioxidant metabolism. *Protoplasma*, *251*(5), 1007-1019.

Kohli, S. K., Handa, N., Sharma, A., Gautam, V., Arora, S., Bhardwaj, R., & Ahmad, P. (2018). Combined effect of 24-epibrassinolide and salicylic acid mitigates lead (Pb) toxicity by modulating various metabolites in Brassica juncea L. seedlings. *Protoplasma*, *255*(1), 11-24.

Leyser, O. (2009). The control of shoot branching: an example of plant information processing. *Plant, Cell & Environment*, *32*(6), 694-703.

Masood, A., Khan, M. I. R., Fatma, M., Asgher, M., Per, T. S., & Khan, N. A. (2016). Involvement of ethylene in gibberellic acid-induced sulfur assimilation, photosynthetic responses, and alleviation of cadmium stress in mustard. *Plant Physiology and Biochemistry*, *104*, 1-10.

Matsuoka, M. (2003). Gibberellin signaling: how do plant cells respond to GA signals?. *Journal of Plant Growth Regulation*, *22*(2), 123-125.

Nam, Y. J., Tran, L. S. P., Kojima, M., Sakakibara, H., Nishiyama, R., & Shin, R. (2012). Regulatory roles of cytokinins and cytokinin signalling in response to potassium deficiency in Arabidopsis. *PLoS One*, *7*(10).

O'Brien, J. A., & Benková, E. (2013). Cytokinin cross-talking during biotic and abiotic stress responses. *Frontiers in plant science*, *4*, 451.

Ouzounidou, G., & Ilias, I. (2005). Hormone-induced protection of sunflower photosynthetic apparatus against copper toxicity. *Biologia Plantarum*, *49*(2), 223.

Pena, L. B., Azpilicueta, C. E., & Gallego, S. M. (2011). Sunflower cotyledons cope with copper stress by inducing catalase subunits less sensitive to oxidation. *Journal of Trace Elements in Medicine and Biology, 25*(3), 125-129.

Per, T. S., Khan, M. I. R., Anjum, N. A., Masood, A., Hussain, S. J., & Khan, N. A. (2018). Jasmonates in plants under abiotic stresses: crosstalk with other phytohormones matters. *Environmental and experimental botany, 145*, 104-120.

Piotrowska-Niczyporuk, A., & Bajguz, A. (2014). The effect of natural and synthetic auxins on the growth, metabolite content and antioxidant response of green alga Chlorella vulgaris (Trebouxiophyceae). *Plant Growth Regulation, 73*(1), 57-66.

Porter, J. R., & Sheridan, R. P. (1981). Inhibition of nitrogen fixation in alfalfa by arsenate, heavy metals, fluoride, and simulated acid rain. *Plant Physiology, 68*(1), 143-148.

Rady, M. M. (2011). Effect of 24-epibrassinolide on growth, yield, antioxidant system and cadmium content of bean (Phaseolus vulgaris L.) plants under salinity and cadmium stress. *Scientia Horticulturae, 129*(2), 232-237.

Ramakrishna, B., & Rao, S. S. R. (2012). 24-Epibrassinolide alleviated zinc-induced oxidative stress in radish (Raphanus sativus L.) seedlings by enhancing the antioxidative system. *Plant growth regulation, 68*(2), 249-259.

Rubio-Wilhelmi, M. M., Sanchez-Rodriguez, E., Rosales, M. A., Begona, B., Rios, J. J., Romero, L., ... & Ruiz, J. M. (2011). Effect of cytokinins on oxidative stress in tobacco plants under nitrogen deficiency. *Environmental and Experimental Botany, 72*(2), 167-173.

Sethy, S. K., & Ghosh, S. (2013). Effect of heavy metals on germination of seeds. J Nat Sci Biol Med 4: 272–275.

Sharma, P., & Dubey, R. S. (2005). Lead toxicity in plants. *Brazilian journal of plant physiology, 17*(1), 35-52.

Sharma, S. S., & Kumar, V. (2002). Responses of wild type and abscisic acid mutants of *Arabidopsis Thaliana* to cadmium. *Journal of plant physiology, 159*(12), 1323-1327.

Shi, Q., & Zhu, Z. (2008). Effects of exogenous salicylic acid on manganese toxicity, element contents and antioxidative system in cucumber. *Environmental and Experimental Botany, 63*(1-3), 317-326.

Siddiqui, M. H., Al-Whaibi, M. H., & Basalah, M. O. (2011). Interactive effect of calcium and gibberellin on nickel tolerance in relation to antioxidant systems in Triticum aestivum L. *Protoplasma, 248*(3), 503-511.

Singh, A., & Prasad, S. M. (2011). Reduction of heavy metal load in the food chain: a technology assessment. *Reviews in Environmental Science and Bio/Technology, 10*(3), 199.

Singh, H. P., Kaur, G., Batish, D. R., & Kohli, R. K. (2011). Lead (Pb)-inhibited radicle emergence in Brassica campestris involves alterations in starch-metabolizing enzymes. *Biological trace element research, 144*(1-3), 1295-1301.

Singh, I., & Shah, K. (2014). Exogenous application of methyl jasmonate lowers the effect of cadmium-induced oxidative injury in rice seedlings. *Phytochemistry, 108*, 57-66.

Singh, N., Ma, L. Q., Vu, J. C., & Raj, A. (2009). Effects of arsenic on nitrate metabolism in arsenic hyperaccumulating and non-hyperaccumulating ferns. *Environmental pollution, 157*(8-9), 2300-2305.

Sneideris, L. C., Gavassi, M. A., Campos, M. L., D'Amico-Damiao, V., & Carvalho, R. F. (2015). Effects of hormonal priming on seed germination of pigeon pea under cadmium stress. *Anais da Academia Brasileira de Ciências, 87*(3), 1847-1852.

Song, W. Y., Yang, H. C., Shao, H. B., Zheng, A. Z., & Brestic, M. (2014). The alleviative effects of salicylic acid on the activities of catalase and superoxide dismutase in malting barley (Hordeum uhulgare L.) seedling leaves stressed by heavy metals. *CLEAN–Soil, Air, Water, 42*(1), 88-97.

Srivastava, S., Srivastava, A. K., Suprasanna, P., & D'sSouza, S. F. (2013). Identification and profiling of arsenic stress-induced microRNAs in Brassica juncea. *Journal of experimental botany, 64*(1), 303-315.

Tassi, E., Pouget, J., Petruzzelli, G., & Barbafieri, M. (2008). The effects of exogenous plant growth regulators in the phytoextraction of heavy metals. *Chemosphere, 71*(1), 66-73.

Wang, B., Wei, H., Xue, Z., & Zhang, W. H. (2017). Gibberellins regulate iron deficiency-response by influencing iron transport and translocation in rice seedlings (Oryza sativa). *Annals of botany, 119*(6), 945-956.

Wani, S. H., Kumar, V., Shriram, V., & Sah, S. K. (2016). Phytohormones and their metabolic engineering for abiotic stress tolerance in crop plants. *The Crop Journal, 4*(3), 162-176.

Wu, F., Zhang, G., Dominy, P., Wu, H., & Bachir, D. M. (2007). Differences in yield components and kernel Cd accumulation in response to Cd toxicity in four barley genotypes. *Chemosphere, 70*(1), 83-92.

Xie, X., Yoneyama, K., & Yoneyama, K. (2010). The strigolactone story. *Annual review of phytopathology, 48*, 93-117.

Xu, Y. X., Mao, J., Chen, W., Qian, T. T., Liu, S. C., Hao, W. J., & Chen, L. (2016). Identification and expression profiling of the auxin response factors (ARFs) in the tea plant (Camellia sinensis (L.) O. Kuntze) under various abiotic stresses. *Plant Physiology and Biochemistry, 98*, 46-56.

Xu, Z., Jiang, Y., & Zhou, G. (2015). Response and adaptation of photosynthesis, respiration, and antioxidant systems to elevated CO2 with environmental stress in plants. *Frontiers in plant science, 6*, 701.

Zhu, X. F., Jiang, T., Wang, Z. W., Lei, G. J., Shi, Y. Z., Li, G. X., & Zheng, S. J. (2012). Gibberellic acid alleviates cadmium toxicity by reducing nitric oxide accumulation and expression of IRT1 in Arabidopsis thaliana. *Journal of hazardous materials, 239*, 302-307.

Use of Algae in Wastewater Treatments

E Lokesh Goud

Department of Agronomy, School of Agriculture

Lovely Professional University, Jalandhar, Punjab, India,144411

Email: lokeshgoud1832@gmail.com

Abstract

Discharge of contaminants after domestic use, industrial and manufacturing remnants, and drains from the cultivated field into the water bodies without any concern about the treatments provided for such wastewater would cause polluted water bodies and scarcity for quality water. Efficient treatments with a cost-effective approach are desirable. Wastewater can be treated by primary, secondary, tertiary and quaternary phases. Safe discharge cannot be ensured even if the performance of the primary and secondary phase do well due to dissolved nutrients. Algae are the photoautotrophic organism which when added to contaminated water, do uptake and assimilate the dissolved nutrient forms like N (Ammonium and Nitrate) and phosphates. They can adsorb heavy metals on their extracellular and intracellular parts for efficient removal otherwise a threat to living organisms. Incorporation of Algae into the tertiary phase will assist in the removal of nutrients. HRAPs (High rate Algal Pond System) are well-equipped treatment system for wastewater by culturing algae using Wastewater nutrients and exploiting the subsequent harvest of biomass for resource recovery like biofuel, feed, food fertilizers etc.

Introduction

Earth should pay for the water being the reason behind the existence of life and biological phenomenon happening on its surface. The requirement of the same cannot be substituted with any other substances. Studies at the early centuries revealed that Earth has its 71 % composed of water and a major part of this, around 96.5% resides in Ocean, 1% in the form of saline groundwater and remaining 2.5 % found as freshwater. The amazing fact is that about 1% of the freshwater on Earth is in the accessible form and remaining trapped as glaciers and snowfield. Water has an irreplaceable and prominent role in each sector and varying aspects of human life. The burgeoning population has a direct impact on all the natural resources because of high demand and misuse. This is not a different case in the condition of water resources as they have the visual pieces of evidence in our surroundings and ultimately, it results in water stress. Water pollution is one of the serious societal concern, which does not have a permanent solution yet but has stricter enforcements of Environmental laws against the issue. Since agricultural sector utilizes 70% of the water considering worldwide (FAO report for G20 presidency of Germany, 2017), the qualitative degradation along with quantitative depletion of water should be a major area of study due to the large dependence of population for food. Any method used to rejuvenate water resources should belong to sustainable water management. A non-proportionate blend of organic and inorganic materials or particles that came as a remnant from the industries, domestic uses, agricultural uses and municipal waste along with the water creates wastewater. This process of contamination existing in the present scenario has the chance for rapid increase because of the demographic forecast on increasing population and if this continues, it will result in water scarcity. Therefore, recycling or reutilizing those contaminated water by applying proper treatment would be a more suitable technique to tackle the problem. These treatment procedures include physical, chemical and biological procedures. This chapter deals with the treatment of contaminated water or the wastewater by proper analysis and studies with the help of algae.

Algae are identified as those organisms exhibiting photosynthetic properties, which reside in both freshwater ecosystem and marine ecosystem also found some terrestrial species, having a high multiplication rate in their biomass production. They have immense application in the industrial sector for food and fuel production. Their capabilities for absorbing certain irrelevant inorganic substances like heavy metals are exploited in the wastewater treatments. BOD (Biological Oxygen Demand, a term that gives the extent of pollution) of the wastewater could be relieved by Oxygen produced by Algal species in photosynthesis (Lee and Lee, 2001). They use eutrophic compounds in water to enhance their carbon sources. Tackling environmental problems using Algae has been emerging as an area of study. Wastewaters from urban region comprised of many compounds prone to eutrophication give a hazardous effect on the ecosystem in water (Table 1).

ALGAE

A diverse group of organisms, which can be said as polyphyletic, that it does not possess a common ancestor and photosynthetic in nature are algae. Algae have resemblance with plants in the way of their response to predators and parasites, photosynthesis etc. while both of them have large differences in many aspects. Here are some characteristics that define algae.

Characteristics of Algae

(i) Photosynthetic organisms: They have pigments in plastids like chlorophyll, the primary photosynthetic pigment (chlorophyll a, b, c, d, and e), Carotenoids (Alpha (α) carotene, Beta (β) carotene, Gamma (γ) carotene, Theta carotene, leucine, lycopene, flvicine, violaxanthin, fucoxanthin, astaxanthin, myxoxanthin and zeaxanthin) and biliproteins also called phycobilins like phycocyanin, allophycocyanin and phycoerythrin.

(ii) They include both unicellular organisms and multicellular organisms. Unicellular thallus characterized by coccoid or rhizopodial and non –

motile while multicellular thallus could be dendroid, colonial, palmelloid, filamentous, siphonous etc.

(iii) The organisation of algal cell is of three levels which are prokaryotic cell (eg: Myxophceae), mesokaryotic cell (Dinophyceae) and eukaryotic cell (other groups) characterized with the cell wall of cellulose.

(iv) The main distinguishing characteristic of an alga from the plant is the well-differentiated body of plants with roots, stem, leaves etc.

(v) Habitat with moisture content is most suitable for algae. There are freshwater algae and marine algae. They can be said as ubiquitous organisms.

(vi) They reproduce through vegetative methods (fragmentation, akinetes, hormogonia), Asexual means (zoospore, aplanospores, exospores, endospore, hypnospores, curpospores etc produced inside sporangia and sexual means via isogamy, anisogamy and oogamy.

(vii) Algae are of free-living types where some exhibit symbiotic association with another group of organisms like Algae and Fungi associated to form Lichen.

Uses of Algae

(i) **Algae for Biofuel Production:** Recent studies show evidence for using algae for the production of biofuels, bio-oils, biodiesel, biohydrogen, biomethane, bioethanol, biobutanol etc, which help in meeting the requirement of energy in various sector. Compared to other fuels from non-renewable resources, algal biofuels are the most cost-effective, nature friendly and clean. Algae are considered as the third generation biofuels where crops like cereals were belonging to first-generation biofuels. Woertz et al., 2009, observed wastewater as a medium for algal growth products give low-cost lipids from which conversion of biofuel is possible. His investigation was on wastewater collected from a dairy farm and water from municipal that has been added of carbon dioxide.

(ii) **Algae in utilizing Wastelands and their role in Reclamation:** Those lands on which normal crops cannot be raised can be used for Algal cultivation. They need some moist conditions and for the same, they can utilize the water that could be detrimental for the growth of the regular crop. They can act as a binding agent because of their extensive growth on soil surface holds the soil tightly to escape them from erosion. Highly alkaline and saline soils can be reclaimed by the growth of algae.

(iii) **Algae used as animal feed, fodder and Fertilizer:** Many of them are rich in protein, minerals, fibre and vitamins since used as feed for cattle give good meat quality and growth of animal (*Spirulina maxima, Spirulina platensis, Schizochytrium*). Seaweeds are a better substitute for Animal feed (*Rhodymenia palmate*). Pisciculture also uses algae as food for fishes. Brown algae, large red algae and liquid fertilizers extracted from seaweed are the fertilizers used as they can repair N levels in the soil.

(iv) **Algae in Foods, Pharmaceuticals and Cosmetics:** They are rich in many micronutrients, which are essential such as Calcium, Potassium, Magnesium,

Iron, Iodine and Zinc. Also enriched with Vitamins A, B (B_1, B_2, and B6), C, niacin, carbohydrates, enzymes, proteins and fibres. This autotrophic organism used for the extraction of Agar, which is used as a preservative agent in canned foods and gelling agent in some food items, Carrageenan for ice creams, toothpaste, lotions, pet foods and stabilizing agent for dairy products and Alginate as a stabilizing agent in malts, milkshakes, mayonnaise etc.

(v) Algae for Bioplastics: As an alternative for the present-day plastics, which are creating hazardous effects to soil and environment, algae are utilized for creating bio-friendly, green plastics that suits the environment. The requirement for such products is the sunlight and contaminated water.

(vi) Algae used to treat wastewater: The capability of huge biomass production in algae is utilized for treating the wastewater along with their property in absorbing inorganic N, P and heavy metals. The tertiary treatment phase is the stage of Algal use. In this chapter, this use of Algae is elaborated.

Table 1: Classification of Algae

Type	Division	Common Names	Examples
Macroalgae	Chlorophycophyta	Green algae	*Chlamydomonas, Chlorella Cosmarium, Ulothrix, Oedogonium, Cladophora, Ulva, Chara*
Macroalgae	Phaeophycophyta	Brown algae	*Ectocarpus, Laminaria, Fucus, Sargassum*
Macroalgae	Rhodophycophyta	Red algae	*Polysiphonia, Batrachospermum, Chondrus, Gracillaria*
Microalgae	Bacillariophycophyta	Diatoms	*Pinnularia, Navicula, Biddulphia, Melosira*
Microalgae	Euglenophycophyta	Euglenoids	*Euglena*
Microalgae	Xanthophycophyta	Yellow-Green Algae	*Botrydium, Listeria*
Microalgae	Chrysophycophyta	Golden – Brown Algae	*Ochromonas, Uroglena, Dinobryon*
Microalgae	Cryptophycophyta	Cryptomonads	*Cryptomonas*
Microalgae	Pyrrophycophyta	Dinoflagellates	*Gymnodinium, Ceratium, Peridinium*
Microalgae	Cyanophycophyta	Blue-green Algae	Nostoc, Oscillatoria, Spirulina, Microcystis, Anabaena

Wastewater and its Composition

Wastewater refers to the contaminated water with remnants after human use. Still, an indefinite number of cases of deposition and discharge of municipal wastewater and domestic wastewater from rural region takes place into the water resources. The arrangements or set up the plan in most villages for the water supply to date play a major role in this increased discharges. A lack of proper and timely investigation on the expanding networks of water supply in many towns along with an abstention in construction of proper sewage system or maintenance of existing sewage plants provoked this problem aiding to pollution in water bodies and health threats to the public. Suspended solids, pathogens, organic and inorganic compounds, materials that demand Oxygen, salts, heavy metals etc. make up the wastewater input towards water bodies.

Unchecked release of human waste is the paramount source of water pollution. Since water is the cardinal requirement among the population, majority of the countries make huge investments in the area of water supply but careless in taking such measures for sanitation projects which reduces those advantages in the health sector as expected from water investments. A consolidated approach with combining technical, manual and economic aspects alleviate the problems for wastewater treatment and safe discharge. There will not be such a condition of prescribing an adequate treatment plant universally as situations have their uniqueness but necessity lies for those standards meeting regional situations. Here, conventional treatment methods are also elucidated with innovative treatment procedures including Algae. Mainly agricultural drains accept domestic discharges, which are bulky pollution load, and this resulted in spoilage of canals (irrigation source) with wastewater pollutants. Even though agriculture owns the state of the largest sector in terms of water consumption, it has the responsibility in polluting water bodies due to contamination by chemical residues such as pesticides, herbicides, as they dissolve in irrigation water and make a hazardous effect on downstream water users. Groundwater is being contaminated due to nitrate from agricultural sources. Organic and inorganic particulates many issues disturbing the quality of water cited as the main reason for eutrophication and sometimes stimulate the growth of Algae forming Algal blooms.

In a society, the lifestyle and technologies adopted by the population determine the composition of wastewater. It is a composite mixture comprising the inorganic, organic and fabricated compound. Carbon present inside this mixture (organic part) has its majority portion as Carbohydrates, volatile acids, amino acids, fats, and proteins. Inorganic portion accounts to elements like Na, Ca, Mg, K, Cl, S, and compounds like Bicarbonate, Phosphates, Ammonium salts, Nitrates and heavy metal compounds. The presence of immense microorganisms (protozoa, bacteria, fungus, etc.) is expected on wastewater as it serves as ideal media for them. These comprise of harmless and harmful pathogenic microorganisms. Harmless one is utilized in sewage treatment biologically and pathogenic should be removed efficiently as it may include bacteria for TB, typhoid, cholera etc. and protozoa making dysentery. Raw sewage received from the urban and rural area accompanied with treated sewage, discharge of effluents from manufacturing units (also industrial plants), runoff water containing all the agrochemical residues from cultivated lands, waste received by leaching from the sites where solid waste has been deposited etc. were some of the areas pointed for the origin of pollutants.

Treatments for Wastewater

Sewage treatment at preliminary stage discards solid and large particles that can cause hindrance inflow path and harm the equipment. Heavier grit particles, rags, wood, faecal material etc. remain floating over water are removed during this phase. At an interval of 20 – 60 millimetres, some bars are arranged through which floating substances are blocked and thus removed. A particular range of velocity for the flow is adopted 0.2 – 0.4 m/s on which grit and silt undergo settling and suspended condition of organic matter.

Primary treatments allow the sewage passage into sedimentation tanks for a gravity settlement to eliminate 70 % of TSS (Total settleable Solids) and besides BOD, removal also happens at 40% if the design of sedimentation tank is well planned. The principle behind the secondary treatment is nothing but the organic matter assimilation by the mediation of heterotrophic bacteria to maintain their growth and meet energy needs eventually abating the BOD. Unit operation largely in a biological way achieves BODs oxidation under aerobic condition while each categorized into fixed film and dispersed growth. Systems with biological oxidation have the provision for removal of > 90% bacteria (i.e., pathogenic) but for adsorption favours viral elimination.

All organic radicals, ions and compounds are the target of tertiary treatment, which can complete chemically and through biological manner. Chemical means are expensive with the threat of secondary pollution. An additional step in this sequence of treatment process always adds up the cost of operation. Nitrate, Phosphates, Ammonium are removed from their dissolved state in this phase that is approximately 4 times costlier than the primary phase. Here is the need for Algal biomass is under study as they have the photoautotrophic condition for survival.

The quaternary phase of treatment could account for 8 to 16 fold expensive from primary treatment as they designed to eliminate heavy metal toxicity and all soluble minerals.

High Rate Algal Ponds; Designed for Wastewater Treatment

To treat and recycle the wastewater with the recovery of water resources, a feasible technology that possesses long-term benefits has to be adopted in the perspective of the water-scarce situation, energy need and necessity of food. HRAPs described as High Rate Algal Ponds are the energy-conserving and energy-producing technology using Algal biomass for wastewater treatment. It has a requirement of low energy for whole processing in the treatment whereas it recovers energy from separated solids inside the wastewater and assists in the biofuel yield from algal biomass that has been harvested. Before starting with HRAPs, contaminated water should undergo a pre-treatment inside some systems like gravity settlers or CAP, Covered Anaerobic Ponds and CDP, Covered Digester Ponds to remove solid particles. The remaining part called effluent is allowed to HRAP which are shallow and where gentle mixing is provided and lead to the efficient breakdown of Organic Matter dissolved within. The main principle behind is the assimilating capability in Algae for nutrients which gives a partial tertiary treatment along with the secondary level of the treatment process. It ensures the disinfection naturally (Craggs et al., 2014). Major communities (farm, industry etc.) used a two-staged pond system for treatment of wastewater like in New Zealand. In terms of cost, ease of maintenance, and removal of organic solids they gave a good performance. Nevertheless, a rethinking is necessary when comes to elimination of Nitrogen and Phosphorus (nutrients) and algal solids, improper disinfection etc. in this system. The drawback of the conventional system is the lack of natural resource recovery from wastewater like biogas for energy, high standard treatment to recover water, fertilizer from the assimilated nutrients by the microorganism biomass, biofuel, feed, fodder etc. In the conventional pond system, mean productivity of algae or bacteria in a year accounts for approximately 2.5 g/m^2/day while the accumulation of the same creates an organic load in received water. Therefore, the need arises for replacement of facultative systems with some cost-effective manners like upgrading the technology of existing infrastructure of the pond technology.

A "massive leap" suits with the HRAP over conventional systems as HRAPs incorporate cordially, the principles of engineering in an ecological way and conglomeration of natural treatment (physicochemical, microbiological) process. This diversity makes them unique, robust and resilient over mechanical means. The unit process that arranged in HRAPs are:

(a) Elimination of solid particles from contaminated water and thereafter-anaerobic digestion at ambient temperature.

(b) Resulted supernatant from the first process is exposed to sunlight with aerobic conditions for promoting algal growth.

(c) Harvest of the biomass of algae and processed to biofuel.

(d) Polishing if necessary for treated effluent.

Covered Anaerobic Pond and Covered Digester Ponds

Before HRAPs treatment, wastewater transmitted to these two systems. Using CAP, solid particles are either settled or subjected to digestion and decomposed to produce methane or with the help of primary clarifier known to be gravity settler, concentration followed digestion is performed in CDP. Thus, ensure a cost-saving and efficient anaerobic digestion. A continuous transfer of solid particles from gravity settler towards CDP observed with the large equipped system but the intermittent transfer of an interval of 1 – 4 hour adopted for a small setup. Compare to CAP, CDP possesses less volume and gradual filling happens over time. When observed that it is filled, removal of the supernatant remnants is done but in the case of solids of wastewater or algae remains there itself for their full digestion that takes many years. Besides, these two products could be used as a liquid fertilizer (supernatant) and amendment materials for soil (digested solids).

The Process Involved in HRAPs

A remaining portion or called supernatant received from gravity settlers is allowed into HRAPs. They are described with (i) Shallow (ii) Paddlewheel mixed (iii) Raceway ponds. In the year 1950, Oswald and co-workers developed this system to recycle resources by treating contaminated water. Like any other system, HRAPs too evaluated with several variables controlling operations that are depth, HRT (Hydraulic Retention Time), Rate of organic loading and horizontal mixing rate. Clarity of the treated water is the scale for measuring the depth of the pond, usually 0.2m to 0.6m. Rate of organic Loading rate in HRAPs has its maximum value range in 100 – 150 kgBOD_5 $ha^{-1}d$ highly varies with climate. In case of HRT, variation occurs season-wise in the temperate climatic condition given as 7 to 9 days suitable for winter and 3 to 4 days preferred for summer. Speed of mixing with paddlewheel ranged from 0.15 – 0.30 m/s enabled turbulence causing the intermittent exposure of algal biomass to solar radiation. Uptake and utilization rate of nutrients like Nitrogen have 24 kg N/ha/day and rate of Phosphorus assimilation is 3 kg P/ha per day based on the composition of nutrients in Algae. Studies show the theoretical possibilities for assimilating N and P completely into biomass yield of algae exhibiting ratios of N: P in 4: 1 to 40: 1. Conversion of solar energy to reside them as algal biomass inefficient manner is involved in HRAPs as well as the energy used in Oxygen release from photosynthetic process enhance decomposition of bacteria in aerobic condition. When photosynthesis continuously happened during the daytime in algal biomass with the light will create a super Saturation point for Dissolved oxygen, i.e. 20 g/m^3. One of the assistance in HRAP, the paddlewheel provides aeration by expending one-tenth of the energy spent by mechanical aeration. The shallow depth of the pond is designed with the specific aim of ease in receiving the sunlight even at the bottom for killing or inactivating the pathogenic microbes resided in faecal matter and oxidation of organic contaminant dissolved in water. The Phosphorus levels could be sometimes declined by addition of a very low level of chemical flocculant to the effluent in HRAPs. The pH is found at higher scale during the daytime due to photosynthesis in HRAPs which helpful in the removal of NH_3 through volatilization and removal of orthophosphates via precipitation. The ultimate biomass received out of HRAPs has its 70 to 90 % algae and remaining part comprised of bacteria, viruses, fungi and some other invertebrates. They perform efficiently if provided with algal strains that adapt to changes in light (diurnal, seasonal), pH, temperature, dissolved O_2. Evaluating the biomass yield per unit area from HRAPs in the absence of carbon dioxide inclusion showed the values of 30 t/ha/year (Mediterranean climates) that seems to be 2-3 fold comparatively higher than that of conventional ponds (Craggs et al., 2014) (Fig. 1).

Fig. 1. Demonstration of High Rate Algal Pond System

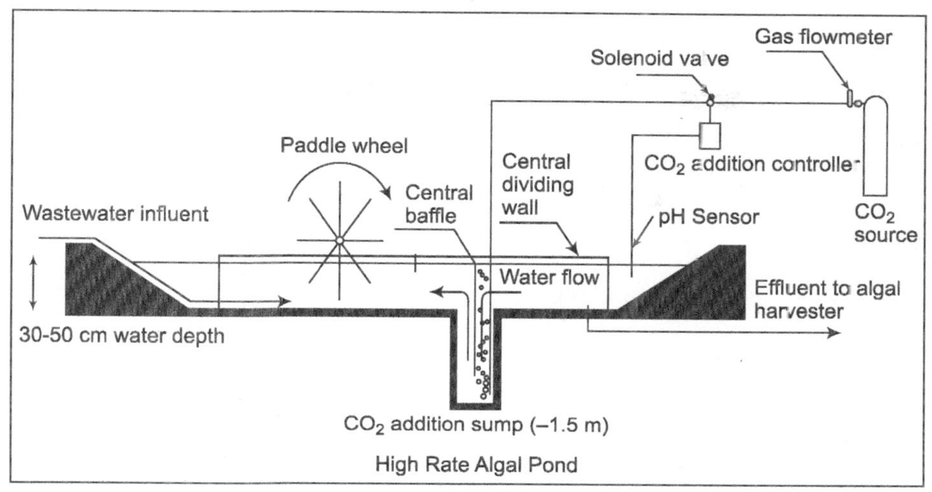

(Source: Sciencedirect.com)

Addition of Carbon Dioxide in HRAPs

Production of algae and their involvement in the removal of nutrient are the performances involved within HRAPs, which have chances of suppression in the limited Carbon availability during daytime specified by pH > 10.0, and the reason is the utilization of both Carbon dioxide and Bicarbonate ions for algal photosynthesis. Low Carbon: Nitrogen (C: N) ratio, i.e., 3: 1 to 4: 1 for wastewater is cited as the reason for the limitation of Carbon when compared to the biomass of algae (6:1). Assimilating N and P directly by biomass of Algae is limited due to the insufficient quantity of C in domestic wastewater, which helps in the removal of the same. Any constraint in C availability hike the pH > 8.5 and this weakens the rate of growth and production but some species continue to exhibit growth in diminished rate even at a pH > 10. At hiked pH level, growth of Algae diminishes with a reason of high concentration of free NH_3. The increased mass of Algae during the daytime conduct photosynthesis and release Oxygen that itself dissolve in the wastewater until it touches the saturation of 200 to 300 %. The condition of supersaturated oxygen level enhances the decomposition of organic contaminant compounds with bacteria but a hindrance for algal growth. Bacteria converting organic matter in the contaminated water into CO_2, which survive as heterotrophic, is inhibited. This condition of C limitation is tackled with CO_2 addition thereby optimizing pH 7.5 to 8.5. With CO_2 addition biomass yield estimated to be 16 to 20 g/m^2/day for wastewater treatment using HRAPs. This assisted more absorption of nutrients by uptake into biomass. CO_2 addition performed by using CO_2 rich flue gas.

Harvest of Algae from HRAPs

HRAP system has a dominated population of some species of microalgae forming colonies like *Micractinium* sp., *Pediastrum* sp., *Scenedesmus* sp., *Coelastrum* sp.,

Actinastrum sp., and *Dictyosphaerium* sp. which can settle down naturally as influenced by gravity under the condition of mixing the High Rate Algal Pond system with simple ponds for algal settling or Algal harvest tanks having low HRT. Competency in the algae removal is better performed by the ability of algal colonies to aggregate each other or undergo bio flocculation as a response to CO_2 addition inside HRAP or this alga that has been settled may recycle to HRAP leading to activated sludge process.

If preferred a natural settling of Algae, adopt Algal-settling ponds as it has the provision for storing the biomass with a view for recovering the settled algae at periodic intervals. However, for Algal harvest tank, the design-enhanced gravity settling efficiently assisted with lamella plates and with 1 – 3 % solids, a secondary thickening has been given to that settled biomass. Removal should perform daily in a continuous manner otherwise chance for deterioration is high in harvested algae even before they exploited beneficially. After recovery, algal biomass has eminent application as fertilizer, feed, fuel, the substrate in any chemical industry as their mass is ample with N, P and K. In 1960, the proposal for Algae as biofuel was put forward by Oswald and Golueke for the first time and considered as a valuable energy resource.

Treatment of Wastewater after Algal biomass harvest

Sometimes there exists a need for polishing the algal treated water or effluent further as per the requirement regarding the quality of water to be discharged. Many further processes are used alone or in the combined form including

- Maturation ponds allowing the effluent for their storage and reuse (disinfect with UV radiation from sunlight)
- Rock filter (used next to maturation pond) to eliminate solids
- UV disinfection (depends on land availability)
- Filtration membrane (UF and MF) providing great quality effluent

The areas where HRAPs dominate the Conventional system inefficient effluent quality are:

(i) The inclusion of anaerobic ponds as pre-treatment phase encouraged the settle down and digestion of Total Soluble Solids and particulate BOD

(ii) Instead of frequent removal of sludge, annual removal is attained through complete digestion in anaerobic condition.

(iii) UV disinfection is provided naturally from sunlight rather than artificial provision.

(iv) The high rate of nutrient elimination especially toxic NH_3 Ammonia.

Economic Analysis of HRAP system

Analysing the costs involved in secondary treatment stage (removal of BOD) of wastewater with electromechanical means and HRAPs, it was observed that costs of capital and operation are only one fourth to one third in HRAPs than other means. For tertiary treatment (removing nutrients) also, give similar ratios or lower ratios for HRAPs compared with electromechanical methods. The function of treating the wastewater alone cover up the costs involved as capital and operating costs expended on the production of algae and their harvesting besides receiving biofuel (minor co-product). However, biofuel has an insignificant effect on these economical values. HRAPs can effectively recover resources in the form of algal biomass that dispensed for fertilizer, feed, fuel etc.

Environmental Benefits

Abating emissions of greenhouse gas and sustainable approach is the marked benefits of HRAPs in comparison to electromechanical treatment. Consumption of energy is low mainly electricity and reduced GHG emitted from fossil fuel. Replacement of fossil fuels and non-renewable resources with renewable biofuels that could meet the energy needs to some extent is another benefit. Mining of rock phosphate and producing the N - fertilizers enhance the emission of greenhouse gases but the use of HRAPs tally the issue by nutrients recovered from wastewater and Carbon that reside with harvested products and another residue can be applied instead of chemical fertilizers. Dry matter of the algae (unlimited culture) is 10% of the biomass harvested and phosphorus accounts to 1% along with some micronutrients (Fig. 2).

Fig. 2. Schematic representation of HRAPs with CDP and CO_2 addition

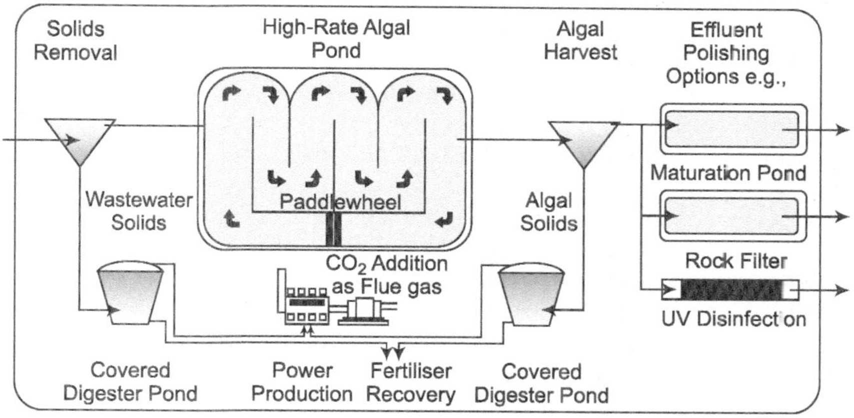

(Source: Craggs et al., 2014)

Removal of Heavy metals by Algae from Wastewater

Water resources in nature and living creatures inside the aquatic ecosystem are prone to threat from haphazard disposal of heavy metals via the supply of domestic sewage and industrial effluents. This serious issue has to be attenuated and several measures (physicochemical) adopted were inclusive of solvent extraction, precipitation in a chemical way, adsorption, membrane separation, ion exchange etc. to strip off toxic metals. Disadvantages regarding this method are: (i) Incomplete process (ii) expensive (iii) requires some monitoring pieces of equipment and system (iv) Energy expensive (v) End products from the processes may become another toxic sludge. Methods become expensive in terms of money and energy if the concentration of metal exceeds 10 – 100 mg/litre. Studies progress to adopt the environmental friendly approach through biological organisms like Algae, bacteria, fungus etc. for the solving of environmental problems and its protection, as they are low cost and strategic approach. The term that entails this approach is 'biosorption' and this replaces accumulation completely as this term has the indication of adsorbing toxic ions of metal on the surface of dead biomass as well as extracellular binding and ligand formation in intracellular.

Accumulation of Metals by Algae

The external concentration of metals make the algae to accumulate them and capability varies from species to species. A case of biomonitoring is proposed to know the pollution extent of water bodies by metals with a concentration of metal in indigenous algae. Load and burden of toxic metals can be alleviated with the inherent characteristics and ability for metal accumulation and recover some very precious and costly metals such as gold, silver from the contaminant containing water. A two-phase of deposition with passive (rapid) and active (slow) ways are part of algal processes. The cell surface of Algae takes up the ion in the short period, which does not rely on metabolic pathways, are involved in passive uptake. Metabolism dependent process that is responsible for taking the metal ion through cell membrane towards the cytosol. Cell membrane if permeable for movement of metals may be induced by the metal itself show passive kind of uptake.

A method prescribed for identifying the metal ions presents intracellularly from that of adsorbed in algae (Bates et.al., 1982) instruct to wash those cells using 2 milli Molar EDTA and this wash off all ions bound to the external surface leaving fractions of ions present within the cell. Quantity of ions within the cell (intracellular) estimated after digesting the above-prepared cells. The metal ions present in water and species allowed in the same surrounding area the factors playing to determine the adsorption and uptake of metals. In comparison with intracellular uptake, metal ions removed in any manner account for its > 80% for adsorption. An alga, *Scenedesmus suspicious* has been analyzed after exposing it to Copper (Cu) for one minute and it showed than > 90percentage got adsorbed rather than intracellular uptake. This adsorbed one only transported into cells at some states. Accumulating the metal through the surface is by holding them with the wall, external membrane or presence of any other polysaccharides then also bind with some other molecules named phytochelatin, cytoplasmic ligands and metallothioneins present intracellularly. It is as much as important to locationalise the ions on the algal cell with X-ray energy-dispersive and electron microscopy. Spectroscopy aims to have a search and investigation on Algal cells about the oxidative state of that metal ion. Cell surface or the membrane of the algae is found negatively charged due to the presence of phosphoryl, hydroxyl, sulphydryl, amino and carboxyl. pKa or Dissociation constant for these functional groups are specific and they subject to dissociation into a proton and Anion when exposed to particular pH. Polysaccharides, peptidoglycan, teichoic acid, protein and

Technotronic acid were components comprising the cell wall constitution and all the above said functional groups remain associated with each. There are variations expected among these components and the type of functional group seems to vary. Majority site for binding the metals seems to be higher with proteins and polysaccharides among all other components.

Role of different Algal Group in heavy metal sorption

The cyanobacterial group have their cell wall composed of Peptidoglycan (N-acetyl glucosamine + β 1, 4-N- Acetyl muramic acid along with peptide chain- Amino group) which makes their carboxyl group (most active sites in cyanobacteria) and Phosphoryl groups mostly remain active for associating with the metal. Membrane proteins (amino group is active), lipids (phosphoryl), lipopolysaccharides (phosphoryl) were other sites meant for combining with metal ions in blue-green Algae. An anionic (presence of uronic acid) capsule or can be said as sheath comprised with polysaccharides on their outer portion made their affinity towards the metal ions at a high level and enhanced their role as an auspicious chelating agent used in removing heavy metals. The cell wall has a part of the chelating matrix to where metal ions after sorption diffuse into and this rate can be slowed down if the capsule is made of gelatine. Those strains do not possess a well-defined or said to be devoid of the capsule has comparatively low affinity for metal ions with that of capsule or sheath covered ones. Heteropolysaccharides are components that sequester heavy metals in green algae those offers Carboxyl functional group and Sulfate functional group. An experiment conducted (in *Ulva sp.*) by extracting the polysaccharides, that accounts for 12% of the dry mass was composed of uronic acid (15 – 19%) and sulphate (16 %). Around 70 % accounts for proteins in the cell wall and among that, 12 % are Aspartic acid + Glutamic acid, which represents 0.15 meq/g, carboxylic group. A proper knowledge should be acquired about metal sorption in marine algae. Constitution of the cell wall in brown algae comprises of cellulose (fibrous skeleton), extracellular mucilage and alginate + fucoidan (Amorphous matrix). Among them, Alginate (chemically NH_4 salt or alkali salt of Alginic acid) have better involvement in metal accumulation. Alginic acid is a polymer made of D – mannuronic acid + L – guluronic acid + residues of both. Carboxyl group (– COOH) left at each component act for metal sorption and binds with cation. Divalent cations (Pb^{2+}, Zn^{2+}, Cu^{2-}, Ca^{2+}, Cd^{2+}) are more special targets for the polyguluronic acid showing more specificity towards them. Alginate has 10 – 40 % dry weight part in brown algae. Fucoidan, a sulphated polysaccharide has its presence with 5 – 20 % in dry weight of algae of which 40 % resides with sulphate esters. Sulfonate group does not have much prominent role in the binding of metal but may become significant with low pH (Fourest and Volesky, 1996). The 10 % of the biomass in *Sargassum* is protein, a part of 10 % of which is Arginine + Lysine contributed to 0.17 meq/g – COOH and 0.07 meq/g amine group respectively. But for brown algae, Alginate carboxyl group is dominant and abundant over the protein carboxyl or protein amine group if metal sorption is considered.

Metal Biosorption by Algae

One of the remedial measures to alleviate discharges containing heavy metals is biosorption with algal biomass. Yu et al., (1999) examine nine macro algae residing in marine ecosystem for analysing the capability in removing heavy metals like Cu, Cd, Pb. Used algae were *Ascophyllum nodosum, Lessonia flvicans, Durvillaea potatorum, Ecklonia Maxima, Laminaria japonica, Lessonia nigresense, Ecklonia radiate and Laminaria hyperbola.* Capacities of biomass observed were 0.8 – 1.2 mmol/g for Cadmium, 1.0 to 1.2 mmol/g for Copper and in case of lead, 1.0 – 1.6 mmol/g which is comparatively high values and benefits for biosorption efficiently. Chaisuksant et al., (2003) used a marine Rhodophytophycea *Gracilaria fisheri* after giving a pretreatment with Calcium chloride to analyse their adsorption ability in metals like Copper and Cadmium from which results show that they are efficient material with biosorbent properties with a maximum adsorption capacity is 0.63 mmol/g in Cadmium and 0.72 mmol/g for Copper. Biosorption in *Spirogyra hyaline* (freshwater and green alga) for metals like Lead (Pb), Cadmium (Cd), Mercury (Hg), Cobalt (Co) and Arsenic (As) keeping parameters like initial concentrations and time of contact. As Hg and Cd adsorbed highly at 40 mg/l of initial concentration and or lead and cobalt amount initially should be 80 mg/l. Dried biomass of algae favoured the adsorptive conditions efficiently and Hg uptake was highest while Co uptake least.

Table 2. Algae and the metals adsorbed by different species

Algal Species	Metals sorbed
Chlamydomonas reinhardtii	K, Mg, Fe, Ca, Co, Cu, Mn, Ni, V, Zn, Mo, Se, Pb, Sr, Cd
Chlorella salina	Co, Zn and Mn
Chlorella sorokiniana	Ni
Chlorella vulgaris	Cd, Pb, Ag and Cu
Chlorella miniata	Chromium
Chlorococcum	Lead, Cadmium, Silver and Copper
Lyngbya taylorii	Nickel, Zn, Cadmium, Lead
Cyclotella cryptic	Lead, Zinc, Al, Cu and Cd
Phaedactylum tricornutum	Aluminium, Copper, Cd, Zn and Pb
Spirulina sp.	Cadmium
Scenedesmus subspicatus	Copper, Zinc, Lead, Aluminium and Cadmium
Stichococcus bacillaris	Se, Pb, K, Mg, Sr, Ca, Mn, Cu, Ni, V, As
Ascophyllum nodosum	Lead, Cadmium, Zinc, Copper
Cladophora crispate	Cu, Ag, Pb, Cd,
Codium fragile	Cadmium
Colpomenia sinuosa	Copper and Nickel
Ecklonia sp.	Chromium
Fucus sp.	Cadmium, Copper, Zn, lead

Gracilaria sp.	Pb, Zn, Ni, Cu, Cd
Laminaria sp.	Zn, Cd, Cu, Pb
Laurencia obtusa	Ni, Co, Cr, Cd, Cu
Padina pavonia	Cd, Ni, Pb
Palmaria palmata	Copper
Petalonia fascia	Nickel and Copper
Pilayella littoralis	Cadmium, Chromium, Aluminium, Cobalt, Nickel, Zinc, Iron
Porphyra columbina	Cadmium
Sargassum asperifolium	Lead
Sargassum hemiphyllum	Nickel and Cu
Sargassum hystrix	Lead
S. natans	Lead
Turbinaria conoides	Lead
Ulva fascia	Ni, Cu
Ulva lactuca	Lead
Ulva reticulata	Zinc
Ulva sp	Cadmium, Nickel, Zinc, Copper Lead

(Source: Brinza et al., 2007)

Growth conditions for Algae about their use in Wastewater treatment

Growth of any living organism at every stage has some influence from its surroundings as both biotic and abiotic factors. Enhancing the study on each factor directly helps in optimizing its growth and development beneficial in many ways. The same analysis is applicable for the algae as they have a prominent role in many sectors like food, pharmaceuticals, cosmetics, fodder, energy, sewage treatment etc. in a cost-effective manner.

Effect of Temperature on Algal Growth

The temperature of the growing condition is the supreme factor that regulates growth. Even microorganism tries to acclimatize with the prevailing atmospheric temperatures by some metabolic changes. On another side, some organisms are themselves thermophilic, mesophilic and cryophilic (Snow algae like *Chlamydomonas, Mesotaenium berggrenii*). In a study conducted by Cho, et al., (2007) on the analysis of different temperatures (15°C, 20°C, 25°C, 30°C) and salinity (10, 20, 30) conditions on the growth of two green algae *Nannochloris occulata* and *Chlorella ellipsoidea,* he was observed highest specific growth rate at 25ºC and 10 salinity while 15°C and 10 salinity for the maximum density in case of *Chlorella ellipsoidea* whereas *Nannochloris occulata* exhibited higher SGR and maximum density at 25°C and 10 and 30 saline conditions. Foy et al. (1976) observed no growth for *Anabaena flos-aquae, Aphanizomenon Flos-aquae, Oscillatoria agardhii* and *Oscillatoria redekei* at 35ºC and a severe inhibition in growth for *Anabaena flos-aquae* observed at 25°C. Konopka and Brock (1978) conducted a study to determine the reason behind the insignificant number of cyanobacteria in Mendota Lake during the spring season and observed that a significant rate of photosynthesis happens at low temperature. Therefore, it cannot be said that low temperature alone is responsible for an insignificant number of a blue-green alga. Progressive growth of algae observed up to the optimum temperature range for algal growth and beyond the range, enhances the respiration in algae and photorespiration, which diminishes overall productivity. At sufficient light and nutrient levels, optimum temperature differs from species to species and a range from 22°C to 35°C is considered as optimum temperature for microalgal growth. Photosynthesis in blue-green algae during June – November enhanced with optimum temperature 0 - 20°C and in summer 20 - 30°C (Singh and Singh, 2015). Sato et al. (1979) found that growth temperature affects lipid composition in *Anacystis nidulans* (blue-green algae) which influenced chain length and caused desaturation. Low temperature declines the efficiency of those treatments depending on microalgae. A review by Munoz and Guieysse, (2006) observed that a range of temperature from 25 - 30ºC enabled a doubling of contaminant removal efficiency assisted by a symbiotic system consisting strains of *C. sorokiniana* (algae) *and R. basilensis* (fungus). The studies show that possibilities for treating wastewater without any obstacles even under the temperature-dependent decline of biological activity in strains adapted to cold could meet with optimized bioreactors. Energy conversion from light to heat by the algal biomass is responsible for the excess temperature in the presence of increased light intensities. Some proposals to alleviate these problems are water spray or external heat exchanger thus guaranteeing a stable population of microalgae but prohibitive costs are the problem. Tackling the temperature-dependent uncertainties in the algal growth can be made possible by combining different strains of algae with similar features and various optimum temperatures.

Effect of pH on Algal Growth

It is essential to study on pH of the growing habitat as it influences biological phenomenon. Each organism has a specific tolerance limit for a particular Ph. Chen, Durbin (1994) examined the growth rate of two marine algal species Thalassiosira pseudonana and Thalassiosira occanica, and a consistent decline in the rate of both growth and photosynthesis observed with increasing pH level (> 8.8). Elevated pH level makes the CO_2 unavailable for marine algae. Wilde et al. (2006) studied the effect of pH (5.5 – 8.0) on the *Chlorella* sp. for the uptake and toxicity of Cu and Zn. Results showed that increasing pH level increased the toxicity of both Copper and Zinc and growth inhibited with the increase in the concentration of Cu and Zn. At some occasion, the presence of algae results in algal bloom, and it would result in water, which is unsuitable for drinking. Besides, this should be treated properly through filtration like ultrafiltration (UF) and Microfiltration (MF). The Ultrafiltration membrane function for removing algal cells along with Extracellular Organic Matter (polysaccharides, proteins, nucleic acids etc) at a moderate level and the fouling of membrane happens due to the same. Zhang et al. (2012) studied the pH effect on the membrane fouling by Algae and observed that membrane fouling is accelerated by low pH. Extracellular Organic Matter (EOM) and cell lysis were increased with reduced pH along with an increase in smaller MW EOM fractions. Water contaminated with metals especially having ecotoxicological concern have a detrimental effect on aquatic organisms. A study conducted on the toxicity of such metals like Copper and Uranium towards freshwater algae, *Chlorella* by Franklin et al. (2000) about the pH levels (5.7 and 6.5) and a comparison of metal concentration which is intracellular and extracellular were done at both pH levels. It was found that pH 5.7 exhibited reduced toxicity due to the reason of less concentration of Cu and U with the intracellular and extracellular condition as compared to 6.5. Xin, et al., (2010) conducted a study on the effect of different sources of Nitrogen on the growth of algae *Scenedesmus* sp. and observed that acidic pH (especially < 6.0) do not promote the growth of microalgae. Acidic pH makes an inhibitory effect on the algal growth leading to a stable phase having low algal density and less efficiency in nutrient removal for Ammonium N source. Also suggested to add a buffer solution to maintain pH value above 7.0 during the wastewater treatment. The pH in a growing medium (here the wastewater) modifies equilibrium between Ammonia (NH_3): Ammonium (NH_4^+) and Carbon Dioxide, Carbonate and Bicarbonate ions or the available Phosphorus and heavy metals. As said earlier, Ammonia volatilization regulates the removal of Nitrogen and precipitation of orthophosphate at elevated pH (9.0 – 11.0) removes Phosphorus from contaminated water under treatment.

Effect of Light and Irradiance on Algal Growth

Light is the major factor controlling the algal growth in terms of accumulation of biomass as they have light-harvesting pigments, chlorophyll a and chlorophyll b, which enable them to remain photosynthetic and autotrophic. Light has such an effect that maximum growth rate in algal species reported at light saturation and beyond this particular limit, the chance of growth inhibition due to photoinhibition happens (Singh and Singh, 2015). The dim light and dark condition will inspire the algae to be heterotrophic. Baalen et al. (1971) conducted a study on two cyanobacteria, the unicellular *Agmenellum quadruplicated* and filamentous multicellular *Lyngbya lagerheimii* to examine the effect of dim light on them. Both of them exhibited an incomplete TCA cycle under the dim light as it blocked the autotrophic growth. Here, a major contribution to the cell constituents was from glucose. Light is the supreme factor as it has an inevitable role in dispensing energy requisite to aid metabolism but surfeit intensity results in the production of ROS (Reactive Oxygen Species) along with oxidative stress. Using a flatbed photobioreactor, which has adjustments to reduce self – shading in cells, Sforza et al. (2012) studied the impact of intensity of light on the growth as well as lipid productivity in the algae *Nannochloropsis salina*. He experimented with continuous illumination as well as alternate light-dark cycles at varying frequencies. The occurrence of dark cycles helped the *Nannochloropsis* for re – oxidizing the electron transporter inside the PS II even though this alga can exploit at very light intensity otherwise in the absence of alternative light-dark cycles, radiation damage occurs. Algae are efficient depositors of lipids in their cellular parts and these are the precursor for biodiesel. These lipids are transformed to FAME (Fatty acid Methyl Esters) which replaces the diesel fuel. Kleinova et al. (2012) found that the microalgae such as *Chlorella* and *Nannochloropsis* deliver valuable triacylglycerols to produce biofuel. In the *Nannochloropsis,* Das et al. (2011) studied the production of biomass and FAME under different wavelengths of light and intensities under phototrophic as well as mixotrophic cultures with blue, red, green and white LED. Blue light gave high specific growth rate. Under the green (550nm) LED, FAME yield in the percentage of biomass dry weight was maximum with 20.45% in phototrophic and 15.11% in mixotrophic conditions while the maximum volumetric yield of FAME observed in blue light with 55.13 mg/l in phototrophic and 111.96 mg/l in mixotrophic because of biomass production at a high level. Besides, the study suggested that incremental exposure rather than continuous illumination saves 20% of input energy. Lee and Lee, (2001) experimented on *Chlorella kessleri* to examine the influential part of light: dark cycle with 12 hours each concerning their ability to treat wastewater. Microalgae exhibited low organic C removal rate because of low cell concentration in light: dark cycle scheme than continuous illumination. During the 12 hours of the dark period, *Chlorella kessleri* do its growth through the chemoorganotrophic mechanism. Algae could remove Nitrate efficiently in continuous illumination system rather than 12 hours each Light and Dark cycles but these diurnal cycles with an optimized photobioreactor used for high algal biomass could save energy and better removal of Organic C source. Light intensity ranging from 200 – 400 $\mu E/m^2/s$ make the saturation of photosynthetic apparatus by which algal activity is promoted up to that particular limit. Sometimes in a sunny day, irradiance up to 4000 $\mu E/m^2/s$ is received during mid hours that can cause photoinhibition (Munoz and Guieysse, 2006). A halt or drastic reduction in photosynthesis happens due to deprived light or dim light condition, which make the anaerobic environment inside the reactor. Removal of pollutants along with photosynthesis recommence if the availability of light is restored. Any sort of natural fluctuation in light availability (seasonal and diurnal) can be adjusted in Water stabilization pond (system for wastewater treatment) by an increase in HRT (Hydraulic retention time) for that moment which in turn avoids the accumulation of toxic pollutants.

Effect of Nutrients on Algal Growth

For the growth in the majority population, nutrients are the substantial limiting factor. Nutrients are a very essential part of an organism to build up himself or herself and operationalize their biological system. Every plant type is different in their dependence on nutrients. Pederson and Borum (1996) suggested that the reason for variation in growth rate and critical Nitrogen concentration in algae as a species-specific manner is that the period of limitation in nutrients varied with algal types. This study was conducted in a context where replacement of slow-growing algae by fast-growing algae happened after the nutrient enrichment in shallow coastal water. Here, a comparison of the effect was made between four macroalgae, which are ephemerals (Cladophora serica, Ceramium rubrum, Ulva lactuca, Chaetomorpha linum), and one perennial (*Fucus vesiculosus*). Fast-growing algae (ephemerals) exhibited a 30-fold higher requirement of N per unit biomass in comparison to slow-growing species. Large algae which are long-living requires low nutrient per unit time and biomass due to their low rate of growth and low concentration of nutrient inside the tissue finally allocating to the meristems. Factors like rate of diffusion, internal nutrient concentration, external concentration, species, boundary layer thickness etc. control the growth of algae concerning nutrient uptake. To achieve a high rate of production of mass culture of microalgae, ensure the optimum supply of nutrients like N and P. Any limitation in its supply declines the crude protein content and productivity. Mostert and Grobbelaar (1987) observed the highest crude protein and production rates when concentration exceeded 25 mg/l Nitrogen and 2 mg/l phosphorus. When nitrogen is supplied with Nitrate – N: Ammoniacal – N in 3:1 ratio, rate of production increased. Schnurr et al. (2013) studied biofilm growth system of two algae *Nitzschia palea* and *Scenedesmus obliquus* about their ability to produce neutral lipid (triacylglycerol – commonly advantageous in the production of biofuel) concentration if a condition of essential nutrients (such as N and Si) starvation is applied. The result showed no increase in neutral lipid amount due to nutrient starvation observed in biofilm system of algal growth. However, in suspension culture of algae, for *N. palea,* lipid productivity was 0.45 $g/m^2/d$ and biomass productivity was 2.8 $g/m^2/d$ while for *S. obliquus,* values were 0.18 $g/m^2/d$ and 2.1 $g/m^2/d$ respectively under nutrient starvation. Nowadays, the application of microalgae in the tertiary treatment of wastewater is worldwide used and it accounts for the inorganic nutrient removal capability of the algae. *Scenedesmus* sp., a freshwater alga was analysed by Xin et al. (2010) on the growth as well as properties of nutrient removal under various sources of Nitrogen (Urea, NO_3-N, and NH_4-N). Fastest growth or SGR (Specific Growth Rate) of the microalgae exhibited with NH_4-N as a Nitrogen source, then Urea –N and NO_3-N (NH_4-N > Urea-N > NO_3-N). Removal of both Nitrogen and Phosphorus were efficiently done with nitrate and Urea nitrogen source. (Total Nitrogen removal with NO_3-N is 90.4%, Urea-N is 87.8% and Total Phosphorus nearly 100%). If NH_4-N is used, the H^+ ions released to create an acidic pH in algal culture thus reduced growth observed in stable phase with a decline in nutrient uptake.

Effect of CO_2 in Algal Growth

On behalf of the National Oceanic and Atmospheric Administration, Carbon dioxide has an average concentration of 407.4 ppm globally in 2018 while it was 315 ppm in 1960 (IPCC report). This release should subject to under control otherwise leads to dramatic climatic changes. Phototrophic algae such as seaweed, seagrass and phytoplankton in the aquatic ecosystem are efficient carbon sequesters that aid carbon fixation than their terrestrial offset and thus alleviating global warming. Kaladharan et al. (2009) studied the carbon sequestration of a few marine microalgae and macroalgae. It was observed that the utilization of dissolved CO_2 by seaweed *Ulva lactuca* is 100% for fixation and 27.7% utilization by microalgae. An estimation is made that biomass of seaweed along the Indian coast can utilize 9052 tonnes CO_2 per day. The Demeter et al. (1995) conducted an experiment on

Chlamydomonas reinhardtii and *Chlamydomonas stellata* by illuminating these green algae with CO_2 – depleted medium for its cultivation under 50 and 150 μmol/m²/s photon flux density respectively. This condition created a complete inhibition for the electron transport from the water towards diaminodurene and a slight influence on electron transport from water to 2,6 – dichlorobenzoquinone. A prolonged condition with CO_2 depletion on these green algal cells results in a loss, which is irreversible for electron transport and variable fluorescence. Nakano et al. (1996) experimented on the *Euglena,* which is a photosynthetic alga to evaluate its photosynthetic conditions under high CO_2 levels. A level of 5 – 40% CO_2 enhanced photoautotrophic growth as compared to bubbling air. Cell proliferation of *Euglena* will not happen above 45% CO_2 level. A considerable increase in chlorophyll and RuBisCO has been noticed in the cell at 5 – 40% concentration. Under high CO_2 concentration PEP, carboxyl kinase and PEP carboxylase were not inhibited. Even if the affinity of RuBisCO towards CO_2 decreases, RuBisCO has high activity under elevated CO_2 concentration. The CO_2 level above 45% leads to inhibition in growth as PSII activities were suppressed where PSI activity remain persistent. One of the limiting substrates for photosynthetic rate is CO_2. More than 50% of photosynthetic production worldwide is the contribution of aquatic microorganisms. So, that any change in the supply of CO_2 or HCO_3^- (dissolved inorganic carbon) could take effect on these organisms. The reasons cited for these supply issues are variability in soil or sediment respiration, the influence of pH on CO_2 and HCO_3^- (dissolved in water) and their relative distribution, diffusion rate in the aquatic environment of CO_2, which is 10^4 times slower than air conditions etc. A mechanism against this condition for their acclimatization in limited CO_2 supply is the CCM (Carbon – concentrating mechanism) to enhance C assimilation by an increase in internal inorganic Carbon. Transcriptome analysis to know the components of CCM is done in *Chlamydomonas reinhardtii* (Yamano and Fukuzawa, 2009). High Rate Algal Ponds abbreviated as HRAP are efficient in treating wastewater in comparison with conventional stabilization ponds. CO_2 addition promotes the N removal and assimilation of algae in HRAPs as it adds to the available CO_2 during the daytime. A decreased pH below 8 due to CO_2 addition (Park and Craggs, 2010) will create a chance of preventing the inhibitory act exerted by Ammonia on algal growth while on another side it restricts the removal of N through some physical and chemical process happened at high pH like Ammonia volatilization but this is counterbalanced by enhanced nutrient assimilation by algae. Park and Craggs (2011) conducted a study on HRAPs with 4 day and 8 day hydraulic retention times to examine the influence of adding CO2 on N removal. The results obtained show that a higher proportion of Algae (80.5%) found in 4 days HRT and in 8 days HRT, it was 55.60% in HRAP has and decreased pH resulted from CO_2 addition. In addition, a reduced Nitrogen loss below 9% obtained with the addition of CO_2 while a 26% loss in Nitrogen happened without CO_2 addition. While wastewater treatment using Algae, if CO_2 level is enhanced, it enables the algae for complete removal of dissolved N and dissolved P over a range of 2.5: 1 to 102: 1 for Nitrogen: Phosphorus mass ratios (Fulton, 2009).

Conclusion

Contaminated water, their treatment and safe disposal or discharge are serious issues to be noticed with societal concern and attention. Even though many mechanical, physicochemical technologies exist for treating the water, they possess drawbacks such as high expenditure, incomplete removal of contaminants, resulting in further toxic waste etc. The biological replacement for these methods by Algae efficiently removes the dissolved compounds within the water and let it for safe discharge. Use of Algae not only provide the benefit of treating wastewater but also have the surplus benefit of resource recovery. Using HRAPs in algal treatment aid the effective dismissal of Phosphates, nitrates, Ammonium heavy metals from water in coordination with the utilization of biomass yield from algae for biofuel production. They should provide with adequate growth condition while used for the treatment of contaminated water like temperature with optimum range 22 - 35ºC, pH of 5.7 to 8.5 blue, red light radiation and CO_2 addition in the condition of limitation enhances the growth of Algae. HRAPs are economically and environmentally feasible. Research work should continue with more areas of applying the benefits of Algae.

References

Abdel – Raouf, N., Al- Homaidan, A.A. and Ibraheem, I.B.M. (2012). Microalgae and wastewater treatment. *Saudi Journal of Biological Sciences, 19*, 257 – 275.

Baalen, C.V., Hoare, D.S. and Brandt, E. (1971). Heterotrophic Growth of blue – green Algae in dim light. *Journal of Bacteriology, 105*(3), 685 – 689.

Bates, S.S., Tessier, A., Campell, P.G.C. and Buffle, J. (1982). Zinc adsorption and transport by *Chlamydomonas variabilis* and *Scenedesmus subspicatus* (Chlorophyceae) grown in semicontinuous culture. *J. Phycology. 18*, 521 – 529.

Brinza, L., Dring, M.J. and Gavrilescu, M. (2007). Marine micro and macro algal species as biosorbents for heavy metals. *Environmental Engineering and Management Journal, 6*(3), 237 – 251.

Butterwick, C., Heaney, S.I. and Talling, J.F. (2005). Diversity in the influence of temperature on the growth rates of freshwater algae, and its ecological relevance. *Freshwater Biology, 50*, 291 – 300.

Chaisuksant, Y. (2003). Biosorption of Cadmium (II) and Copper (II) by pretreated biomass of marine algae *Gracilaria fisheri. Environmental Technology, 24*, 1501 – 1508.

Chen, Y.C. and Durbin, E.G. (1994). Effects of pH on the growth and Carbon uptake of marine phytoplankton. *Marine Ecology Progress Series, 109*, 83 – 94.

Cho *et al.* (2007). Optimum temperature and salinity conditions for growth of green algae *Chlorella ellipsoidea* and *Nannochloris oculata. Fisheries Science, 73*, 1050 – 1056

Craggs, R.J., Heubeck, S., Lundquist, T.J. and Benemann, J.R. (2011). Algal biofuels from wastewater treatment high rate algal ponds. *Water Science & Technology, 63*(4), 660 – 665.

Craggs, R., Park, J., Heubeck, S. and Sutherland, D. (2014). High rate algal pond systems for low – energy wastewater treatment, nutrient recovery and energy production. *New Zealand Journal of Botany, 52*(1), 60 – 73.

Das, P., Lei, W., Aziz, S.S. and Obbard, J.P. (2011). Enhanced algae growth in both phototrophic and mixotrophic culture under blue light. *Bioresource technology, 102*, 3883 – 3887.

Foy, R.H., Gibson, C.E. and Smith, R.V. (1976). The influence of daylength, light intensity and temperature on the growth rates of planktonic blue – green Algae. *British Phycological Journal, 11*(2), 151-163. DOI: 10.1080/00071617600650181.

Fourest, E. and Volesky,B. (1996). Contribution of Sulfonate groups and Alginate to heavy metal biosorption by the dry biomass of *Sargassum fluitans. Environ. Science. Technology, 30*, 277 – 282.

Franklin, N.M., Stauber, J.L., Markich, S.J. and Lim, R.P. (2000). pH- dependent toxicity of copper and uranium to a tropical freshwater alga (*Chlorella* sp.). *Aquatic toxicology*, 48, 275 – 289.

Kleinova *et al.* (2012). Biofuels from Algae. *Procedia Engineering, 42*, 231 – 238.

Konopka, A. and Brock, T.D. (1978). Effect of temperature on blue – green Algae (Cyanobacteria) in Lake Mendota. *Applied and Environmental Microbiology, 36*(4), 572 – 576.

Kumar, J.I.N. and Oomen, C. (2012). Removal of heavy metals by biosorption using freshwater alga *Spirogyra hyaline.J. Environ. Biol. 33*, 27 – 31.

Lee, K. and Lee, C. (2001). Effect of Light/Dark cycles on Waste water Treatments by Microalgae. *Biotechnol. Bioprocess Eng., 6*, 194 – 199.

Mehta, S.K. and Gaur, J.P. (2005). Use of Algae for removing heavy metal ions from wastewater: Progress and Prospects. *Critical Reviews in Biotechnology, 25*(3), 113 – 152.

Mostert, E.Z. and Grobbelaar, J.U. (1987). The influence of Nitrogen and Phosphorus on Algal Growth and Quality in outdoor mass Algal cultures. *Biomass, 13*, 219 – 233.

Nakano *et al.* (1996). Growth of Photosynthetic Algae *Euglena* in high CO_2 conditions and its photosynthetic characteristics. *Proc.IntSym.Plant Production in closed ecosystems. Acta Hort., 440*

Park, J.B.K. and Craggs, R.J. (2010). Wastewater treatment and algal production in high rate algal ponds with carbon dioxide addition. *Water Science and Technology, 61*, 3.

Park, J.B.K. and Craggs, R.J. (2011). Nutrient removal in wastewater treatment high rate algal ponds with carbon dioxide addition. *Water Science and Technology, 63*, 8.

Park, J.B.K., Craggs, R.J. and Shilton, A.N. (2011). Wastewater treatment high rate algal ponds for biofuel production. *Bioresource Techology, 102*, 35 – 42.

Pederson, M.F. and Borum, J. (1996). Nutrient control of algal growth in estuarine waters. Nutrient limitation and the importance of nitrogen requirements and nitrogen storage among phytoplankton and species of macroalgae. *Marine Ecology Progress Series 142*, 261 – 272.

Sato, N., Murata, N., Miura, Y. and Ueta, N. (1979). Effect of Growth Temperature on lipid and fatty acid compositions in the blue – green algae, *Anabaena variabilis* and *Anacystis nidulans. Biochimica et Biophysica Acta., 572,* 19 – 28.

Schnurr, P.J., Espie, G.S. and Allen, D.G. (2013). Algae biofilm growth and the potential to stimulate lipid accumulation through nutrient starvation. *Bioresource technology, 136,* 337 – 344.

Sforza, E., Simionato, D., Giacometti, G.M., Bertucco, A. and Morosinotto, T. (2012). Adjusted light and dark cycles can optimize photosynthetic efficiency in Algae growing in Photobioreactors. *PLoS ONE, 7*(6), e38975. Doi: 10.1371/ journal.pone.0038975.

Singh, S.P. and Singh, P. (2015). Effect of temperature and light on the growth of algae species: A review. *Renewable and Sustainable Energy Reviews, 50,* 431 – 444.

Wilde, K.L., Stauber, J.L., Markich, S.J., Franklin, N.M. and Brown, P.L. (2006). The effect of pH on the uptake and toxicity of Copper and Zinc in a tropical freshwater Alga (*Chlorella* sp). *Archives of Environmental Contamination and Toxicology, 51,* 174 – 185.

Woertz, I., Feffer, A., Lundquist, T. and Nelson, Y. (2009). Algae grown on dairy and municipal wastewater for simultaneous nutrient removal and lipid production for biofuel feedstock. *Journal of Environmental Engineering, 135*(11). 1115 – 1122.

Xin, L., Hong-Ying, H., Ke, G. and Jia, Y. (2010). Growth and nutrient removal properties of a freshwater microalgae *Scenedesmus* sp. LX1 under different kinds of nitrogen sources. *Ecological Engineering, 36,* 379 – 381.

Yamano, T. and Fukuzawa, H. (2009). Carbon – concentrating mechanism in a green alga, *Chlamydomonas reinhardtii,* revealed by transcriptome analyses. *Journal of Basic Microbiology,* DOI 10. 1002/jobm.200800352.

Yu, Q., Matheickal, J.T., Yin, P. and Kaewsarn, P. (1999). Heavy metal uptake capacities of common marine macro algal biomass. *Wat. Res., 33*(6), 1534 – 1537.

Zhang, Y., Tang, C.T. and Li, G. (2012). The role of hydrodynamic conditions and pH on algal – rich water fouling of ultrafiltration. *Water Research, 46,* 4783 – 4789.

Significance of Microbial Volatiles in Agro-Ecological Health

E Lokesh Goud

Department of Agronomy, School of Agriculture

Lovely Professional University, Jalandhar, Punjab, India, 144411

Email: lokeshgoud1832@gmail.com

Abstract

Microbial Volatiles include both organic and inorganic compounds released by the soil microbes like bacteria and fungi, characterized with the physicochemical properties such as lipophilic, compounds with high vapour pressure (evaporate at 0.01 kPa at 20^{0}C), low molecular weight (100 – 500 Da) with small size (up to 20 carbon atoms), low polarity and low boiling point. Lipophilic property of volatiles enables their membrane free movement and exhibits accelerated diffusion through soil pores and atmosphere. Chemical classes of this volatiles encompass hydrocarbons, alcohols, alkenes, terpenes, ketones, organic acids, aldehydes, esters, pyrazines, cyclohexanes, benzene derivatives and inorganic compounds. Microbial volatiles is the side products of complex biosynthetic and metabolic pathways like aerobic, fermentation, amino acid catabolism, heterotrophic carbon metabolism, sulphur reduction, fatty acid degradation, protein degradation, glycolysis, the anaplerotic pathway of Kreb's cycle and terpenoid biosynthesis inside the microbes. Volatile compounds mediate ecological interaction in the soil – atmosphere system such as bacteria – bacteria, bacteria – fungi, fungi – fungi, fungi – bacteria, bacteria – insects, fungi – insects and with plants in above ground and below ground environment. Microbial Volatile Compounds have the function to maintain agroecological health by exhibiting antibacterial and antifungal properties against phytopathogenic organisms, enhancement of plant growth, a defence mechanism against insect pests, attractant and repellent to certain social insects and making the plant tolerant for certain abiotic stress condition like salinity and drought. Microbial Volatiles can confer Induced Systemic Tolerance or Induced Systemic Resistance in plants against biotic and abiotic stresses. This chapter discusses the extent to which chemical inputs can be substituted by microbial volatiles in promoting plant growth, controlling pathogenic and damaging pests as well as inducing tolerance to abiotic stress in plants to protect the agro-ecological system on a sustainable manner.

Introduction

Agricultural practices emphasizing on sustainability have a high demand for alternatives for chemical inputs that have been used in current agricultural production. Ease in use and results in shorter time influenced the producers to use more chemical inputs like pesticides, fungicides, herbicides, synthetic fertilizers etc. Even though chemical inputs can augment the crop yield, their continuous use creates environmental pollution. Agricultural production should be increased to meet the food requirements of a growing population. Abating the number of chemical inputs and encouraging the environmentally friendly and cost-effective strategies can promote the conservation of ecosystem and natural resources through sustainable management practices. A large number of compounds including both organic and inorganic chemicals released by microbes are discussed in this chapter. Microbial Volatile Compounds (MVCs) are the metabolic side products obtained from microorganisms like bacteria and fungi. Recent studies have put forward the evidence of various applications of MVCs like improved crop productivity and yield, plant protections against pathogenic and insect attack (Beck and Vannette, 2017) as well as induce drought tolerance in plants (Cho et al., 2008). Microorganisms have continuous interaction with the plant roots. The rhizosphere is a narrow zone but has an intense microbial population and highly influenced by plant roots (Sarsan, 2017). The rhizosphere is a complex ecosystem that resides bacteria and fungi as a source of biodiversity where exudates from the plant roots attract the microbes (Barea et al., 2005). This helps in obtaining the bacterial cell number of 10^8 / gm of fresh root. Besides, symbiotic fungi (mycorrhiza) has a major cover (>95%) of short roots of terrestrial plants which again surrounded by mycorrhiza helper bacteria (Penuelas et al., 2014). Interactions between the plant and microbial populations may be detrimental or beneficial as they can influence crop health, nutrition and plant growth at above and below the soil surface. Even though microbes release both inorganic and organic volatile compounds, organic compounds are vastly discussed and abundantly present in the soil. Microbial Volatile Organic Compounds (MVOCs) have a multidimensional role in agroecosystems. They can act as an attractant or repellent for many insect species and invertebrates called semiochemicals for chemical ecologists. For agricultural scientists, MVOCs has the role of biofertilizer and biocontrol agent. Plant diseases after harvesting can be controlled by the biocontrol property of MVOCs in food industries. Recent studies reveal the application of MVOCs as biofuel. In this chapter, microbial volatiles is discussed with emphasis on their significance in maintaining agro-ecological health, the extent of substitution of chemical inputs with microbial volatile organic compounds, interaction among inter and intra organisms through volatile signalling and various applications of MVOCs.

Microbial Volatile Compounds (MVCs)

A wide range of lipophilic compounds with low polarity characterized by low molecular weights (100-500 Da), low boiling point, small compounds with high vapour pressure (0.01 kPa at 20^{0}C) occur as a complex mixture and derived through various biosynthetic pathways are called volatile compounds (Schmidt et al., 2015; Kanchiswamy et al.,2015). These metabolic products are the side products of either primary metabolism (eg. Amino acids, Fatty acids and DNA synthesis) or secondary metabolism as intermediate products of primary metabolism (eg. Oxidation of glucose in metabolic pathways)(Korpi et al., 2009). Above-mentioned characteristics of volatiles enable properties like evaporation and diffusion. So they can move from the site of production via soil pores (water or air-filled) and atmosphere that suits them as an ideal communicative mediator that can communicate in an inter organism and intra organism manner by mediating short and long-distance interactions. Therefore, they are called info chemicals (Wheatley, 2002). Biosynthetic pathways responsible for volatile synthesis include aerobic, fermentation, amino acid catabolism, heterotrophic carbon metabolism, sulphur reduction, fatty acid degradation and terpenoid biosynthesis (Schmidt et al., 2015). Living organisms ranging from microorganisms to well-developed plants and animals can secrete volatile organic compounds, which influence each other. Even human activities and biogenic sources could evolve volatile compounds. Biogenic volatile compounds come under esters, alcohols, aldehydes, terpenoids, thiols, fatty acid derivatives, ketones, benzenoids, alkenes, pyrazines (Schulz-Bohm et al., 2017). It has been reported by Schenkel et al., 2015 that 853 volatiles were listed as released from plant roots and residing microbes of soil which are shared among 5 groups, that are, roots of a plant producing 39 compounds, rhizosphere bacteria producing 209 volatile compounds, soil bacteria giving out 483 compounds, rhizospheric fungi and soil fungi producing 261 and 187 volatile compounds respectively. Lemfack et al., 2014 reported that the study conducted till that date shows that 69 fungi and 349 bacteria emitted volatile compounds. Lemfack et al., 2018 identified around 1860 microbial volatiles released from 340 fungi and 604 bacterial species through manual approach and literature studies.

Gas chromatography – Mass spectrometry can be used as a reliable analytical tool for volatile compounds in which gas chromatography have separation capability and mass spectrometry can identify and quantify each compound (Bitas et al., 2013). Synthesis of MVOCs in the rhizosphere is influenced by the availability of nutrients, oxygen, growth stages of microbes, pH, temperature and soil moisture (Schulz-Bohm et al., 2017). Volatile compounds can mediate the interactions among biotic environments like plant-to-plant, microbe to microbe and plant-microbe. The formation and development of an ecosystem are facilitated by biogenic VOCs as the interactions mediated by them can balance the ecosystem.

Chemistry of Volatile Compounds from Microbes

The microbes as volatiles into the soil or atmosphere release a wide range and diverse chemical compounds, which can be included in different chemical classes as discussed below.

Hydrocarbons: These are the compounds resulting from the biosynthetic pathways of fatty acid metabolism. In Bacteria, alkanes of short-chain ranging from decane to tetra decane were found while cyanobacteria claimed for synthesising longer chains of hydrocarbon (eg: hexadecane) as well as branched ones also. Fungi can produce short-chain hydrocarbons. As part of the defence mechanism in

A. thaliana against *Pseudomonas syringae*, long-chained hydrocarbon C13 tridecane was released from the bacterium *Paenibacillus polymyxa* E681 by augmenting ISR. Applying pure tridecane as a priming material at a dose of 10mM exhibited 4.7 times of transcription of PR1 gene (Pathogenesis Related Gene). Volatile blends of same microbes consist of dodecane, acetylene, isoprene, 1 – undecane and cyclohexane. Biocontrol of *Sclerotinia sclerotiorum* is done using n – decanal volatile emitted from *Pseudomonas* sp. Under salt stress, pure tridecane can enhance soybean germination.

Ketones / Alcohols: Products of fatty acid decarboxylation belong to this group. Acetoin is a prominent microbial volatile in terms of plant growth and disease resistance. The chemical name of acetoin is 3 – hydroxyl -2- butanone. 2, 3 – butanediol is the oxidised form of acetoin. These two compounds resulted from anaerobic fermentation of pyruvic acid. If sucrose has been used as the nutrient source, 2, 3 – butanediol can be produced in – vitro favourably. SA dependent mechanism is a drought-mitigating strategy in the plant that can be stimulated through 2R, 3R- butanediol that is a volatile released from *P. chlororaphis* strain O6. *Bacillus amyloliquefaciens* are capable to release a group of alcohols, which are branched and short-chain like 2 – methyl – 1 – butanol and 3 – methyl -1- butanol. These alcohols are products of the Ehrlich pathway in which an enzyme catalysed the transformation of leucine and isoleucine (Amino acids and branched-chain) takes place. Fatty acids under α oxidation and β oxidation can derive certain long-chain aliphatic alcohols which mostly a part of Enterobacteriaceae (gram-negative bacteria) which includes following compounds 1 – decanol, 1 – octanol, 1 – dodecanol etc. Short-chain alcohols (eg: 2, 3 – butanediol) are the group of volatile compounds from a microbial group, for example, *Firmicutes, Proteobacteria.* 1 – hexanol is an alcoholic volatile compound with antifungal property released from *Bacillus* and *Pseudomonas* sp. Besides, it can enhance the growth of plants.

Acids: Acetic acid is a well-known short-chain carboxylic acid that is a product among the blend of VOCs released from *Burkholderia Tropica* (bacterium) for inhibiting fungi, which are phytopathogenic like *Fusarium oxysporum, F. Culmorum, C. gleosporioides* and *Sclerotium rolfsii.* Other acids belonging to microbial volatile compounds are phthalic acid, oxalic acid, hexadecanoic acid, stearic acid and myristic acid from *Pseudomonas simiae* AU that is a plant growth-promoting rhizobacteria with effect on salinity tolerance in soybean plants. These acids are the side products of anaerobic metabolism like the fermentation of carbohydrates in bacteria.

Sulfur Compounds: Volatile compound belonging to the chemical classes of sulfur compounds includes the example of DMDS and DMTS (Dimethyl Disulphide and Dimethyl Trisulpide). These compounds are claimed for odour in cheese and wine. In soybean, a sulfur-containing compound named Benzathiazole, which is then released volatile from Pseudomonas spp, can inhibit the activity of sclerotia. The characteristic odour of this group of volatiles is due to methionine decomposition in a catabolic pathway.

Inorganic Compounds: Cyanide (HCN) production by rhizospheric microorganism is an example for the inorganic volatile compound from bacteria like *Chromobacterium or Pseudomonas,* which is capable to inhibit metal-containing enzymes like cytochrome c oxidase. Besides, it helps in growth promotion. Other examples are nitric oxide, hydrogen sulphide, Ammonia (Piechulla et al., 2017).

Types of Microbial Volatile Compounds

1. **Bacterial Volatile Compounds:** Bacteria are a cosmopolitan group of microorganism that has a higher cell number in the soil (10^{11} cells per gram of soil) and present in colonies around the root and other organic substances. They produce volatiles whose potential has been exploited in food and beverage industries like yoghurt, cheese, wine, sauerkraut etc. Besides, they act as an interaction agent with other living organism resulting in either beneficial or harmful impact and to overcome competition inside the same niche. Fungal pathogens are negatively affected by albaflavenone and dimethyl disulphide while plant growth is positively influenced by acetoin, geosmin, 2, 3-butanediol and tridecane. Bacterial phenotypes exhibiting resistance against stress and antibiotics can also produce volatile compounds. Bacterial cells engaged in biofilm formation produce volatiles to link each other. In humans, immunization against certain pathogens is the result of the changes caused in the physiology of the gut cell by some bacterial volatiles like ammonia and trimethylamine. Bacterial volatiles produced mostly belong to classes of amines and pyrazines. Apart from the benefits, these volatiles is responsible for pathogenesis in different strains. Catabolic pathways produce bacterial volatiles in certain processes like glycolysis, lipid degradation and protein degradation. The pathways involved in the biosynthesis of some common bacterial volatiles and their chemical classes are discussed below.

   (i) Methyl ketones are derived from the decarboxylation process of fatty acids

   (ii) Two pyruvic acid molecules are condensed into acetolactate in the presence of the enzyme acetolactate synthase. Further decarboxylation of the product gives acetoin. Oxidation of acetoin gives 2, 3 butanedione. The entire process is anaerobic or can be said as fermentation of pyruvate.

   (iii) Short chained Organic acids (eg: Acetic acid, Butyric acid, propionic acid) side products of carbohydrate fermentation in bacteria.

   (iii) TCA (Tricarboxylic Acid) cycle under an anaplerotic pathway encourages the cell in using the carbon and its compounds, which ultimately produces glyoxylic acid.

   (v) Bacteria mediate the biosynthesis of dimethyl sulphide and 1 – (methyl thio) -3- pentanone. These are volatile compounds derived from methionine. Cleavage of the compound, 3 – dimethylsulfoniopropionate lead to L – methionine biosynthetic pathway of volatiles of sulfur compound.

   (vi) Reduction of TMAO (Trimethyl Amine Oxide) results in the release of trimethylamine, which is a tertiary amine and has the odour of spoiled fish, occurs in anaerobic condition. Conversion of this compound by bacteria can adopt the aerobic and anaerobic pathway.

(vii) Indole compounds released by more than 85 bacterial species have been reported. In *E. coli,* tryptophan is modified into indole, ammonia, pyruvate etc with the help of *tnaA* (tryptophanase-encoded gene) reversibly. This biosynthetic process of indole is regulated by environmental factors pH, temperature, cell population density, catabolite repression, the population density of microbes etc.

(viii) Deoxyxylulose phosphate pathway or mevalonate pathway would give terpenes group of volatiles. Building units for terpenes are dimethylallyl pyrophosphate and isopentenyl pyrophosphate by which bacteria produce sesquiterpenes, monoterpenes and other derivatives. Other examples are albaflavenone (antibiotic from *Streptomyces*), geosmin (a degraded sesquiterpene producing earthy odour) etc.

(ix) Degradation of the amino acid, cysteine produces hydrogen sulphide. Nitric Oxide synthase is an enzyme that produces NO from L – arginine. In *E. coli,* conversion of aspartic acid to fumaric acid catalysed by aspartate ammonia-lyase gives Ammonia as a side product. Hydrogen Cyanide is a product from the biosynthetic pathway catalyzed by HCN synthase from Glycine in bacterias encoded by gene *hcnABC* under oxygen-deprived condition (Piechulla et al., 2017).

Diversity and complexity are more in case of bacterial volatiles than that of plants and fungi. As the volatile compounds can distribute in an aerial environment with a specific odour, it can act as an identification mark for the presence of a specific organism. For example, *Pseudomonas aeruginosa* releases a volatile compound named 2- AA (2 –amino–acetophenone) which has a grape-like odour and helps in identifying their infections. They can play the roles of carbon release valve, growth-promoting agent, and growth-inhibiting agent and enhances survival and establishment of bacterial colonies in the ecological niche. Their site of action is above and below ground.

2. **Fungal Volatile Compounds:** Diverse microorganisms with complexity in structure and competency to other microorganisms with a saprozoic and heterotrophic nature can be characterized as fungi, which also can form colonies. Most analysis and studies conducted on fungal volatiles were related to their application as an indirect indicator of fungal presence, food and flavour properties, info chemical or semi or chemical against insects, food spoilage monitoring, chemotaxonomical purposes, detection of diseases and ailments in large organisms including animals and plants, biodiesel and biofilters as well as mycofumigation (Hung et al., 2015). Fungi are capable to synthesise mixture of carbon-based and gas-phase compounds termed as volatile organic compounds. Movement and diffusion of these particles are quite easy through soil and atmosphere due to their small size. The volatile compounds released from fungi are effective as a green pesticide. Fungi

exhibit diversified reaction pathways to synthesise secondary metabolites that are volatile aromatics. These compounds from fungi mostly belong to the chemical class called polyketides. An interaction established between the polyketide synthases (a multidomain protein) and acetyl CoA and Malonyl CoA (both are short-chained carboxylic acids) would give polyketides of different Carbon chain lengths eg. Terpenes. Techniques used to analyze volatiles from fungi include (i) Steam distillation coupled with Liquid-Liquid Extraction (ii) Gas Chromatography-Mass Spectrometry (GC-MS) (iii) Proton Transfer Reaction – Mass Spectrometry (iv) Selected Ion Flow Tube – Mass Spectrometry (v) Gas Chromatography – Time of Flight Mass Spectrometry (vi) Solid-Phase microextraction. Mycofumigation is the characteristic property of fungus exhibited through their volatile emissions by which they can protect the crop during its growth and after harvest from bacterial and fungal diseases (Selim et al., 2017). Example for this property is *Muscodor albus* and *Oxyporus latemarginatus,* whose volatile increase shelf life of transported agricultural commodities. Since fungi are most efficient decomposers, the hydrolytic enzymes and others can degrade the exogenous chemicals. There are chances for the misconception that all VOCs analyzed as part of a particular fungal species are emitted because of anabolic pathways in their biosynthesis. However, in actual, they are the products or byproducts of some extracellular substrates. The exposure of fungal growth on human from damp environments and mouldy residential cause certain clinical conditions and illness termed as 'mould-related illness' or 'sick building syndrome' (Morath et al., 2012). However, they also can produce certain fungal VOCs, which can be used as antifungal, nematicides and insecticides. Eg. 1- Octen -3- ol, Sabinene, 1, 8 - Cineole, 3 - methyl-butanol etc.

Fig. 1. Application of microbial volatiles in different sectors

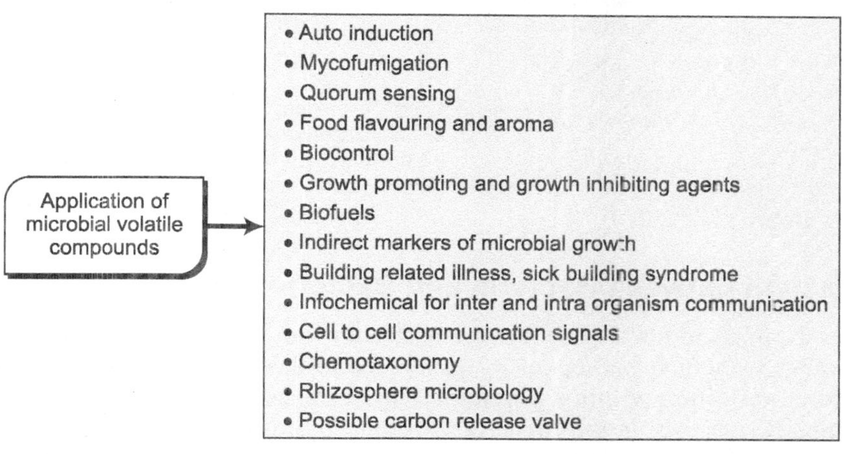

(Source: Enespa and Chandra, 2017)

Inter – Organism and Intra – Organism Interactions through Microbial Volatiles

It is essential to analyze and study various interactions between organisms while studying an ecosystem. Microbes are an inevitable part of the agroecosystem as they can affect the crops in a beneficial and harmful manner. The diverse microbial population are present in soil and constantly interact with each other. Besides, they interact with plants and insects. Either volatiles, it from plant or microbe, is the best signalling agent in intercommunication than root exudates (Schenkel et al., 2015). Microbial Volatile Organic Compounds are the mediator for signalling those interactions largely.

Fig. 2. Microbial Volatiles interaction with organisms and the exhibited response of the organisms

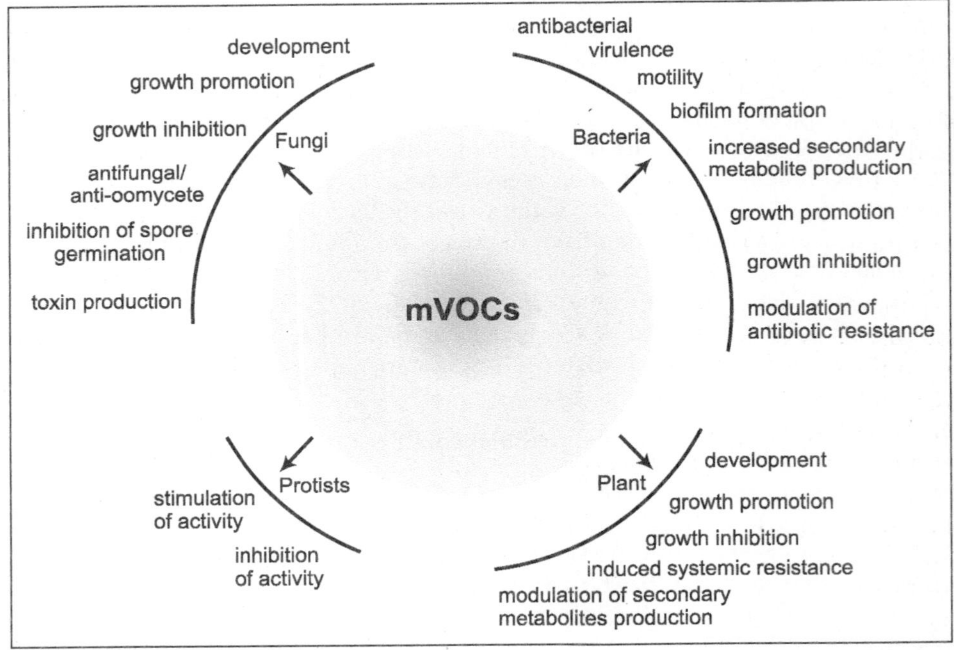

(Schulz-Bohm et al., 2017)

Interaction between Bacteria and Bacteria

Volatile compounds produced by one bacterium can have an antagonistic effect on another by which pathogenicity can be controlled. For a bacteria to enable their communication within a population, quorum sensing (a process to monitor cell density or population density through the cell-to-cell communication) and to coordinate the behaviour in a group is made possible through volatiles. Chemistry of communicative agents differs in gram-positive and gram-negative bacteria. According to Kai et al., 2008, gram-positive bacteria use gamma butyrolactones, modified peptones and amino acids whereas gram-negative bacteria use cyclic dipeptides, fatty acid derivatives (eg: *cis*- unsaturated fatty acids, 3 –hydroxy palmitic acid methyl ester) and *N*-AHLs (*N*- acyl-homoserine lactones) for communication. *Bacillus subtilis* is affected by the volatile compound sesquiterpene albaflavenone, which is released by *Streptomyces albidoflavus*. Two bacterias residing in rhizospheric zones *Serratia plymuthica* and *Pseudomonas fluorescens* produced the volatile compound, dimethyl disulphide exhibited bacteriostatic effect on phytopathogens *Agrobacterium vitis* and *A. tumefaciens* (Sarsan, 2017). *Ralstonia solanacearum* is the causal organism of bacterial wilt in tomato, which a bacteriostatic effect is produced by the organic compounds 1-methyl naphthalene and benzathiazole released by *Pseudomonas fluorescens* WR-1. Wilt disease is decreased by the property of volatiles to change the expressions of certain genes controlling pathogenicity (eg: Type III secretion system, EPS production – extracellular polysaccharide, Global virulence Regulator PhcA) so that plants exhibit ISR (Induced Systemic Resistance).

Some compounds can cause bacterial virulence eg, compounds like acetoin and 2,3 butanediol induce virulence in the bacterias *Pectobacterium carotovorum* and *Pseudomonas aeruginosa*. There is bacterial volatiles that enhances the growth of other bacterias within the rhizosphere. *Collimonas pratensis* is a bacterium that produces the VOCs responsible for inducing the growth in *Pseudomonas fluorescence* PFO – 1 by regulating the gene expression for mobility within the bacteria along with stimulation for their antibacterial properties against *Bacillus*. Those bacteria consuming the exudates from plant roots released certain volatiles that can stimulate the growth in bacteria surviving at a distant location and nutrient-limited conditions (Schulz – Bohm et al., 2015). The characteristics of bacteria to move from one place to next location in need of resources and less competency is motility and this can be affected by the volatile compounds released by bacteria like H_2S, NH_3, trimethylamine, 2- amino- acetophenone and nitric oxide. These compounds can also alter the formation of biofilm. Exploration is a new process evolved in certain bacteria like *Streptomyces venezuelae* (non- motile bacteria) which helps them to access nutrient-rich regions. To escape from the nutrient stress condition, vegetative hyphae exhibit a hydrophilic growth in a fast manner in *S. venezuelae, which* is triggered by increased pH and glucose depletion. For facilitating long distance communication, explorer cells release certain signals like trimethyl amine that can also induce antibacterial activity by increasing pH against *Micrococcus luteus* and *Bacillus subtilis*.

Interaction between Fungi and Bacteria

There exists the effect of fungal volatiles on bacteria and bacterial volatiles on fungi. There exists specificity in this interaction, as any fungal or bacterial volatile cannot cause harmful actions on any other organism at the same extent. Butyrolactonas, Phenazine -1- carboxylic acid, Ammonia, Alcohols, HCN, acids are some group of volatiles that may affect various fungi to inhibit their mycelial growth (Ortiz-Castro et al., 2009). However, long-distance communication with phenotypical responses is obtained through volatile interaction. Gene expression (phlA) in *Pseudomonas fluorescens* that is encoded for the synthesis of 2,4- diacetylphloroglucinol has been increased by certain VOCs produced by the fungus *Trichoderma atroviride. Pleurotus ostreatus* (oyster mushroom) produce certain fungal VOCs that are growth inhibitors in bacterias like *B. subtilis* and *B. cereus.* Phenotypical responses in bacteria against the oomycetal volatiles and fungal volatiles included changes in growth, the formation of biofilm, antimicrobial properties, bacterial motility (Swarming / Swimming) which can be positively or negatively regulated. In fungus, volatile compounds can be used for attracting mutualistic bacteria that have a positive interaction while they can also exclude the competitive category. The study on transcriptomics and proteomics in the bacteria *S. plymuthica* PRI-2C shows that a change in the gene expression and protein expression has been observed when exposed to certain fungal volatile compounds with observed effect on cell envelope biogenesis, signal conduction, motility, energy metabolism and metabolite production. Schmidt et al., 2017 inferred in a study that the bacteria, *S. plymuthica, when exposed to the VOCs synthesised by Fusarium culmorum,* produced an unusual terpene compound Sodorifen.

Soil fungistatis is a characteristic phenomenon induced by certain rhizospheric bacterial volatile compounds by which fungal propagules loss their ability to germinate or further grow. Fungi causing rice sheath blight, called *Rhizoctonia solani* can be controlled by VOCs produced by the bacteria *Streptomyces* sp. This can reduce the disease intensity. In addition, it has been analysed by Kai et al., 2008 that volatiles from *Streptomyces* sp. are promoters for mycorrhizal formation (*Glomus mosseae*) as well as arbuscular mycorrhizal formation. Structural changes occur in fungal mycelium with the exposure to bacterial volatiles. Some examples listed by Kai et al., 2008 are thick walled, irregular shaped conidia and hyphae with disappeared longitudinal and traverse septae in *Alternaria alternata* under volatile fumigation from *B. subtilis,* Stunted and vegetative conidiophores in *Cladosporium oxysporum* and granulation and vacuolization in mycelium with fungal hyphae lysis occurs in *F. oxysporum.* The property of volatile compounds, i.e., antifungal, antagonistic or non-antagonistic depends on the growth conditions of the bacteria along with nutrient availability.

Interaction between Fungi and Fungi

Volatile compounds produced by one fungus can act as the interactive mediator within the fungal species. 1-Octen-3-ol is a volatile organic compound that has mushroom odour also can be called as mushroom alcohol, produced by certain group of fungi can develop other fungal growth. 1-octen-3-ol act as self-inhibitor of spore germination in *Penicillium paneum.* They exhibit antagonistic effect on other fungal species. Volatile compounds like dimonene, 3, 4 – 2H – dihydropyran, 5 – hexenoic acid and Octanoic acid are released from *Fusarium oxysporum* CanR – 46 (non – pathogenic fungus) can cause inhibition in another pathogenic fungi - *Verticillium dahlia.*

Detoxification of antifungals can also be performed by some VOCs produced by fungus itself. An example cited for this is 6 – pentyl – alpha – pyrone (a toxic compound) released by *Trichoderma harzianum* can be detoxified by *Fusarium graminearum.* It will be difficult for the fungus to survive in carbon-limited environment but certain fungal VOCs help them to colonize in such condition, as VOCs can be an efficient carbon source.

Interaction between Bacteria and Insect

Bacterial volatiles can induce certain developmental and physiological alterations within the insect. A bacterium, *Pantoea aggolomerans* can induce the formation of guaiacol (aggregation pheromone) within the guts of the desert locust (*Schistocerca gregaria*) with the help of vanillic acid as the substrate that ingested through diet. *Dendroctonus valens* is the most dangerous pest that attacks coniferous forest also called as 'bark beetle'. Microbes can perform certain biotransformations. Here, the gut bacterial community converts the monoterpenoids (toxic compounds from conifers) to verbenone (pheromone with the multifunctional role). The high concentration of verbenone signals for anti – aggregation thus ensures coniferous forest management. Signals for oviposition in some insects eg: *Anopheles gambiae* can be transducted by the bacteria residing within their gut through the bacterial volatiles like 2-methyl-3-decanol 24, 2-phenyl ethanol 11, diisopropyl pyrazine 26, and isopropyl- secbutyl pyrazine 28, 2 – phenyl ethanol 31, which can be said as semiochemicals. Mutualistic interaction between aphids (*Acyrthosiphon pisum*) and ants (*Lasius niger*) are mediated by certain bacterial VOCs associated with the aphids. Aphids are the sucking pest that damages the crop and ants in this interaction maintain and protect the Aphid nest. Two bacteria named *Staphylococcus xylosus* and *S. sciuri* were present in the honeydew secreted by aphids. These bacteria emit a blend of volatile compounds (semiochemical) like butanoic acid, 3–methyl –2 – butenal, limonene, 2– methyl –2 – buten-1-ol, 3– methyl –1-butanol, 3–methyl–3–buten–1–ol, 3 – methyl butanoic acid, 2 – methyl butanoic acid, 2,3 – butanedione, 2 – methyl butanal, 3 – methyl butanal, 2,5 – dimethyl pyrazine and propanone – propane that can attract the ants. The ant – aphid mutual system with two bacterias *Acinetobacter calcoaceticus* and *S. sciuri* releases certain semiochemicals capable for inducing attraction and oviposition for the *Episyrphus balteatus, which* is the predator or natural enemy of the aphid, which can efficiently and effectively act as aphidophagous predators. *E. blateatus* is a pollinator in adult stage and predator during larval stage. Semiochemicals involved for these parasitic interactions are 2 – methyl butanoic acids, 3 – methyl – 2 – butenal and butanoic acid. Bacterial volatiles can make regulations over the social behaviour of certain social insects also.

Interaction between Fungus and Insect

Even though both fungi and insect are responsible for crop losses largely, there are some beneficial aspects too. Interaction between two organisms can alter the purpose of their existence in an ecosystem. Fungi interacting with the insects through volatile compounds can alter their damaging nature, reproductive habits, nutritional requirements and another physiological process. Fungi residing in symbiotic association with some insects can produce aggregation pheromones. Yeast is unicellular fungi. Association of yeast (*Hansenula capsulate and Pichia pini*) with the insect *Dendroctonus ponderosae* (mountain pine beetle) can synthesise the verbenone, which is an anti-aggregation pheromone. This is produced by the involvement of yeast in converting the *cis*- verbenol and *trans*- verbenol that emitted by pine beetles. This prevents the beetle attack of pine trees massively. In the same beetle, symbiotic relation with *Grossmania clavigera* (fungus) that produces sterols induces aggregation pheromone by female beetles. The volatile compound 1-octen-3-ol produced by mushroom is attractive to certain insect species. Oviposition of the insects is determined by the volatiles of fungus as if volatile compounds from the yeast inhabiting on the grape vine can positively affect the oviposition of the moth *Lobesia botrana.* In *Conogethes punctiferalis* (gravid yellow peach moths), inhibition of oviposition happens by fungal volatiles. Some fungi can cause diseases in certain insects like *Metarhizium anisopliae, Beauveria bassiana* (entomopathogenic fungi). They are eminent in releasing volatile compounds and certain insects like termites can identify such odours, which help them to avoid direct contact. Pine weevil (*Hylobius abietes*) can be repelled from pine twigs by the volatile compounds named styrene and 3 – methylanisole produced by the fungus *Penicillium expansum.* Yeast (*Ogataea pini*) in association with bark beetle can produce certain volatiles that functions against the growth of *B. bassiana* while the same can induce the growth of another fungus *Entomocorticium sp.* Yeast named *M reukaufii* produces a volatile blend that can attract the honey bees, so they are called to be nectar specialist (Rering et al., 2018).

Microbial volatiles for Agro-Ecological Health

Disturbances in the proper functioning of an ecosystem may result in unhealthy condition. The health of an agroecosystem depends on the plant growth and development, soil health, tolerance to biotic and abiotic stresses and co – the existence of all components of the ecosystem. Microbial Volatiles act as info chemical or mediator for Intra – and inter – organisms communication as well as can change the physiological responses in plants

Plant Growth induced by Microbial Volatiles

Exudates are released from plant roots into the soil for which bacteria get attracted towards them and live in association with plant roots in the rhizospheric zone. Bailly and Weisskopf (2012) that bacterial volatiles can make a 5-fold increase in plant growth have reported it. *Pseudomonas* is one among such microbes that can resist the pathogen attack on plant and maintain soil health. 2 – heptanol, 2 – pentanone, 2 – pentadecanone, 2 – undecanone, 4 – heptanone and 2 – tridecanone are the bacterial originated volatiles commonly seen as well as a bicyclic oligomethyl octadiene compound named sodorifen (produced by *Serratia odorifen*) can interfere with plants. Microbial volatiles such as 2 – pentyl furan, 2,3 – butanediol and acetoin are efficient in activating the growth of a targeted plant in leaf-level and root level along with alterations in metabolic processes, phytohormone signalling and nutrition levels (Fincheira and Quiroz, 2018). *Bacillus cereus* synthesised one of the most bioactive compound named DMDS (Dimethyl Disulphide) which can protect *Nicotiana tabacum* plant from *Botrytis cinerea* and *Zea mays* from *Cochliobolus heterostrophus.* The same compound DMDS can negatively regulate the gene expression for methionine biosynthesis and sulphur assimilation in *N. attenuate.* In *Arabidopsis thaliana,* acetoin (3 – hydroxyl – 2 – butanone and 2, 3 – butanediol released from B. subtilis and B. amyloliquefaciens, have enhanced induced systemic resistance and total leaf area. The 2, 3 - butane diol produced by *P. chlororaphis* through GacS kinase-dependent method when exposed to tobacco plants induced their growth promotion (Kanchiswamy et al., 2015) and the same GacS kinase can control the production of AHL (Acyl – homoserine Lactones) The filamentous fungus, Trichoderma has its application as both biofertilizer and bio fungicide that can enhance the growth of plants (Lee et al., 2016). Both the monocotyledonous and dicotyledonous have shown an effect on its carbohydrate metabolism by microbial volatiles. Fungal volatiles treated potato leaves showed ADP glucose, sucrose, 3 – phosphoglycerate and UDP glucose at higher levels along with starch accumulation (Ezquer et al., 2010). Volatile compounds from *Bacillus* can promote plant growth and it can provide value-added service of pathogenic fungal control also (Asari et al., 2016). Influence of volatiles from *Trichoderma* was observed in biomass of root and shoot, chlorophyll content and root growth of *A. thaliana* (Nieto – Jacobo et al., 2017). In *Populus* and *A. thaliana,* some ectomycorrhizal fungi synthesis sesquiterpenes that prominently induce lateral root growth. *Bacillus subtilis* strain SYST2 is a PGPR (Plant Growth Promoting Rhizobacteria) producing VOCs that can promote growth and regulate hormones in Tomato. Albuterol and 1,3 – propanediol are two volatile compounds produced by *B. subtilis* that stimulate plant growth significantly compared to all other compounds in terms of fresh and dry weights, length of root and shoot, stomatal conductance rate and photosynthesis enhanced with VOCs of SYST2 strains (Tahir et al., 2017). Even though volatile compounds can promote plant growth, it largely depends on the age of both microbe and plant that revealed in a study conducted by Lee et al. (2015) in which *T. Atroviride* of 5 days released some volatiles to which plants are exposed resulted in inhibited growth while other conditions showed enhanced growth. Case studies on some microbial volatiles such as β – caryophyllene (25 – 100 μM) inferred that they enhanced length of root and shoot, the weight of fresh mass and chlorophyll content in seedlings, as well as dimethyl hexadecyl amine at a dose level of 8 – 32 μM, can promote the root density, length of root and stem as well as fresh weight (Fincheira and Quiroz, 2018). The same studies also showed that in *Mentha piperita,* volatiles from *B. subtilis* can increase 3.5 times of root dry weight and twice of shoot fresh weight.

Fig. 3. Different MVCs and the conditions influencing their production with current knowledge and future proposals (Arrows in green colour)

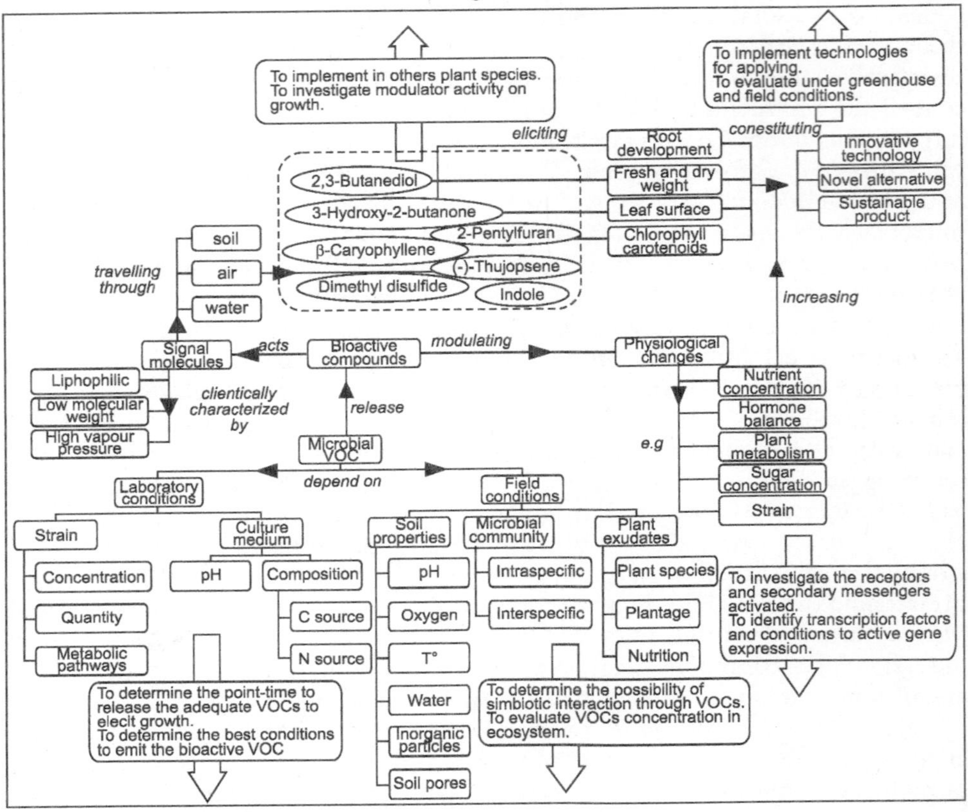

(Source: Fincheira and Quiroz, 2018)

Tolerance of plants against Abiotic Stresses through Microbial Volatiles

Abiotic stress tolerance is also a feature of microbial volatiles for the plants. A useful group of bacteria residing in the rhizosphere that enhances the yield of the crop, growth of the plant, the emergence of seedlings, and resistance against pathogenic diseases are PGPRs. They can emit both non – volatile compounds (eg: Hormones like auxin, cytokinins, siderophores, 1 – aminocyclo propane -1- carboxylase deaminase) and volatile organic compounds. Plant growth is suppressed and yield is reduced for those plant cells under the ionic stress or osmotic stresses because of excessive Na^+. Volatile organic compound treatment from the GB03 bacterial strains of *B. amyloliquefaciens* for *A. thaliana* will produce greater biomass and accumulate only fewer sodium ions. Under salinity conditions, *Pseudomonas simiae* strain AU releases VOCs that can confer Induced Systemic Tolerance (IST) by producing VSP (Vegetative Storage Proteins) and other proteins. Volatile compounds from

GB03 strains of *Bacillus amyloliquefaciens* are capable of acidifying the rhizospheric environment due to which a proton gradient is produced which allows the Na^+ export from roots of *A. thaliana* (Liu and Zhang, 2015). HKT1 (High-Affinity K^+ Transporter) has a major role among other transporters in regulating the Na^+ ion entry (Vaishnav et al., 2017). *Bacillus subtilis* (soil bacterium) GB03 strain released certain VOCs capable to regulate the expression of HKT1 by which salinity stress tolerance is exhibited by the *A. thaliana.* HKT1 is expressed according to the tissue, i.e., whether root or shoot. GB03 strains can down-regulate the expression of HKT1 in roots while up regulating in shoots which lower down the accumulation of sodium ions in plants. Vaishnav et al., 2015 studied in Soybean that AU strain of *Pseudomonas simiae* released some putative VOCs enhanced salt stress tolerance by reducing Na^+ ion accumulation in roots and shoots, stimulating proline content in the root, expressions of VSPs in an indirect way as well as affect the RuBisCO large chain. Gamma – Glutamyl Hydrolase (GGH) is also produced in addition to VSPs and RuBisCO proteins in soybean to maintain its growth in abiotic stress condition in response to VOCs from *Ps. simiae* AU strains. When soybean has made exposure towards AU strain, shoot and root growth promoted due to IAA concentration in the root, low Na^+ uptake rate, elevated chlorophyll content due to upregulation of RuBisCO protein. Drought is an adverse condition that can reduce crop productivity and food availability in the whole world and plant itself tries to adapt, mitigate and tolerate the constraint. PGPR can assist the drought tolerance mechanism by releasing EPS (Exo Polysaccharides), 1 – aminocyclo propane –1– carboxylate deaminase, plant hormones (Auxins, Cytokinins etc.), VOCs, up and downregulation of genes responding to stress, root morphology alteration and osmolyte accumulation. IST is the microbial induced tolerance enhancement of the plant by making physical changes and chemical changes against abiotic stresses (Vurukonda et al., 2016). *Pseudomonas chloraphis* O6 synthesises 2R, 3R – butanediol that can make the plant tolerant to drought stress by inducing closure of stomata and thus prevents water loss in *A. thaliana* (Cho et al., 2008). The presence of sodium nitoprusside in the soybean under salinity stress act as a donor of nitric oxide which can induce the bacterium *Pseudomonas simiae* to release quinolone and 4 – nitro guaiacol which was found helpful in seed germination of the soybean. As a relief from the harmful effects of salinity stress, volatile compounds influenced to induce nitrate reductase gene expression and antioxidative enzymes. Some volatiles named 7 – hexanol, 2 – undecanone, DMDS and 3 – methyl butanol released from *Paraburkholderia phytofirmans* PsJN can bring together bacterial colonies, enhanced plant growth with salinity stress tolerance. MVCs from *F. oxysporum* and *V. dahliae* enhanced chlorophyll content and growth in the plant under abiotic stress and decreased symptoms of diseases when exposed to pathogenic microbes (Li and Kang, 2018). Antioxidation activity is increased in plants as a part of their self-protection strategies from osmotic stress. Before sowing, if seeds are primed with *Paenibacillus polymyxa* B and *Bacillus thuringiensis* AZP2, it would help provide tolerance against drought stress as they exhibited increased survival, dry weight, the activity of antioxidant enzymes and WUE (Water Use Efficiency). This happens due to the volatile compounds emitted such as benzaldehyde, geranyl acetone and b – pinene. In *Arabidopsis thaliana,* expression of an enzyme PEAMT (Phospho ethanol amine N- methyltransferase) for the choline synthesis had increased when exposed to VOCs from *B. subtilis* GB03 and this leads to enhance the glycine – betaine and choline pool in the plants exposed to osmotic stress (Vaishnav et al., 2017). Another mechanism of the VOCs is to induce the formation of biofilm that has EPS as their main constituent. EPS can maintain moisture levels in the soil to make the plants drought tolerant. Salinity stress of high salt concentration can abate the solubility of Iron (Fe) due to which it is a limiting nutrient. This makes the Fe deficient in plants with metabolic changes. In plants, GB03 strains of *B. subtilis* can activate the mechanism for Fe acquisition with increasing metal ion assimilation in the presence of VOCs. It is necessary for the induction of FRO2 and IRT1 (Iron transporters) which is made possible through up-regulation of FIT1 gene expression (Iron deficiency-induced transcription factor 1). Microbial Volatiles can acidify the media due to which protons from roots are released that enhances the Fe mobility within plants. The same conditions happen for Sulphur (S) availability and its deficiency abnormally affects amino acid synthesis, productivity and photosynthesis. The plant *Nicotiana attenuate* grown in Sulphur deprived soils with impairment in uptake, metabolism and assimilation of S can do uptake of Sulphur from the DMDS volatile compound released by the soil bacterium *Bacillus* sp. B55.

Microbial Volatiles used as a Defence Mechanism against Pathogenic Microorganisms

Volatile compounds from the microorganism act as a defensive tool against the pathogenic organism that helps them directly through inhibition of disease-causing organism or indirectly by stimulating the plants to give defence response against those pathogens or Induced Systemic Resistance (ISR). Volatile compounds elicit their actions depending upon the released quantity. A single colony of *B. subtilis* strain GB03 is not sufficient to inhibit disease-causing fungi *F. oxysporum* and *Botrytis cinerea* and inhibitive action increases as the number of bacterial colonies increases. This shows that the inhibition of a pathogen is directly proportional to the concentration of the volatile compound. *Ps. syringae* is a bacterium that causes canker on stems of various fruit trees like Almond. Plants in the presence of bacterial volatiles tri – decane and hexadecane that can confer ISR., resist this pathogen. Each volatile compound requires different doses to stimulate their action in a specific manner. Bautanediol at a low level can inhibit *Microdochium nivale* while the same level of acetoin can induce resistance against *P. syringae* pv. 100mM concentration level of some volatiles like 3 – pentanol, acetoin, methyl jasmonate and 2 – hydroxyl – 3 – pentanone do not inhibit *B. cinerea* but the same dosage of same compounds can induce defence response against the pathogen. Ryu et al., 2004 in their study in *Arabidopsis thaliana* revealed that GB03 strain of *B. subtilis* and IN937a strain of *B. amyloliquefaciens* emitted a blend of volatile organic compounds that can significantly reduce the severity of diseases from the pathogen *Erwinia carotovora* subsp. carotovora. Fungal pathogens, which are soil-borne, and a threat to irrigated crops with broad host range that causes white mold in many cops is *Sclerotinia sclerotiorum*. Volatile compounds belonging to sulphides, aldehydes, alcohols and ketones have antifungal properties against mycelia growth, sclerotia and germination of ascospores of *S. sclerotiorum*. DMDS, methanethiol and 1 – undecene are bacterial volatile compounds that exhibit bioactivity against *Agaricus bisporus, Pleurotus ostreatus* and *P. eryngii* produced from *Pseudomonas tolaasii* which are identified through chromatography. Sulphur metabolism, uptake and assimilation are improved by VOCs of *B. amyloliquefaciens* that also have the stimulating effect on increasing glucosinolates, which is protective agent from beet armyworm (*Spodoptera exigua*). Hydrophobicity is a characteristic feature of volatile compounds. For example, monoterpene changes membrane permeability in cells of *E. coli* and *Staphylococcus aureus*. More than 250 volatile compounds have been identified till date which exhibited remarkable features like 8 – Carbon alcohol, aldehyde, short-chain alcohol and ketones. *Mucor, Aspergillus, Ulocladium, Alternaria* and *Penicillium* are some fungi engaged in volatile compounds. Each volatile has a definite pathway for their synthesis and derived from certain precursors. For example, amino acids are the origin for isoprene N, lactones and S containing VOCs. Styrene, 3 – methyl -1- butanol and 2 – methyl – 1 – propanol derived from fatty acid metabolism. Metabolic pathway of Acetyl CoA derives monoterpenes, sesquiterpenes and isoprenes. *Talaromyces* sp. is a growth-promoting fungus isolated from the agricultural field and inferred from its evaluation that it can emit β – caryophyllene, a volatile compound, promoting the growth and confers ISR against *Colletotrichum higginsianum*. ISR against the disease-causing *Pseudomonas syringae* is enhanced due to volatiles from *Cladosporium* sp. The appearance of symptoms by diseases can be reduced in plants with the exposure of volatile compound 6PP (6 – pentyl – 2H pyran -2- one) released by fungus. Chemical fumigation is done in soil using methyl bromide against the soil-borne pathogen. However, it can be substituted by some microbial volatiles released from *Muscodor* like naphthalene, azulene, tetrahydrofuran, α-phellandrene, 2 – pentyl furan and β – caryophyllene. *Muscodor crispans* releases volatile compounds that inhibit the growth of *Mycosphaerella fijiensis* (Black Sigatoka disease in Banana) and *Xanthomonas axonopodis* pv *citri* (bacteria causing citrus canker). *Guignardia citricarpa* is a pathogen responsible for citrus black spot whose growth is inhibited by the volatile compounds released by *Saccharomyces cerevisiae* (yeast) like methyl butanol, ethyl acetate and phenyl ethanol (Monika et al., 2017). Phytopathogenic microbes like *Verticillium, Cercospora* and *Sclerotinia* are controlled by VOCs from *Phoma* sp. Some previous works mentioned by Nadarajah (2017) shows that exogenous application of some microbial volatiles is possible with examples like disease impact of soilborne pathogens like *M .nivale, S. homeocarpa* and *R. solani* have reduced when applied with 2, 3 butanediol.

Above Ground and Below Ground Action of Microbial Volatiles

Functions of volatile compounds spread in both atmospheric (aerial) and soil (below ground) ecosystem. Complex volatile blends are released from the endophytic fungal group, which has a beneficial part on plant growth and defence responses. An endophytic fungus is nothing but that group of fungi having no harmful impact on the host during the association. VPEs (Volatile Producing

Endophytes) are industrially important due to the release of odorous compounds having physical, biological and chemical properties and occur in both below and above ground ecosystems. Members of the Xylariaceae are prominent in emitting MVOCs rather than Ascomycota, even though they are capable of producing such volatiles. As an example of VPEs acting above ground, *Muscodor* is considered. Disease-causing microbes including fungi and bacteria can be controlled by the volatiles of *Muscodor albus* as well as other two species of *Muscodor* also exhibit same property, *M. fengyangensis, M. yucatanensis* (Selim et al., 2017). *Xanthomonas axonopodis* pv *citri* which is pathogenic to citrus and *Mycosphaerella fijiensis* that infects banana are suppressed by volatiles from *M. crispans*. Another example is the *Phoma sp.*that releases volatiles that make some plant capable to survive in desert areas. Under the optimum growth conditions, if a fungal hyphae deprive to grow in soil because of competing microorganisms or negatively affecting microbial volatiles can be termed as fungistasis. Fungal dominance in the soil can be identified by the presence of DMDS, trimethyl – amine, 3 – methyl -2- pentanone, methyl pyrazine, N – dimethyl octyl amine and nonadecane because soil having no fungistasis does not produce these compounds. Soil pores are filled with VOCs and they spread depending on the surroundings. In soil, pathogenic fungus *F. oxysporum* is inhibited with its associated bacteria, which are non – pathogenic from the released volatiles. *Phytophthora capsici, the* causal organism of root rot of bell pepper and *R. solani,* causal organism for broccoli seedling death are naturally managed from the volatile compounds of *Muscodor albus*. Some volatiles released by a particular microbe can induce the biocontrol feature of another microbe. The gene expression is upregulated in *P. fluorescens* to exhibit its biocontrol property at an enhanced level if the microbe is exposed to the volatiles from *T. Atroviride.* Volatile compounds from the soil-borne fungus can be used to treat seed or seedlings to decrease their susceptibility against a pathogen. Defence gene is produced against *B. cinerea* if *A. thaliana* has been treated with mushroom alcohol. The same property exhibited when treatment is done with a 6-carbon aldehyde and allo – ocimene.

Table No. 1: Summary of Microbial Volatiles, Organism releasing the volatile and their influence on Plant

BACTERIA			
Bacterial Species	**Volatile Compounds**	**Plant associated**	**Effect on Plant**
Bacillus subtilis GB03	2, 3 – Butanediol (Acetoin)	*Arabidopsis thaliana*	Promotes the growth Induced Systemic Resistance
Bacillus megaterium XTBG34	2 - Penthylfuran	*A. thaliana*	Growth promotion

Bacillus amyloliquefaciens T-5	Compounds including benzenes, ketones, organic acids, alkanes and aldehydes	*Lycopersicon esculentum*	Enhancement of growth
B. amyloliquefaciens SQR-9	Volatile compound blends	*Lycopersicon esculentum*	Control plant pathogen
B. amyloliquefaciens IN937a	2, 3 - Butanediol	*A. thaliana*	ISR and Growth promotion
Serratia plymuthica ICI270	DMDS (Dimethyl Disulphide)	*Lycopersicon esculentum*	Plant Growth is promoted through inhibition of *Agrobacterium tumefaciens* and *A. vitis*
Pseudomonas fluorescens B-4117	DMDS	Do -	
P. chlororaphis O6	2, 3 - Butanediol	*A. thaliana, Nicotiana tabacum*	Desiccation tolerance, ISP, Growth is enhanced
P. fluorescens SS101	13 – Tetradecadien-1-ol, 2 – butanone, 2 – methyl –n-1-tridecene	*Nicotiana tabacum*	Plant growth Promotion
P. palleroniana R43631	Volatile mixture	*Solanum tuberosum* subsp. *andigenum*	Increased yield
Arthrobacter agilis UMCV2	N, N –Dimethyl - hexadecanamine	*Medicago sativa*	Promotion of growth
Rhizobacteria strains	VOCs mixture	*Arabidopsis thaliana*	Architecture of root system and biomass production
FUNGI			
Fungal Species	**Volatile Compound**	**Effective on**	**Exhibited Effect**
Muscodor yucatanensis	Volatile blend	Fungus and Plants	Allelochemical effect against other endophytic fungi, plants and plant disease-causing pathogen

M. albus	Isoamyl acetate 2 – methyl butanol and Isobutyric acid	Fungus, Bacteria	Synergistic action against a wide range of plant pathogenic bacteria and fungi. Control post-harvest plant diseases.
M. crispans	Volatile mixtures	Fungus, Bacteria	Effective against *M. fijiensis* causing black sigatoka disease in banana and X. *axonopodis* pv. *citri* as well as *Pythium ultimum and S. sclerotiorum*
Tuber melanosporum, T. indicum, T. borchii	2 - Octenal	Plants	Growth inhibition in plants
Trichoderma virens	β elemene, β – caryophyllene, germacrene, Gamma - cadinene	Plants, Fungus	Growth promotion, defence response against *B. cinerea* in *A. thaliana.*
Mold Fungi	1 – Octen -3-ol	Plants	Induced defence Protection from *Botrytis cinerea*
Fusarium oxysporum MSA 35	β- caryophyllene	Plants	Induced shoot and root length. Increase fresh mass in lettuce.
Phomopsis sp.	Isoamyl alcohol 2-propanone 2 – methyl propanol Sabinene	Fungi	Antifungal properties and intensely bioactive against *Pythium, Fusarium, Botrytis, Colletotrichum, Sclerotia, Verticillium, Rhizoctonia and Phytophthor*

Trametes gibbosa	1 – Octen -3-ol	Fungi	Attract the beetles that eat fungi
T. versicolor	⊚- Patchoulene, isoledene, β-guaiene	Fungi	Do -
Phoma sp.	VOCs mixture that includes naphthalene derivatives along with some alcohols and sesquiterpenes.	Fungi	Antifungal and Fuel property

(Source: Kanchiswamy et al., 2015; Altaf et al., 2017)

Table No. 2: Summary of the Microbe – Insect Communication through Microbial Volatiles

BACTERIA		
Species	**Volatile Compound Produced**	**Insect Responding**
Citrobacter freundii	2-pentanone, 3-methyl butanamine, Pyrazine	*Anastrepha ludens*
Klebsiella pneuomoniae	3-methyl-1-butanol, 2-methyl-1-butanol, DMDS, 2- hexanone, 2- ethyl hexanol, 2-butanone, phenol	Do -
Pantoea agglomerans	3- hydroxyl butanone, 3-methyl-1-butanol, Ammonia, DMS, Indole, 2- phenyl ethanol	*A. Ludens, A. suspense, Rhagoletis pomonella, S. gregaria*
Klebsiella oxytoca: Bacillus spp	Dialkyl sulphides	*Diadromus pulchellus*
Staphylococcus aureus	Trimethyl amine, Ammonia, Isoamyl amine, 2, 5 –dimethyl pyrazine, 2 – methyl-butyl amine, Acetic acid	*A. ludens*
S. epidermidis	1 – butanol, 2- methyl butanoic acid, 3- hydroxyl -2- butanone, 3- methyl -1- butanol, 3 – methyl butanoic acid	*Anopheles gambiae*
S. sciuri	Propanone Benzene ethanol, 3- methyl butanoic acid, 3-methyl -1- butanol, 3 – methyl -2- butenal, 3- hydroxyl -2- butanone	*Episyrphus balteatus*
Proteus mirabilis	2 – methylpropan -1- amine	*Lucilia sericata*
FUNGI		
Species	**Volatile Compound**	**Insect giving response**
Amylostereum sp.	Acetaldehyde	*Ibalia leucospoides* *Rhyssa perscusoria*

Aspergillus versicolor *A. sydowii*	1-octen-3-ol	*Lariophagus distinguendes*
Beauveria bassiana	Not determined	*Anopheles stephensi*
Cladosporium cladosporio-desC. herbarum *Penicillium spinulosum*	Decanal, dipentene, Nonanone, decane, phenyl ethanone	*Orchesella cincta* *Tomocercus flavescens*
Phoma sp. *Rhizopus sp.* *Fusarium sp.*	DMTS, 2 – phenyl ethanol	*Musca domestica*
Pencillium expansum	3 - methylanisole	*Hylobius abietis*
Trichothecium roseum	Cis – and trans- Octa – 1,5-dien-3-ol	*Tyrophagus putrescentiae*
Sclerotium rolfsii	Not determined	*Spodoptera exigua*
Saccharomyces cerevisiae	Ethanol, Acetaldehyde, 1 – butanol, 2-methyl-1-propanol Ethyl decanoate	Carpophilus lugubris C. hemipterus Drosophila melanogaster

(Davis et al., 2013

Conclusion

This chapter summarizes that microbial volatile can be an ecofriendly substitute for chemical inputs as it does not harm the ecosystem in addition to value-added services. As they have the property to transform into gaseous state directly, diffusion of these substances into the soil pores and atmosphere is not difficult. They have diverse application in agro-ecologies like biocontrol agent for pathogenic and disease-causing organisms, mycofumigation, semiochemical, biochemical, quorum sensing, biofuels, chemotaxonomy, mediator for Intra and inter organism communication, food flavouring agent and plant growth promoter. Volatiles released from the microbes have complex biosynthetic pathways and included in different chemical classes like organic acids, short-chain and long-chain hydrocarbons, alkenes, aldehydes, esters, pyrazines, terpenes, ketones and benzene derivatives. They are the byproducts of metabolic processes like degradation of fatty acids, fermentation of carbohydrates, decarboxylation of fatty acids, sulphur reduction, glycolysis etc. Further studies have to be promoted with improved technology for better identification of the volatiles in deeper soil layers and extreme conditions with more specificity in their action concerning soil and climatic conditions.

References

Altaf, M.M., Khan, A.S.A. and Ahmad, I. (2017). Bacterial Volatiles: Potential Applications in Plant Growth and Health. In: Choudhary, D.K., Sharma, A.K., Agarwal, P., Varma, A. and Tuteja, N. (Eds), *Volatiles and Food Security: Role of Volatiles in Agro – Ecosystem,* 199 – 217. DOI 10.1007/978-981-10-5553-9_10

Asari, S., Matzen, S., Petersen, M.A., Bejai, S. and Meijer, J. (2016). Multiple effects of *Bacillus amyloliquefaciens* volatile compounds: plant growth promotion and growth inhibition of phytopathogens. *FEMS Microbiology Ecology, 92.* doi: 10.1093/femsec/fiw070

Bailly, A. and Weisskopf, L. (2012). The modulating effect of bacterial volatiles on plant growth; Current knowledge and future challenges. *Plant Signaling & Behaviour, 7*(1), 79–85. DOI: 10.4161/psb.7.1.18418

Barea, J., Pozo, M.J., Azcon, R. and Azcon-Aguilar, C. (2005). Microbial co-operation in the rhizosphere. *Journal of Experimental Botany. 56*(417), 1761 – 1778. Doi: 10.1093/jxb/eri197

Baysal, O. and Silme, R.S. (2017). Bacterial Volatiles n Rhizosphere. In: Choudhary, D. K., Sharma, A.K., Agarwal, P., Varma, A. and Tuteja, N. (Eds), *Volatiles and Food Security: Role of Volatiles in Agro – Ecosystem,* 219 - 226. DOI 10.1007/978-981-10-5553-9_11

Beck, J.J. and Vannette, R.L. (2016). Harnessing Insect- Microbe Chemical Communications to control Insect Pests of Agricultural Systems. *Journal of Agricultural and Food Chemistry, 65,* 23 – 28.

Bitas, V., Kim, H. S., Bennett, J. W. and Kang, S. (2013). Sniffing on Microbes: Diverse Roles of Microbial Volatile Organic Compounds in Plant Health. *MPMI, 26* (8), 835 – 843.

Calcagnile, M., Tredici, S. M., Tala, A. and Alifano, P. (2019). Bacterial Semiochemicals and Transkingdom Interactions with Insects and plants. *Insects,* 10.3390/insects10120441

Cho, S. M., Kang, B. R., Han, S. H., Anderson, A. J., Park, J. Y., Lee, Y. H., ... & Kim, Y. C. (2008). 2R, 3R-butanediol, a bacterial volatile produced by Pseudomonas chlororaphis O6, is involved in induction of systemic tolerance to drought in Arabidopsis thaliana. Molecular plant-microbe interactions, 21(8), 1067-1075.

Davis, T. S., Crippen, T. L., Hofstetter, R. W. and Tomberlin, J. K. (2013). Microbial Volatile Emissions as Insect Semiochemicals. *Journal of Chemical Ecology.* DOI 10.1007/s10886-013-0306-z

Enespa and Prem Chandra. (2017). Microbial Volatiles as Chemical Weapons. In: Choudhary, D.K., Sharma, A.K., Agarwal, P., Varma, A. and Tuteja, N. (Eds), *Volatiles and Food Security: Role of Volatiles in Agro – Ecosystem,* 227 – 254. DOI 10.1007/978-981-10-5553-9_12

Ezquer, I., Li, J., Ovecka, M., Baroja-Fernández, E., Muñoz, F. J., Montero, M., ... & Pozueta-Romero, J. (2010). Microbial volatile emissions promote accumulation of exceptionally high levels of starch in leaves in mono-and dicotyledonous plants. Plant and Cell Physiology, 51(10), 1674-1693.

Fincheira, P. and Quiroz, A. (2018). Microbial volatiles as plant growth inducers. *Microbiological Research, 208,* 63 – 75.

Hung, R., Lee, S. and Bennett, J.W. (2015). Fungal Volatile Organic Compounds and their role in ecosystem. *Appl Microbiol Biotechnol., 99,* 3395 – 3405

Kai, M., Haustein, M., Molina, F., Petri, A., Scholz, B. and Piechulla, B. (2008). Bacterial volatiles and their action potential. *Appl Microbiol Biotechnol.* DOI 10.1007/s00253-008-1760-3

Kanchiswamy, C.N., Malnoy, M., Maffei, M.E. (2015). Bioprospecting bacterial and fungal volatiles for sustainable agriculture. *Trends Plant Sci.* doi: 10.1016/j.tplants.2015.01.004

Kanchiswamy, C.N., Malnoy, M. and Maffei, M.E. (2015). Chemical diversity of microbial volatiles and their potential for plant growth and productivity. *Front. Plant Sci., 6,* 151. https://doi.org/10.3389/fpls.2015.00151

Korpi, A., Jarnberg, J. and Pasanen, A.L. (2009). Microbial Volatile Organic Compounds. *Critical Reviews in Toxicology, 39*(2), 139 – 193.

Lee, S., Hung, R., Yap, M. and Bennett, J.W. (2015). Age matters: the effects of volatile organic compounds emitted by *Trichoderma atroviride* on plant growth. *Arch Microbiol.* DOI 10.1007/s002023-015-1104-5

Lee, S., Yap, M., Behringer, G., Hung, R. and Bennett, J.W. (2016). Volatile organic compounds emitted by *Trichoderma* species mediate plant growth. *Fungal Biology and Biotechnology, 3*(7). DOI: 10.1186/s40694-016-0025-7

Lemfack, M.C., Nickel, J., Dunkel, M., Preissner, R. and Piechulla, B. (2014). mVOC: a database of microbial volatiles. *Nucleic Acids Research.* 42, D744-D748, database issue: doi: 10.1093/nar/gkt1250

Lemfack, M.C., Gohlke, B, Toguem, S.M.T., Preissner, S., Piechulla, B. and Preissner, R. (2018). mVOC 2.0: a database of microbial volatiles. *Nucleic Acids Research, 46. Database issue,* D1261 – D1265. doi: 10.1093/narlgkx1016.

Li, N. and Kang, S. (2018). Do volatile compounds by *Fusarium oxysporum* and *Verticillium dahlia* affect stress tolerance in plants? *Mycology, 9*(3), 166–175. DOI: 10.1080/21501203.2018.1448009

Liu, X. and Zhang, H. (2015). The effects of bacterial volatile emissions on plant abiotic stress tolerance. *Frontiers in Plant Science, 6, 774.* doi: 10.3389/fpls.2015.00774.

Monika, Sarim, K.M., Arya, S.S., Devi, S., Kaur, V. and Singla, A. (2017). Microbial Volatiles in Defense. In: Choudhary, D.K., Sharma, A.K., Agarwal, P., Varma, A. and Tuteja, N. (Eds), *Volatiles and Food Security: Role of Volatiles in Agro – Ecosystem,* 45 – 60. DOI 10.1007/978-981-10-5553-9_4

Morath, S.U., Hung, R. and Bennett, J.W. (2012). Fungal volatile organic compounds: A review with emphasis on their biotechnological potential. *Fungal Biology Reviews, 26,* 73 – 83.

Nadarajah, K.K. (2017). Microbial Volatiles and Plant Defence. . In: Choudhary, D.K., Sharma, A.K., Agarwal, P., Varma, A. and Tuteja, N. (Eds), *Volatiles and Food Security: Role of Volatiles in Agro – Ecosystem,* 355 – 373. DOI 10.1007/978-981-10-5553-9_19

Nieto-Jacobo, M. F., Steyaert, J. M., Salazar-Badillo, F. B., Nguyen, D. V., Rostás, M., Braithwaite, M., ... & Mendoza-Mendoza, A. (2017). Environmental growth conditions of Trichoderma spp. affects indole acetic acid derivatives, volatile organic compounds, and plant growth promotion. Frontiers in plant science, 8, 102.

Ortiz-Castro, R., Contreras-Cornejo, H. A., Macias-Rodriguez, L. and Lopez-Bucio, J. (2009). The role of microbial signals in plant growth and development. *Plant Signalling & Behaviour, 4(8),* 701 – 712. DOI: 10.4161/psb.4.8.9047

Panpatte, D.G., Shukla, Y.M., Shelat, H.N., Vyas, R.V. and Jhala, Y.K. (2017). Bacterial Volatile Organic Compounds: A new Insight for Sustainable Agriculture. *Microorganisms for Green Revolution.* https://doi.org/10.1007/978-981-10-6241-4_8

Penuelas (2014). Biogenic volatile emissions from the soil. *Plant, Cell and Environment, 37,* 1866 – 1891.

Piechulla, B. and Degenhardt, J. (2014). The emerging importance of microbial volatile organic compounds. *Plant, Cell and Environment, 37,* 811 – 812.

Piechulla, B., Lemfack, M.C. and Kai, M. (2017). Effects of discrete bioactive microbial volatiles on plants and fungi. *Plant, Cell and Environment, 40,* 2042 – 2067. doi: 10.111/pce.13011

Rering, C.C., Beck, J.J., Hall, G.W., McCartney, M.M. and Vannette, R.I. (2018). Nectar – inhabiting microorganisms influence nectar volatile composition and attractiveness to a generalist pollinator. *New Phyologist, 220,* 750 – 759. doi: 10.1111/nph.14809

Ryu, C.M., Farag, M.A., Hu, C., Reddy, M.S., Kloepper, J.W. and Pare, P.W. (2004). Bacterial Volatiles Induce Systemic Resistance in *Arabidopsis. Plant Physiology, 134,* 1017 – 1026.

Sarsan, S. (2017). Bacterial Volatiles for Plant Growth. . In: Choudhary, D.K., Sharma, A.K., Agarwal, P., Varma, A. and Tuteja, N. (Eds), *Volatiles and Food Security: Role of Volatiles in Agro – Ecosystem,* 335 – 354. DOI 10.1007/978-981-10-5553-9_18

Schenkel, D., Lemfack, M. C., Piechulla, B. and Splivallo, R. (2015). A meta – analysis approach for assessing the diversity and specificity of belowground root and microbial volatiles. *Frontiers in Plant Science, 6,* 707. doi: 10.3389/fpls.2015.00707

Schmidt, R., Cordovez, V., de Boer, W., Raaijmakers, J. and Garbeva, P. (2015). Volatile affairs in microbial interactions. *The ISME Journal, 9,* 2329 – 2335

Schmidt, R., de Jager, V., Zühlke, D., Wolff, C., Bernhardt, J., Cankar, K., ... & Garbeva, P. (2017). Fungal volatile compounds induce production of the secondary metabolite Sodorifen in Serratia plymuthica PRI-2C. Scientific Reports, 7(1), 1-14.

Schulz- Bohm, K., Zweers, H., de Boer, W. and Garbeva, P. (2015). A fragrant neighbourhood: Volatile mediated bacterial interactions in soil. *Frontiers in Microbiology, 6,* 1212. doi:10.3389/fmicb.2015.01212

Schulz-Bohm, K., Martin-Sanchez, L. and Garbeva, P. (2017). Microbial Volatiles: Small Molecules with an important role in Intra- and Inter Kingdom Interactions. *Frontiers in Microbiology, 8,* 2484. doi: 10.3389/fmicb.2017.02484

Selim, K.A., El Ghwas, D.E., Selim, R.M. and Hassan, M.I.A. (2017). Microbial Volatles in Defense. . In: Choudhary, D. K., Sharma, A. K., Agarwal, P., Varma, A. and Tuteja, N. (Eds), *Volatiles and Food Security: Role of Volatiles in Agro – Ecosystem,* 135 – 170.

Siddiquee, S. (2017). Fungal Volatile Organic Compounds: Emphasis on Their Plant Growth- Promoting. In: Choudhary, D.K., Sharma, A.K., Agarwal, P., Varma, A. and Tuteja, N. (Eds), *Volatiles and Food Security: Role of Volatiles in Agro – Ecosystem,* 313 – 334. DOI 10.1007/978-981-10-5553-9_17

Tahir, H.A.S., Gu, Q., Wu, H., Raza, W., Hanif, A., Wu, L.,Colman, M.V. and Gao, X. (2017). Plant Growth Promotion by Volatile Organic Compounds Produced by *Bacillus subtilis* SYST2. *Frontiers in Microbiology, 8, 171.* Doi: 10.3389/fmicb.2017.00171.

Vaishnav, A., Kumari, S., Jain, S., Varma, A. and Choudhary, D.K. (2015). Putative bacterial volatile – mediated growth in soybean (*Glycine max* L. Merrill) and expression of induced proteins under salt stress. *Journal of Applied Microbiology, 119,* 539 – 551.

Vaishnav, A., Varma, A., Tuteja, N., Choudhary, D.K. (2017). Characterization of Bacterial Volatiles and their impact on Plant Health under Abiotic Stress. In: Choudhary, D.K., Sharma, A.K., Agarwal, P., Varma, A. and Tuteja, N. (Eds), *Volatiles and Food Security: Role of Volatiles in Agro – Ecosystem,* 16 – 24. DOI 10.1007/978-981-10-5553-9_2

Vurukonda, S.S.K.P., Vardharajula, S., Shrivastava, M. and SkZ, A. (2016). Enhancement of drought stress tolerance in crops by plant growth promoting rhizobacteria. *Microbiological Research, 184,* 13 – 24.

Wheatley, R.E. (2002). The consequences of volatile organic compound mediated bacterial and fungal interactions. *Antonie van Leeuwenhoek, 81*, 357 – 364.

Avenue of Phytoremediation Strategies for Fluorine Toxicity

Lalit Saini

Department of Agronomy, School of Agriculture

Lovely Professional University, Jalandhar, Punjab, 144411, India

Email: sainilalit994@gmail.com

Abstract

The element fluorine (F) is highly abundantly present on earth. It behaves as natural xenobiotics in even a very low concentration. The content of fluorine in soil, water, and the air is continuously increasing by various uncontrolled anthropogenic activities. Irrigation with groundwater, which is heavily contaminated with fluoride and various regional agricultural-practices in fluorine contaminated soil influence negatively the biochemical and physiological parameters in vegetables and crops, which further results in an inhibition of productivity and growth. In a few plants, toxic ions are translocated from the root part to shoot and gather in fruits and edible parts of plants. There are harmful consequences of fluorine bioaccumulation; also, the amount is greater than its tolerable concentration in eatable parts of plants. Studies show that plants have inherent ability to tolerate fluorine and some cases show hyper-accumulation of fluorine. This potent nature of plants toward fluorine is highly useful in phytoremediation. Scientists often go after few bio-indicators for measuring the amount of contaminated fluorine in the specimens of interest. Trials in the field via robust, research in the future should be pointed on the strategies of phytoremediation with proper executions. In this detailed review, we tried to highlight the loci trait of quantitative determinants in genetic studies with large throughput, which has been utilizing to forming suitable characters in a crop variety that is tolerance to fluorine.

Introduction

Fluorine is the 13th essential element present in earth crust in forms of minerals having cryolite, fluorspar, and fluorapatite. The exceptionally rise in electronegativity is responsible for the excessive reactivity in fluorine (Hong et al., 2016). The 0.2 to 0.9 mg dm^{-3}amount of NaF concentration is used to control the osteoporosis and caries of dental, which are having due to contaminated drinking and groundwater. Uptake of fluorine above its secure limit caused invariant damage i.e. deformities of skeletal, fluorosis of non-skeletal fluorosis and in animals, it forms neurologic complications in humans also including animals (Choubisa 2013). The contamination of fluorine has adverse consequences on the environment. Plants may absorb F from the polluted soil through roots and from the atmosphere (Bhat et al., 2015). Anthropogenic activities are the main regions for releasing fluorine in the atmosphere, which is hydrolyzed, by corrosive hydrogen fluorine. The vaporous and aerosols are reacted with it to forming fluorines having stable non-volatile (Hong et al., 2016).

In Poland atmosphere, high amount of fluorine 2.1 mg kg^{-1}have recorded in the precipitations of the atmosphere (Walna et al. 2013). The minute concentration of hydrogen fluorine (1μg kg^{-1}) also causes damage in plants (Gupta et al., 2009). During the study of 2 years in Poland was measured a huge deposition of fluorine 52mg m^2 (soil) of the year (Walna et al., 2013). Low concentration of fluorine exposure was sufficient to originate different physiological problems like damage of leaves, burning of lesions due to necrosis, development inhibition and capacity of productivity insentient plant. In plants, the stress-related symptoms are observed slowly and more noticeably on sustained exposure of fluorine (Hong et al., 2016). Bioaccumulation of fluorine in eatable parts of the plant causing problems in many countries having Africa, Asia, Southern America, and Australia was facing intense regional fluorosis (Maheshwari and Meenakshi 2006, Susheela 1999). Different researches have examined multiple stresses of abiotic like drought, salinity, heat, light, toxicity due to heavy metals and cold (Banerjee and Roychoudhary 2017), In the plants (slow poisonous) very few amounts of literature are accessible on its effects. We have discussed adverse consequences of F as an agent of biohazardous also discussed prospects related to phytoremediation of fluorine.

Whenever the plants are exposed to the F contaminated soil and atmosphere in which the atmospheric F is often absorbed by the stomata whereas in the soil it is taken up by the roots and then accumulated F is translocated to stems as well as leaves by the apoplastic route. Within the cell, the accumulation of fluoride exact in localization is still not clear.

Translocation of Fluorine from Soil to Plants

Physico-chemical characters of fluoride in soil: In various states of India content of fluorine recorded in water having. In various states of India, the content of fluorine in water was reported as 48 mg dm^{-3}. A large content of fluorine has recorded in alkaline lakes and ponds of north Tanzania (59-689 mg $dm^{-3)}$, (14-62 mg dm^{-3}) in spring and rivers having (11-25mg dm^{-3}) (Nanyaro et al., 1984). In Pakistan, region researchers tested 146 samples of groundwater. The concentration of fluorine is going beyond its secure range (1.4 mg dm^{-3}) and the increase in their amount concentration is toxic and causes damage in plants, humans as well as animals (Farooqi et al., 2007). It also recorded that the content of fluorides (100-3500 mg kg^{-1}) reaches to soil (clay soil) than it is heavily contaminated (Hong et al., 2016). Normally the range of fluorine content in soil is lies between (149-399 mg kg^{-1}) (Bhat et al., 2015). However, in soil fluorine exceeded its critical limit are absorbed by parts of plant and translocation it through root to shoots were increased in specific plant species. Fluorine is mainly a universal toxin that highly adsorbed by the soil particles and in solution soil, very fewer percentage oxy-hydroxides are dissolved (Marier, 1971). Hence, the mobility of fluorine is varying in between permeability of soil and capacity of sorption. The capacity of sorption is dependent upon sorbent types, pH of soil and salinity of soil (Tylenda, 2011). Content of fluorine in silt, soil having non-calcareous and in clay loam compare to sandy particle soil because Al^{3+}, Ca^{2+}, Fe^{2+}, and $Mg^{2+}$are combined with the stable formation of bonds (Hong et al., 2016). CaF_2 is the form of fluorine that is fixed effectively by increasing the content of Ca^{2+} in soil (Abugri 2010). The result shows that in free state low percentage of negative fluorine ions are present in soil which containing more aluminum (Vasudevan et al., 2003). For producing a large concentration of free negative fluorine ions, the soil must be alkaline. The bioavailability of plants is enhanced by releasing the fluorine from the surface of the soil. The retention of fluorine is reduced when the pH of the soil is having alkaline. Negative ion of fluorine bond is displaced by OH^- and with the potential of electrostatic emergence (Saxena and Rani, 2012). Consequently, the soil having Physico-chemical and soil pH properties highly dependent on the bioavailability of fluorine to plants. Fluorine solubility in water has increased in the acidic state with Ale_x formation, although free fluorine desorption is promoted by the alkaline situation (Wehr et al., 2014) (Fig. 1).

The entry of Fluoride in Plants

Plants from the soil, water, and air absorb fluoride. Uptake of fluoride amount depends on the species of plant, characteristics of soil and fluoride amount present in irrigation water, soil and air (Anshumali, 2014). The potentiometric technique is used to analyze the content of fluoride in *Triticum aestivum* tissues by using a discriminating electrode of fluoride. Various fluoride content was a notice in leaves, roots, and stems (Agarwal and Chauhan 2014). In cultivars, different types of fluoride accumulation have recorded in biomass of shoot. Due to passive fluoride uptake through roots of spite. The different translocation of root to shoot having a pivotal role in determining the accumulation of fluoride through shoots (Mondal 2017). The sensitivity of fluoride is mainly determined by the content of Ca^{2+} in the cell wall of the root. In the cell wall of roots, Ca^{2+} ions are present that scavenging the F^- ions. F^- Entry is controlled in the species having more Ca^{2+}ions. It is considered that channels of chloride are responsible for the absorption of F^- and uptake of fluoride enhances the deficiency of chloride (Baunthival and Ranghar, 2013). Because fluoride having more electronegativity, it does not give permission to transport through symplastic that is slightly obvious. Permeability in ions of fluoride excludes in the membrane of plasma can be fixed by the negative charge. Therefore, spaces in between the intercellular and in cell walls are common via the transport of apoplast (Agarwal and Chauhan, 2014) (Fig. 1).

Fig. 1. Fluorine translocation from soil to plant

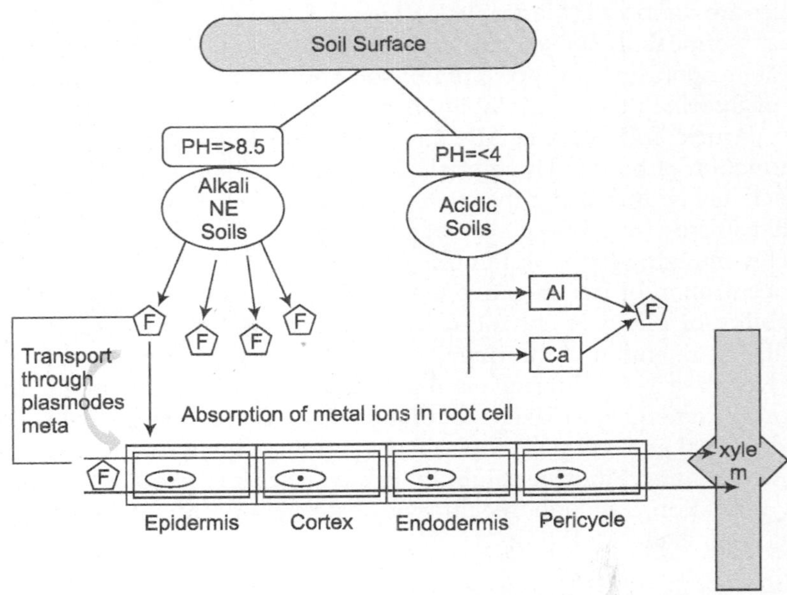

The main role of endodermis is to prevent the transport of effective fluoride through tissues of vascular and behaves as a shielding barrier. It helps to decreases the translocation from root-shoot. In the leaves, flow of fluorine carry out by xylem sap (Panda, 2015). In species to species, the accumulation of fluorine in shoots may vary and the functions having a concentration of fluorine in the medium of roots and flow rate of water (Pitman, 1985). The latest research shows that cadmium and fluorine having a pessimistic effect on the growth of vegetable plants. The accumulation of cadmium and fluorine are noticed in the tip of leaves and the inside part of the leaves.

Phytotoxicity of Fluoride

Anatomical and Morphological Symptoms

The 20mM sodium is fluoride-treated with Arabidopsis thaliana suspension of cells in 2 hours for decreasing loss in viability of cells. Therefore, the unveiling to only 1 mM sodium fluoride at 10d has specifically hindered cellular growth (Pulyaevskaya et al., 2011). In the foliage of the plant, fluoride may accumulate and normal damage is minimal, with necrosis of tip slowly extending inward. With the help of vascular tissue, fluoride is taken and accumulated in the margins of the leaves (fig. 2). Based on the Cell Sap material. The epidermis of the lower part and spongy mesophyll firstly destroyed by fluoride secondly cells of a palisade in chloroplast and then damage the upper part of the epidermis (Panda, 2015). *Pinaceae* members are shows destructive patches that spread from base to tip of the needle (Hong et al., 2016).

Fluoride Stress and Preparatory Development in Plants

The survival of plants is crucial in stages of growing the seedlings and reasonable germination of seeds. When fluoride enters the metabolism of the plant then the cell partition rate is decreased. The stress of fluoride is induced in the beginning development of seedling and the germination rate was decreases (Ram et al., 2014). Aleurone of gibberellic acid degenerated when the stress of fluoride is induced and slightly decreases the germination of seeds. Due to this unusual metabolism in saccharides endosperm (Gadi et al., 2016). Seedling having inappropriate development and uptake of the nutrient is unbalanced when fluorine is intervention and it decreases the length of root and shoots, fresh and dry masses and also lower down the vigours index of *Abelmoschus esculentus, Triticum æstivum, Oryza sativa, Cicer arietinum* and *Citrullus lanatus* (Gupa et al., 2016, Khan and Iram, 2016).

Fig. 2. Translocation of Fluorine in plants

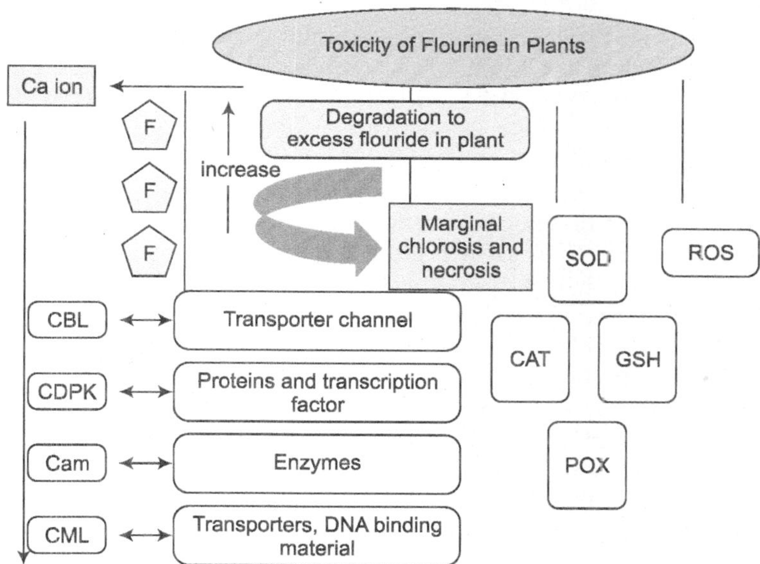

Fluoride Impact on Photosynthesis

As a messenger in the system of plant, Ca^{2+} plays an important role in transferring the signals (Banerjee and Roychoudhury, 2017). Chemically interaction of Ca^{2+} and fluoride help in altering the permeability of membrane by CaF_2 (Stevens et al., 1998). In the response of different abiotic stresses, protein kinase behaves as the nerve centre of signalling when calcium is depending on it (Banerjee and Roychoudhury, 2017). Due to the stress of fluoride the intracellular content of Ca^{2+} is decreased then it damages the process of signalling when the plant is having under abiotic and non-living stresses (Fig. 2). Inside the plants, fluoride behaves as a resistor of metabolic processes (Khan and Iram, 2016). The normal physiological activities i.e. photosynthesis are deteriorated and chlorosis occurs due to a prolonged period of fluoride stress. $Fe^{2+/3+}$ and Mg^{2+} are actively associated with fluoride and in leaves; it decreases translocation to the upward side (Elloumi et al., 2005). That is responsible for the inhibition of carotenoids and chlorophyll biosynthesis. The involvement of acid having α-aminolevulinic in the biosynthesis of the chlorophyll pathway is prevented when the stress of fluoride is induced (Wallis et al., 1994). The complexions of magnesium fluoride degraded the pigments of the plant (Abdallah et al., 2006). When *Triticum aestivum* and *Oryza sativa* are having in contact with the stress of fluoride significantly decreases chlorophyll a and b content Gupta et al., 2009, Bhardwaj and Bhargava, 2010). The result shows that two different varieties of rice are exposed to various concentrations (20, 10 and 5 dm^{-3}) with sodium fluoride degraded the carotenoid and chlorophyll (Chakrabarti and Patra, 2015). Likewise (Mondal, 2017) find out that degradation of pigments when various concentrations (20, 10 and 5 dm^{-3}) NaF exposed with different four varieties of rice. On the new leaves of plant having *Populus deltoides* Severe necrosis and chlorosis were noticed when 500.01 mg dm^{-1} amount of fluoride present in irrigated water (Verma and Singh, 2013). Anthocyanins are involved in tolerance of stress and properties of antioxidants are having in anthocyanin (Paul et al., 2017). The decline in Content of carotenoid and chlorophyll and increases in the content of anthocyanin in the plant of Salicornia brachiate when it exposed to fluoride stress (Kaur and Reddy, 2008). In the cultivation of radish plant absorption of Mn^{2+} and Zn^{2+} is high in the soil of which is contaminated with cadmium and fluoride. Still, the absorption of Cu^{2+} was reduced drastically (Chen et al., 2017). For the important enzymes such micronutrients are also used as a cofactor, for the plant development and their growth important pathway of signaling is used that altered by influencing their profile (Roychoudhury and Banerjee, 2018). The stress of fluoride affecting the rate of photosynthesis and form sever types of chlorosis (Mondal, 2017, Roychoudhury and Banerjee, 2018). Reducing the associated enzyme activities within the cycle of Calvin. The enzymes that are affected by the exposure of fluoride are ribulose, synthetase of sucrose, ATPase chloroplast (Ranghara and Baunthiyal, 2014). When the fluoride stress is induced in conductance of stomata resulting in lower down the atmospheric diffusion of CO_2 in the cell, carbon dioxide assimilation during photosynthesis is reduced (Singh et al., 2013). Stress-induced by fluoride in the known plants reduced the fluorescence of chlorophyll (CHL) and efficiency of photosynthesis (Fv/Fm) (Verma and Singh 2013). The efficiency of the plant analyzer has monitored the yield of the fluorescence. There is a decline in the photosystem second (PS) II, which occurs when subunits antenna of photosystem second binds with the proteins of chlorophyll (CP47 and CP43) and their activities, decreased. Water stress is induced by fluorine decreases the relative content of water (RWC) of plants i.e. *Triticum asetivum, Brassica oleracea, Beta vulgaris, Oryza sativa,* and *Pisum sativum* when irrigated water is contaminated with fluoride (Gupta and Mondal 2015). The normal growth of the plant is also inhibited by less RWC.

Fluoride Induced by the Alteration of Biochemical

Parameters of biochemical play an important role in determining the sensitivity of plant against specific stress. More than 50% of respiration is suppressed by the stress of fluoride and preventing the various activities of glycolytic and mitochondrial enzymes i.e. dehydrogenase of succinate, enolase, and maltose (Miller, 1993). In chloroplast and mitochondria having the acidic pH within organelles of cells that accumulate the fluoride and stop the activities of ATPase in the membrane (Struglics et al., 2000). Transportation of phosphorous is inhibited due to an analogue of phosphate ions complexes with aluminium fluoride (AlF^{-4}) ions. The fluoride content is exposed to the protein present in the membrane of potato inhibiting the transportation of phosphorus (Struglics et al., 2000, Golicnik, 2010). The various process is prohibited like glycolysis, signalling, and synthesis of nucleotide due to phosphorylation inhibition (Barbier et al., 2010). In various four varieties of rice leakage of ions in root and lipid peroxidation (H_2O_2) having due to the rise in the amount of fluoride content (Mondal, 2017). The fluoride stress is induced in plant affected the metabolism of total N and amino acids. The fluoride is induced in tea plant leaves soluble protein is degraded due to increases in the content of amino acid. Large numbers of content having aspartic, glutamic acid, serine and asparagine are accumulated in the leaves (Cia et al., 2016). Those plants, which are sensitive and tolerant to fluoride a correlation, will find in between the fluoride stress and metabolism of nitrogen has detected (Szostek and Ciecko, 2017). In the manner of specific-species total nitrogen present in the biomass of shoots accumulated the stress of fluoride. Possibly developing the strategies to preventing the degradation of protein and the plant of maize being tolerant of fluoride. The lower content of nitrogen would be observed and does not require regulating the synthesis of protein in fluoride stressed plant (Ciecko and Szostek, 2017). The spite of the *Lupinus luteus* plant being tolerated to fluoride and a large quantity of total N is accumulated in the shoots (Ciecko and Szostek, 2017). In-plant tolerance due to fluoride is exhibited by increasing the synthesis of protein and degradation of protein keep away from equilibrium tilting. In the fluoride sensitive plants, viz., *Lupinus angustifolius* noticed that the content of nitrogen is present is a large amount by unregulated degradation of the protein. The reductase of nitrogen is inhibiting by the assimilation of nitrogen deteriorates when it induced in the stress of fluoride (Rao *et al.* 2013). For handling the stress and producing the energy through the process of metabolic towards the degradation of product through channelization, it all explained when the content of total protein is decreases by induced fluoride stress (Szostek and Ciecko, 2017). The content of protein in the mulberry plant was decreased by up to 59% under the stress of fluoride (Rao et al., 2013). Pal et al. also observed in an experiment that the content of protein is decreased in different 13 species of the plant when irrigation is done by contaminated fluoride water (Pal et al., 2012). Seedlings of *Abelmoschus esculentus* irrigated by contaminated fluoride water, as a comparison with the controlled plant it shows that the percentage of protein and nitrogen was decreased (Khan and Iram, 2016). If the plant of potato was grown in the soil contaminated with fluoride shows that increasing the content of total protein and make a balance in loss of protein (Saleem et al., 2015). The stress of fluoride-induced than in roots the conductivity of water and uptake of water is inhibiting by decreases in the potential of water and the result shoes that increase in osmotic potential rate (Gadg et al., 2016). In those conditions, Biological hydration is maintained in different inorganic ions and solubilized sugar by accumulated them into plants (Paul et al., 2017). The rise in contents of proline, phenol and amino acids shows in table 2. The structure of enzymes and protein is stabilized and membranes are protected with the help of suitable solute. The solubilize sugar is accumulated properly by different species of plants which exposed in the stress of fluoride. They behave like as osmotica for the transport of water through roots or stabilized the protein, structures of the membrane (Banerjee and Roy Choudhury, 2016; Ram et al., 2014).

Antioxidant System and Fluoride Stress

The previous discussions resulted that the severe effect of fluoride decreases the physiological activities in plants. In this type of inhibition production in the species of reactive oxygen (ROS) i.e. radical of hydroxyl, radicals of superoxide and hydrogen peroxide are increased (Roychoudhury and Banerjee, 2018) (Fig. 2). The ROS increases lipid peroxidation and degrades the nucleic acid and protein. When the index of membranes stability is decreased than leakage of electrolytes were increases. ATPases membrane activities and interaction of protein-lipid seriously hampered (Roychoudhury and Das, 2014). To detoxify the effect of ROS, plant variety increases the antioxidative activity of enzymes i.e. (CAT) catalase, (POX) peroxide and (SOD) superoxide dismutase (Paul et al., 2017). *Spinaceaoleracea, Coriandrumsativum, Brassica Juncea,* and *Raphanussativus*plants are exposed in the stress of fluoride and it increases the activities of catalase, peroxide and superoxide dismutase (Gadi et al., 2016, Singh et al., 2013). A report shows that the activity of catalyzing is maximum in rice plant that irrigated by contaminated fluoride water (Mondal 2017). In the maize plant when fluoride stress is induced than the activity of glutathione peroxidase (GPX) and CTA decreases prominently, thus the genotype shows susceptibility to fluoride (Mutahir et al., 2016). After the exposure of fluoride in various species of plants has observed accumulation of phenolics and (AsA) ascorbic acid that is non-enzymatic antioxidants. Increases in the content of AsA indicate that AsA efficiency is higher in the glutathione cycle (Roychoudhury and Das, 2014). Under this study encoded, the expression of the gene pattern in antioxidants is not reported.

Accumulation of Fluoride in Edible Plant Parts

Accumulation of F is the direct risk in the food chain by fluoride bioaccumulation in grains of different crops mainly in legumes as well as cereals and different vegetables edible parts (Mondal and Gupta, 2015). There is a high amount of F accumulate in different popular plants named as *Chihorium* sp., *Spinacia* sp., *Brassica* sp., Apiumgraveolensetc (Hong et al., 2016). The main reason for the accumulation is the efficient translocation from contaminated soil is rooted in shoots and from the atmosphere; there is direct absorption (Ambuvel et al., 2015). The experiment was conducted and reported that in rice seedlings there was an accumulation of F 5000 to 4000 mg/kg which was irrigated with the contaminated water which has sodium fluoride concentration 20 and 30 mg/dm^{-3} for 15 days(Gupta et al., 2009). The plants, which are having a high concentration of fluoride, are not suitable for animal feeding which also acts as the indicator for the accumulation of F in the grains (Gupta et al., 2009). When the rice plant was exposed to the metal stress by giving the sodium fluoride of concentration 5 mg/dm-3 and reported that accumulation of F translocated in the part of the shoot was maximum (Mondal, 2017). In Tamilnadu, the rice seedling and the black gram were irrigated with groundwater which results in the accumulation of F which was higher than the safe limit i.e 1.5 mg/kg^{-1} by WHO (Ambuvel et al., 2015). When camellia sp. was grown in the fluoride, contaminated soil found that the leaves contain the highly accumulated F i.e. 214 mg/ kg^{-1} (Bauthiyal et al., 2013).

Phytoremediation

The phytoremediation approach is a cheap and eco-friendly strategy used for detoxification metal-contaminated infected soils. Phytoremediation technique is associated with the use of plants, which are having better translocation efficiency with tolerance. It was identified that species of *Apiumgraveolens, Spinaceaoleracea, Brassica oleracea,* and *Cichorium spp.* can accumulate a huge amount of F (Hong et al., 2016). Accumulation of F is done in two phases first directly absorbed from the atmosphere and second after translocation of ions from roots to shoot. The selection of plants that are suitable for phytoremediation of F can be easily identified because in areas be easily identified in areas where F infected. The species that are selected for remediation can accumulate F ions and does not have any phytotoxic effect. (Yoon et al., 2006). The Fabaceae most famous family used as phytoremediators. F tolerant plants have been screened from the site of phosphate fertilizer manufacturing industries because there is a synergistic (Mezghani et al., 2005) relationship between fluoride and phosphate. An interesting fact noticed arsenic-contaminated groundwater also contains a very high concentration of fluoride. *Peteris vittata* species were identified best for decontamination of As and F. (Boukhris et al., 2015). Some species can be used as in situ phytoremediators in F contaminated arid and semiarid region soils such as *Rhanterium suaveolens, Erodium* glaucophyllum for arid regions and *Accasia toetilis, Prosopis juliflora, Cassia fistula* for semiarid regions. However, *C. fistula* shows sensitivity in later stages of growth. There is some aquatic flora that can be used for purpose of detoxification of F .Aquatic algae to increase F tolerance through the process of increasing Ca ions to confiscate free ions of F and convert into the form of CaF2 in roots of the plant. The efficiency of these aquatic plants are described in the table (Karmakar et al., 2011). The visible effect of F can be monitored on leaves of *Eucalyptus rostrata* and the leaf tip of *Pinus radiate.* Near about 6-3652 mg (f) kg/dry matter is accumulated on leaf tissue of *E. rostrata.* External F present on the leaf surface can be easily removed by washing which leads to minimizing total F content up to 24-51% in *E. rostrate* and *P. hybridus.*

Table 1: Aquatic plants and phytoremediation

S.No	Phytoremediators	Efficiency to remove F ions %
1	*P. stratiotes*	19.87
2	*E. crassipe*	12.17
3	*S. polyrhiza*	19.23

Table 2: Fluoride effect on various species of plant (Carotenoid - Car, Chlorophyll - Chl, Preoxidase – POX, Relative Water Content – RWC, Ascorbic Acid – AsA, Superoxide Dismutase – SOD, Proline – Pro, Electrolyte Leakage – EL)

Type/ family	Plant	Pigment	Growth parameters	Photosynthesis	Osmolytes	Biochemical	Reference
Gramineae	Rice	Car, Chl a and Chl b degradation	Vigorous index and germination reduction	The efficiency of photosynthesis is reduced	Cat, phenol, soluble sugar is increased	Low protein, RWC, increases lipid oxidation	Gupta et al., 2009; Chakrabarti and Patra 2015 and Mondal, 2017
	Wheat	Damage in Chl	Root and shoot length is reduced	Photosynthetic efficiency is effected	Soluble sugar, AsA and proline increase	RWC is low	Bhargava and Bhardwaj, 2010; Mondal and Gupta, 2015
	Triticale	Not Studied	Very sensitive to fluoride	Not Studied	Not Studied	In roots content of N is less	Szostek and Ciecko, 2017
	Maize	Not Studied	Biomass of shoot is increases and root is decreases	Not Studied	Not Studied	Protein and N is low	Szostek and Ciecko, 2017
Leguminosae	Moong Bean	Chl a Chl b and Car degradation	Germination growth is less, abnormalities, vigour index is low	The efficiency of photosynthesis is reduced	Soluble sugar and proline is increased	The protein content is low and EL is increased	Gadi et al., 2012

	Lentil	Chl content is decreased	Not Studied Efficiency of photo-synthesis is reduced	The efficiency of photosynthesis is reduced	Less increase in Pro, AsA and soluble sugar	RWC content in protein is low	Mondal and Gupta, (2015)
	Lupinus-culinaris	Not Studied	Root and shoot biomass increases	Not Studied	Not Studied	A large content of N in biomass of shoot	Szostek and Ciecko, (2017)
	Lupin	Not Studied	Growth is reduced	Not Studied	Not Studied	A large content of N in biomass of shoot	Szostek and Ciecko, (2017)
	Chick Pea	Not Studied	Less % of germination	Not Studied	Accumulation of soluble sugar and oxidation of AsA, POX activities is increases	Increased toxicity % in shoot and roots	Datta et al. (2012)
	Pea	Chl will not damage	Not Studied	Not Studied	AsA, soluble sugar content Increases	Decreases RWC	Mondal and Gupta (2015)
	Alfalfa	Not Studied	Slightly growth is reduced	Not Studied	Not Studied	Increase the content of N	Szostek and Ciecko (2017)
Cruciferae	Cabbage	Content of Chl decreased	Not Studied	Photosynthetic efficiency is reduced	Content of AsA is increases	Decreases RWC and content of Protein	Mondal and Gupta (2015)

	Mustard	Content of Chl decreased	Not Studied	Photosynthetic efficiency is reduced	The minimal rise in AsA, soluble sugar and activities of SOD, CAT, and POX are increased.	RWC is low	Mondal and Gupta (2015)
	Rapeseed	Not Studied	Increase in biomass of root and shoot.	Not studied	Not studied	Content of Total nitrogen in roots in high	Szostek and Ciecko (2017)
	Radish	Not Studied	Length of shoot and root decreases significantly	Not Studied	Activities of POX, CAT, and SOD	Moderately rise in content of total nitrogen	Szostek and Ciecko (2017)
Amaryllis	Onion	Content of Chl decreased	Not Studied	Photosynthetic efficiency is reduced	Content of sugar is decreased and the content of Pro and AsA increases	RWC is low	Mondal and Gupta (2015)
	Garlic	Content of Chl decreased	Not Studied	Photosynthetic efficiency is reduced	The rise in content of AsA and soluble sugar	RWC is low	Mondal and Gupta (2015)
Chenopodiaceae	Pickleweed	Growth is arrested	Degradation of Car and Chl	Photosynthetic efficiency is reduced	The rise in content of anthocyanin	Not Studied	Reddy and Kaur (2008)
	Spinach	Content of Chl decreased	Not Studied	Photosynthetic efficiency is reduced	Rise in the content of Pro and AsA, Activities of CAT, POX, and SOD	RWC is low	Mondal and Gupta (2015)

	Beetroot	Damage in Chl	Not Studied	Photosynthetic efficiency is reduced	Decreases in the content of proline	RWC is low	Mondal and Gupta (2015)
Umbelliferae	Celeriac	Content of Chl decreased	Not Studied	Photosynthetic efficiency is reduced	The minimal rise in content of Pro, AsA and soluble sugar	RWC is low	Mondal and Gupta (2015)
	Wild carrot	Damage in Chl	Not Studied	Photosynthetic efficiency is reduced	The rise in content of soluble sugar	RWC is low	Mondal and Gupta (2015)
Cucurbits	Watermelon	Index of Chl stability is reduced	Vigour index is low and growth is arrested	Photosynthesis is retarded	Accumulation of phenolics and soluble sugar	Membrane stability is lower	Ram *et al.* (2014)
	Cucumber	Content of Chl is decreases	Not Studied	Photosynthetic efficiency is reduced	Increase the content of soluble sugar	RWC is low	Mondal and Gupta (2015)
Borago	Phacelia Juss.	Not Studied	Less sensitivity towards fluoride	Not Studied	Not Studied	In the biomass of aerial content of total N is high	Szostek and Ciecko (2017)
Mallows	Okra	Chl a Chl b are decreases	Reduces the length of shoot and root	Photosynthetic efficiency is reduced	Not Studied	Content of protein and nitrogen is reduced	Iram and Khan (2016)
Trees	Poplar	Reduction in yield of chlorophyll inflorescence and degradation of Chl	Expansion in leaf and biomass of plant are decreases	Assimilation of CO_2 and stomatal conductance is reduced	Not Studied	Harvest index is low	Singh and Verma (2013); Singh *et al.* (2013)

	Mulberries	Not Studied	Decline in growth	Not Studied	Not Studied	decreases the content of protein in leaf	Rao *et al.* (2013)

Table 3 Suppl. Bioaccumulation of fluoride in eatable plants is greater than 1.50 mg kg^{-1} (Safe limit WHO)

Type	Plant	Conc.of fluoride (mg dm^{-3})	Stress of fluoride duration	Accumulation of fluoride [mg kg^{-1} (d.m.)]	Reference
Crops	Rice	20.1	16 days	2000.1	Gupta *et al.* (2009)
	Mustard	4.81	90 days	43.7	Mondal and Gupta (2015)
	Wheat	4.81	90 days	28.5	Mondal and Gupta (2015)
	Lentil	4.81	90 days	25.3	Mondal and Gupta (2015)
	Tea plant	If soil is contaminated	Whole life cycle	214.1	Baunthiyal and Ranghar (2013)
Vegetables	Okra	51	21 days	9.31	Iram and Khan (2016)
	Spinach	4.81	90 days	55.1	Mondal and Gupta (2015)
	Wild cabbage	4.81	90 days	29.9	Mondal and Gupta (2015)
	Celery	4.81	90 days	42.1	Mondal and Gupta (2015)
	Cucumber	4.81	90 days	18.7	Mondal and Gupta (2015)
	Pea	4.81	90 days	27.1	Mondal and Gupta (2015)
	Carrot	4.81	90 days	23.8	Mondal and Gupta (2015)
	Beetroot	4.81	90 days	20.7	Mondal and Gupta (2015)
	Onion	4.81	90 days	26.2	Mondal and Gupta (2015)
	Garlic	4.81	90 days	27.4	Mondal and Gupta (2015)

Table 4 Suppl. Candidate plant feature for phytoremediation of fluoride

Type	Plant	Characteristics	References
Phytoremediation in soil by trees	Prosopis	Present in regions of semi-arid, recapture the fluoride in groundwater and soil, huge accumulation of fluoride in roots, highly tolerant to fluoride	Banthiyal and Sharma (2012, Saini *et al.* (2012)
	Olive	300.1 [mg kg^{-1} (d.m.)] of fluoride accumulates sans sowing phytotoxicity	Mezghaniet *al.* (2005)
	Mulberry	Tolerant to fluoride, exhibited less inhibition in the parameter of photosynthesis.	Kumar and Bhaska-ra (2008)
	Tulip tree	Accumulation of fluoride is high in the tissues of the leaf	Wang-Cahill (2007)
	Grapevine	No inhibition in metabolic, sequestered of fluoride in by increasing the content of calcium	Abdallah *et al.* (2006)
Phytoremediation in soil by trees	Guinea grass, Rhodes grass	15.1[mg KF(dm^{-3})] conc.is tolerant due to sequestration induced by calcium, nope translocation via the phloem	Chaves *et al.* (2002).
Phytoremediation by aquatic plants	Camellia, Japanese *pittosporum*	Accumulation of fluoride in contaminated H_2O	Santos-Diaz and Zamora-Pedraza (2010)
	Green algae	Producing CaF_2 by use of $CaCO_3$ to detoxified the aquatic fluoride	Baunthiyar and Ranghar (2013)
	Field mustard	Boron remediation and sediments of contaminated fluoride	Asada *et al.* (2005)
	Duckweed	Efficiently defluoridated the water	Shirke and Chandra (1991)

Bioindicators of fluoride	Eucalyptus rostrata	The plant leaves deposited 3651[mg KF kg^{-1} (d.m.)] near to Al smelters.	Rodriguez *et al.* (2012)
	Populus hybridus	In the plant leaves an accumulation of fluoride neat to AL smelters.	Rodriguez *et al.* (2012)

Conclusion

The phytotoxic effects of F toxicity in plants are inadequately studied. The content of fluoride is observed in a different form of soil like in silt clay, loam soil and phosphatic fertilized soils. There are different factors upon which the accumulation of fluoride depends like pH of the soil, capacity of sorption as well as permeability. The F related phytotoxicity is demonstrated by the burning of leaf tip, scorching of leaves and chlorosis. Many mechanisms within the plants are affected by the accumulated F like photosynthetic production, metabolism of nitrogen, synthesis of nucleic acid hence increase the production of ROS. When the ROS is increased in the plants there is membrane damage and structure of enzymes but non-enzymatic antioxidants accumulate as AsA and polyphenols result in increasing the enzymatic activities antioxidants like, CAT, SOD, and POX that helps in mitigating the ROS toxicity. In the future, the gene encoding of the profiles expression of antioxidants should be studied. In human and animal, the effect of accumulated F in the plants is the serious risk, which needs to be solved. For controlling such problems different F hyperaccumulators and tolerant sp. need to be identified for the phytoremediation. For contaminated soils and water with fluoride metals, the best technique for removal of F metal is phytoremediation because it is cost-effective and easy to adopt. However, this may take more time and sometimes act as partially effective. In any situation where the field is contaminated with F toxic metals so the first thing is to remove the metal by using the F hyperaccumulators then only suitable for the cultivation. With the low F accumulation, it is very much important to produce genetically modified crops, which are tolerant of the F-tolerant genotypes.

Reference

Abdallah, F.B., Elloumi, N., Megfani, I., Garrec, J.P., Sfax, M.B., Tunisia, M.: Fluoride pollution of Jerbi grape leaves and the distribution of F, Ca, Mg and P in them. - Fluoride 39: 43-48, 2006

Abugri, D.A., Pelig-Ba, K.B.: Assessment of fluoride content in tropical surface soils used for crop cultivation. - Afr. J. environ. Sci. Technol. 5: 653-660, 2010.

Agarwal, R., Chauhan, S.S.: Bioaccumulation of sodium fluoride toxicity in Triticum aestivum var. - Int. J. Food Agr. Vet. Sci. 4: 98-101, 2014.

Ambuvel, D., Kumaresan, S., Jothibai, M.R.: F accumulation on paddy (Oryza sativa) and black gram (Phaseolus mungo Linn.) in cultivated areas of Kanyakumari district, Tamil Nadu, India. - Indian J. Fundam. Appl. Life Sci. 5: 280-285, 2015.

Anshumali, B.K.: Fluoride in agricultural soil: a review on its sources and toxicity to plants. - Global sustainability transitions: impacts - Innovations 3: 29-37, 2014.

Asada, M., Parkpian, P., Sumio, H.: Remediation technology for boron and fuoride contaminated sediments using green plants. - J. ASTM Int. 3: 304-310, 2005.

Banerjee, A., Roychoudhury, A.: Group II late embryogenesis abundant (LEA) proteins: structural and functional aspects in plant abiotic stress. – Plant Growth Regul. 79: 1-17, 2016.

Banerjee, A., Roychoudhury, A.: Abscisic-acid-dependent basicleucine zipper (bZIP) transcription factors in plant abiotic stress. - Protoplasma 254: 3-16, 2017.

Banerjee, A., Roychoudhury, A.: Strigolactones: multi-level regulation of biosynthesis and diverse responses in plant abiotic stresses. - Acta Physiol. Plant. 40: 86, 2018a.

Banerjee, A., Roychoudhury, A.: The gymnastics of epigenomics in rice. - Plant Cell Rep. 37: 25-49, 2018b.

Banerjee, A., Roychoudhury, A.: Effect of salinity stress on growth and physiology of medicinal plants. - In: Ghorbanpour, M., Varma A. (ed.): Medicinal Plants and Environmental Challenges. - Springer, Cham 2018c.

Banerjee, A., Roychoudhury, A.: Abiotic stress, generation of reactive oxygen species, and their consequences: an overview. - In: Singh, V.P., Singh, S., Tripathi, D., MohanPrasad, S., Chauhan, D.K. (ed.): Revisiting the Role of reactive Oxygen Species (ROS) in Plants: ROS Boon or Bane for Plants? Pp. 23-50. John Wiley & Sons, Chichesterconcentrations in waters in parts of northern Tanzania. -Geochem. J. 41: 213-234, 2007.

Gadi, B.R., Bhati, K., Goswami, B., Rankawat, R., Kumar, S.: Sources and phytotoxicity of fluorides in the environment. -Environ. Impact. Biodivers. 2016: 251-266, 2016.

Gadi, B.R., Verma, P., Amra, R.: Influence of NaF on seed germination, membrane stability and some biochemical content in Vigna seedlings. - J. Chem. biol. Phys. Sci. 2: 1371-1378, 2012.

Golicnik, M.: Metallic F complexes as phosphate analogues for structural and mechanistic studies of phosphoryl group transfer enzymes. - Acta Chim. slov. 57: 272-287, 2010.

Gupta, S., Banerjee, S., Mondal, S.: Phytotoxicity of fluoride in the germination of paddy (Oryza sativa) and its effect on the physiology and biochemistry of germinated seedlings. - F 42:142-146, 2009.

Hong, B.D., Joo, R.N., Lee, K.S., Lee, D.S., Rhie, J.H., Min, S.W., Song, S.G., Chung, D.Y.: Fluoride in soil and plant. -Korean J. Agr. Sci. 43: 522-536, 2016.

Iram, A., Khan, T.I.: Effect of sodium fluoride on seed germination, seedling growth and biochemistry ofAbelmoschusesculentus. - J. Plant Biochem. Physiol. 4: 170, 2016.

Jha, S.K., Nayak, A.K., Sharma, Y.K.: Fluoride toxicity effects in onion (Allium cepa L.) grown in contaminated soils.Chemosphere 76: 353-356, 2009.

Kanduti, D., Sterbenk, P., Artnik, B.: Fluoride: a review of use and effects on health. - Mater. Sociomed. 28: 133-137, 2016.

Karmakar, S., Mukherjee, J., Mukherjee, S.: Removal of fluoride contamination in water by three aquatic plants. - Int. J.Phytoremediation 18: 222-227, 2016.

Kumar, A., Bhaskara, V.R.: Physiological responses to fluoride into cultivars of mulberry. - World J. Agr. Sci. 4: 463-466, 2018.

Lorestani, B., Cheraghi, M., Yousefi, N.: Phytoremediation potential of native plants growing on a heavy metals contaminated soil of copper mine in Iran. - World Acad. Sci. Eng. Technol. 77: 377-382, 2011.

Marier, R.: F Pollution. - http://Falert.org/ articles/growth-1975/,1971.Meenakshi, Maheshwari, R.C. Fluoride in drinking water and its removal. - J. Hazard. Mater. 137: 456-463, 2006.

Mezghani, I., Elloumi, N., Abdallah, F.B., Chaieb, M., Boukhris, M.: Fluoride accumulation by vegetation in the vicinity of a phosphate fertilizer plant in Tunisia. - Fluoride 38: 69-75, 2005.

Miller, G.W.: The effect of fluoride on higher plants with special emphasis on early physiological and biochemical disorders. -F 26: 3-22, 1993.

Mondal, D., Gupta, S.: Influence of fluoride contaminated irrigation water on biochemical constituents of different crops and vegetables with an implication to human risk through diet. - J. Mater. Environ. Sci. 6: 3134-3142, 2015.

Mondal, N.K.: Effect of fluoride on photosynthesis, growth and accumulation of four widely cultivated rice (Oryza sativa L.) varieties in India. - Ecotoxicol. Environ. Safety 144: 36-44, 2017.

Mutahir, A., Bisht, M., Baunthiyal, M.: Comparative assessment of fluoride tolerance in two genotypes of Zea mays. - J. med.Plant Studies 4: 253-258, 2016.

Nanyaro, J.T., Aswathanarayana, Mungure, J.S. A geochemical model for the abnormal fluoride concentrations in waters in parts of northern Tanzania. - J. Afr. Earth Sci. 2: 129-140, 1984.

Pal, C.K., Mondal, K.N., Bhaumik, R., Banerjee, A., Datta, K.J.: Incorporation of fluoride in vegetation and associated biochemical changes due to fluoride contamination in water and soil: a comparative field study. - Ann. Environ. Sci. 6:123-139, 2012.

Panda, D.: Fluoride toxicity stress: physiological and biochemical consequences on plants. - Int. J. Bioresour. Environ. Agr. Sci.1: 70-84, 2015.

Paul, S., Roychoudhury, A., Banerjee, A., Chaudhuri, N., Ghosh, P.: Seed pre-treatment with spermidine alleviates oxidative damages to a different extent in the salt (NaCl)-stressed seedlings of three indica rice cultivars with contrasting level of salt tolerance. - Plant Gene 11: 112-123, 2017.

Pitman, M.G.: Fluoride transport across plant roots. - Rev. Biophys. 15: 481-554, 1985.

Pulyaevskaya, M.A., Varakina, N.N., Gamburg, K.Z., Rusaleva, T.M., Stepanov, A.V., Voinikov, V.K., Rikhvanov, E.G: Sodium fluoride inhibits HSP synthesis in heat-stressedcultured cells of Arabidopsis thaliana. - Russ. J. Plant Physiol. 58: 589-596, 2011.

Ram, A., Verma, P., Gadi, B.R.: Effect of fluoride and salicylic acid on seedlings growth and biochemical parameters of watermelon (Citrullus lanatus). - F 47: 49-55, 2014.

Rao, B.V.A., Kumar, A.K., Nagalaksmamma, K., Vidyunmala, S.: Effect of fluoride on protein profiles in two cultivars of mulberry leaves. - J. Agr. Environ. Sci. 13: 957-960, 2013.

Reddy, M.P., Kaur, M.: Sodium fluoride-induced growth and metabolic changes in Salicornia brachiataRoxb. - Water Air Soil Pollut. 188: 171-179, 2008.

Rodriguez, J.H., Wannaz, E.D., Pignata, M.L., Fangmeier, A., Franzaring, J.: Fluoride biomonitoring around a large aluminium smelter using foliage from different tree species. -Clean Soil Air Water 40: 1315-1319, 2012.

Roy, M., Giri, A.K., Dutta, S., Mukherjee, P.: Integrated phytobial remediation for sustainable management of arsenic in soil and water. - Environ. Int. 75: 180-198, 2015.

Roychoudhury, A., Banerjee, A.: Abscisic acid signalling and involvement of mitogen-activated protein kinases and calcium-dependent protein kinases during plant abiotic stress.- In: Pandey, G. (ed.) Mechanism of Plant Hormone Signaling under Stress. Vol. 1. Pp. 197-241. John Wiley & Sons, Hoboken 2017.

Roychoudhury, A., Banerjee, A., Lahiri, V.: Metabolic and molecular-genetic regulation of prolinesignalling and its cross-talk with major effectors mediate abiotic stress tolerance in plants. - Turk. J. Bot. 39: 887-910, 2015.

Saini, P., Baunthiyal, M., Sharma, V.: Organ-wise accumulation of F in Prosopis juliflora and its potential for phytoremediation of fluoride contaminated soil. -Chemosphere 5: 633-635, 2012.

Saleem, M., Ahmad, N.M., Khan, A.B., Zia, A., Ahmad, S.S., Shah, U.H., Khan, A.N., Qazi, M.I.: Effects of soil fluoride on the growth and quality of two tomato varieties grown in Peshawar, Pakistan. - F 48: 174-178, 2015.

Santos-Díaz, M.S., Zamora-Pedraza, C.: Fluoride removal from water by plant species that are tolerant and highly tolerant of hydrogen fluoride. - Fluoride 43: 150-156, 2010.

Saxena, S., Rani, A.: Fluoride ion leaching kinetics for alkaline soils of Indian origin. - J. Sci. Res. Replicat. 1: 29-40, 2012.

Shirke, P.A., Chandra, P.: Fluoride uptake by duckweed Spirodelapoyrrhiza. - Fluoride 24: 109-112, 1991.

Singh, M., Verma, K.K.: Influence of fluoride-contaminated irrigation water on physiological responses of poplar seedlings (Populusdeltoides L. clone-S7C15). - F 46: 83–89, 2013.

Singh, M., Verma, K.K., Verma, C.L.: An approach to develop a model for describing the influence of fluoride-contaminated irrigation water on physiological responses in poplar (Populus deltoids clone S7C15). - Acta Physiol. Plant. 35: 3357-3364, 2013.

Sinha, S., Saxena, R., Singh, S.: Fluoride removal from water by Hydrilla verticillata (l.f.) Royle and its toxic effects. - Bull. Environ. Contam. Toxicol. 65: 683-690, 2000.

Stevens, D., McLaughlin, M.J., Alston, A.M.: Phytotoxicity of the fluoride ion and its uptake from solution culture by Avena sativa and Lycopersicon esculentum. - Plant Soil 200: 119-129, 1998.

Struglics, A., Fredlund, K.M., Konstantinov, Y.M., Allen, J.F., Müller, I.M.: Protein phosphorylation/ dephosphorylation in the inner membrane of potato tuber mitochondria. – FEBS Lett. 475: 213-217, 2000.

Susheela, A.K.: Fluorosis management programme in India. -Curr. Sci. 77: 10, 1999.

Szostek, R., Ciecko, Z.: Effect of soil contamination with fluoride on the yield and content of nitrogen forms in the biomass of crops. - Environ. Sci. Pollut. Res. 24: 8588-8601, 2017.

Tylenda, C.A.: Toxicological Profile for Fluorides, HydrogenFluoride and Fluorine. - DIANE Publishing, Darby 2011.

Vasudevan, D., Harrington, L.F., Cooper, E.M.: Fluoride sorption and associated aluminium release in variable charge soils. - J. Colloid. Interface Sci. 267: 302-313, 2003.

Wallis, W.J., Miller, G.W., Psenak, M., Shieh, J.: Fluoride effects on chlorophyll biosynthesis in Nicotianatabacum. – Fluoride7: 69-77, 1974.

Walna, B., Kurzyca, I., Bednorz, E., Kolendowicz, L.: Fluoride pollution of atmospheric precipitation and its relationship with air circulation and weather patterns (WielkopolskiNational Park, Poland). - Environ. Monit. Assess. 185: 5497–5514, 2013.

Wang-Cahill, F.: Investigation of fluoride distribution indeciduous trees at a hazardous waste landfill. - In: The Annual International Conference on Soils, Sediments Water land water. University of Massachusetts, Amherst 2007.

Wehr, J.B., Kopittke, P.M., Scholz, L., Menzies, N.W.: Understanding the Movement of Fluoride from Soils into the broader Ecosystem. - The University of Queensland, School of Agriculture and Food Sciences, Gatton 2014.

Yoon, J., Cao, X., Zhou, Q., Ma, L.Q.: Accumulation of Pb, Cu, and Zn in native plants growing on a contaminated Florida site. - Sci. Total Environ. 368: 456-464, 2006.

Zhao, J., Guo, H., Ma, J., Shen, Z.: Effect of fluoride on arsenic uptake from arsenic-contaminated groundwater after using Pteris vittata L. - Int. J. Phytoremediation 17: 355-362, 2015.

Phytohormones: Master Mediators in Abiotic Stress Tolerance in the Crop Plants

Priyanka Devi

Department of Agronomy, School of Agriculture

Lovely Professional University, Jalandhar, Punjab, 144411, India

Email : parichib17@gmail.com

Abstract

Due to abiotic stress, there are global losses in the crops. Abiotic stress comprises the flood, drought, heavy metal stress, heat and many more, which causes a huge loss in crop production. In this modern world, the human population is increasing day by day so preventing these types of losses to meet the demand of a growing population by producing more food is a matter of concern. Due to the various anthropogenic activities, there is variation in a climate that results in the section of agricultural land towards the abiotic stress. There is great importance to know about the different mechanisms under which plants go and develop to mitigate abiotic stresses to withstand under severe conditions. In the recent studied it was found that there are various phytohormones like Cytokinins, gibberellins, auxins, ethylene and some are new members named as Jasmonates, strigolactones and brassinosteroids are important in metabolic activities which help in producing the tolerant plants against the abiotic stress. Phytohormones are well known for the vital functioning in the capability of plants to adapt under the various environmental stresses by facilitating plant growth as well as development, distribution of nutrients and source or sink transitions. Inside the plants, various types of molecules signal are produced which is known as growth regulators of plants. Present review at physiological as well as at molecular level described the role of various phytohormones against the stress and progress concerning the response of plants towards the environmental stress. In different stress conditions, we also define the current attainments in classifying the main role of various phytohormones.

Introduction

The rate of the human population nowadays increasing very fast which requires a sustainable increase in productivity of agricultural products globally. Conversely, there are different factors, which limit the productivity of crop, and these factors are biotic as well as abiotic (Wani *et al.*, 2014). By the mean of 2050, there is an additional increase in 2.3 billion people which require to increase productivity by 70% overall (Tilman *et al.*, 2011). Whenever the plants experienced environmental stress then plants undergo different mechanisms for stress response and tolerance against the stress (Qin *et al.*, 2011). To know and classifying the principles is one of the critical features for scientists in which plants retort to numerous environmental stresses. Among all other abiotic stresses, the most extensive and significant stress are salt stress, drought as well as high temperature (Wani *et al.*, 2013). There is a decrease in average yield by 50% of major crops due to the abiotic stress. (Wang *et al.*, 2001) Due to the complexity of various tolerance characters, techniques of conventional breeding termed as limited or less successful to mitigate these types of stresses. By considering these points there is a great need for effective advances in the production of crops to feed the whole nation. One of the most important advanced techniques i.e. phytohormones engineering method to produce the crops with high yield under the various stress conditions. In stressful environment plants having the many strategies that respond to abiotic, stress which results in enhancing the growth of the plants as well as crop productivity. Plants undergo different phenomena, which include the growth of plasticity, physiological as well as biochemical processes in contrast to numerous stresses (Saud *et al.*, 2014; Tuteja, 2007). Whenever plants experience any type of stress they try to adapt themselves in that particular conditions which supplement the metabolic modifications which result to the accretion of various organic solutes mainly named as betaines, proline, sugars, polyols, cellular protection, free radicals scavenging, ionic homeostasis maintenance, protein expression as well as phytohormones induction (Parida and Das, 2005; Tuteja, 2005; Munns and Tester, 2008). The concentration of produced phytohormones molecules is very less but the again able to control different cellular processes in plants. In higher plants phytohormones work like chemical messenger which interconnects the cellular activities (Bishopp *et al.*, 2014). During the abiotic stress, phytohormones play an important role in coordinating the different signal transduction pathways and help in regulating the external and internal stimuli (Kazan *et al.*, 2015) but there are some phytohormones, which act as the stress hormones named as ABA. It has a different role in plant development, dormancy of seed, growth regulation, germination inhibition, the closing of stomata, abscission of fruit besides facilitating the environmental stress (Yang *et al.*, 2010). Even though the response of plants to various stresses like salinity stress influenced by numerous factors and relation of these can vary with space and time, but still there are most important endogenous elements named as phytohormones, which help in moderating the physiological responses, which ultimately result in adaptation to unfavourable environment conditions (Khan and Khan, 2013). To predict about mechanisms of plant susceptibility as well as tolerance against these stresses endogenous compounds and their level are responsible (Velitcukova and Fedina, 1998) because under stress many proteins are expressed by the plants are induced by phytohormones (Hamayun *et al.*, 2010). When different phytohormones are applied by exogenous application it was suggested that it is one of the best practice approaches to deal with these type of stresses (Iqbal *et al.*, 2012) and it also reduced the growth as well as emergence attributes which leads to decrease the level of endogenous phytohormones (Jackson, 1997). Sharma *et al.*, 2013 investigated the application of exogenous phytohormones helps in alleviating the adverse effect of salinity stress. In this review article, we try to summarize the whole impression of phytohormones and their importance in the plant's response to abiotic stress and different engineering metabolic activities to converse the abiotic stress, which results in increasing the production by mean of quantity-wise as well as quality point of view. It includes recent studies and their successes as well as prospects.

Abiotic Stress

Due to abiotic stress, there are huge and extensive losses in the production of crops all over the world so, it is necessary to know or examining how this is going to alter or affect the growth of plants and its development at various level molecular, physiological as well as biochemically (Kazan *et al.*, 2015). Many many environmental factors are responsible for limiting the productivity of crops mainly named drought, heat, ozone, salinity, chilling, pathogens, heat, UV radiation, and freezing. The stress response is mainly influenced by the stress period, biotic as well as abiotic features and different plant growing stages (Feller *et al.*, 2015). As we, all know about the three mechanisms of stress that are adaptation, escape and resistance and depending upon the species and crop genotype this mechanism may perhaps differ because some crops may be affected by stress at early stages, however, recover themselves for survival. Drought stress is one the most drastic stress which is liable for decreasing the agricultural production globally by affecting the plants in different ways; growth of a plant, degrading the integrity of the membrane, by altering the osmotic adjustment, degrading the content of pigment, water relations and activity of photosynthesis (Sanghera *et al.*, 2011; Pathak *et al.*, 2014). By Postel (2006) it is correctly said that water is an oil of the twenty-first century and several rivers of the world are dehydrating every day which results in less discharge to the sea (Postel *et al.* 2006). Salinity stress is also considered as one of the severe stress, which causes less productivity of crops after the drought stress. There are various hormones of the plants, which have a positive function and role in mitigating the stress, which result plant to withstand in those harsh conditions. After much investigation, a different attempt was made to allocate a particular hormone against the particular stress and some hormones having multiple resistance functions against different environmental stress.

Significance of Plant Stress Combination

Earlier if we want to study the abiotic stress situations then normally investigation was done for specific or particular stress conditions, which includes heat, drought, salinity, and acclimation of varied molecular features were examined. However, in the natural environment, the conditions and circumstances are different in which plants suffer a different combination of stresses. The experiment was conducted to know about the molecular majority in the controlled condition that is laboratory as well as greenhouse and actual conditions are maintained which does not reflect as that arise in the field. Only one stress is executed in the greenhouse as well as in the laboratory because conditions are controlled but in case of natural or field conditions, many other environmental stresses are imposed in the amalgamation. Hence, investigation was done to assessing the stress combination effects factors other than of single stress, which may ascertain valuable. In the USA combination of heat and drought, stress was responsible for causing of loss of two hundred dollars (Suzuki *et al.*, 2012).

Phytohormones; Basic Conciliator of Different Responses of Plant Against Abiotic Stress

Different growth and development must be regulated in the plants to respond to the different interior as well as exterior stimuli (Walters *et al.*, 2009). Plant hormones that belong to the diverse group responsible for signalling of molecules that are present in the lead concentration the cells and facilitate the different responses. Phytohormones help the plants by promoting the acclimatization of plants to these environment stress by enhancing the plant growth, maintaining the source and sink as well as by providing the adequate amount of nutrients to plant inappropriate time(Fahad *et al.*, 2015). Phytohormones can help the plants at either their synthesis site or somewhere else in the plants by the transporting through different pathways (Peleg *et al.*, 2015). Phytohormones have significant importance in the development of the plants as well as in the plasticity of plant growth. There are different type of plant hormones, which mainly includes abscisic acid, gibberellin, Cytokinins, auxin, ethylene, salicylic acid, Jasmonates, and brassinosteroids (figure 1).

Fig. 1. Cross talk of hormones at the time of Embryo development and maturation in flowering plants

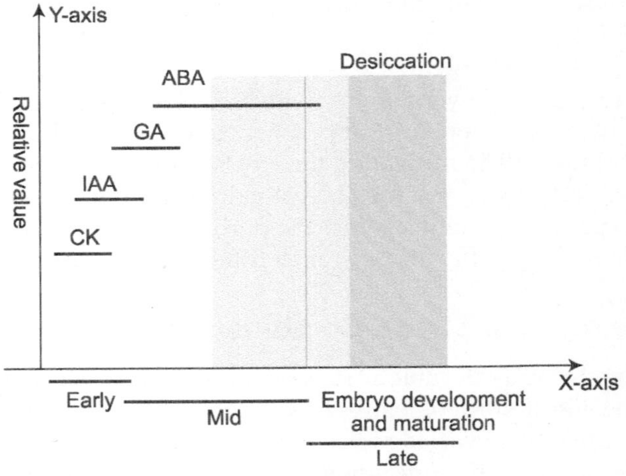

Abscisic Acid (ABA)

From its name, it indicates that responsible for the plant leaves abscission and maximum studied plant hormones for the role in the adaptation of plants and its responses against the abiotic stresses for that reason it is known as the stress hormone. ABA is produced in the plastidial 2-C methyl-D-erythritol-4-phosphate pathway and termed as isoprenoid hormone of plants. ABA having the significant role in abiotic stress which help the plants in several physiological processes and different stages of development which also includes the dormancy of seed, the opening of stomata, synthesis storage of proteins as well as lipids and regulating the morphogenesis of embryo(Sreenivasulu *et al.*, 2010). In abiotic stress, the plants adapt the various responses due to the ABA, considered as a necessary messenger in plants and having ample attention due to its function in stress tolerance (Figure 2). Whenever plants experienced any type of environmental stress the level of ABA within the plants increased quickly, the pathway of particular signalling activated and also responsible for modifying the levels of expression of genes(Brein *et al.*, 2013).

Fig. 2. Reservoir of Calcium and potassium solutes and ABA Receptors

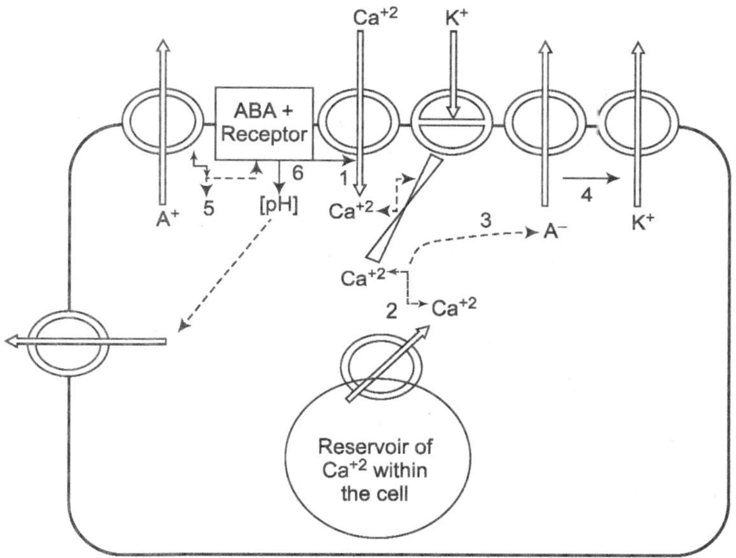

The investigation was done and reported that 10%of genes protein-encoding was regulated transcriptionally (Nemhauser *et al.*, 2006). It also helps the plant in the harsh environmental conditions by acting as the internal signals (Keskin *et al.*, 2010). Whenever there is less water infield, it provides the ability to plants which sent the signal to shoots that plants are experiencing the stress which causes the stressful circumstances around the roots, ultimately to withstand in that particular conditions activity of water-saving activity, production of antitranspirants, reduction in the expansion of leaf and closing of stomata which reduce the rate of transpiration (Wilkinson *et al.*, 2012). In the deficiency of nitrogen and drought conditions, it is responsible for the formation of vigorous roots as well as some other architectural alterations (Giuliani *et al.*, 2005; Zhang *et al.*, 2007). ABA is also responsible for regulating the many stress-responsive genes and synthesis of many protective proteins named as dehydrins and LEA proteins (Verslues *et al.*, 2006; Sreenivasulu *et al.*, 2015). It is also involved in many processes like maintaining the turgor of cells, production of antioxidants, which helps in conversing dehydration tolerance and Osmoprotectant synthesis (Chaves *et al.*, 2003). When the plant is exposed to salinity stress, the concentration of ABA increased proportionally (Zhang *et al.*, 2006). In the cytoplasm, the concentration of the calcium ions increased due to the induction of ABA when the stomata are closed in salinity stress. When the activation of the plasmalemma in the ion channels and guard cell loses the turgor pressure also associated with induction of ABA which increases the production of hydrogen peroxide (H_2O_2) that assists as a transitional signal in the closing of stomata process(Kim and Wang 2010). In salt stress level of the transcript, quantification revealed that hormones responsible for the expression of genes for adaptation in a barley crop. Keskin *et al.*, 2010 stated that induction of $MAPK4^-$ is faster like genes of (TIP 1 and GLP 1) due to the treatment of ABA in wheat crops. For understanding, the mechanism in better way upregulation of biosynthesis of ABA in the salinity stress is still required clarification and supplementary research is expected to recognize the functions of all types of responsive genes of ABA. Application of exogenous ABA, which increases the K+ and Na+ encouraging the tolerance against the salinity stress in rice (Bohra *et al.*, 1995). Recently it was studied that there is an increase in the yield of rice grain due to an increase in the accumulation of soluble sugars, homeostasis of K^+ and Ca^+ and proline by the application of ABA (Gurmani *et al.*, 2013).

Fig. 3. Signal perception and transduction of ABA

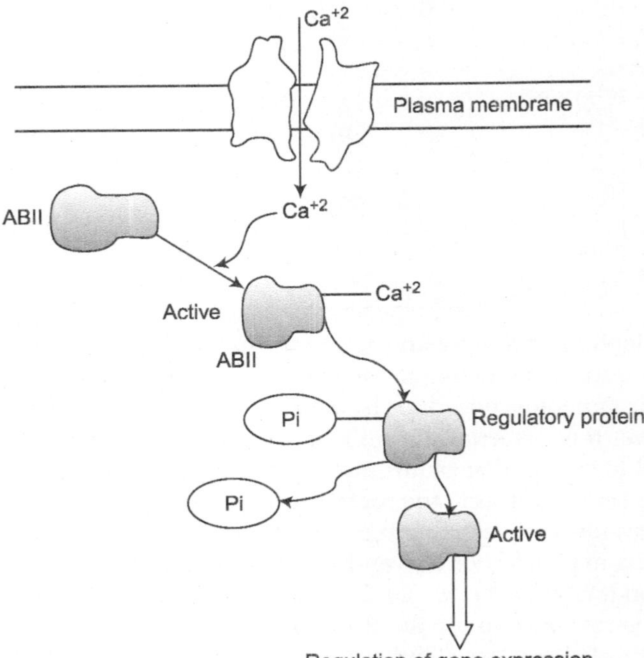

Auxins (IAA)

Over a hundred-year, it has been studied but still the pathway for signalling, transportation and biosynthesis in not understood (Wang *et al.*, 2015). There is some interconnected pathway are known for the biosynthesis of auxin named as tryptophan dependent and trp-independent pathway (Mano *et al.*, 2014). Under the stress conditions, IAA acts as the multifunctional plant hormones, which help the growth of plants and its development as well as also responsible for the coordinating and governing the growth of plants (Kazan *et al.*, 2013). Under the diverse conditions and environment, the biosynthesis of auxin and transport in green algae, which help the plants for adaptation, is a clear sign of evolutionary function played by IAA (Smet *et al.*, 2010). IAA plays an important function in the adaption of the plants to the various stress mainly salinity stress with clear evidence (Iqbal *et al.*, 2014). In heavy metal stress and salinity stress, it can increase the growth of shoots as well as the roots in maize plants (Sheng *et al.*, 2006; Egamberdieva 2009). When maize plants under the salinity stress decrease the concentration of IAA but when the application of salicylic acid is done then concentration increased effectively which creates the balance in hormones and serious crosstalk to perception signal as well as mediation response against the stress. A large number of genes are produced due to the transcription stimulation of auxin known as primary response genes of auxin and they are present in the different plant species like soybeans, rice and Arabidopsis (Javid *et al.*, 2011). Auxin is considered as significant constituent's resistance responses with the help of regulation in the several genes and able to mitigate the abiotic as well as biotic stress (Fahad *et al.*, 2016).

Cytokinins

Cytokinins are considered as one of the important phytohormones, which acts as the master regulators in the growth of the plants as well as in development (Kang et al., 2012; Nishiyama *et al.*, 2011). In salinity and drought stress, it responds against the stress when there is an alteration in the level of endogenous cytokinins (Nishiyama *et al.*, 2011). Alteration in the activity of the cytokinins enzymes metabolism by transgenic tissues and mutants towards the vital contribution in numerous crops, which includes the crop productivity as well as the tolerance against the stress (Zalabak *et al.*, 2013). This investigation has been done to estimate the responses of plants to cytokinins when the application is done externally in stress conditions, which also increase the level of endogenous cytokinins through uptake and biosynthesis (Pospisilova *et al.*, 2003). It is also responsible for the inhibition of seed germination and helps in releasing the seed from dormancy similar to ABA (Fahad *et al.*, 2015). There is an increase in the ratio of ABA and CK due to decrease content of CK and ABA accumulated under the water stress conditions (Pospisilova *et al.*, 2003).

Ethylene

Ethylene is phytohormone, which is in the form of gaseous and involved in many plant growth phases as well as in the development of plants. It is mainly responsible for senescence of flower, the ripening of fruits, abscission of petal and leaf; moreover act as essential stress response regulators (Gamalero *et al.*, 2012; Groen *et al.*, 2014). When plants experienced abiotic stress mainly chilling and salt stress which changes the endogenous level of ethylene in plants and with the higher concentration of ethylene, it enhances the tolerance against the stress (Tian *et al.*, 2012). In heat stress, also ethylene acts as a major function in plant response tolerance against stress (Lerkindale *et al.*, 2005). Whenever there is environmental stress, which helps in the induction of accumulation of ethylene due to which there is an increase in survival of the plants against the heat stress (Lerkindale *et al.*, 2005). Ethylene having the function through modulating the expression of a gene, which is considered ethylene signal effect (Klay *et al.*, 2014). Ethylene acts cooperatively when the combination is done with other phytohormones like SA as well as JA. It also involved in different regulations in which plants are protected against the pathogen and pest. It was reported that when ethylene and ABA are in combination they are having the positive interaction and negative interaction to control the growth of the plant and different development stages (Yin *et al.*, 2015).

Gibberellins

Gibberellins are the member of large groups named tetracyclic diterpenoid carboxylic acid, on the other hand in the higher plant, only a few of them are act as growth hormones (Sponsel *et al.*, 2004). The gibberellins having the many helpful functions like seed germination, elongation of the stem, initiation of flower, expansion of leaf, and development of fruit (Yamaguchi 2008). They act as the essential function in the stimulation of plant growth throughout the life cycle of plants and endorse phase of developmental transitions (Colebrook *et al.*, 2014). This investigation was done to evaluate the function of ethylene in the response of osmotic stress in *Arabidopsis sp.* Seedlings (Skirycz *et al.*, 2011; Claeys *et al.*, 2012). It also interacts with other all phytohormones for the stimulation of many developmental as well as the different responses of plants against the stress (Munteanu *et al.*, 2014).

Brassinosteroids

Brassinosteroids are able for strong plant growth and a high potential for promoting the development, which relatively comes under a new group, named polyhydroxy steroidal hormones of plants. In mustard plants, brassinosteroids were isolated in the pollen of the rape plants. Around seventy brassinosteroids or more than that are isolated till now from the plants. The three most active brassinosteroids are widely used in the studies of experiment and physiological studies named brassinolide, 24 epibrassinolides and 28 homobrassinolide (Vardhini *et al.*, 2006). Brassinosteroids is present in plants almost in every plant parts like a flower bud, seeds, fruits, pollen, vascular cambium, shoots, leaves, and roots (Bajguz *et al.*, 2009). Brassinosteroids play a significant function in many processes of developments such as the growth of roots, flower development and initiation of floral and fruits (Bajguz *et al.*, 2009). In recent research shown, the brassinosteroid is potential in modulating the antioxidant in response to the plants against the abiotic stress (Varadhini *et al.*, 2015).

Jasmonates

When the metabolic activity takes place in fatty acids membrane this includes the primary methyl jasmonate and jasmonic acid which is free acid comes together to form the jasmonates present in the plants. It has a multifunctional property, which plays a crucial role in associated with the development of plants as well as in the processes of reproduction, senescence, fruiting and involvement on direct and indirect responses (Fahad *et al.*, 2015; Pauwels *et al.* 2009). Jasmonate helps the plant in two ways like protecting it from any type of pathogen attack as well as provide the strength to plants against the different abiotic stresses (Pauwels *et al.*, 2009; Seo *et al.*, 2011). When the jasmonate is applied in the field by exogenous application, it was found that able to reduce the salinity stress in seedlings of soybean (Yoon *et al.*, 2009). Application of jasmonate able to produce the antioxidants in the plants, which result in mitigating the heavy metal stress in, plants (Jan *et al.*, 2013). It also helps the plants by the phytochelatins accumulation, which provides the tolerance to plants withstand against the stress mainly Cu as well as Cd stress (Maksymiec *et al.*, 2007).

Signalling Interconnection between different Phytohormones

Whenever plants experienced any type of stress, they are required to observe and respond to the different signals, which are extremely coordinated as well as interactive. The plant should have the capability to maintain the growth plasticity and to withstand harsh conditions by maintaining the network of signalling (Harrison *et al.*, 2012). Under the stress condition stress, hormone-responsive fluctuates which helps in changing the dynamics and regulating the plant growth co-ordinately (Kholi *et al.*, 2013). The conjunction of the different transduction hormone signals considered as the crosstalk and in the combined form, they form the network of signalling (Harrison *et al.*, 2013). In the recent studied it was that biosynthesis and essential signalling the major components of phytohormones which include IAA, GA, ABA, JA and BRs (Singh *et al.*, 2013). When all the factors considered together creates more confusion rather than finding any solution (Pinherio *et al.*, 2011). Among all plant hormone JA, ABA and SA are the most important in regulating the signalling network pathway. The interaction between the phytohormones can mitigate the abiotic stress and added novel attention. It is important to understand the crosstalk among different phytohormones for the signalling pathway defending and result in the novel potential for targeting the host resistance development (Grant *et al.*, 2009).

Fig. 4. Interconnection between the phytohormones in different stresses

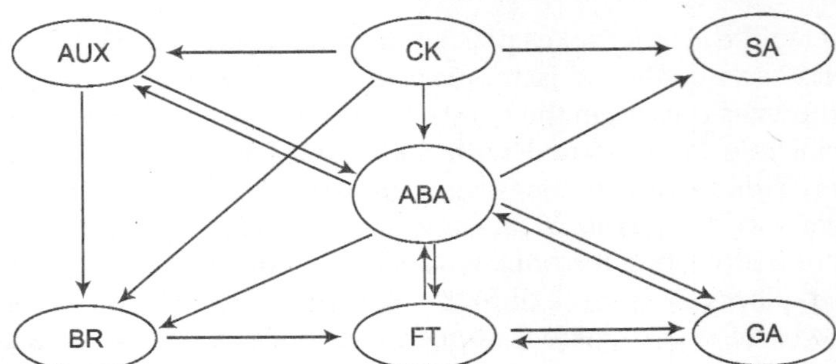

The different colour arrow shows different stresses;-

High Light AUX- auxin,

High Salt ABA- Abscisic acid

Cold Stress ET- ethylene

Heavy Metal GA- gibberellic acid

Excess Water **SA**- salicylic acid, **BR**-brassinosteroids

Phytohormones Role in Plants under Abiotic Stress

There are lot of responses off to the plant hormones, which help the plant to withstand in that harsh environment. Some key function/role are followed as-

Under cold stress development of pollen: stress damage any stage of development, but the utmost critical is the reproductive stage. If the plant experience the stress in the reproductive stage it may damage the whole plant, which results in a huge loss in crop yield. In the reproductive stage, the development of pollen is the most critical stage can be affected by various abiotic stress hence there is a loss in yield (Zinn *et al.*, 2010). Hormones like ABA and GA considered as the most important for understanding the mechanism of molecular development of pollen under the harsh as well as no stress conditions result in the production of hybrid seed through producing the sterile plants (figure 3).

Hormonal balance in cold stress: reported that hormones like JA, SA, and ABA involve in phosphoprotein pathways and activate the genes of expression, which are associated with tolerance against cold stress, and reported concentration of JA is high in three different populations in controlled temperature with reduction of tenfold in sensitive plants (Kolasazov *et al.*, 2013).

Fig. 3. Cold stress and Mechanism for tolerance to cold stress

Cold Stress

Membrane fluidity alterations

Calcium ion produce

Protein kiase/phosphatase cascade

Induction or repression of TFs expression

Cold regulated gene expressed

Development of freezing tolerance

Zn Finger, AREBs/ABFs, ICEs, CBFs/DREBs, MYC/MYB

LEA proteins, Osmolyte synthesis, Detoxifications enzymes

Fig. 4. Engaged transcriptional regulatory network of the genes involved in cold stress response

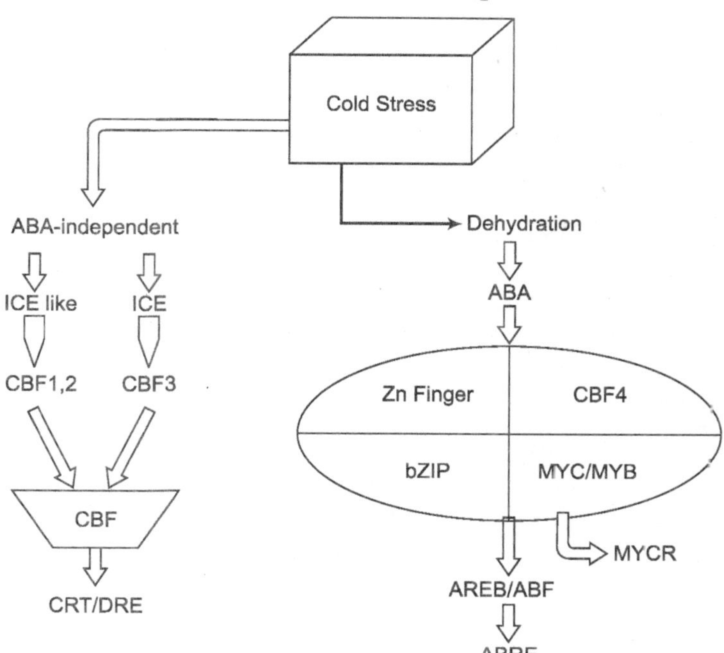

Role of salicylic acid: In bean seedlings, SA is responsible for stimulation of various plant growth, growth-promoting biosynthesis, inhibitions of compounds and adverse effect of cold stress in bean (Gharib and Hegazi, 2010). The experiment was conducted with six bean varieties with temperature 25°C and with 15°C as chilling stress. They found that germination is slowed in chilling stress as compared to optimal temperature but a seed, which is treated with salicylic acid, increased the germination rate as compared to the untreated seed under chilling stress as well as control conditions (Gharib and Hegazi, 2010).

Role of Auxin Carrier Against the Response of Drought Stress and Tolerance of Drought Stress

To mitigate the abiotic stress it also deeds as an essential signal in different reactions (Zhang *et al.*, 2012) and reported that efflux of auxin in rice carrier gene *osPIN3t* that responsible for the transport of polar auxin against the response of drought stress. They also examined that when there is the degradation of a gene named *osPIN3t* results in the losses to seedlings stage i.e. abnormalities in crown root and its expression increased the tolerance against the drought stress.

Conclusion and Prospects

Phytohormones' function in regulating the plants against various abiotic stresses plays an important role, which provides the new prospects in maintaining sustain production of crops, helps in feeding the nation, which is growing at a very fast rate in terms of population. By involving in the different responsive genes for stress has a major function in plants by the tolerance of stress as well as by adaptation to the various stresses. Due to the development of technology in genomic rapidly from past years most of the research mainly focused on understanding the responses of plants against abiotic stress. However, the point is that due to the complexity behind the transduction of stress signal is still not covered and has many challenges. To understand the different responses of plants against the abiotic stress in a better way then work needs to be done at the level of genetic to know about the pathway of hormones biosynthetically like IAA. From recent studies until the date designates that initiation of engineering of phytohormones started. The response of plant hormones to the changing environment plays a crucial role in the regulation of response to plant stress. In this review, we discussed the major role of phytohormones, interconnection among the hormones, and the significance of studying the hormones in combination. Phytohormones having the direct as well as an indirect role in defending themselves against the various abiotic stresses and interactions between the plants and environment. In the direction of the goal, the focus should be on a combination of responses of stress in those harsh field conditions, because in the field environment the different stresses mainly occur concurrently.

References

B.C. Keskin, A.T. Sarikaya, B. Yuksel, A.R. Memon, Abscisic acid regulated gene expression in bread wheat, Aust. J. Crop Sci. 4 (2010) 617–625.

B.V. Vardhini, N.A. Anjum, Brassinosteroids make plant life easier under abiotic stresses mainly by modulating major components of the antioxidant defense system, Front. Environ. Sci. 2 (2015) 1–16.

B.V. Vardhini, S. Anuradha, S.S.R. Rao, Brassinosteroids: a great potential to improve crop productivity, Indian J. Plant Physiol.11 (2006) 1–12.

Bajguz, S. Hayat, Effects of brassinosteroids on the plant responses to environmental stresses, Plant Physiol. Biochem. 47 (2009) 1–8.

Bohra JS, Dorffling H, Dorffling K (1995) Salinity tolerance of rice (Oryza sativa L.) with reference to endogenous and exogenous abscisic acid. J Agron Crop Sci 174:79–86

C.C. Yin, B. Ma, D.P. Collinge, B.J. Pogson, S.J. He, Q. Xiong, J.S. Zhang, Ethylene responses in rice roots and coleoptiles are differentially regulated by a carotenoid isomerase-mediated abscisic acid pathway, Plant Cell 27 (2015) 1061–1081.

Chaves, M. Manuela, J.P. Maroco, J.S. Pereira, Understanding plant responses to drought from genes to the whole plant, Funct. Plant Biol. 30 (2003) 239–264.

D. Egamberdieva, Alleviation of salt stress by plant growth regulators and IAA producing bacteria in wheat, Acta Physiol. Plant. 31 (2009) 861–864.

D. Tilman, C. Balzer, J. Hill, B.L. Belfort, Global food demand and the sustainable intensification of agriculture, Proc. Natl. Acad. Sci. U.S.A. 108 (2011) 20260–20264.

D. Zalabák, H. Pospíšilová, M. Šmehilová, K. Mrízová, I. Frébort, P. Galuszka, Genetic engineering of cytokinin metabolism: prospective way to improve agricultural traits of crop plants, Biotechnol. Adv.31 (2013) 97–117.

De Smet, U. Voss, S. Lau, M. Wilson, N. Shao, R.E. Timme, R. Swarup, I. Kerr, C. Hodgman, R. Bock, M. Bennet, G.Jurgens, T. Beeckman, Unraveling the evolution of auxin signaling, Plant Physiol. 155 (2010) 209–221.

E. Gamalero, B.R. Glick, Ethylene and abiotic stress tolerance in plants, in: P. Ahmed, M.N.V. Prasad (Eds.), Environmental Adaptations and Stress Tolerance of Plants in the Era of Climate Change, Springer, New York, 2012, pp. 395–412

E.H. Colebrook, S.G. Thomas, A.L. Phillips, P. Hedden, The role of gibberellin signaling in plant responses to abiotic stress, J. Exp. Biol. 217 (2014) 67–75.

F. Qin, S. Kazuo, Y.S. Kazuo, Achievements and challenges in understanding plant abiotic stress responses and tolerance, Plant Cell Physiol. 52 (2011) 1569–1582.

F.A. Gharib, A.Z. Hegazi, Salicylic acid ameliorates germination, seedling growth, phytohormone and enzymes activity in bean (*Phaseolus vulgaris* L.) under chhold stress, J. Am. Sci. 6 (2010) 675–683.

G.S. Sanghera, S.H. Wani, W. Hussain, N.B. Singh, Engineering cold stress tolerance in crop plants, Curr. Genom. 14 (2011) 30–43.

Gurmani AR, Bano A, Ullah N, Khan H, J-ahangir M, Flowers TJ (2013) Exogenous abscisic acid (ABA) and silicon (Si) promote salinity tolerance by reducing sodium (Na?) transport and bypass flow in rice (Oryza sativa Indica). Aust J Crop Sci 7:1219–1226

H. Claeys, A. Skirycz, K. Maleux, D. Inze, DELLA signaling mediates stress-induced cell differentiation in *Arabidopsis* leaves through modulation of anaphase-promoting complex/cyclosome activity, Plant Physiol. 159 (2012) 739–747

H. Du, H.B. Liu, L.Z. Xiong, Endogenous auxin and jasmonic acid levels are differentially modulated by abiotic stresses in rice, Front. Plant Sci. 4 (2013) 397. doi: 10.3389/fpls.2013.00397.

H. Wolters, G. Jurgens, Survival of the flexible: hormonal growth control and adaptation in plant development, Nat. Rev.Genet.10 (2009) 305–317.

H.S. Seo, J.T. Song, J.J Cheong, Y.H. Lee, Y.H. Lee, I. Hwang, J.S. Lee, D.C. Yang, Jasmonic acid carboxyl methyltransferase: a key enzyme for jasmonate regulated plant responses, Proc. Natl. Acad. Sci. U.S.A. 98 (2001) 4788–4793.

Hamayun M, Khan SA, Khan AL, Shin JH, Ahmad B, Shin DH, Lee IJ (2010) Exogenous gibberellic acid reprograms soybean to higher growth and salt stress tolerance. J Agric Food Chem 58:7226–7232

Iqbal N, Masood A, Khan NA (2012) Phytohormones in salinity tolerance: ethylene and gibberellins cross talk. In: Khan NA, Nazar R, Iqbal N, Anjum NA (eds) Phytohormones and abiotic stress tolerance in plants. Springer, Berlin, pp 77–98

J. Larkindale, D.J. Hall, M.R. Knight, E. Vierling, Heat stress phenotypes of Arabidopsis mutants implicate multiple signaling pathways in the acquisition of thermo-tolerance, Plant Physiol. 138 (2005) 882–897.

J. Pospíšilová, Interaction of cytokinins and abscisic acid during regulation of stomatal opening in bean leaves, Photosynthetica 41 (2003) 49–56

J. Zhang, W. Jia, J. Yang, A.M. Ismail, Role of ABA in integrating plant responses to drought and salt stresses, Field Crops Res. 97 (2006) 111–119

J.A. O'Brien, E. Benkova, Cytokinin cross talking during biotic and abiotic stress responses. Front. Plant Sci. 4 (2013)451. doi: 10.3389/fpls.2013.00451

J.A. O'Brien, E. Benkova, Cytokinin cross talking during biotic and abiotic stress responses. Front. Plant Sci. 4 (2013)451. doi: 10.3389/fpls.2013.00451

J.L. Nemhauser, F. Hong, J. Chory, Different plant hormones regulate similar processes through largely non-overlapping transcriptional responses, Cell 126 (2006) 467–475.

J.S. Seo, J. Joo, M.J. Kim, Y.K. Kim, B.H. Nahm, S.I. Song, J.J. Cheong, J.S. Lee, J.K. Kim, Y.D. Choi, OsbHLH148, a basic helix-loop-helix protein, interacts with OsJAZ proteins in a jasmonate signaling pathway leading to drought tolerance in rice, Plant J. 65 (2011) 907–921.

J.Y. Yoon, M. Hamayun, S.K. Lee, I.J. Lee, Methyl jasmonate alleviated salinity stress in soybean, J. Crop Sci. Biotechnol. 12 (2009) 63–68.

Jackson M (1997) Hormones from roots as signals for the shoots of stressed plants. Trends Plant Sci 2:22–28

K. Kazan, Auxin and the integration of environmental signals into plant root development, Ann. Bot. 112 (2013) 1655–1665.

K. Kazan, Diverse roles of jasmonates and ethylene in abiotic stress tolerance, Trends Plant Sci. 20 (2015) 219–229

K. Kazan, Diverse roles of jasmonates and ethylene in abiotic stress tolerance, Trends Plant Sci. 20 (2015) 219–229

K.E. Zinn, M. Tunc-Ozdemir, J.F. Harper, Temperature stress and plant sexual reproduction: uncovering the weakest links, J. Exp. Bot. 61 (2010) 1959–1968.

Keskin BC, Sarikaya AT, Yuksel B, Memon AR (2010) Abscisic acid regulated gene expression in bread wheat. Aust J Crop Sci 4:617–625

Khan MIR, Khan NA (2013) Salicylic acid and jasmonates: approaches in abiotic stress tolerance. J Plant Biochem Physiol 1:e113. doi:10.4172/2329-9029.1000e113

Kim T-W, Wang Z-Y (2010) Brassinosteroid signal transduction from receptor kinases to transcription factors. Annu Rev Plant Biol 61:681–704

Klay, J. Pirrello, l. Riahi, A. Bernadac, A. Cherif, M. Bouzayen, S. Bouzid, Ethylene response factor *Sl-ERF.B.3* is responsive to abiotic stresses and mediates salt and cold stress response regulation in tomato, Sci. World J. 2014 (2014) 167681. doi: 10.1155/2014/167681.

Kohli, N. Sreenivasulu, P. Lakshmanan, P.P. Kumar, The phytohormone crosstalk paradigm takes center stage in understanding how plants respond to abiotic stresses, Plant Cell Rep. 32 (2013) 945–957

L. Pauwels, D. Inze, A. Goossens, Jasmonate-inducible gene: what does it mean? Trends Plant Sci. 14 (2009) 87–91. doi: 10.1016/j.tplants.2008.11.005.

M. Kolaksazov, F. Laporte, K. Ananieva, P. Dobrev, M. Herzog, E.D. Ananiev, Effect of chilling and freezing stresses onjasmonate content in *Arabis Alpina,* Bulg. J. Agric. Sci. 19 (2013)15–17.

M.A. Harrison, Cross-talk between phytohormone signaling pathways under both optimal and stressful environmental conditions, in: N.A. Khan, R. Nazar, N. Iqbal, N.A. Anjum (Eds.), Phytohormones and Abiotic Stress Tolerance in Plants, Springer-Verlag, Berlin Heidelberg, 2012, pp. 49–76.

M.A. Matilla-Vazquez, A.J. Matilla, Ethylene: Role in plants under environmental stress, in: P. Ahmad, M.R. Wani (Eds.), Physiological Mechanisms and Adaptation Strategies in Plants Under Changing Environment, Vol. 2, Springer Science + Business Media, New York, 2014, pp. 189–222.

M.G. Javid, A. Sorooshzadeh, F. Moradi, S.A.M.M. Sanavy, I. Allahdadi, The role of phytohormones in alleviating salt stress in crop plants, Aust. J. Crop Sci. 5 (2011) 726–734.

M.R. Grant, J.D. Jones, Hormone (dis)harmony moulds plant health and disease, Science 324 (2009) 750–752

M.R. Pathak, J.A. da Silva Teixeira, S.H. Wani, Polyamines in response to abiotic stress tolerance through transgenic approaches, GM Crops Food 5 (2014) 87–96.

Munns R, Tester M (2008) Mechanisms of salinity tolerance. Ann Rev Plant Physiol 59:651–681

N. Iqbal, S. Umar, N.A. Khan, M.I.R. Khan, A new perspective of phytohormones in salinity tolerance: Regulation of proline metabolism, Environ. Exp. Bot. 100 (2014) 34–42.

N. Sreenivasulu, V. Radchuk, A. Alawady, L. Borisjuk, D. Weier, N. Staroske, J. Fuchs, O. Miersch, M. Strickert, B. Usadel, U. Wobus, B. Grimm, H. Weber, W. Weschke, De-regulation of abscisic acid contents causes abnormal endosperm development in the barley mutant *seg8*, Plant J. 64 (2010) 589–603.

N. Sreenivasulu, V.T. Harshavardhan, G. Govind, C. Seiler, A. Kohli, Contrapuntal role of ABA: does it mediate stress tolerance or plant growth retardation under long-term drought stress? Gene 506 (2012) 265–273.

N. Suzuki, R.M. Rivero, V. Shulaev, E. Blumwald, R. Mittler, Abiotic and biotic stress combinations, New Phytol. 203 (2014) 32–43.

N.Y. Kang, C. Cho, N.Y. Kim, J. Kim, Cytokininreceptor-dependent and receptor-independent pathways in the dehydration response of *Arabidopsis thaliana*, J. Plant Physiol. 169 (2012) 1382–1391.

P.E. Verslues, M. Agarwal, S. Katiyar-Agarwal, J. Zhu, J.K. Zhu, Methods and concepts in quantifying resistance to drought, salt, and freezing, abiotic stresses that affect plant water status, Plant J. 45 (2006) 523–539.

Parida AK, Das AB (2005) Salt tolerance and salinity effects on plants: a review. Ecotoxicol Environ Saf 60:324–349

Pinheiro, M.M. Chaves, Photosynthesis and drought: can we make metabolic connections from available data? J. Exp. Bot.62 (2011) 869–882.

Q. Ke, Z. Wang, C.Y. Ji, J.C. Jeong, H.S. Lee, H. Li, B. Xu, X. Deng, S.S. Kwak, Transgenic poplar expressing Arabidopsis YUCCA6 exhibits auxin-overproduction phenotypes and increased tolerance to abiotic stress, Plant Physiol. Biochem. 94 (2015) 19–27.

R. Nishiyama, Y. Watanabe, Y. Fujita, L.D. Tien, M. Kojima, T. Werner, R. Vankova, K. Yamaguchi-Shinozaki, K. Shinozaki, T. Kakimoto, H. Sakakibara, T. Schmuelling, P.T. Lam-Son, Analysis of cytokinin mutants and regulation of cytokinin metabolic genes reveals important regulatory roles of cytokinins in drought, salt and abscisic acid responses, and abscisic acid biosynthesis, Plant Cell 23 (2011) 2169–2183.

R. Singh, N.S. Jwa, Understanding the responses of rice to environmental stress using proteomics, J. Proteome Res.12 (2013) 4652–4669.

S. Fahad, A. Bano, Effect of salicylic acid on physiological and biochemical characterization of maize grown in the saline area, Pak. J. Bot. 44 (2012) 1433–1438.

S. Fahad, L. Nie, Y. Chen, C. Wu, D. Xiong, S. Saud, L. Hongyan, K. Cui, J. Huang, Crop plant hormones and environmental stress, Sustain. Agric. Rev. 15 (2015) 371–400.

S. Fahad, S. Hussain, A. Bano, S. Saud, S. Hassan, D. Shan, F.A. Khan, F. Khan, Y.T. Chen, C. Wu, M.A. Tabassum, M.X. Chun, M. Afzal, A. Jan, M.T. Jan, J.L. Huang, Potential role of phytohormones and plant growth-promoting rhizobacteria in abiotic stresses: consequences for changing environment, Environ. Sci. Pollut. Res. 22 (2015) 4907–4921.

S. Fahad, S. Hussain, A. Matloob, F.A. Khan, A. Khaliq, S. Saud, S. Hassan, D. Shan, F. Khan, N. Ullah, M. Faiq, M.R. Khan, A.K. Tareen, A. Khan, A. Ullah, N. Ullah, J.L. Huang, Phytohormones and plant responses to salinity stress: a review, Plant Growth Regul. 75 (2015) 391–404

S. Giuliani, M.C. Sanguineti, R. Tuberosa, M. Bellotti, S. Salvi, P. Landi, Root-ABA1 a major constitutive QTL affects maize root architecture and leaf ABA concentration at different water regimes, J. Exp. Bot. 56 (2005) 3061–3070.

S. Wilkinson, G.R. Kudoyarova, D.S. Veselov, T.N. Arkhipova, W.J. Davies, Plant hormone interactions: innovative targets for crop breeding and management, J. Exp. Bot. 63 (2012) 3499–3509.

S. Yamaguchi, Gibberellin metabolism and its regulation, Annu. Rev. Plant Physiol. 59 (2008) 225–251.

S. Zhang, J. Hu, Y. Zhang, X.J. Xie, A. Knapp, Seed priming with brassinolide improves lucerne (*Medicago sativa* L.) seed germination and seedling growth in relation to physiological changes under salinity stress, Aust. J. Agric. Res. 58 (2007) 811–815.

S.C. Groen, N.K. Whiteman, The evolution of ethylene signaling in plant chemical ecology, J. Chem. Ecol. 40 (2014) 700–716.

S.H. Wani, N.B. Singh, A. Haribhushan, J.A. Mir, Compatible solute engineering in plants for abiotic stress tolerance—the role of glycine betaine, Curr. Genomics 14 (2013) 157–165.

S.H. Wani, S.K. Sah, Biotechnology and abiotic stress tolerance in rice, J. Rice Res. 2 (2014) e105.doi: 10.4172/jrr.1000e105.

S.L. Postel, For our thirsty world, efficiency or else, Science 313 (2006) 1046–1047.

Saud S, Li X, Chen Y, Zhang L, Fahad S, Hussain S, Sadiq A, Chen Y(2014) Silicon application increases drought tolerance of kentucky bluegrass by improving plant water relations and morphophysiological functions. Sci World J. doi:10.1155/2014/368694

Skirycz, H. Claeys, S. De Bodt, A. Oikawa, S. Shinoda, M. Andriankaja, K. Maleux, N.B. Eloy, F. Coppens, S.D. Yoo, K. Saito, D. Inze, Pause-and-stop: the effects of osmotic stress on cell proliferation during early leaf development in *Arabidopsis* and a role for ethylene signaling in cell cycle arrest, Plant Cell 23 (2011) 1876–1888.

Tuteja N (2007) Mechanisms of high salinity tolerance in plants. Methods Enzymol 428:419–438

U. Feller, I.I. Vaseva, Extreme climate events: impacts of drought and high temperature on physiological processes in agronomically important plants, Front. Environ. Sci. 39 (2014) 1–17.

U. Vob, A. Bishopp, E. Farcot, M.J. Bennett, Modelling hormonal response and development, Trends Plant Sci. 19 (2014) 311–319.

V. Munteanu, V. Gordeev, R. Martea, M. Duca, Effect of gibberellin cross-talk with other phytohormones on cellular growth and mitosis to endoreduplication transition, Int. J. Adv. Res. Biol. Sci. 1 (6) (2014) 136–153.

V.M. Sponsel, P. Hedden, Gibberellin, biosynthesis, and inactivation, in: P.J. Davies (Ed.), Plant Hormones Biosynthesis, Signal Transduction, Action! Springer, Dordrecht, 2004, pp. 63–94.

Velitcukova M, Fedina I (1998) Response of photosynthesis of Pisum sativum to salt stress as affected by methyl jasmonate. Photosynthetica 35:89–97

W. Maksymiec, M. Wojcik, Z. Krupa, Variation in oxidative stress and photochemical activity in Arabidopsis thaliana leaves subjected to cadmium and excess copper in the presence or absence of jasmonate and ascorbate, Chemosphere 66 (2007) 421–427.

Wang Y, Mopper S, Hasentein KH (2001) Effects of salinity on endogenous ABA, IAA, JA, and SA in Iris hexagonal. J Chem Ecol 27:327–342

X.F. Sheng, J.J. Xia, Improvement of rape (*Brassica napus*) plant growth and cadmium uptake by cadmium-resistant bacteria, Chemosphere 64 (2006) 1036–1042.

X.J. Li, M.F. Yang, H. Chen, L.Q. Qu, F. Chen, S.H. Shen, Abscisic acid pretreatment enhances salt tolerance of rice seedlings: proteomic evidence, BBA-Proteins Proteom. 1804 (2010) 929–940

X.J. Li, M.F. Yang, H. Chen, L.Q. Qu, F. Chen, S.H. Shen, Abscisic acid pretreatment enhances salt tolerance of rice seedlings: proteomic evidence, BBA-Proteins Proteom. 1804 (2010) 929–940.

Y. Mano, K. Nemoto, The pathway of auxin biosynthesis in plants, J. Exp. Bot. 63 (2012)2853–2872.

Y. Shi, S. Tian, L. Hou, X. Huang, X. Zhang, H. Guo, S. Yang, Ethylene signaling negatively regulates freezing tolerance by repressing expression of *CBF* and Type-A *ARR* genes in Arabidopsis, Plant Cell 24 (2012) 2578–2595.

Y. Wang, S. Mopper, K.H. Hasentein, Effects of salinity on endogenous ABA, IAA, JA, and SA in *Iris hexagona,* J. Chem. Ecol. 27 (2001) 327–342.

Z. Peleg, E. Blumwald, Hormone balance and abiotic stress tolerance in crop plants, Curr. Opin. Plant Biol. 14 (2011) 290–295

Z. Yan, J. Chen, X. Li, Methyl jasmonate as a modulator of Cd toxicity in *Capsicum frutescens* var. *fasciculatum* seedlings, Ecotox. Environ. Safe. 98 (2013) 203–209.

Foliar Heavy Metal Uptake,Toxicity and Detoxification in Plants

Priyanka Devi
Department of Agronomy, School of Agriculture
Lovely Professional University, Jalandhar, Punjab, 144411, India
Email: parichib17@gmail.com

Abstract

From the diverse sources of heavy metal like human activities, which changes the cycles of heavy metal worldwide by biogeochemically hence, emitting the different concentration of heavy metal in the atmosphere. With the increase in the industrialization and urbanization nowadays leads to increase the heavy metal pollutants in the atmosphere, which have toxic effects on the human as well as on the plants. Plants adsorbed these atmosphere heavy metals by the foliar organs of the plants after the wet-dry deposition on the canopy of plants. The transfer of metal from the roots has been studied already in much research but the foliar uptake of the metal by the plants in the atmosphere still needs great research. For the best understanding regarding the heavy metal uptake by foliar is significant research. This review indicates the heavy metal behaviour in the atmosphere plant system as biogeochemically. This review abridges of all the mechanisms of the heavy metal uptake as foliar, translocation, toxicity uptake compartmentation and detoxification of plants. In this chapter, I described the environmental factors as well as biological, which effects the heavy metal uptake as foliar as compared to that of the biogeochemical behaviour (translocation, uptake, toxicity and detoxification) uptake of heavy metal in roots and foliar uptake. The effects of heavy metal and health risks on consumption and entering the food chain are also discussed.

Introduction

For the base of socio-economic, mineral resources are the important material to raise the economic progress. Nowadays the extraction of the mineral and its continuous use in the different industries leads to the pollution in the environment mainly results in the heavy metal pollution (Li *et al.*, 2014; Naizi *et al.*, 2016). Due to their toxicity of heavy metal, bioaccumulate tendency and high persistence in the ecosystem in the environment is the big threat for the human health as well as for the ecosystem (Leveque *et al.*, 2014). Some of the heavy metals are considered as the required micronutrients such as manganese, chromium, copper, nickel and zinc but at the higher level, they are toxic in nature (Shahid *et al.*, 2015). There are some non-essential metals, which are toxic even in the small quantities like mercury, arsenic, cadmium and lead (Harguinteguy *et al.*, 2016). As we all know that there is a continuous increase in the industrialization, as well as urbanization along with modern farming, has put the environment under the pressure, which faces the lot many problems. Through the numerous human activities and modern development result in the release of the toxic pollutants in the atmosphere (Su *et al.*, 2013). The heavy metal emissions in the atmosphere are much higher by human activities than that of the natural emissions. Increase in the growing industry mainly the mining industry is one the main reason for the emission of the heavy metal in the atmosphere, which affects the human health as well as the plant ecosystem (CA *et al.*, 2016). By both activities like anthropogenic as well as natural, it emits the metals in the atmosphere (Shahid *et al.*, 2015) which can transport for the long distance. The area near to the urban area and near to the industrial units have the metal released in the air, soil, crop and water reservoirs from the natural as well as by the anthropogenic activities. The dust particles of the heavy metal in the atmosphere have a persistent and toxic effect on the health of human (Norouzi *et al.*, 2016). Within the high-density population is mainly in the urban area, the dust particles of heavy metal are inhaled and ingestion of the polluted crops may lead to the hazards in human health. The accumulation of the heavy metal in the crops and the vegetables are examined only from the root system accumulation this is due to the build-up of metal in the soil system and the plants adsorbed only from the root system. Apart from the roots of plants, the aerial plant's parts like fruits, leaves and flowers can absorb the heavy metal. The aerial plant parts organ absorbing structure for the metal uptake is similar to that of the root structure of the plants. After the deposition of the atmospheric particles on the surface of the leaf, the metal accumulates in the leaves of the plant parts by the transfer of foliar (Xiong *et al.*, 2014). The concentration of heavy metal is increased in those plants, which are near the mining area, as well as in the smelting area. The study was done near those industry areas to know about the heavy metal contamination and deposition of these heavy metal receive larger attention. The less research was done on the heavy metal uptake by the plants leaves from the atmosphere (Pourrut *et al.*, 2011). Those metal which is essential for the metabolic activities of plants their absorption pathway was known and evaluated for the biochemical reaction of plants. The various example in which studies was done to know the absorption by foliar like iron, zinc and copper (Fernendez *et al.*, 2013). This metal absorbs by the plants through the cuticles penetration and finally accumulates in the tissues of the plant leaves. The leaves of the plants by the foliar transfer can absorb the metal, which is non-essential like cadmium, chromium, lead, arsenic and mercury. This review comprises of the different mechanism involved in heavy metal uptake by plants, factors that affect the heavy metal by foliar, transportation and compartmentation of the foliar heavy metal, toxic and detoxification of heavy metal inside the plant after the foliar uptake and the root and foliar uptake comparison.

Heavy Metal Pollution in the Atmosphere: Need Great Attention

The global concerns regarding the increasing heavy metal pollution in the atmosphere as the atmosphere contains only oxygen, carbon dioxide and nitrogen. From the last two-three decades, the increasing industrialization, as well as urbanization, leads to deteriorating the atmosphere by the emissions of the different pollutants. There are the common organic and inorganic pollutants named as ozone, sulphur dioxide, carbon dioxide, hydrogen fluoride and nitrogen oxides (Cruz *et al.*, 2015). From the different anthropogenic activities of the human various heavy metal are emitted into the atmosphere. Heavy metal also emitted in large quantities into the atmosphere by the metal processing of ores heating smelters, which acts as the air contaminated. The particles, which are suspended in the air, termed as the particulate matter, which is liquid and solid whereas the aerosol also contributes the threat for the environmental health and there are some volatile compounds, which are released from the chimneys of different industry (Csavina *et al.*, 2011). From the different industry, the different metal is released such as chromium, lead, arsenic, nickel, zinc, cadmium and vanadium termed as carcinogenic and high in the toxic potential exposure to the living organisms. Due to their ecotoxicity, bioaccessibility of humans, the potential of oxidative and cytotoxicity metals emitted to the atmosphere and their hardness in the atmosphere. The heavy metal in the atmosphere is emitted by the incineration of waste, burning of domestic oil, power generation plants, industrial units, and traffic of vehicle and different remediation of contaminated sites (Manno *et al.*, 2006). Among all the sources for the heavy metal pollution in the atmosphere, the most important source is industrial and the traffic activities lead to atmospheric pollution. The heavy metal emission in the atmosphere by the various process of industry like fusion, crushing, reduction, refining and processing (Zuhang *et al.*, 2009). The mobility of heavy metal is high in the atmosphere then the pollutants can travel for several kilometres along with the wind design (Table 1).

Heavy Metal Atmospheric Pollution and Particulate Matter

Heavy metal may be attached to the various atmospheric particles after the pollution of the atmospheric heavy metal and the emissions of the particulate matter (Equani *et al.*, 2016). It was evaluated that heavy metal majority co-exist with atmospheric particles due to the metal integration and adsorption on to the surface of ferri-magnetics found in the atmosphere. The particulate matter of the heavy metal, which have the various complex mixtures of particles in the atmosphere, which are harmful. The reactivity of particulate is higher than coarse emissions concerning toxicity and bioavailability so this is the major concern for the environmental. The various study has been done to know about the composition of metal chemically, spatial and temporal particulate variations and different level of human exposure (Belis *et al.*, 2013). From this different studies, it was indicated that the colloids of organic as well as inorganic of different size present in the atmosphere of environments compartments like soil, streams, lakes, oceans and rivers have the vital role in the pollutants biogeochemical cycles(Chen *et al.*, 2016). The sorption capacity of nanoparticles is high for the arsenic, zinc and lead in the environment. The nanoparticles of size 12 nm magnetite were more efficient almost 200 folds than that of 20 and 300 nm particles in removing of arsenic from the water (Mayo *et al.*, 2007). Due to the small size of the particulate matter, it is considered as one of the serious health hazards and easily inhaled by the lungs and can enter into the blood circulation. In various studies, it was reported that exposures of the atmospheric heavy metal in term of both short term and long term has the severe effect on the human as well as on the local habitat, which leads to the morbidity and mortality (Dockery *et al.*, 2009). In 1952 it was that in London the 4000 people died due to the illness in the respiratory system like bronchitis and as pneumonia and along with that 8000 people were dead in next few months due to the effects of smog in the atmosphere. Then the samples were collected from the infected people and it was found that the lungs were contaminated with the fine particles of higher level including the heavy metals like lead, zinc and iron (Table 2).

Foliar Uptake of Heavy Metal and Deposition

By the wet and dry deposition of the heavy metal that is emitted by the smelters of industrial are transported to several kilometres away from their sources (Douay *et al.*, 2009). To the targeted organism, the presence of heavy metals can create many risks in the health of human. Health risk guidelines for the heavy metals in the air are established by the world health organization (WHO, 2009). After the deposition of the atmospheric particles in the leaf, surfaces by the foliar transfer heavy metal can be accumulated in the leaves of the plants (Sehreck *et al.*, 2012). The metal uptake by the roots of plants has been studied many times whereas the uptake of metal by the leaves of the plants from the atmosphere is behind in research (Toma *et al.*, 2005). In context with pollution, the foliar transfer of metal is neglected which is the major pathway of the pollution mainly when the fine ultra-particles interact with leaves of the plants (De *et al.*, 2012). The arrangement of the leaves or canopy of plants acts as the filtration of heavy metal from the atmosphere efficiently. On the foliar parts, particulate matter is captured and adsorb or reduce the particulate matter in the atmosphere by the plant canopy (Prusty *et al.*, 2005). From the various study, it was found that pine forest could retain the airborne particles up to 36.4 tonnes hectare per year (Feng *et al.*, 1992). The shelterbelt of the tress can retain the dust particles 38.9 % and it was reported that plants, as well as growing vegetables, has a high level of the foliar heavy metal. The concentration of the heavy metal in the foliar plant organ easily defined the risk in the environment, which represent the pollution of environment load. The concentration of the heavy metal in the crop plant tissue was more than that of their permissible limits (Dashmani *et al.*, 2000). The vegetables, which are grown in the urban areas, are enhanced to accumulate the lead and cadmium concentration of the heavy metal (Mombo *et al.*, 2015). The uptake of the heavy metal by the foliar surfaces by the cuticular cracks, stomata, lenticels and through the ectodermal which is the main channels for the subsidiary cells and guard cells or epidermal cell wall. Above the guard cell, the cuticle is present which are more permeable compared to the epidermal cell. From the various research, it was found that most of the particulate matter is retained by the waxes of cuticular and trichomes on the surfaces of leaves but some particulate matter enters inside the leaf tissue of plants. Metal uptake by the foliar is considered as the surface phenomenon where the adaxial cuticular is the key role in high absorption of metal. From the different studies, it was reported that copper and nickel transfer rich particles might enter the leaves of the plants through the stomata (Kozlov *et al.*, 2000). Due to the consumption of the polluted plants with heavy metal result in risk in the human health which needs great attention to be solved and there are rare data regarding the health risk in the kitchen garden near to the contamination area with the heavy metal(Mombo *et al.*, 2015).

Table 3. Guidelines of health risk values in ambient air for various heavy metals (Source WHO)

Metal	Time-Weighted average	Average time
Cd	5.0ng/m^3	One year
Pb	0.4μg/m^3	One year
Mn	0.16μg/m^3	One year
Hg	1μg/m^3	One year
Ni	0.4μg/m^3	many years
Ar	5.0μg/m^3	-

Bioindicator of Atmosphere Heavy Metal Pollution; Lichen and Mosses

Due to increase in the heavy metal in the atmosphere, the best method to reduce the pollution is biomonitor means the use of the lichens for the level of the atmospheric pollutants, which also includes the heavy metal pollutants (Garty *et al.*, 2001). To maintain the air quality the diversity of the epiphytic lichens are most suitable and acts as the strong indicators. The mostly macrolichens are used for the air quality because it has the foliose and fruticose lifeform. Biomonitoring of the lichens method used which includes the abundance species, diversity and health, isotopic and the composition chemically. Transplant lichens may be used in that area which is not having the natural lichens. It improves the absorption of the pollutants from the atmosphere due to its physiology. There are the two mechanisms under which the lichen goes to accumulate the heavy metals are cation binding by the extracellular and the thallus surface adsorption of particulate matter and intercellular space penetration(Garty *et al.*, 2001). Around the industrial sites and near about the urban areas the mosses have been used for the atmosphere metal pollution (Szczpaniak *et al.*, 2003). In the European countries, the use of the mosses as the bioindicators are used for the purification of the deposited pollution in the atmosphere. Terrestrial mosses are more effective as compared to the sorbents in the high adsorption of the pollutants from the atmosphere all over the world. The most commonly used method is moss-bag methods for the biomonitoring of the atmosphere (Giordano *et al.*, 2013). From the various study, it was found that the extraction of the mineral elements from the soil mosses is used. For the higher accumulation of the heavy metal from the atmosphere, the mooses are the best method for the accumulating of the higher concentrations of heavy metal from the atmosphere.

Fig. 1. Foliar heavy metal uptake by mosses and lichens

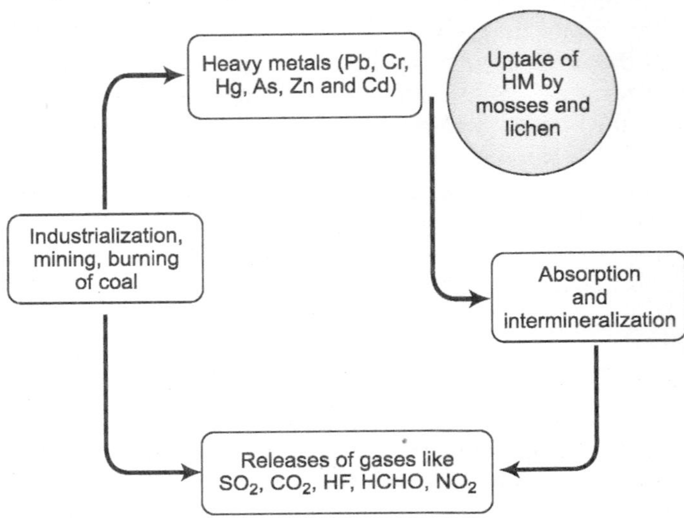

Table 4. Heavy metal concentration in bioindicators plants (mosses and lichen) of air quality

Species of plant	Plant	Metal	Concentration	References
Lichen	*Lecanora muralis*	Lead	4428	Rola *et al.*, 2016
	Parmelia carperata	Zinc	59.84	Ite *et al.*, 2016
	Phaeophyscia hispidula	Arsenic	8	Bajpai *et al.*, 2011
	Hypogymnia physodes.	Coer	68.2	Klimek *et al.*, 2015
	Pyxine cocoes	Cadmium	34.6	Bajpai *et al.*, 2011
	Hypogymnia physodes	Lead	16.8	Kos *et al.*, 2011
	Hypogymnia physodes	Nickel	1247	Kos *et al.*, 2011
	Phaeophyscia hispidula	Zinc	2	Bajpai *et al.*, 2011
Mosses	*Fabriona ciliaris*	Zinc	15.7	Macedo-Miranda *et al.*, 2016
	Hylocomium splendens	Nickel	136.7	Lequy *et al.*, 2016
	Racopilum africanum	Nickel	6.5	Ite *et al.*, 2016
	Sphagnum papillosum	Lead	23.77	Lequy *et al.*, 2016
	Sphagnum papillosum	Cadmium	0.45	Salo *et al.*, 2012
	Thuidium tamariscinum	Cadmium	4.30	Lequy *et al.*, 2016
	Pleurozium schreberi	Zinc	457	Kos *et al.*, 2011
	Hylocomium splendens	Lead	171	Ford *et al.*, 2001

Mechanism of Foliar Uptake of Heavy Metals

Albeit partially assumed, Knocking down is important in the mechanism of metal translocation all over the plant, which takes place since in form of foliar the heavy metals have been absorbed by leaves of the plant. The different writer gives a various mechanism for the entrance of nutrients/metals in the plants via transfer as foliar. Generally, the content of metal in tissue was increased by foliar absorption, but that was difficult to make a distinction in between the sages of transferring and foliar absorption of metals in the plants (Schreck et al.). Saumel (2012) explained the two important steps involved in the uptake of metal by foliar as (a) internalization and adsorption by cuticle and (b) metal penetration via pores of stomata. Biological, physical and chemical are three mechanisms that explained about the decomposition of metals in the leaves of plants (Kinnersley et al., 2001). PM through the foliage containing heavy metals was mechanically captured that related to physical uptake of heavy metals. Note down the measured morphological parameters in plant vegetation. Metals initial decomposition/retention by cuticle of the plant was noted down and measuring the biological and chemicals factors for the uptake of metals. In addition, the heavy metals (HM) chemistry and cuticle permeability were proven uptake of HM on the leaves of plants. Birbaum et al. explained that small particle was penetrated in the inner part of leaves, as particles having large size attached on the wax of the surface. Generally, cuticular was trapped the metals present in particles of dust and after that outspreading in plant leaves. In cuticle, the compounds of soluble are directly interacted with it, which are having a waxy surface layer and provided the hydrophobic characteristic on the plant surface. The degree of plant surface hydrophobicity and polarity vary according to species of plants. The cuticle of the plant contained operative moieties like as carboxylic moieties and that is why it behaves as polyelectrolyte (Fernandez *et al.*, 2013). The arrangement of the biopolymer matrix differs by genotypes of plants, organs, types of species, development stages and growing conditions (Kerstiens *et al.*, 2010). Furthermore, treatment of the surface with silicon and addition of low weight compounds of organic can improves the reactivity and characteristic of the phyllosphere. According to the study of Chamel *et al.*, (1991) mainly 4 steps are involved for penetration of heavy metals in the cuticle of leaf: Cuticle adherence, Penetration by cuticle, Adsorption through cells of subjacent, Apoplast desorption.

Fig. 2. Entrance of heavy metal to the plants by foliar pathways

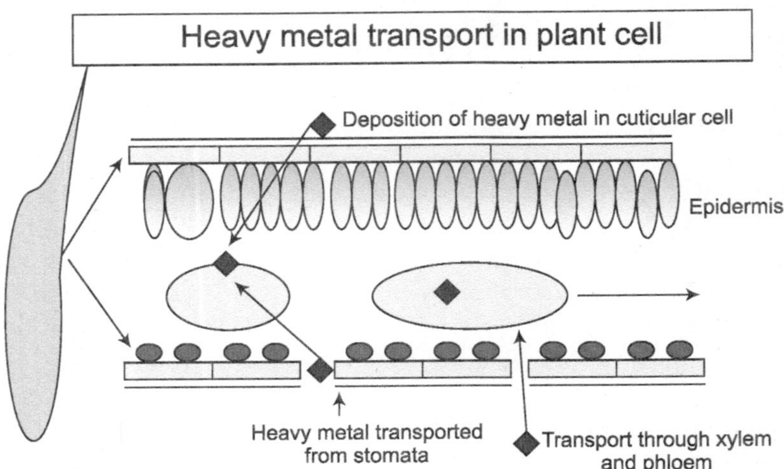

Different Factors which affect the Heavy Metal Absorption

There are the different factors upon which the absorption the heavy metal vary like the physicochemical character of the cuticle and metals, leaf surface area and the morphology of the plant leaves, chemical and physical form of the adsorbed metal, leaf texture, habitat of the plant, duration of exposure, conditions of environment and exchange of the gas (Beckett *et al.*, 2000). The level of the heavy metal present in the atmosphere decided the adsorption and deposition of heavy metal on the leaf of the plant. From the recent study, it was found that the speciation of the heavy metal represents the behaviour of the biogeochemical of metals in the ecosystem that are uptake, translocation, toxicity, compartment and detoxification inside the plants(Shahid *et al.*, 2014). As the variation of the metal occurs in the atmosphere which also results in the variation in the capacity in entering inside the plants by the foliar and root uptake. When the vegetables are grown in the urban area, it is very important to understand the speciation role of the metal because the ingestion of these foods leads to human health risk. Most of the study has been done to regarding with foliar metal uptake with the total quantity of the metal present in the air but the few of them focused on the quality of the air concerning the speciation of the metal. Even under the similar climatic conditions, the different plant species having a variation in the retention and adsorption capabilities. The foliar uptake of the heavy metal varies with the change in the plant anatomical features of the plant like a canopy of plants, leaf angle, no. of branches, area of the leaf, lamina of leaf and structure of the leaves (Chai *et al.*, 2002). Among all the characters of the plant's size and stomata density and the surface area of the leaf are the most important factors, which result in the adsorption the heavy metal. It was also reported that the morphological factors play a vital role in the adsorption of heavy metal like an index of stomata, the density of the trichome and length affect the efficiency of the collection of the dust particles from the atmosphere by the plants (Rao *et al.*, 1992). It was also found that the deposition amount of the particles on the surface of leaves dependent upon the type of the plant species because it has the epidermis characteristics, are different (Liu *et al.*, 2011). It was also found that lead concentration was increased with the four times more in the loiumperenne as compared to that of Lactuca sativa under the transfer of foliar metal. Due to the high density of the stomata in the Acer palmatum having the more capacity to adsorb the foliar heavy metal than that of the other plant species. It was also reported that the index of the stomata is a more important factor for the transfer of the atmospheric plant heavy metal (Ferretti *et al.*, 2005).

Meteorological Conditions for Heavy Metal Uptake

As we, all know that the variation in the climate has the direct effect in the physiochemical properties of the plant and the surface of the leaf, which result in the alterations in the potential of foliar heavy metal uptake. The conditions of the climate also influence the metabolic and the biological processes inside the plants, which lead to effect the foliar uptake as well as the metal compartmentation. The climatic conditions like humidity, temperature and the period of light for the foliar deposition alter the metabolic activities and the process of penetration across the surface of the leaf and the inside movement of leaf spaces. The relative humidity is also one of the most important factors, which also alter the foliar uptake of the heavy metal. The permeability potential of the surface of the leaf is altered by the relative humidity. In terms of the solubility and redox of the particulate matter by the alteration in the physiochemical response of plants through the change in the relative humidity. When the relative humidity is high, the potential of permeability of plant leaf surface is improved due to the hydration of the cuticle. During the high relative humidity, the particulate matter deposited and the heavy metal remains wet, which increases the penetration of metal inside the leaves of plant this happens due to the high humidity in the phyllosphere heavy metal remains as the solutes, which increase the leaf penetration of plants. Rainfall is another most important factor which manipulating the deposition of the metal on the plants as well on the foliar uptake of the metal. From the study, it was found that the accumulation, as well as the deposition of the applied lead as foliar in the Lactuca sativa, was correlated with the measurements of rainfall (Shereck *et al.*, 2012). During the high rainfall in that particular period, it was found that lead accumulation in Lactuca sativa was higher in the first and last weeks of foliar application of the lead. It was also reported that the lead accumulation, as well as the deposition in rainfall, was equal to or greater than that of dried deposition which results in the increase in the penetration and deposition in rainy period (Lawman *et al.*, 2001). As we, all know that during the rainy season the opening of stomata and the cuticle expansion was more, which favour the higher level of metal penetration from the atmosphere.

Fig. 3. Emission of the heavy metal from the different anthropogenic activities

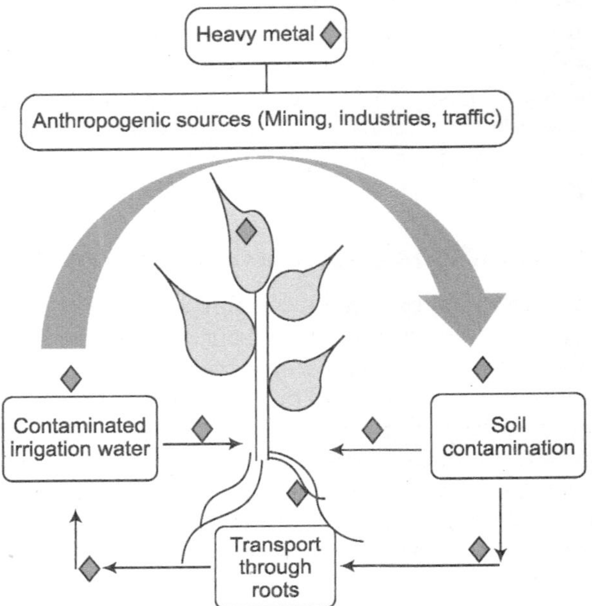

Uptake of the Heavy Metal by Foliar and Root Comparison

The uptake of metal and accumulation of metal can be by both roots as well as by the foliar uptake. The major pathway for the heavy metal uptake is through the soil root translation. From the various study, it was found that there are two ways for the metal uptake by the plants. The different mechanism of the metal uptake, compartmentation, toxicity, translocation and detoxification differ with the pathway of the metal applied (Zandi *et al.,* 2012). Plants near the urban area and industrial area are very difficult to differentiate the concentration of the metal in the tissues of the plants that whether it comes from the uptake of the metal by roots from the soil or either by the absorption of the atmospheric metal through the surface of leaves. There are very fewer data from which it can indicate that the foliar application of the from the uptake of roots. The experiment was conducted to investigate the difference between the foliar and root metal transfer (Schrek *et al.,* 2014). For the transfer of the metal by the roots the experiment was conducted in the earthen pots in which lettuce plants are grown under the lead-contaminated soil and for the uptake of the foliar, the earthen pots are kept near the smelter in which lettuce is grown. To avoid the atmospheric fallouts, contamination of soil and transfer of metal by the roots a geotextile membrane was placed on the top of the soil.

Entrance Comparison of Metal to Plants by Foliar and Root Pathway

In the soil, the metal first uptake the metal from the soil solution and adsorbed on the surfaces of roots after that it binds with polysaccharides in the rhizosphere surfaces of the cell onto the root surfaces. Heavy metal penetrates the roots passively after the absorption of heavy metal and finally enter through the streams of water translocation. Due to the presence of the wax cover on the leaf and hairs on the surface of leaves, the particulate adsorbed in case of the foliar transfer of the heavy metal from the atmosphere (Eichert *et al.*, 2008). There are mainly two pathways for the pollutants to enter one is the cuticle diffusion for the lipophilic elements and through the stomata aqueous pores and hydrophilic compounds (Loure *et al.*, 2014).

Uptake of the Heavy Metal by Foliar and Root Pathway Comparison

Under the root and foliar metal uptake, the different factors, which affect the metal uptake and the accumulation, index by the plants. The uptake of the metal by plants alter with the soil and plant type metal speciation in soil chemically, particle size of soil, ph of soil, cation exchange capacity content of organic matter and also the activity of the microbial in the soil. The foliar uptake of the metal depends upon the physical, biological and chemical factors (Kinnersley *et al.*, 2000). Physical factors, which include the chemistry and type of the meatal also the plant type where the chemical factors include the cuticle composition and metal speciation. The biological stages, which include the stages of the growth in which the metal deposited on the surfaces of the plants.

Fig. 4. Foliar and root metal uptake comparison

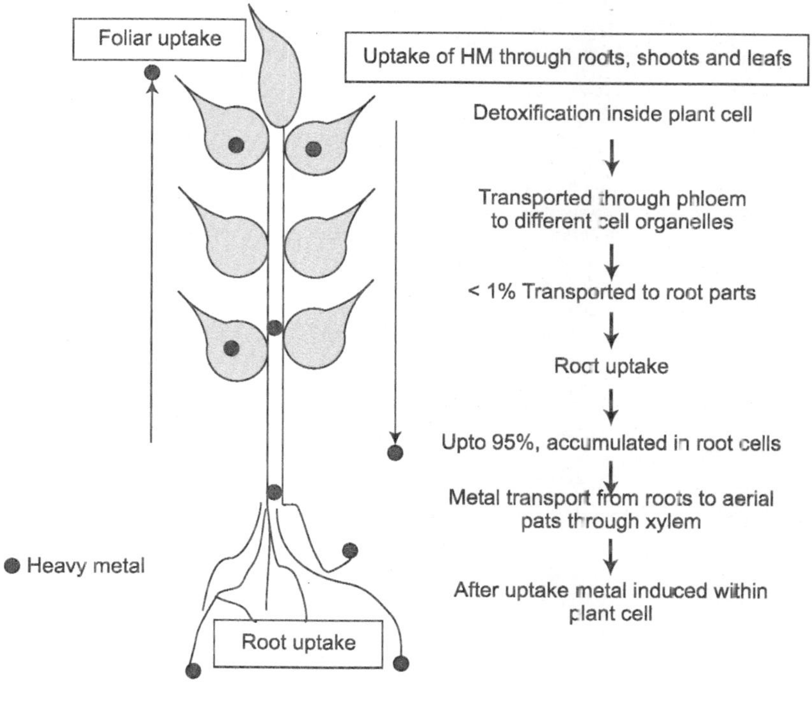

Heavy Metal Foliar Uptake Health Risk Assessment

To understand the risk in human health it is important to estimate the exposure level of heavy metal and the tracing routes to target the contamination of heavy metal. Due to the ingestion, inhalation and consuming the polluted food with this atmospheric contamination have the risk related to human health. From the above, it is stated that due to the accumulation, translocation, compartmentation, toxicity and detoxification alter with the pathways, which in turn affected the metal accumulation in the edible parts of plants and hence enters in the food chain resulting in the health risk. By keeping these things in mind, it is important to evaluate the assessment of health risk due to the foliar-deposited heavy metal. Heavy metal accumulation in the edible plant parts after the deposition by the foliar transfer used to estimate the health risk potential, daily intake, hazard index, tolerance daily intake and the carcinogenic risk (Shakoor *et al.*, 2015). From recent research, it was found that there are various vegetables, which accumulate the different levels of the metals in their edible plant parts (Khalid *et al.*, 2015). The human, not just one plant species, takes up the mixtures of vegetables and the crops. Under the high metal foliar deposition and transfer, the cultivation of vegetables in those urban areas is the best technique to manage that urban agriculture. The accumulation of the heavy metal by the root as well as by the foliar transfer differ with the different plant species.

Table 5: Foliar uptake of heavy metals by vegetables and associated risks to health near industrial areas

Metal	Reference	Vegetables	Concentration in atmospheric fallouts (mg/cm)	Concentration in plant shoot (mg/kg)	Concentration in grains	GEF	DIM (mg kg/ day)	HRI
Cadmium	Xiong *et al.*, 2014	Spinacia oleracea	-	317.2	-	396.5	0.126	25.492
Cadmium	Liu *et al.*, 2009	Oryza sativa	-	30	1.9	0.1	0.011	2.417
Cadmium	Uzu *et al.*, 2010	Lactuca sativa	0.8	1.6	-	1.8	0.001	0.136
Lead	Schreck *et al.*, 2012	Lactuca sativa	-	334.9	-	300	0.134	26.914
Lead	Schreck *et al.*, 2015	Lactuca sativa	-	171.4	-	248.6	0.068	13.778
Lead	Uzu *et al.*, 2010	Lactuca sativa	456.1	121.9	-	0.2	0.048	9.801

Lead	Xiong *et al.*, 2016	Spinacia oleracea	-	484.9	-	79.4	0.194	38.965
Zinc	Uzu *et al.*, 2010	Lactuca sativa	6.8	29	-	4.1	0.011	2.337
Zinc	Wang *et al.*, 2012	Triticum aestivum	-	31.67	43.60	1.3	0.012	2.544
Zinc	Xiong *et al.*, 2016	Spinacia oleracea	-	144.1	-	5.6	0.057	11.584
Zinc	Li *et al.*, 2015	Triticum aestivum	-	86.7	43.3	0.4	0.034	6.973
Stibium	Uzu *et al.*, 2010	Lactuca sativa	1.8	1.3	-	0.6	0.001	0.111
Stibium	Xiong *et al.*, 2016	Spinacia oleracea	-	276.2	-	50.1	0.110	22.1198
Nickle	Kozlov *et al.*, 2000	Betula pendula	58.1	4.7	-	0.1	0.001	0.385
Copper	Uzu *et al.*, 2010	Rye grass	1.7	6.9	-	4.0	0.002	0.561
Arsenic	Uzu *et al.*, 2010	Lactuca sativa	0.2	1	-	5.4	0	0.037

Contaminated Plant Residue Management

When the concentration of metals in the plant biomass is increased than that of the threshold level then it is considered as the hazardous material. Form the recent study it was reported that when the concentration of heavy metal is high it may be disposed of off to minimize the risk associated with the human health (Sas *et al.*, 2004). The disposed of the contaminated biomass plants, treatment and storage are acts as the secondary pollution, which needs to be solved. From the few studies, it was found that the heavy metal-containing biomass can be converted to the energy which termed as the heavy metal recovery by the radical approach (Schuetzle *et al.*, 2015) and this is considered as one of the most important approaches in terms of economical as well as environmental safety to utilize the plant biomass contamination and converted to the thermal and electric energy. On the pyrolysis of the plant, biomass, which is contaminated by heavy metal, involves in the thermochemical process and likewise, the bio-ores can be produced by the ashing of the contaminated biomass of plants.

Conclusion

From the research which were done in the recent about the heavy metalfoilar absorption by the plants still there exist some areas which need to resolve including the foliar absorption mechanism by the plants. Through the stomata of the plants and the cuticle, the penetration of the heavy metal occurs to result in the deposition of the metal in the leaves of the plants. Plants adsorbed these atmosphere heavy metals by the foliar organs of the plants after the wet-dry deposition on the canopy of plants. The transfer of metal from the roots has been studied already in much research but the foliar uptake of the metal by the plants in the atmosphere still needs great research. The heavy metal emissions in the atmosphere are much higher by human activities than that of the natural emissions. Increase in the growing industry mainly the mining industry is one the main reason for the emission of the heavy metal in the atmosphere, which affects human health as well as the plant ecosystem. By both activities like anthropogenic as well as natural, it emits the metals in the atmosphere, which can transport for the long distance. The area near to the urban area and near to the industrial units have the metal released in the air, soil, crop and water reservoirs from the natural as well as by the anthropogenic activities. The dust particles of the heavy metal in the atmosphere have a persistent and toxic effect on the health of human. From the recent study, it was found that the speciation of the heavy metal represents the behaviour of the biogeochemical of metals in the ecosystem that are uptake, translocation, toxicity, compartment and detoxification inside the plants. As the variation of the metal occurs in the atmosphere which also results in the variation in the capacity in entering inside the plants by the foliar and root uptake. There are the common organic and inorganic pollutants named as ozone, sulphur dioxide, carbon dioxide, hydrogen fluoride and nitrogen oxides. From the different anthropogenic activities of the human various heavy metal are emitted into the atmosphere. Heavy metal also emitted in large quantities into the atmosphere by the metal processing of ores heating smelters, which acts as the air contaminated. The particles, which are suspended in the air, termed as the particulate matter, which is liquid and solid whereas the aerosol also contributes the threat for the environmental health and there are some volatile compounds, which are released from the chimneys of different industry. Due to the ingestion, inhalation and consuming the polluted food with this atmospheric contamination have the risk related to human health. From the above, it is stated that due to the accumulation, translocation, compartmentation, toxicity and detoxification alter with the pathways, which in turn affected the metal accumulation in the edible parts of plants and hence enters in the food chain resulting in the health risk.

Prospects and research gap needs to be explored

- The behaviour of heavy metal inside the plant by the foliar uptake is yet not well known so proper study on the foliar uptake of heavy metal in plants need more focus.

- A detailed study on the "Metal speciation" its influence with foliar uptake and detoxification of the heavy metal inside the plant needs to be more exposed.
- Deep studies on the "entry of heavy metal" through the leaf tissue and its further translocation inside the plants to different parts are required to be explored at the molecular level.
- Effect of the heavy metals on the human health is well known but there is need of deep study on the mechanism of heavy metal inside the human body so that it can minimize the effects of the heavy metal.
- Among all the micronutrients except the boron, the plants in the trace level require all. Meanwhile, the proper mechanism of the metals tends to bought toxicity and disturb physiology of plants, which required the great need to be exposed.
- Foliar uptake of the heavy metal leads to produce the more reactive oxygen species in plant apart from this there are major effects on the metabolism of the plants. There are not sufficient approaches has been observed to minimize ROS production

Reference

A. Sas-Nowosielska, R. Kucharski, E. Małkowski, M. Pogrzeba, J. Kuperberg, K. Kry′nski, Phytoextraction crop disposal—an unsolved problem, Environ. Pollut. 128 (2004) 373–379

A.M.J. Cruz, S. Sarmento, S.M. Almeida, A.V. Silva, C. Alves, M.C. Freitas, H.Wolterbeek, Association between atmospheric pollutants and hospital admissions in Lisbon, Environ. Sci. Pollut. Res. 22 (2015) 5500–5510.

B. Pourrut, M. Shahid, C. Dumat, P. Winterton, E. Pinelli, Lead uptake, toxicity, and detoxification in plants, Rev. Environ. Contam. Toxicol. 213(2011) 113–136.

B. Prusty, P. Mishra, P. Azeez, Dust accumulation and leaf pigment content in vegetation near the national highway at Sambalpur, Orissa, India, Ecotoxicol. Environ. Saf. 60 (2005) 228–235.

C. Belis, F. Karagulian, B. Larsen, P. Hopke, Critical review and meta-analysis of ambient particulate matter source apportionment using receptor models in Europe, Atmos. Environ. 69 (2013) 94–108.

C. Feng, The Research of Greening Environment, Environmental SciencePress of China, Beijing, 1992.

C. Larue, H. Castillo-Michel, S. Sobanska, L. Cécillon, S. Bureau, V. Barthès, L.Ouerdane, M. Carrière, G. Sarret, Foliar exposure of the crop Lactuca sativato silver nanoparticles: evidence for internalization and changes in Agspeciation, J. Hazard. Mater. 264 (2014) 98–106.

C. Luo, C. Liu, Y. Wang, X. Liu, F. Li, G. Zhang, X. Li, Heavy metal contamination in soils and vegetables near an e-waste processing site, south china, J. Hazard. Mater. 186 (2011) 481–490.

C.A. Harguinteguy, M.N. Cofré, A. Fernández-Cirelli, M.L. Pignata, Themacrophytes Potamogeton pusillus L. and Myriophyllum aquaticum (Vell.)Verdc. as potential bioindicators

D. Schuetzle, R. Schuetzle, S.K. Hoekman, B. Zielinska, The effect of oxygen on formation of syngas contaminants during the thermochemical conversion of biomass, Int. J. Energy Environ. Eng. 6 (2015) 405–417.a

D.W. Dockery, Health effects of particulate air pollution, Ann. Epidemiol. 19(2009) 257–263.

E. Manno, D. Varrica, G. Dongarra, Metal distribution in road dust samplescollected in an urban area close to a petrochemical plant at Gela, Sicily, Atmos. Environ. 40 (2006) 5929–5941.

E. Schreck, V. Dapper, G. Sarret, S. Sobanska, D. Nowak, J. Nowak, E.A.Stefaniak, V. Magnin, V. Ranieri, C. Dumat, Foliar or root exposures to smelter particles: consequences for lead compartmentalization and speciation in plant leaves, Sci. Total Environ. 476–477 (2014) 667–676.

E. Schreck, Y. Foucault, G. Sarret, S. Sobanska, L. Cécillon, M. Castrec-Rouelle, G. Uzu, C. Dumat, Metal and metalloid foliar uptake by various plant species exposed to atmospheric industrial fallout: Mechanisms involved for lead, Sci. Total Environ. 427–428 (2012) 253–262

E. Schreck, Y. Foucault, G. Sarret, S. Sobanska, L. Cécillon, M. Castrec-Rouelle, G. Uzu, C. Dumat, Metal and metalloid foliar uptake by various plant species exposed to atmospheric industrial fallout: Mechanisms involved for lead, Sci. Total Environ. 427–428 (2012) 253–262.

F. Douay, C. Pruvot, C. Waterlot, C. Fritsch, H. Fourrier, A. Loriette, G. Bidar, C.Grand, A. De Vaufleury, R. Scheifler, Contamination of woody habitat soilsaround a former lead smelter in the North of France, Sci. Total Environ. 407(2009) 5564–5577.

H. Dahmani-Muller, F. Van Oort, B. Gelie, M. Balabane, Strategies of heavy metal uptake by three plant species growing near a metal smelter, Environ.Pollut. 109 (2000) 231–238.

J. Csavina, A. Landázuri, A. Wonaschütz, K. Rine, P. Rheinheimer, B. Barbaris,W. Conant, A.E. Sáez, E.A. Betterton, Metal and metalloid contaminants inatmospheric aerosols from mining operations, Water Air Soil Pollut. 221(2011) 145–157..

J. Garty, Biomonitoring atmospheric heavy metals with lichens: theory and application, Crit. Rev. Plant Sci. 20 (2001) 309–371.

J. Mayo, C. Yavuz, S. Yean, L. Cong, H. Shipley, W. Yu, J. Falkner, A. Kan, M.Tomson, V. Colvin, The effect of nanocrystalline magnetite size on arsenic removal, Sci. Technol. Adv. Mater. 8 (2007) 71–75.

K. Szczepaniak, M. Biziuk, Aspects of the biomonitoring studies using mosses and lichens as indicators of metal pollution, Environ. Res. 93 (2003)221–230.

K.P. Beckett, P.H. Freer-Smith, G. Taylor, The capture of particulate pollution by trees at five contrasting urban sites, Arboriculture. J. 24 (2000) 209–230.

L. De Temmerman, A. Rutten's, N. Waegeneers, Impact of atmospheric deposition of As, Cd and Pb on their concentration in carrot and celeriac, Environ. Pollut. 166 (2012) 187–195.

M. Ferretti, J. Bucher, A. Bytnerowicz, W. Prus-Glowacki, D. Karnosky, K.Percy, State of science and gaps in our knowledge concerning air pollution, Dev. Environ. Sci. 3 (2003) 437–446.

M. Kozlov, E. Haukioja, A. Bakhtiarov, D. Stroganov, S. Zimina, Root versus canopy uptake of heavy metals by birch in an industrially polluted area: contrasting behaviour of nickel and copper, Environ. Pollut. 107 (2000)413–420.

M. Rao, P. Dubey, Occurrence of heavy metals in air and their accumulation by tropical plants growing around an industrial area, Sci. Total Environ. 126(1992) 1–16.

M. Shahid, C. Dumat, B. Pourrut, J. Silvestre, C. Laplanche, E. Pinelli, Influence of EDTA and citric acid on lead-induced oxidative stress to Vicia faba roots, J.Soils Sed. 14 (2014) 835–843.

M. Shahid, S. Khalid, G. Abbas, N. Shahid, M. Nadeem, M. Sabir, M. Aslam, C.Dumat, Heavy metal stress and crop productivity, in K.R. Hakeem (Ed.), Crop Production and Global Environmental Issues SE – 1, SpringerInternational Publishing, 2015, pp. 1–25.

M. Shakoor, N. Niazi, I. Bibi, M. Rahman, R. Naidu, Z. Dong, M. Shahid, M.Arshad, Unraveling health risk and speciation of arsenic from groundwater rural areas of Punjab, Pakistan, Int. J. Environ. Res. Public Health 12 (2015)12371–12390.

M. Tomaˇsevi´c, Z. Vukmirovi´c, S. Rajˇsi´c, M. Tasi´c, B. Stevanovi´c, Characterization of trace metal particles deposited on some deciduous tree leaves in an urban area, Chemosphere 61 (2005) 753–760.

N.K. Niazi, E.D. Burton, Arsenic sorption to nanoparticulate mackinawite(FeS): an examination of phosphate competition, Environ. Pollut. 218 (2016)111–117.

N.M. Lawson, R.P. Mason, Concentration of mercury, methylmercury, cadmium, lead, arsenic, and selenium in the rain and stream water of two contrasting watersheds in western Maryland, Water Res. 35 (2001)4039–4052

P. Zhuang, M.B. McBride, H. Xia, N. Li, Z. Li, Health risk from heavy metals via consumption of food crops in the vicinity of Dabaoshan mine, South China, Sci. Total Environ. 407 (2009) 1551–1561.

R. Zangi, M. Filella, Transport routes of metalloids into and out of the cell: a review of the current knowledge, Chem. Biol. Interact. 197 (2012) 47–57.

S. Giordano, P. Adamo, V. Spagnuolo, M. Tretiach, R. Bargagli, Accumulation of airborne trace elements in mosses, lichens and synthetic materials exposed at urban monitoring stations: towards harmonisation of the moss-bag technique, Chemosphere 90 (2013) 292–299.

S. Khalid, Evaluation of wastewater suitability for irrigation in Vehari city: Environmental consequences and health risk assessment, in MS Thesis, COMSATS Institute of Information Technology, Vehari-Pakistan, 2015, pp.1–82.

S. Mombo, Y. Foucault, F. Deola, I. Gaillard, S. Goix, M. Shahid, E. Schreck, A.Pierart, C. Dumat, Management of human health risk in the context of kitchen gardens polluted by lead and cadmium near a lead recycling company, J. Soils Sediments (2015).

S. Norouzi, H. Khademi, A.F. Cano, J.A. Acosta, Biomagnetic monitoring of heavy metals contamination in deposited atmospheric dust, a case study from Isfahan Iran, J. Environ. Manage. 173 (2016) 55–64

S.A.M.A.S. Eqani, A. Kanwal, A.K. Bhowmik, M. Sohail, R. Ullah, S.M. Ali, A.Alamdar, N. Ali, M. Fasola, H. Shen, Spatial distribution of dust–bound trace elements in Pakistan and their implications for human exposure, Environ.Pollut. 213 (2016) 213–222.

T. Eichert, A. Kurtz, U. Steiner, H.E. Goldbach, Size exclusion limits and lateral heterogeneity of the stomatal foliar uptake pathway for aqueous solutes and water-suspended nanoparticles, Physiol. Plant. 134 (2008) 151–160.

T. Leveque, Y. Capowiez, E. Schreck, T. Xiong, Y. Foucault, C. Dumat, Earthworm bioturbation influences the phytoavailability of metals released by particles in cultivated soils, Environ. Pollut. 191 (2014) 199–206.

T. Xiong, C. Dumat, A. Pierrot, M. Shahid, Y. Kang, N. Li, G. Bertoni, C.Laplanche, Measurement of metal bioaccessibility in vegetables to improve human exposure assessments: a field study of soil-plant-atmosphere transfers in urban areas, South China, Environ. Geochem. Health (2016).

V. Fernández, P.H. Brown, From plant surface to plant metabolism: the uncertain fate of foliar-applied nutrients, Front. Plant Sci. 4 (2013) 289.

WHO, Health Risks of Heavy Metals from Long-range Transboundary air pollution, WHO Regional Office for Europe, Copenhagen, Denmark, 2007. [90] V. Fernández, T. Eichert, Uptake of hydrophilic solutes through plant leaves: current state of knowledge and perspectives of foliar fertilization, Crit. Rev.Plant Sci. 28 (2009) 36–68.

Y. Chai, N. Zhu, H. Han, [Dust removal effect of urban tree species in Harbin], Ying Yong sheng tai xue bao = The journal of applied ecology/Zhongguosheng tai xue hui, Zhongguo ke xue yuan Shenyang ying yong sheng tai yanjiu suo zhu ban 13 (2002) 1121–1126.

Y. Su, Y. Liang, The foliar uptake and downward translocation oftrichloroethylene and 1,2,3-trichlorobenzene in air-plant-water systems, J.Hazard. Mater. 252–253 (2013) 300–305.

Y.-M. Chen, J.-b. Gao, Y.-Q. Yuan, J. Ma, S. Yu, Relationship between heavy metal contents and clay mineral properties in surface sediments: implications for metal pollution assessment, Cont. Shelf Res. 124 (2016)125–133.

Z. Li, Z. Ma, T.J. van der Kuijp, Z. Yuan, L. Huang, A review of soil heavy metal pollution from mines in China: pollution and health risk assessment, Sci. Total Environ. (2014) 843–853.

Food Security and Assessment with Special Reference to Metal Toxicity

Christy Ann Philip*, E Lokesh Goud
Department of Agronomy, School of Agriculture
Lovely Professional University, Jalandhar, Punjab, India, 144411
**Email: christyannphilip16@gmail.com*

Abstract

Heavy metals are a well-defined set of elements that occur naturally in earth crust which are at least 5 times denser than water (i.e,$\geq$5gcm^{-3}) including metals and metalloids. Numerous health hazards at neurological, haematological, genetic, nephrological, cardiovascular, carcinogenic and respiratory levels are the after-effects of the heavy metal release into the environment by either natural or anthropogenic processes. Food security ensures the safe, nutritious and sufficient food to the population both quantitatively and qualitatively. The released heavy metal enters the food chain in aquatic as well as terrestrial ecosystem. Plants exposed to metal toxicity exhibit phytotoxicity symptoms that result in heavy yield loss. Plants can sequester the metals in edible plant parts which when consumed by human creating health hazards. This chapter elucidates the impact of non - threshold toxic metals and metalloids like Arsenic, Mercury, Lead, Cadmium and Chromium on major food crops and the after-effects of their consumption in human. Assessment of the transfer of toxic metals and pollutants from soil toward plant and health risks caused by transfer from plant to human is done with some indices like BCF, PLI, HQ, HI, PERI, DIM and HRI. This provides the extent of contamination and risks. Remediation strategies for the heavy metal alleviation should cost effective, eco friendly, safe and rapid. It includes reduction of the source of contaminants, phytoremediation by which plants accumulate contaminants within their tissue, chemical methods and nanotechnology. This chapter elucidates the impact of heavy metal toxicity on major food crops, assessment of the health risks and soil-plant transfer of heavy metal and remediation measures adopted to overcome the negative impact of metal toxicity on food security.

Introduction

Million years ago, hunter-gatherers were curious about the growth and life cycle of a plant at the time they decided to settle down with agriculture. Curiosity surpassed the challenges raised in food production and agriculture throughout these years. Nobody expected this quantum leap of the agricultural sector in producing food for human consumption, feed for livestock population and other commodities in sufficient quantity for the whole world. The world witnessed great famines, pest attacks, climate changes and natural disasters that significantly made an impact on food production. Green revolution implemented in the world achieved self-sufficient nations in terms of food production by combining high yielding varieties (HYVs), chemical fertilizers and irrigation. The IFPRI forecasted the world population around 9.8 billion in the year 2050. Increasing population raise more demand for food based on the regional dietary pattern. These necessities should be met with proper planning that ensures food security. Sufficient and surplus food production is an attainable goal for the majority of nations under proper strategic plans. On behalf of World Food Security, the definition of the committee set up by United Nations for food security summarized that nutritious, safe and sufficient food according to the dietary need and food preferences should be socially, physically and economically accessible for all people during all time to lead a healthy and active life. Researches and breeding technology could enhance nutritional quality in food while proper planning, schemes, along with research could produce sufficient food but the problem lies with the safety of the produced food.

Polluted environment hinders the growth and development of an organism healthily and this is not a different case in plants. Toxicity experienced by plants against contaminants such as heavy metals is a serious threat to the quality and productivity of the crop. Earth crust consists of several heavy metals naturally (Singh et al., 2011) but their presence due to excess release resulted from anthropogenic actions, industrial wastes etc make adverse effects. Cultivation and growth of plants in soil contaminated with heavy metals could uptake and accumulate metals at high concentration due to which contaminated food chain occur (Singh et al., 2016). This serves as the primary pathway that makes an entry of toxic metals into human and animal tissues, which can be the reason for serious health issues from skin disorders to cancer (McLaughlin et al., 1999).

Sources of Heavy Metals and their Trophic Transfer

Unwanted and excess accumulation of toxic substances in an area due to human involvement make the way of metal into animal tissues by several means like intake of food and water (diet), inhalation and improper and unsafe handling of such particles gradually leads their binding with cellular components. Heavy metals are not artificially made but have natural occurrence, which is five or more times denser than water and possess high atomic weight (Tchounwou et al., 2012). Their natural occurrence itself is the biggest threat for their alleviation as they are non-biodegradable particles (Alluri et al., 2007). Recent definitions show that heavy metals are a set of elements with Atomic No. > 20 and specific gravity (density) $\geq$ 5gcm^{-3} (Ali et al., 2019). Sectors responsible for the release of heavy metals are agriculture, domestic, industry, technology and medical due to the wide range of applications they perform. The priority of toxicity categorised some metals and metalloids into high degrees toxic grades such as Lead, Chromium, Cadmium, Mercury and Arsenic as they create damage to multiple organs even when exposure occurs at very low concentration (Rahman and Singh, 2019). Some metal ions, which are essential in regulating the metabolic functions in plants and other living organisms (Ni, Fe, Zn, Cu, Mo, Mn, Co), will exhibit toxic reactions in excess uptake by plants and resulted accumulation at different organizational level.

Fig. 1. Transfer of Heavy metals from different sources to plants to human

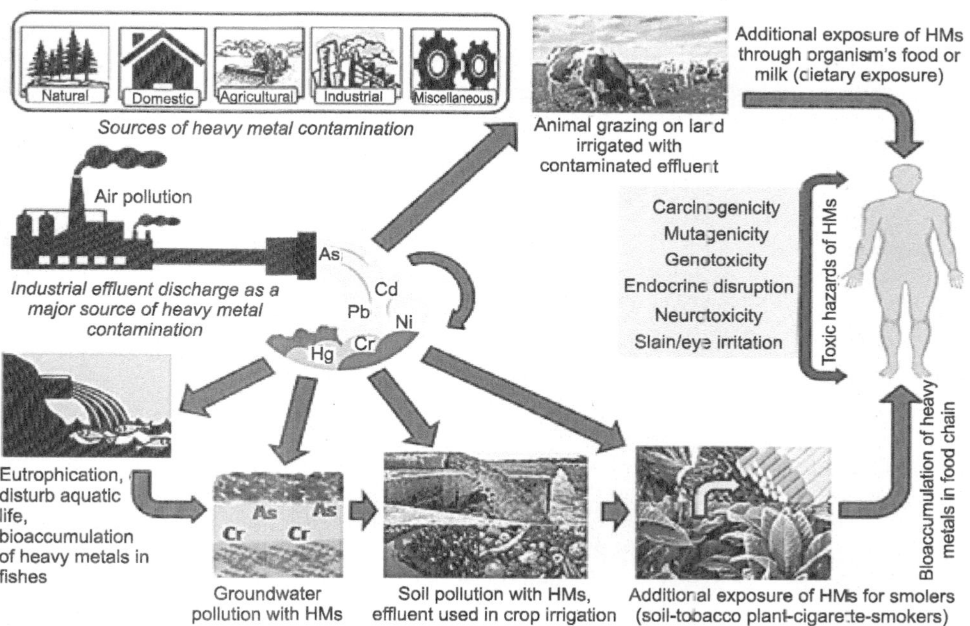

(sciencedirect.com)

Non-threshold toxins include the most toxic and problematic heavy metals (Cd, Pb, Cr, Hg and As) that their presence itself is a harmful condition while the threshold toxins at low concentration (Co, Cu, Fe, Zn, Mn, Mo) is beneficial for plants but exceeding the threshold level produces toxic impact (Ali et al., 2013). Food chains in the terrestrial ecosystem are contaminated due to the presence of Cadmium and Selenium (McLaughlin et al., 1999). Plants tend to transport and deposit the contaminants towards their edible and harvested parts. This process depends on some factors like soil, climate, agronomical management practices, phenotype and genotype of the plant. Inclusion of metal-deposited plant parts and contaminated water in the daily diet of human threatens the safety of food intake and further health issues. When a toxic contaminant moves in the food chain from one trophic level to next, it is referred to as biotransference or trophic transfer. Accumulation of toxic substance from the abiotic environment in an organism is called bioconcentration while bioaccumulation includes accumulation from the dietary intake too. In a food chain, if the successive trophic levels have an increased concentration of the toxic metal, biomagnification occurs while a decrease in concentration with trophic levels is biodilution.

Fig. 2. Trophic transfer of heavy metals

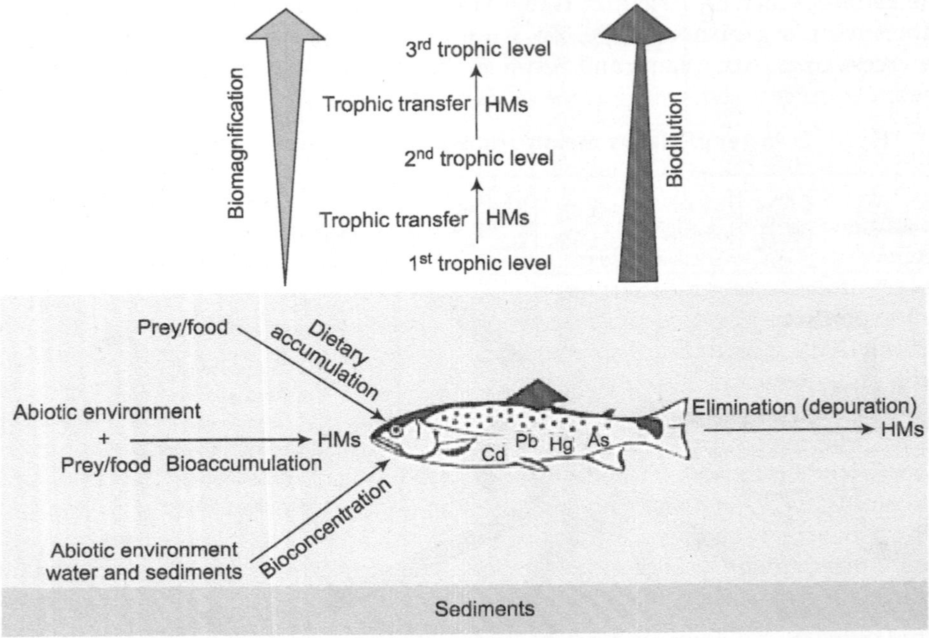

(Source: Ali and Khan, 2018)

Since food security defined by the authorised committee includes safety food for a healthy life, an assessment is necessary for the context of heavy metal toxicity to evaluate the influence on food security. Natural sources of heavy metals are the erosion, mineral weathering and volcanic explosions that make the exposure of underlying heavy metals to surroundings while human-induced processes are discharged from industries, smelting, mining, use of phosphatic fertilizers, pesticides, electroplating, biosolids, dumping of sludge, manufacturing batteries etc (Ali et al., 2013). Accumulation of such toxic metals in soils for agricultural use creates ecological effects in plants growing over that place, microbes residing in particular soil and other flora and fauna. The extent of toxicity depends on the chemical nature of the toxic metals as it varies from one another and the vulnerability of the exposed individuals. Non-threshold toxic metals are elucidated in detail with their impact on major food crops.

Arsenic

Arsenic (As), a ubiquitous element arranged in the periodic table (Atomic no. 33) within the same group of N and P and exhibits itself with metallic and non-metallic properties, so-called metalloid. It has organic and inorganic forms for the existence where inorganic part manifests more toxicity. Arsenic species display two oxidation states in inorganic form i.e. Arsenite, As(III) and Arsenate, As(V) while organic forms include Arsenosugars, Arsenobetaine, Arsenolipids etc that commonly found in seafood with less toxicity. Total As was found to be responsible for As concentration in 98% of results while analyzing some food commodities such as seafood, Algae-based supplements and food (hijiki), fish, cereal-based foods (especially rice-based) to detect As concentration by EFSA Panel on Contaminants in the food chain (CONTAM, 2010). This report enlisted the most used methodologies for analysing and determining As forms in foods. Inorganic forms were chemically separated from organic forms and each was quantified separately.

Arsenic species were separated using chromatography along with on – line detection followed by quantification using As-selective detector. Many kinds of research and studies give evidence that plants can accumulate Arsenic in their tissues. Humans consuming the food crops and vegetables cultivated in soil contaminated with As and drinking water also pave the way for the entry of Arsenic. In the present scenario of climatic changes and unpredictable precipitation, irrigation sources are groundwater and surface waters that are prone to chances of Arsenic contamination in a natural way. This can prominently make an impact on crops grown under a flooded condition like Rice as it being the staple food for a large population. Growth of the plant reduced with unfilled florets of rice plant ultimately result in 'straight head' condition due to As toxicity (panicles devoid of grains). Amount of As was found higher in grains of rice grown in Boro season than that of Aman since Boro season require more water provided from groundwater (Duxbury et al., 2003).

In a day, if average consumption of rice by an adult is 400 – 650 gram (regions where rice is the staple food) in Arsenic affected area, then 0.20 – 0.35 mg arsenic has been reported to intake through food in a day and a water intake of 4 litres daily could take 0.2 mg Arsenic per day (Rahman et al., 2008). Byproducts after rice harvesting like husk, paddy straw grown under Arsenic containing soils along with As contaminated water when given to cattle make As deposition in its body. South Asia and S-E Asia face problems in the quality of food, food security and sustainability in agriculture due to accumulation of As in plant parts inducing yield reduction as an observed yield reduction of 16% occurred over 8 ha (Panaullah et al., 2008). Arsenic toxicity in wheat has a significant decrease in root length and height of shoot with more effective reduction observed with Arsenite than Arsenate (Liu et al., 2005). The yield of grain per plant and dry weight of a spike in wheat lowered with As contamination while it varies with wheat varieties and plant tissue. Arsenic levels in a plant varied with Awns < Glumes < Grains < leaf & rachis < Stem < Root (Zhang et al., 2009). Inorganic P applied as fertilizer can hinder the Arsenic translocation towards shoot and grain (Pigna et al., 2009). Arsenite forms were found in food items (*nori* and *yakinori*) made from edible red algae *Porphyra tenera* (Shibata et al., 1990). *Hijikia fusiforme* is a seaweed that has been recommended as a healthy food included in the diet. Recently, studies reveal the findings that intake of *Hijikia* could make the same adverse effects as an Arsenic poisoning due to a large amount of inorganic As present in it (Nikajima et al., 2006). One of the inorganic species of Arsenic, Arsenate has much chemical similarity to that of phosphate, which has been exploited in the marine ecosystem by which algal population absorbs arsenate in place of phosphate requirement mostly in limited supply conditions of phosphate that results in the hindrance of

ATP formation by forming ADP- Arsenate (Landner, 1998).

Changes in the physiological parameters happen in Maize with As contamination like reduced pigment concentration for Chlorophyll and Carotenoid, decreased protein content, increase in activity of peroxidase, reduced growth, dry weight and leaf area, induction of lipid peroxidation (Stoeva et al., 2003). An analysis performed for proteomes in the root of Maize (*Zea mays* L.) in response towards its exposure on arsenic compounds (Na arsenite and Na arsenate) and observed the down-regulation and up-regulation of about 10% of proteins (Requejo and Tena, 2005) detected totally in roots of Maize (67 out of 700). Among these proteins, seven enzymes, which have functional homogeneity existing for redox perturbation by cellular homeostasis, and four other enzymes having different functions gave high responses. This study suggested that indication for Arsenic toxicity is given by inducing oxidative stress in plants.

According to WHO / FAO, there is a Provisional Tolerable Weekly Intake abbreviated as PTWI whose value should not be more than 15 μg / kg body weight for inorganic Arsenic in coordination with Expert Committee on Food Additives (as cited in Rosas – Castor et al., 2014). In China, a maximum permissible concentration has been fixed for Arsenic as 0.7 mg/kg, which exceeds this limit when Maize crop exposed with > 50 mg Arsenic per kg soil. The same study also revealed that low levels of As (12.5 – 25 mg/kg) have the improved effect on growth and nutritional quality of Maize kernels and high levels abated the yield and yield attributes. Arsenic form found in plant parts like (root, grain, shoot) but the grain has organic As species. (Xiao - ke et al., 2012). According to an experiment conducted in West Bengal, by Roychowdhury et al., 2002 observed the As a presence in vegetable leaves, potato skin, papaya, arum leaf, wheat, rice, cumin, spices, vegetables, cereals and bakery products, turmeric powder etc. Here, As was detected mostly in vegetable skin rather than fleshy part and cooked items in comparison to raw. The study also mentioned As uptake by adults as 171.20 – 189.13 μg and 91.89 – 101.63 μg in children through food items in two different blocks – Jalangi and Damkal. Samples of plant tissues have an As conc. of 107 ppb – 2000 ppb in commonly used vegetables like gourds, kalmi and arum having high levels of water-soluble Arsenic (Huq et al., 2001).

Mercury

Mercury (Hg) hold a rank of 16 among the abundance list of elements on earth crust even though ubiquitous (present in hydrosphere, lithosphere, biosphere and atmosphere at trace amounts). Mercury at a level of 20-150 ppb is found in normal soils according to WHO, 1976 and USEPA,1984. Mercury is capable to strongly bind with soil organic matter making the minimum rate of leaching due to mobility and they cannot be responsible for groundwater contamination unless leaching happens from municipal waste (as cited by Patra and Sharma, 2000). Erosion promotes the transport of mercury towards surface water and the process called degassing in a natural way occurring from oceans and earth's crust enhance its circulation in the atmosphere. Mercury input in the environment accounts for 1/3rd by anthropogenic actions and 2/3rd lies with natural emission (NRC, 1978).

The 48% of Mercury emitted to land, water with 7% and 45% released to air (Patra and Sharma, 2000). Three forms for the existence of mercury have to be known. They are (CICAD50, 2003; Patra and Sharma, 2000).

Elemental mercury, which does exist in gaseous form (atmosphere) and liquid form (as a metal). This category has a prominent impact on toxic nature due to high vapour pressure and ability to soluble in both lipid and water.

Inorganic Hg compounds/ionic form consisting Hg^{+} (mercurous. eg: Mercurous chloride) and Hg^{2+} (mercuric. eg: Mercuric sulphide, Mercuric arsenate, Mercuric chloride).

Organic Hg forms: Make a covalent bond with C. (eg: Phenyl mercury, Ethyl mercury etc)

Apart from present-day studies on mercuric toxicity, some incidents happened earlier reveals the danger of Hg toxicity. Central Asia and South Asia witnessed the death of many hundreds of people by accidentally consuming seeds treated with a fungicide of Hg – based type (1960) and Japan witnessed the aftereffect of entry of Hg into the food chain of the aquatic ecosystem while effluents with elemental Hg converted to bioavailable methyl Hg as Minamata disaster. Exposure of human toward Hg poisoning creates health problems, disorders related to immunological, neurological, cardiac, nephrological, reproductive, motor, and hereditary (Zahir et al., 2005). Water (be it from rain, sea, and groundwater), when contaminated with mercuric compounds, make no one safe due to exposed meat, fish, vegetables etc. Inorganic Hg is the common form emitted and once it meets water bodies are transformed to methyl mercury which is lipophilic (organic). This creates a condition of biomagnifications in the food chain. Diet of a major population includes seafood or freshwater originating foods and this lead to Hg exposure (Zahir et al., 2005).

As methyl mercury finds its way mainly through rice consumption in human, mitigation methods like the careful selection of varieties along with water management (Peng et al., 2012) could tackle the problem by a reduction in methyl mercury and Total Hg in rice grain. In China, analysis of rice plants at the different mining site for total Hg and Methyl Hg accumulation (Meng et al., 2014) were done and it was observed that greater accumulation of methyl mercury is in seeds and water-soluble mercury has a prominent role in it. Hg ions have low Hg translocation from roots towards above-ground portion because the root has a role of potential Hg barrier so that the inorganic Hg found in stalks and leaves are directly absorbed from the atmosphere and the same cannot be moved towards seeds as same as methyl mercury (Meng et al., 2012). Wang et al. (2015) concluded in a study that mercury uptake and deposition in roots of rice could be correlated with the formation of iron plaque and organic acids. In root endodermis, the formation of Apoplastic barriers hinders the mercury uptake.

The increased concentration of inorganic mercuric ions will make a negative impact on the growth of coleoptiles and elongation of the root. *Brassica oleracea* has been found sensitive towards mercury contamination while *B. campestris* have resistance to some extent (Ling et al., 2010). Application of untreated wastewater will create an accumulation of mercury in low concentration in vegetables

(Shariatpanahi and Anderson, 1986). Uptake of mercury by vegetables depends on some factors like Bioconcentration factor of vegetables, soil pH and soil organic matter (Yu et al., 2018). It was also observed that Hg uptake slows down with increasing pH (alkaline condition). Leafy vegetables can highly enrich the Hg concentration while cucurbits are less capable. Mercury stress can do inhibition of root growth and shoot growth along with protein synthesis in wheat. Wheat seedlings prone to lipid phosphorylation and water deficit condition that enhanced protein phosphorylation in leaves under mercury stress (Cailin et al., 2009). Besides, Hg stress makes deviations in Calcium inside wheat leaves (Liu et al., 2010). Edible parts of maize can be accumulated with Hg when grown under Hg polluted soils (Zhao et al., 2017) and order varies for mercury levels are Kernels < Stems < Bracts < Leaf.

Cadmium (Cd)

No nutritional benefits are obtained from cadmium and they are seen in association with phosphate rocks as well as ores of metals. They possess a wide range of industrial applications such as electroplating, manufacturing of chemical fertilizers, manufacturing of Nickel-Cadmium batteries, stabilizer in PVC, in electrodes, pigments in plastic industries (Furness, 1996; Ganesan, 2008). Diversity maintained by cadmium in its toxic effects from other heavy metals are: (i) In humans, Cd possesses a half-life of 20 to 30 years (ii) Body excrete out Cd at a very low rate (iii) Soft tissues (liver, kidney etc) in the body are vulnerable for Cd storage (Goering et al., 1995). Exposure of Cd happens in many ways. According to Furness (1996), Cadmium is omnipresent over land, water and air. Terrestrial surroundings close to smelters receive Cd deposition and vehicle tyres getting wears and tears make Cd deposition on land (50 mg/kg). Low volatile property of cadmium accounts for their low deposits in the air as compared to Hg and Pb. 10-30 mg/kg is the Cadmium level in sewage and sludge. Cd toxicity affects seedling growth but not on germination (An, 2004). Exposure of Cd on humans happens with edible parts of plants (vegetables, cereals, fruits), water, other foods including meat (Gallego et al., 2012). Cadmium is available to plant under some regulatory factors like redox potential (Eh), pH, parameters of Physico-chemical concern etc (Prasad, 1995). Apart from diet, cigarette consumption, welding etc can create Cd exposure (Bernhoft, 2013). An estimated value of 30000t per year is the release of Cd into the atmosphere by which 13-43% accounts for industrial release according to ATSDR, 2005. Cadmium exposure has prominent evidence for its toxicity in humans as a disease named *Itai-itai*. In Japan, paddy fields irrigated from a period of 1900-1960, which had been polluted due to slag deposition from the mine nearby is the source of this hazardous disease. Rice grains harvested from such field were containing high Cd levels (Aoshima, 2016). Alleviation of the polluted fields was achieved through soil replacement.

In tomato, growth reduction, browning of roots, the appearance of necrotic spots and chlorosis results due to the presence of Cd. At low concentration level, tomato roots can accumulate Cd, which is cited due to increase in activity of metabolic pathway consisting of PEP carboxylase-Malate dehydrogenase-

Citrate synthase and high concentration, make negative effects on photosynthetic apparatus (Lopez-Millan et al., 2009). Cadmium toxicity promotes the H_2O_2, superoxide and Reactive Oxygen Species in epidermal cells, vascular tissues of pea plant (Rodriguez-Serrano et al., 2009). Dias et al. (2013) found a reduction in the activity of RuBisCO that eventually makes impaired assimilation rate of CO_2, reduce photochemical efficiency in Photosystem II, and low plant dry weight with increasing concentration of Cd on lettuce. Nutrient uptake is also impossible with increased Cd level. Stress due to Cd depresses the exchange of gases, growth of the seedling, chlorophyll amount as well as chlorophyll fluorescence particulars (especially high Cd conc.) in wheat (Ci et al., 2010). Spring wheat has an accumulation of Cd in its grain depending on the N requirement and fertilization of the crop along with Cd level in soil (Li et al., 2011). A chance for remobilization of Cd cannot be excluded in durum wheat from their stem and leaf towards maturing grains, which do accumulate over the grain (Harris et al., 2001). Those crops whose leafy portion and roots are edible were more vulnerable to Cd concentration rather than grains, fruits etc but oilseed has an exception. Maize cultivar named Liyu no.6 could accumulate Cd at high levels due to their proliferated roots and high biomass, so-called hyperaccumulator of Cd (Wang et al., 2007). Ju et al., 2012 concluded that the consumption of seafood along with the inhalation of Cd through cigarette smoking could create adverse health impacts. Among seafood, molluscs have rich Cd content (Sirot et al., 2008). At the early growth stages of Chickpea (*Cicer arientinum*), increasing Cadmium concentration induced proportionate reduction in length of root and shoot, dry weight and fresh weight of root and shoot as well as the enzyme activity in the leaf named Carbonic anhydrase but an increase in proline amount observed in roots and leaf (Hasan et al., 2007). Growth, biomass, RWC (relative water content), stomatal conductance, rate of transpiration and photosynthesis of mustard crop has a detrimental effect due to cadmium stress along with the induction of oxidative stress (Ahmad et al., 2011). A decline in total lipids with Cd levels was observed but the higher Cd level significantly decreased glycolipids and phospholipids (Gaur and Gupta, 1994). In *Brassica juncea* L., Cd stress can reduce growth, pigment content (Carotenoids, Chlorophyll) protein content to 95% (John et al., 2009). Plant in response to Cd toxicity generated some cellular mechanism for self-protection like production of metallothioneins, phytochelatins, proline etc and accelerated rate of antioxidant enzyme activity as well as metal compartmentalization inside vacuole (Choppala et al., 2014).

Lead (Pb)

Lead is a potential pollutant gaining much attention with concern towards the environment as it can enter into the food chain creating multiple health hazards. A reduction of the toxic impacts of Lead is not expected shortly as it has a wide range of applications in daily life. Apart from natural release of lead via weathering, anthropogenic activities and chemical products like smelting, mining, effluents from the battery manufacturing industry, fertilizer, urban soil waste, factory chimneys, automobile exhaust, pesticides, an additive in gasoline, metal plating, Pb containing paints and pigments etc are responsible for Lead release into the environment. Pb is not a necessary element for plant growth, nutrition and metabolism but it gets deposited in various parts of the plant creating a toxic effect on plants and transfer into the next level in the food chain. The particle size of the soil, Cation Exchange Capacity (CEC) of soil, pH, plant species and cultivar, amendments used in soil, exudations from the roots, physic-chemical factors control the lead uptake by plants (Sharma and Dubey, 2005). Lead has both occupational and environmental exposure to human (Gidlow, 2004). In children, the route of Pb intake is through diet and it elevates when Ca, Fe, Zn, P are low in uptake (Papanikolaou et al., 2005). A value above 1.93 μmol/L for lead level in blood executes symptoms and lead to gastrointestinal problems, neurological and haematological dysfunctions (Lockitch, 1993). Many crops can accumulate Lead in their plant parts. It varies with its concentration inside the soil. In rice, the concentration level of Pb varied among different plant parts observed with a decrease from the root towards the grain and the distribution is not uniform inside the grain also as 32.88% lead has been present in polished rice (Liu et al., 2003). Within the rice grain, caryopsis centre has less Pb concentration and gradually increased from the outer surface of rice glume, caryopsis surface, aleuronic layer and inner part of rice glume(Chen et al., 2008). Because of less mobility, Pb remain mostly associated with roots and shoot uptake along with total plant uptake of Lead correlates with redox potential and pH. Increase in pH reduces the Pb uptake in rice plants (Reddy and Patrick, 1977). Application of Selenium can lower down the bioavailability of heavy metals like Cd and Pb in soil and rice grains. However, it cannot reduce Pb in brown rice (Hu et al., 2014). In China, examining the paddy near to mine area found that both soil and plants have higher levels of Pb than their tolerable levels. Besides, the local population who consumes the same in daily diet intake 2.6 mg Lead by adults and 1.2 mg for children, which were far exceeding allowed levels (Yang et al., 2004). Lead stress induces oxidative damages in rice plant but it could overcome by the effective role of antioxidant defence mechanism including peroxidises, SODs etc (Verma and Dubey, 2003). Lead in excess inside plants results in low biomass yield, photosynthesis, respiratory mechanism, chlorophyll biosynthesis and mineral uptake in rice (Ashraf et al., 2015). In aromatic rice, lead toxicity created quality reduction and low yield attributes along with an increase in peroxidation of lipid, leakage of electrolyte, and increase in H_2O_2 that abated the pigment production required for photosynthesis (Ashraf and Tang, 2017). Increased level of Pb reduced chlorophyll (a, b), carotenoids, K^+, Ca^{2+} ions and inhibited growth of rice (Awan et al., 2015). An experiment conducted with wheat and spinach under different Pb levels showed that the Tolerance Index is higher for spinach than wheat (Lamhamdi et al., 2010). Increasing Pb levels in external media has a positive correlation with Pb levels in wheat seedlings but the negative impact on morphological factors (Lamhamdi et al., 2011). Evaluation of wheat components of 25 varieties near lead smelters observed that lead exceeded the MPC for human consumption (maximum permissible concentration) in the wheat grain of 16 varieties (Xing et al., 2016).

Chromium (Cr)

Chromium is a serious pollutant having environmental concern resulting from the extensive industrial uses. It contaminates soil and water. Toxicity is governed by the valency of element i.e., Cr (VI) has high toxicity and mobility as compared to Cr(III) which is less toxic. Carrier ions of some essential elements like sulphur (SO_4^{2-}) and iron help in the transport of Chromium in plants, as they do not have a specific transport system (Shanker et al., 2005). Changes in the germination process, physiological process (mineral nutrition, uptake of water and photosynthesis), growth of shoot, leaf and stem ultimately result in low yield and dry matter. Applications of chromium at industrial level include alloying, textile dyes, plating, mordants, inhibiting water corrosion, refractory bricks, pigments, pressure-treated lumber, ceramic glazes due to which the metal has high risk of environmental contamination. The toxic form of Chromium Cr(VI) exists as dichromate(CrO_4^{2-}), CrO_3 and CrO_4^{2-} (chromate) which has a toxic impact on the microorganism, aquatic organisms and plants while Cr(III) acts as a necessary micronutrient in human required for lipid and sugar metabolism. A concentration level in the range of 0.05 – 1 mg/L for chromium can increase the growth and yield of plants (Oliveria, 2012). It can enter the human body through occupational exposure and dietary intake. Intake of Cr through food, especially vegetables happens with the human at 30-100 µg per day (Smith et al., 1989). Plants growing in soil condition that has been amended with tannery remnants, sewage and sludge, ultrabasic soil, PF-ash (Pulverised firing ash) get accumulated with Chromium due to its high concentration in these substances. Mostly, Cr is restricted to plant roots even if the pH, plant species, valence state of Cr, humic acids and fulvic acids can alter the uptake and translocation of Cr. Again, grains have less quantity than leaves. Plants under the stress of Chromium do exhibit three modifications at the metabolic level. (a) Photosynthetic pigment production (Chlorophyll, anthocyanin) that sustain plant life. (b) Ascorbic acid, glutathione is some metabolites produced largely that can make damage to plants. (c) Metabolites like histidine; phytochelatins etc, which have biochemical functions, can provide tolerance against Cr stress are highly produced (Nagajyoti et al., 2010). Among the rapeseed, wheat and soybean under Chromium stress, it was observed that rapeseed is having hyper tolerating and accumulating capacity than others due to well organized antioxidant defence system and enzymatic activities (Diwan et al., 2012). Chromium can reduce the length of shoot and root as well as the fresh weight of the plants. In maize plant, that has both physiological alterations like lesions on a leaf with interveinal chlorosis, the leaf has a papery appearance with clear veins, pale green leaf with curling of margins, reduction in photosynthetic pigments, impediment of soluble proteins, reduced activity of amylase can ultimately decline the quality and production of grain (Sharma et al., 2003). Increasing Cr levels decline the mineral nutrient availability inside shoots of the mustard crop. This creates a nutrient imbalance within the crop and thus reduce the yield. As Chromium level in the growth media increases, the pH level of rhizosphere also increases favouring the release of malic acid, oxalic acid and citric acids and these factors together promote the accumulation of Cr in rice (Zeng et al., 2008). If Si has applied exogenously for those plants suffering Cr stress, it can encourage Cr tolerance of the plant even at the seedling stage, by reducing Cr accumulation, Chromium transport from root to shoots and increasing mineral nutrient level along with the antioxidant mechanism by which plant has been protected from drastic losses. Chromium can lower down net photosynthesis, stomatal conductance, growth attributes (height, no.of leaves, area of the leaf, the diameter of the stem, dry weight, fresh weight) and yield attributes (no.of ears per plant, 100 kernel weight, grain yield, no.of kernels per cob) in maize (Anjum et al., 2016). If glycine betaine is applied through the foliar application, the plant becomes capable to tolerate Cr stress as it can detoxify heavy metals and lower down C accumulation in wheat grains. According to USEPA (1998), Maximum Concentration Level (MCL) ascribed for Chromium in drinking water is 100μg/l but the Indian standards allowed 0.05 mg/L Cr for water supplies. Chromium level ranging from 20-500 μg/kg is a common value for most of the food items but the highest level seen in molluscs, meat, crustaceans etc. Ingestion of Cr through oral intake accounts for well water, exposure through inhalation of Cr dust and skin contact make serious health issues.

Assessment of the Transfer of Metal in Soil Plant system (Rai et al., 2019)

Indices representing Soil-Plant Transfer

Analysing the impact of pollutant or contaminant to the environment is necessary as it helps in understanding the extent of exposure, survival and tolerance capability of a living organism, accumulation of pollutant and permissible limits of the pollutant. Assessment is done with some indices that govern the effects of eco-toxicology by contaminated food intake. Contamination of food chain with heavy metal is the aftereffect of cultivating food crops in polluted soil as they can uptake, translocate and accumulate the metal within various parts of the plant body. Values obtained through these assessments help to adopt necessary preventive measures ensuring food security.

Bio-Concentration Factor (BCF)

The intake and accumulation of any chemical/pollutant by the plant or any other organism from their surrounding environment are called bio-concentration while the ratio of chemical concentration in the organism to that of its surrounding is indicated by BCF. It was a term proposed for toxicology in the aquatic ecosystem, uptake is from the soil and for aquatic organisms, uptake is from polluted water. In the case of plants, BCF measured in two systems is Plant-Soil and Plant-Air that assisted in evaluating metal concentration in different vegetative tissues. During initial times, heavy metal transfer in the soil-plant system is quantified by EF (Enrichment factor) while environmental biogeochemistry was indicated by plant uptake factor and BCF. Earlier studies implicated that BCF values changed in the order leafy vegetables > Tuber crops > horticultural crop.

$$\text{BCF} = \frac{\text{Concentration of the heavy metal in crop (mg kg}^{-1}\text{)}}{\text{Concetration in soil (mg kg}^{-1}\text{)}}$$

In the aquatic ecosystem, a unit of BCF is Lkg^{-1}.

Pollution Load Index (PLI)

To measure the degree of heavy metal contamination in the soil of a particular site, PLI has been used. The comparison of the contamination status of various places is possible with this index. The determination of the intensity of pollution along with changes occurring with different sites can be done using PLI. PLI represents how many times the metal concentration in sediments is more than that of background concentration. For the calculation of PLI, the required factor is CF (Concentration Factor/Contamination Factor). The ratio between the concentration of a particular metal (C_{metal}) to the concentration in a control treatment/ unpolluted soils or can be said as the background value of the particular metal ($C_{background\ value}$) gives the concentration Factor (CF). PLI can be summarized as the geometric mean of the CF of different metals.

$$CF = \frac{C_{metal}}{C_{background\ value}}$$

$$PLI = (CF_1 \ X \ CF_2 \ X \ CF_3 \ X .... X \ CF_n)^{1/n} \qquad \text{n: No: of metals}$$

PLI values <1 indicate the condition of no pollution while >1 indicates pollution. Again, a scale of 0–6 entirely evaluates the extent of pollution.

0 = No pollution
1 = No pollution-Medium pollution
2 = Moderate pollution
3 = Moderate-Strong pollution
4 = Strong pollution
5 = Strong-Very strong
6 = Very strong Pollution

Assessment of Health Risk

The dangerous part of the heavy metal pollution is that it can transfer from contaminated soil to plant and plant to animal/human body through food crop consumption. This can create numerous health hazards that eventually cause death. Safety of consuming food is a serious threat to food security. Some indices represent the health risk due to intake of contaminated food and drinking water.

Hazard Quotient (HQ) (Ogunlaja et al., 2019; Rai et al., 2019)

Evaluation of chronic but non-carcinogenic health hazard has been evaluated by HQ, which was proposed by USEPA to evaluate health risk through toxic substances in the air. The ratio between the concentration of potential exposure or means daily intake of toxic metals and reference concentration at which no toxic impact is expected is called Hazard Quotient. Reference dose (RfD) is nothing but a specific dose of heavy metal whose daily oral intake will not take any risk to even most sensitive population.

$$HQ = \frac{[DW_{plant}] \times [M_{plant}]}{RfD \times BW} = \frac{ADI\ (\text{mg kg}^{-1}\ \text{day}^{-1})}{RfD\ (\text{mg kg}^{-1}\ \text{day}^{-1})}$$

DW_{plant} : Consumed Dry Weight of Contaminated Plant sample per day (mg day^{-1})
M_{plant} : Metal Concentration within vegetables (mg kg^{-1})
BW : Human Body Weight (kg)

ADI : Average Daily Intake

The value obtained for HQ categorizes the level of safety.

HQ < 1: Safe; 1 < HQ ≤ 5: Low risk; 5 < HQ ≤ 10: Medium risk; HQ > 10: High risk

a. **Hazard Index:** The additive value of HQ that represents the combined toxic effect from the exposure towards more than one toxic metal is called the Hazard Index. Hazard Index > 1 represents that pollutants consumed through the food intake can cause significant non- carcinogenic health hazard. Calculate the Hazard Quotient for each metal and find the arithmetic sum of all values to obtain HI.

$$HI = HQ_1 + HQ_2 + HQ_3 \ldots HQ_n$$

b. **Daily Intake of Metals:** To assess the quantity of heavy metal intake by a human through the consumption of contaminated vegetables or food, DIM is used.

$$DIM = \frac{M_c \times C.F \times DI}{BW_{avg}}$$

M_c : Concentration of heavy metal in plants (mg kg^{-1})

$C.F$: Conversion Factor (0.085) to convert fresh weight into dry weight of a vegetable

DI : Vegetable intake daily

BW_{avg} : Average bodyweight of human

c. **Health Risk Index:** The proportion of daily intake of metal to the reference dose through the oral ingestion of vegetables or any other food is called HRI.

$$HRI = \frac{DIM}{RfD}$$

The health risk is mentioned when the value of HRI>1.

d. **Potential Ecological Risk Index (RI or PERI):** The concept of PERI put forward by Hakanson (1980) evaluates the risk due to heavy metal at the ecological level and considers the level of toxicity, synergic effects, heavy metal concentration and ecological sensitivity. Three basic modules for the formulation of PERI are C_D (Degree or Extent of Contamination), E_R (Ecological Risk Factor) and T_R (Toxic-Response Factor).

$$C^i_f = \frac{C_n}{C_{nr}}$$

$$E^i_r = T^i_r\, C^i_f$$

$$RI = \Sigma E^i_r$$

C_n : Metal concentration in a given sample

C_{nr} : Background value / Reference value

T^i_r : Toxic factor for each metal (~6)

C^i_f : Single-metal pollution factor

RI < 50 : Low risk; RI(50-100) : Moderate risk; RI(100-200) : Considerable risk; RI > 200 : Intense / high risk.

Remediation of Heavy Metal Toxicity in Soil-Plant system to avoid Health Hazards

Concern over the human health prioritize food safety globally because it has been facing large threats by some anthropogenic actions like the dumping of sewage-sludge without proper treatment, irrigation through wastewater and disposal of effluents from industries into the water bodies. Remediation of the process of entry of heavy metal from soil to food crops should incorporate eco-friendly, cost-effective and rapid methods. The existing methods cover alleviation through ecological, physical, biological and chemical means. In addition, nanotechnology with innovative techniques can do heavy metal removal from the environment. A new concept called H-G concept that is an integration of assessment of human health risk and geospatial technologies can assess the problematic soil sites rapidly and accurately. Conversion of the agricultural practices to an organic concept can reduce the heavy metal release into agricultural soils largely. Some remediation methods are:

Reduction of Source of Pollutants

Dealing with the source of a problem can effectively manage further escalation. The usage of untreated sewage- sludge and effluent up to the prescribed standards should be completely avoided in agricultural land as it lowers down the heavy metal accumulation. Soil is vulnerable for deposition of Particulate Matter from the polluted air and this can be checked with proper air quality measures. Avoid the use of food crops from roadside cultivation as urban traffic creates an accumulation of dust and PM on edible parts of plants. The controlled environment like greenhouses for vegetable production is also vulnerable to heavy metal exposure from the applied chemicals and provided condition. Maintain the organic matter content and pH of the soil in greenhouses.

Fig. 3. Plant Response to Heavy metal stress

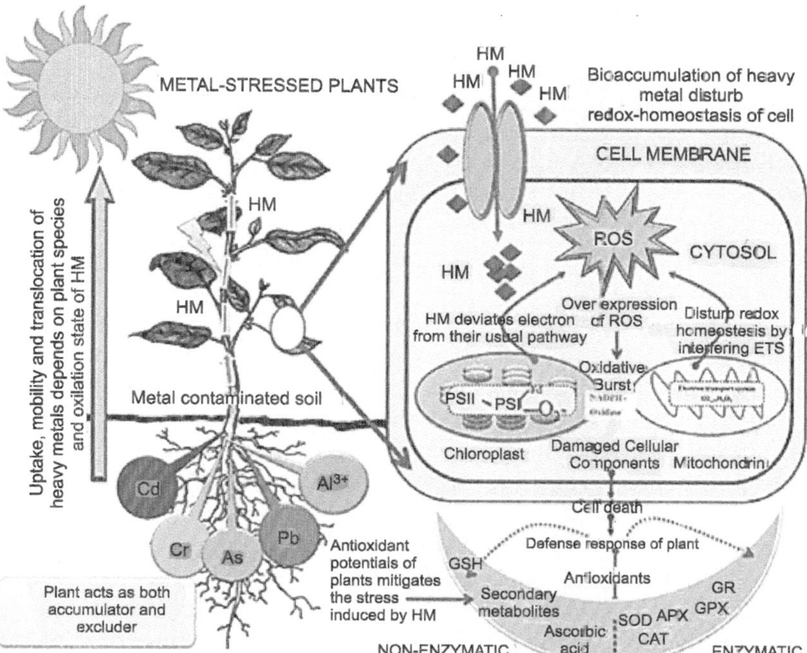

(Source: link.springer.com)

Phytoremediation

The removal of heavy metals from the soil using plant or root zone microorganism is called phytoremediation. Polluted environments can be restored with this efficient, environmentally friendly and inexpensive method. Degree of contamination and amount of contaminant as well as the ability of the plant determines the effective conduct of the process. Hyper accumulator plants having low biomass efficiency and non-hyper accumulator plants with high biomass yield as well as fast-growing species are used for phytoremediation. In addition, it is the sum of several processes. They are:

Phytoextraction: The intake and translocation of pollutant (heavy metal) present within the soil by roots of the plant towards the plant components above ground level by the principle of hyperaccumulation are called phytoextraction also called phytoaccumulation. 100-1000 fold increased accumulation is observed in hyperaccumulator than that of non-hyperaccumulator plant species without exhibiting any phytotoxicity. Some plants belonging to families such as *Asteraceae, Brassicaseae, Poaceae, Euphorbiaceae, Lamiaceae, Cyperaceae, Caryophylacae, Cunouniaceae and Violaceae* can accumulate heavy metals. This process has an economic benefit as along with the removal of pollutant, they also add value by biomass yield. Specific features for plants performing phytoextraction are: (i) Highly capable of removing heavy metals in soil (ii) Metal ions can translocate from root to shoot very easily (iii) Plants are able for detoxification and sequestration of metals at large rates in shoots (iv) Rapid growth (v) Profuse root.

Phytofiltration: It is performed in three forms: (a) Using seedlings – blastofiltration (b) Using roots of the plant – rhizofiltration (c) Using excised shoot of plants – caulofiltration. These methods assist in the removal of pollutants from aqueous wastes.

Phytostimulation: Exudates coming from plant roots can enhance the microbial activity for the degradation of organic pollutants. The high concentration of ethylene can inhibit DNA synthesis and cell division while low concentration stimulates elongation of the root. To balance the level of ethylene in plants, an enzyme named 1-aminocyclopropane-1-carboxylase deaminase is used that can manage abiotic stresses in the plant. This enzyme can degrade the metal pollutants along with plant exudates.

Phytovolatilization: This process occurs if the plant is capable of converting the contaminants in the soil by releasing them as vapour into the atmosphere. From the mercury-contaminated sites, Tobacco plants are capable of depositing methyl mercury which is highly toxic and eventually it gets converted to a less toxic form of volatile elemental Hg that escape through the leaf of the plant. This process is the integrated result of microbial activity in the rhizosphere as well as the metabolic potential of a particular plant.

Phytostabilization: After absorption of the metals by roots, they were allowed to sequester in rhizosphere for further separation and stabilization to converting them as harmless to the environment. Immobilization of metal is attained by root sorption, complexation, precipitation near to root and reduction of metal valency.

Plants capture the metals without permitting into their tissues and denying their spread into the environment.

Chemical Methods: Proper treatments through chemical methods applied to industrial effluents make them non-hazardous to dispose of water bodies. Chemical precipitation makes the easy removal of inorganic pollutants by making insoluble precipitates of carbonates, sulphides, phosphates and hydroxide form of heavy metals. Coagulation-Flocculation is another method based on zeta potential. Reduced net surface charge make stabilization of colloidal particles through electrostatic repulsion process in coagulation while increased particle size obtained through flocculation can easily be separated. Ion exchange methods can replace the ions of any charge from the electrolyte solution by absorbing with ion-exchange resins and releasing the ions of the same charge at same amount into that solution. Strongly contaminated soils are effectively treated by soil washing (ex-situ method) and remove toxic inorganic contaminants like heavy metals, radionuclides etc. Soil flushing is an in situ method by which contaminants are extracted out from the soil by injecting aqueous solution based on leachability of heavy metals.

Nanotechnology (Singh and Prasad, 2015; Rai et al., 2019): Nanoparticles (NP) can reduce the bioavailability of contaminant metals to ensure soil security. To analyze food safety, mostly in food crops, nano-sensors had been used. Carcinogenic metal can be adsorbed on biochar nanosheets to lower bioavailability. Silica Nanoparticles are capable of inhibiting gene expression for OsHMA3 (Cadmium transporters) synthesis in rice by which metal toxicity is alleviated. Bioaccessibility and leachability of Pb^{2+} in soils can be reduced through its immobilization with the application of Vivianite NP (Iron-Phosphate). Removal of Pb^{2+}, Cd^{2+}, $Cu^{2+}$and Zn^{2+} can be done with 10-80% efficiency with multiwalled Carbon Nanotubes (CNT). Chromium is removed at 99% using zero-valent Fe NPs.

Conclusion

The chapter summarizes that heavy metal toxicity harms food security. Heavy metals enter the food chain from contaminated soil and water bodies. Roots allow the entry of heavy metal species by passive absorption or with the help of carriers like Fe or SO_4^{2-}. Concentration level of the toxic metal will be highest in roots as compared to shoot, leaves and lowest in grains. Nevertheless, cultivation of the food crops near the mines and smelters as well as irrigation with untreated sewage-sludge and toxic effluents results in metal accumulation exceeding the permissible limits for human consumption. The intake of such food creates carcinogenic and non-carcinogenic health hazards at neurological, haematological, genetic, nephrological, cardiovascular and respiratory levels. Exposure of plants to heavy metals exhibits phytotoxicity that reduces the growth attributes like plant height, leaf area and root-shoot length as well as yield attributes. Changes are noticed at physiological levels like alteration in photosynthesis, respiratory mechanisms, pigment production and oxidative damages. The combined effect of these changes drastically reduces the crop yield. Assessment of the soil-plant movement of toxic metals and health risk is done with certain indices whose values summarize the extent of pollution. Plants itself have some modifications in response to heavy metal stress like increased proline, ascorbic acid, metallothioneins, phytochelatins, Superoxide dismutase activity and antioxidant defence system to alleviate it. In addition, some remediation techniques used are source reduction, phytoremediation, chemical and nanotechnological approaches.

References

Ahmad, P., Nabi, G. and Ashraf, M. (2011). Cadmium – induced oxidative damage in mustard [*Brassica juncea* (L.) Czern.& Coss.] plants can be alleviated by salicylic acid. *South African Journal of Botany, 77*(1), 36–44.

Ali, H. and Khan, E. (2018). Trophic transfer, bioaccumulation, and biomagnification of non-essential hazardous heavy metals and metalloids in food chains/ webs-concept and implications for wildlife and human health. *Human and Ecological Risk Assessment: An international Journal,* https://doi.org/10.1080/10807039.2018.1469398

Ali, H., Khan, E. and Ilahi, I. (2019). Environmental Chemistry and Ecotoxicology of Hazardous Heavy Metals: Environmental Persistence, Toxicity, and Bioaccumulation. *Journal of Chemistry, 2019,* https://doi.org/10.1155/2019/6730305

Ali, H., Khan, E. and Sajad, M.A. (2013). Phytoremediation of heavy metals-Concepts and Applications. *Chemosphere, 91*, 869-881.

Alluri *et al.* (2007). Biosorption: An eco-friendly alternative for heavy metal removal. *African Journal of Biotechnology, 6*(25), 2924-2931.

Anjum, S.A., Ashraf, U., Khan, I., Saleem, M.F. and Wang, L.C. (2016). Chromium toxicity induced alterations in growth, photosynthesis, gas exchange attributes and yield formation in maize. *Pakistan Journal of Agricultural Sciences, 53*(4).

An, Y. (2004). Soil ecotoxicity assessment using cadmium sensitive plants. *Environmental pollution. 127*(1), 21-26

Aoshima, K. (2016). *Itai-itai* disease: Renal tubular osteomalacia induced by environmental exposure to cadmium- historical review and perspectives. *Soil Science and Plant Nutrition, 62*(4), 319-326. https://doi.org/10.1080/00380768.2016.1159116

Ashraf, U., Kanu, A. S., Mo, Z., Hussain, S., Anjum, S. A., Khan, I., ... & Tang, X. (2015). Lead toxicity in rice: effects, mechanisms, and mitigation strategies—a mini review. Environmental Science and Pollution Research, 22(23), 18318-18332.

Ashraf, U. and Tang, X. (2017). Yield and quality responses, plant metabolism and metal distribution pattern in aromatic rice under lead (Pb) toxicity. *Chemosphere, 176*, 141-155.

Awan, S., Jabeen, M., Imran, Q. M., Ullah, F., Mehmood, Z., Jahngir, M., & Jamil, M. (2015). Effects of lead toxicity on plant growth and biochemical attributes of different rice (Oryza Sativa L.) varieties. Journal of Bio-Molecular Sciences, 3(1), 44-55.

Bernhoft, R.A. (2013). Cadmium toxicity and treatment. *The Scientific World Journal 2013.*

Cailin, G. E., Yan, D. I. N. G., Zegang, W. A. N. G., Dingzhen, W. A. N., Yulong, W. A. N. G., Shang, Q., & Shishi, L. U. O. (2009). Responses of wheat seedlings to cadmium, mercury and trichlorobenzene stresses. Journal of Environmental Sciences, 21(6), 806-813.

Chen, G., Sun, G., Liu, A. and Zhou, W. (2008). Lead enrichment in different genotypes of rice grains. *Food and chemical toxicology, 46*(3), 1152-1156.

Choppala *et al.* (2014). Cellular mechanisms in higher plants governing tolerance to cadmium toxicity. *Critical reviews in plant sciences, 33*(5), 374-391.

CICAD50 (WHO, UNEP, ILO). (2003). Elemental Mercury and inorganic Mercury compounds: Human Health Aspects.

Ci, D., Jiang, D., Wollenweber, B., Dai, T., Jing, Q., & Cao, W. (2010). Cadmium stress in wheat seedlings: growth, cadmium accumulation and photosynthesis. Acta Physiologiae Plantarum, 32(2), 365-373.

Dias *et al.* (2013). Cadmium toxicity affects photosynthesis and plant growth at different levels. *Acta physiologiae plantarum, 35*(4), 1281-1289.

Diwan, H., Ahmad, A. and Iqbal, M. (2012). Characterization of Chromium Toxicity in Food Crops and their Role in Phytoremediation. *Journal of Bioremediation & Biodegradation, 3*(8). DOI:10.4172/2155-6199.1000159

Duxbury, J.M., Mayer, A.B., Lauren, J.G. and Hassan, N. (2003). Food Chain Aspects of Arsenic Contamination in Bangladesh: Effects on Quality and productivity of Rice. *Journal of Environmental Science and Health. Part A- toxic /Hazardous Substances & environmental Engineering, A38*(1), 61-69.

EFSA. (2009). Scientific opinion on arsenic in food. EFSA panel on Contaminants in Food Chain (CONTAM). European Food Safety Authority, Parma, Italy. *EFSA J, 7*(10), 1351.

Francesconi, K., Visoottiviseth, P., Sridokchan, W. and Goessler, W. (2002). Arsenic species in an arsenic hyperaccumulating fern, *Pityrogramma calomelanos*: a potential phytoremediator of arsenic-contaminated soils. *The Sciences of the total Environment, 284*, 27-35.

Gallego, S. M., Pena, L. B., Barcia, R. A., Azpilicueta, C. E., Iannone, M. F., Rosales, E. P., ... & Benavides, M. P. (2012). Unravelling cadmium toxicity and tolerance in plants: insight into regulatory mechanisms. Environmental and Experimental Botany, 83, 33-46.

Gaur, A. and Gupta, S.K. (1994). Lipid components of mustard seeds (*Brassica juncea* L.) as influenced by cadmium levels. *Plant foods for human nutrition, 46*(2), 93-102.

Gidlow, D.A. (2004). Lead toxicity. *Occupational medicine, 54*(2), 76-81.

Goering, P.L., Waalkes, M.P. and Klaassen, C.D. (1995). Toxicology of cadmium. *Toxicology of metals*, 189-214.

Gunatilake, S.K. (2015). Methods of Removing Heavy Metals from Industrial Wastewater. *Journal of Multidisciplinary Engineering Science Studies, 1*(1).

Harris, N.S. and Taylor, G.J. (2001). Remobilization of Cadmium in maturing shoots of near isogenic lines of durum wheat that differ in grain cadmium accumulation. *Journal of Experimental Botany, 52*(360), 1473-1481.

Hasan, S.A., Ali, B., Hayat, S. and Ahmad, A. (2007). Cadmium-induced changes in the growth and carbonic anhydrase activity of chickpea. *Turkish Journal of Biology, 31*(3), 137-140.

Huq, S.M.I., Ara, Q.A.J., Islam, K., Zaher, A. and Naidu, R. (2001). The possible contamination from arsenic through food chain. *Groundwater Arsenic Contamination in the Bengal Delta Plain of Bangladesh.* In: Jacks, G., Bhattacharya, P., and Khan, A.A. (Eds)., 91-96.

Hu, Y., Norton, G.J., Duan, G., Huang, Y. and Liu, Y. (2014). Effect of selenium fertilization on the accumulation of cadmium and lead in rice plants. *Plant and soil, 384*(1-2), 131-140.

John, R., Ahmad, P., Gadgil, K. and Sharma, S. (2009). Heavy metal toxicity: Effect on plant growth, biochemical parameters and metal accumulation by *Brassica juncea* L. *International Journal of plant Production, 3*(3), 66–75.

Ju, Y., Chen, W. and Liao, C. (2012). Assessing human exposure risk to cadmium through inhalation and seafood consumption. *Journal of hazardous materials, 227*, 353-361.

Lamhamdi, M., Bakrim, A., Aarab, A., Lafont, R. and Sayah, F. (2010). A comparison of lead toxicity using physiological and enzymatic parameters on spinach (*Spinacia oleracea*) and wheat (*Triticum aestivum*) growth. *Moroccan Journal of Biology, 6*(7), 64-73.

Lamhamdi, M., Bakrim, A., Aarab, A., Lafont, R. and Sayah, F. (2011). Lead phytotoxicity on wheat (*Triticum aestivum* L.) seed germination and seedlings growth. *Comptes rendus biologies, 334*(2), 118-126.

Landner, L. (1998). Arsenic in the aquatic environment-speciation and biological effects. *Swedish National Chemicals Inspectorate.* Report No. 2. Exemption Substances Project. Stockholm, Sweden: Printgraf

Ling, T., Fangke, Y. and Jun, R. (2010). Effect of Mercury to seed germination, coleoptiles growth and root elongation of four vegetables. *Res. J. Phytochem, 4*(4), 225-233.

Liu, D., Chen, W.Z., Xu, H. and Wang, Y. (2010). Influence of mercury on chlorophyll content in winter wheat and mercury bioaccumulation. *Plant, Soil and Environment, 56* (3), 139-143.

Liu, J., Li, K., Xu, J., Zhang, Z., Ma, T., Lu, X., ... & Zhu, Q. (2003). Lead toxicity, uptake, and translocation in different rice cultivars. Plant Science, 165(4), 793-802.

Liu, X., Zhang, S., Shan, X. and Zhu, Y. (2005). Toxicity of arsenate and arsenite on germination, seedling growth and amylolytic activity of wheat. *Chemosphere, 61*, 293-301.

Li, X., Ziadi, N., Belanger, G., Cai, Z. and Xu, H. (2011). Cadmium accumulation in wheat grain as affected by mineral N fertilizer and soil characteristics. *Canadian Journal of Soil Science, 91*(4), 521-531.

Lockitch, G. (1993). Perspectives on lead toxicity. *Clinical biochemistry, 26*(5), 371-381.

Lopez-Millan, A., Sagardoy, R., Solanas, M., Abadia, A. and Abadia, J. (2009). Cadmium toxicity in tomato (*Lycopersicon esculentum)* plants grown in hydroponics. *Environmental and Experimental Botany, 65*, 376-385.

McLaughlin, M.J., Parker, D.R. and Clarke, J.M. (1999). Metals and micronutrients-food safety issues. *Field Crops Research, 60*, 143-163.

Meng, B., Feng, X., Qiu, G., Wang, D., Liang, P., Li, P., & Shang, L. (2012). Inorganic mercury accumulation in rice (Oryza sativa L.). Environmental toxicology and chemistry, 31(9), 2093-2098.

Meng, M., Li, B., Shao, J. J., Wang, T., He, B., Shi, J. B., ... & Jiang, G. B. (2014). Accumulation of total mercury and methylmercury in rice plants collected from different mining areas in China. Environmental Pollution, 184, 179-186.

Nagajyoti, P.C., Lee, K.D. and Sreekanth, T.V.M. (2010). Heavy metals, occurrence and toxicity for plants: a review. *Environ Chem Lett., 8*, 199-216.

National Research Council (NRC). (1978). An Assessment of Mercury in the Environment: A report. *National Academies.*

Ogunlaja, A., Ogunlaja, O.O., Okewole, D.M. and Morenikeji, O.A. (2019). Risk assessment and source identification of heavy metal contamination by multivariate and hazard index analyses of a pipeline vandalised area in Lagos State, Nigeria. *Science of the total Environment, 651*, 2943-2952.

Ojuederie, O.B. and Babalola, O.O. (2017). Microbial and Plant-assisted Bioremediation of Heavy Metal Polluted Environments: A review. *International Journal of Environmental Research and Public Health. 14.*

Oliveira, H. (2012). Chromium as an environmental pollutant: insights on induced plant toxicity. *Journal of Botany, 2012*. https://doi.org/10.1155/2012/375843

Panaullah *et al.* (2008). Arsenic toxicity to rice (*Oryza sativa* L.) in Bangladesh. *Plant soil.* DOI 10.1007/s11104-008-9786-y

Papanikolaou, N.C., Hatzidaki E.G., Belivanis, S., Tzanakakis, G.N. and Tsatsakis, A.M. (2005). Lead toxicity update. A brief review. *Med Sci Monit., 11*(10), RA329-336.

Patra, M. and Sharma, A. (2000). Mercury Toxicity in Plants. *The Botanical Review. 66*(3).

Peng, X., Liu, F., Wang, W. and Ye, Z. (2012). Reducing total mercury and methylmercury accumulation in rice grains through water management and deliberate selection of rice cultivars. *Environmental pollution, 162*, 202-208.

Pigna, M., Cozzolino, V., Violante, A. and Meharg, A.A. (2009). Influence of Phosphate on the Arsenic uptake by Wheat (*Triticum durum* L.) Irrigated with Arsenic Solutions at Three Different Concentrations. *Water, air and soil pollution, 197*(14), 371-380.

Prasad, M.N.V. (1995). Cadmium toxicity and tolerance in vascular plants. *Environmental and Experimental Botany, 35*(4), 525-545.

Rahman, M.A., Hasegawa, H., Rahman, M.M., Miah, M.A.M. and Tasmin, A. (2008). Arsenic accumulation in rice (*Oryza sativa* L.): human exposure through food chain. *Ecotoxicology and Environmental Safety, 69*(2), 317-324.

Rahman, Z. and Singh, V.P. (2019). The relative impact of toxic heavy metals (THMs) (arsenic (As), cadmium (Cd), chromium (Cr) (VI), mercury (Hg) and lead (Pb)) on the total environment: an overview. *Environmental monitoring and assessment, 19*(7), 419. https://doi.org/10.1007/s10661-019-7528-7

Rai, P.K., Lee, S.S., Zhang, M., Tsang, Y.F. and Kim,K. (2019). Heavy metals in food crops: Health risks, fate, mechanisms and management. *Environment International, 125*, 365-385.

Reddy, C.N. and Patrick, W.H. (1977). Effect of redox potential and pH on the uptake of cadmium and lead by rice plants. *Journal of Environmental Quality, 6*(3), 259-262.

Requejo, R. and Tena, M. (2005). Proteome analysis of maize roots reveals that oxidative stress is a main contributing factor to plant arsenic toxicity. *Phytochemistry, 66*, 1519-1528.

Rodríguez-Serrano, M., Romero-Puertas, M. C., Pazmino, D. M., Testillano, P. S., Risueño, M. C., Luis, A., & Sandalio, L. M. (2009). Cellular response of pea plants to cadmium toxicity: cross talk between reactive oxygen species, nitric oxide, and calcium. Plant Physiology, 150(1), 229-243.

Rosas – Castor, J.M., Guzman – Mar, J.L., Hernandez – Ramirez, A., Garza – Gonzalez, M.T. and Hinojosa – Reyes, L. (2014). Arsenic accumulation in Maize crop (*Zea mays*): A review. *Science of the Total Environment, 488-489*, 176-187.

Roychowdhury, T., Uchino, T., Tokunaga, H. and Ando, M. (2002). Survey of arsenic in food composites from an Arsenic – affected area of West Bengal, India. *Food and Chemical Toxicology, 40*(11), 1611-1621.

Singh, A. and Prasad, S.M. (2015). Remediation of heavy metal contaminated ecosystem: An overview on technology advancement. *Int.J.Environ.Sci. Technol., 12*, 353-366.

Singh, R., Gautam, N., Mishra, A. and Gupta, R. (2011). Heavy metals and living systems: An overview. *Indian Journal of pharmacology, 43*(3), 246.

Singh, S., Parihar, P., Singh, R., Singh, V.P. and Prasad, S.M. (2016). Heavy Metal Tolerance in plants: Role of Transcriptomics, Proteomics, Metabolomics, and Ionomics. *Front. Plant Sci.* https://doi.org/10.3389/fpls.2015.01143.

Sirot, V., Samieri, C., Volatier, J. and Leblanc, J. (2008). Cadmium dietary intake and biomarker data in French high seafood consumers. *Journal of exposure science & environmental epidemiology, 18*(4), 400-409.

Shanker, A.K., Cervantes, C., Loza-Tavera, H. and Avudainayagam, S. (2005). Chromium toxicity in plants. *Environment international, 31*(5), 739-753.

Shariatpanahi, M. and Anderson, A.C. (1986). Accumulation of cadmium, mercury and lead by vegetables following long – term land application of wastewater. *Science of the total environment, 52*(1-2), 41-47.

Sharma, D.C., Sharma, C.P. and Tripathi, R.D. (2003). Phytotoxic lesions of chromium in maize. *Chemosphere, 51*(1), 63-68.

Sharma, P. and Dubey,R.S. (2005). Lead toxicity in plants. *Braz. J. Plant Physiol, 17*(1), 35-52.

Shibata, Y., Jin, K. and Morita, M. (1990). Arsenic compounds in the edible red alga, *Porphyra tenera,* and in *nori* and *yakinori*, food items produced from red algae. *Applied Organometallic Chemistry, 4*(3), 255-260.

Smith, S., Peterson, P.J. and Kwan, K.H.M. (1989). Chromium accumulation, transport and toxicity in plants. *Toxicological & Environmental Chemistry, 24*(4), 241-251.

Stoeva, N. Berova, M. and Zlatev, Z.I. (2003). Physiological response of maize to arsenic contamination. *Biologia Plantarum, 47*(3), 449-452.

Tchounwou, P.B., Yedjou, C.G., Patlolla, A.K. and Sutton, D.J. (2012). Heavy metals toxicity an the Environment. *National Institute of Health – Public Access, 101,* 133-164. doi:10.1007/978-3-7643-8340-4_6.

Verma, S. and Dubey, R.S. (2003). Lead toxicity induces lipid peroxidation and alters the activities of antioxidant enzymes in growing rice plants. *Plant Science, 164*(4), 645-655.

Wang, M., Zou, J., Duan, X., Jiang, W. and Liu, D. (2007). Cadmium accumulation and its effects on metal uptake in maize (*Zea mays* L.). *Bioresource Technology, 98*, 82-88.

Wang, X., Tam, N,F., He, H. and Ye, Z. (2015). The role of root anatomy, organic acids and iron plaque on mercury accumulation in rice. *Plant and Soil, 394*(1-2), 301-313.

CI, X. K., LIU, H. L., HAO, Y. B., ZHANG, J. W., Peng, L. I. U., & DONG, S. T. (2012). Arsenic distribution, species, and its effect on maize growth treated with arsenate. Journal of Integrative Agriculture, 11(3), 416-423.

Xing,W., Zhang, H. Scheckel, K.G. and Li, L. (2016). Heavy metal and metalloid concentrations in components of 25 wheat (*Triticum aestivum*) varieties in the vicinity of lead smelters in Henan province, China. *Environmental monitoring and assessment, 188*(1), 23.

Yang, Q.W., Shu, W.S., Qiu, J.W., Wang, H.B. and Lan, C.Y. 2004. Lead in paddy soils and rice plants and its potential health risk around Lechang Lead/Zinc Mine, Guangdong, China. *Environment International, 30*(7), 883-889.

Yu, H., Li, J. and Luan, Y. (2018). Meta-analysis of soil mercury accumulation by vegetables. *Scientific reports, 8* (1),1-10.

Zahir, F., Rizwi, S.J., Haq, S.K. and Khan, R.H. (2005). Low dose mercury toxicity and human health. *Environmental toxicology and Pharmacologyi.* doi:10.1016/j.etap.2005.03.007

Zeng, F., Chen, S., Miao, Y., Wu, F. and Zhang, G. (2008). Changes of organic acid exudation and rhizosphere pH in rice plants under chromium stress. *Environmental Pollution,155*(2), 284-289.

Zhang, W.D., Liu, D.S., Tian, J.C. and He, F.L. (2009). Toxicity and accumulation of arsenic in wheat (*Triticum aestivum* L.) varieties of China. *Phyton (Beunos Aires)*, *78*(2), 147-154.

Zhao, Z., Fu, Z., Lin, Y., Chen, H., Xing, X., Liu, Z., ... & Tang, J. (2017). Genome-wide association analysis identifies loci governing mercury accumulation in maize. Scientific reports, 7(1), 1-11.

Non-Enzymatic Based Defending Mechanism of Toxicity in Plants

M Lalitha*, D Vasundhara, E Lokesh Goud

Department of Agronomy, School of Agriculture

Lovely Professional University, Jalandhar, Punjab, India, 144411

**Email: lalithasonu124@gmail.com*

Abstract

A variety of abiotic stresses, such as salt stress, heavy metal toxicity, nutrient deficiency, drought, extreme temperatures, UV radiation, etc., are persistently disputed by plant changes, resulting in significant losses in the quality and yield of the crop. A major signal of such molecular stress is accelerated by the regeneration of ROS such as O2, O2-, H2O2 and OH, which is immensely reactive in nature and reacts with metabolites and cellular molecules, resulting in irrecoverable metabolic breakdown and leading to plant death. To overcome these plants, well-developed defence mechanism systems, Enzymes and non-enzymes are used to detoxify ROS. Ascorbic acid, osmolytes (proline, GB, trehalose) Vitamin E is included in the non-enzymatic defence mechanisms, which reduce the extreme effects of oxidative damage by ROS extinguishing. Additionally, the mechanisms of tolerance and detoxification in plants by gene expression against abiotic stress have been reviewed. The non-enzymatic way to defend plant mechanisms against toxicity is briefly discussed in this chapter.

Introduction

Plants undergo various and unavoidable abiotic and biotic stress in their natural habitat, which is unfavourable for their growth and development and triggers a series of physiology, morphology, biochemical and molecular changes in plants. To survive under these adverse conditions, inherent mechanisms are developed to maintain minimum growth potential. Fungi, viruses and herbivores are factors that undergo biotic stress. Abiotic stress includes salinity, waterlogging, drought, temperature variations (heat, chill and frost), heavy metal toxicity, and nutrient imbalances. Based on the severity of stress, these factors limit plant growth and productivity. Due to global climate change, the periodicity and duration of such abiotic stress will soon increase.

Plants designed to protect themselves from the damage caused by certain defence mechanisms. Includes constitutive (continuous defence) and inducible defences

- Constitutive defences: includes cell wall, waxy epidermal cuticles and bark. It helps to protect plants from attacks and also provides strength and maintains rigidity.
- Inducible defences: includes toxic chemicals, enzymes that degrade pathogens and mediate cell suicide.

Toxicity by Heavy Metals

Heavy metals are inorganic chemical compounds that are non-degradable and persistent in nature with an atomic weight of more than 20 grams and a density of more than 5 grams per cm3. They have the toxic effects of living cells and mutagenic effects on plants and animals by affecting and polluting food chains, soil, irrigation water and the surrounding atmosphere. In soils, metals are found in 2 types as essential micronutrients (Fe, Mn, Zn, Cu, Mg, Mo and Ni) and non-essential elements (Cd, Sb, Cr, Pb, As, Co, Ag, Se and Hg) plants require essential elements in small quantities because they play an important role in the structure of the enzyme and proteins as well as in growth and metabolism. If plants have a high concentration of these elements, it leads to a reduction and inhibition of plant growth. These at toxic levels inhibit normal plant functioning and act as a barrier to metabolic processes in various ways, including a change in the building block of protein structure due to the formation of bonds between heavy metals and sulphydryl groups. It disrupts the functions of essential metals in biomolecules, such as pigments or enzymes, and adversely affects the cytoplasmic membrane. It restricts the essential events in plants such as photosynthesis, respiration and enzyme activity and, on the other hand, the high concentration of these heavy metals increases the generation of ROS (Reactive Oxygen species) such as O_2 (superoxide free radicles), OH-(Hydroxyl free radicles) or molecular forms such as O_2 and H_2O_2 and cytotoxic compounds such as MG (Methylglyoxal). This condition shows the source of several disintegrated disorders such as ion leakage, protein and lipid oxidation, redox imbalance, oxidative DNA attack and cell membrane alteration, and the structure finally leads to the activation of PCP pathways (programmed cell death)(Fig. 1).

Defence Mechanism of Plants for Heavy Metal Toxicity

To tolerate and detoxify heavy metals, the plants naturally develop several extra and intracellular mechanisms. They focus mainly on avoiding the build-up of toxic concentrations and preventing damage. Conflicts against high external metal concentrations are divided into two categories. They are A) Avoidance B) Tolerance

Avoidance

In this mechanism, the entry of heavy metals into plants is limited by preventing their entry into root cells. This prevention strategy is known as the first line of defence. Works at the extracellular level by mycorrhizal association, root exudate complexion, metal-binding organic acid secretion and rhizosphere pH change.

1. Immobilization of metals by mycorrhizal associations: mycorrhizal associations between fungi and roots of host plants protect plants from the accumulation of metals and increase their tolerance when grown in heavy metal contaminated soils. This mycorrhiza adopts the absorption of metals, the chelation of metals and inhibits their entry into the host plant. ECM (Ectomycorrizas) and AM (Arbuscular mycorrhizas) are the most common associations in plants.
2. Root Secretions: Root releases various organic acids or amino acids sugars, inorganic ions and extracts such as protons, CO2, bicarbonates and certain secretions such as mucilage, siderophores and allelochemicals. These are collectively referred to as root exudates. These secretions help the plant to survive and provide stable HM ligand complexes surrounding the roots, reduce their toxicity, and make them unavailable. Few of these exudates increase the pH near the rhizosphere, which accelerates metals and limits their availability.

Fig. 1. Defense mechanism induced by plants in heavy metal contaminated soil

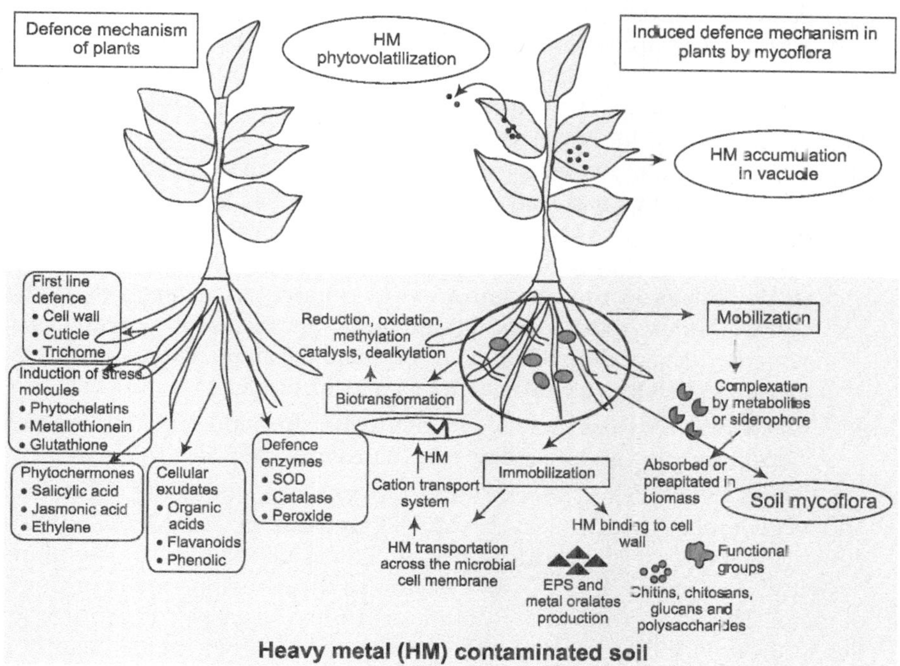

(Source: P.C Singh, 2018, springer.com.)

Tolerance

This mechanism is effective for the accumulation, storage and immobilization of heavy metals by binding them to peptides, amino acids. This is also known as the second line of defence for intracellular detoxification of metals. It has two strategies.

I. By modifying these metal ions near the plasma membrane, the entrance of metals is prevented.

II. When metal ions enter the plant, they can be detoxified by inactivating metals or by covering them in a less toxic form.

   i. Binding of metal ions to the cell wall: the cell wall consists of pectic sites, histidyl groups and carbohydrates, which immobilize heavy metals and prevent them from entering the cytoplasm. Pectins in the cell wall consist of polygalacturonic acid, which prevents metals from being absorbed by binding of metals to carboxylic groups of polygalacturonic acid, which act as cation exchange. This mechanism is efficient for accumulation, storage and immobilization of heavy metals by binding them with peptides and amino acids. This is also known as the second line of defence works at intracellular detoxification of metals. It has two strategies.

   ii. Active efflux /Discharge pumping at the plasma membrane: This mechanism is reported in bacteria and animal cells. In plants, P-type ATPase superfamily and ATP binding cassette (ABC) transporters are worked for the active discharge of heavy metal. The active sequestration of chelated heavy metal includes2 subfamilies of ABC transporters. They are MRP (Multidrug Resistance Associated protein) and PDR (Pleiotropic Drug Resistance)

   iii. Organic acids: In cells, organic acids for detoxification of metals by complexing and makes them unavailable to plant. In nitrogen metabolism, ionic balance and for the formation of ATP from carbohydrates these organic acids act as metabolic intermediates. Any changes in the concentration of organic acids lead to metabolic abnormalities. Excess amount of metal increases the organic acids, which direct the detoxification or on the other hand, causes disturbance in metabolism due to excess metal concentration.

   iv. Inactivation of toxic metals: The phytochelatins and metallothioneins at cytoplasm play an important role to provide tolerance to metals.

a. Pcs (Phytochelations): Phytochelations are derived from GSH (Glutathione and as general structure n-Gly (n=2 to 11). In presence of HMS PC, synthase enzyme activates and produces Pcs by Tran's peptization of ⊚-glutamyl-cysteinyl dipeptides. In Pcs, the cysteine part carries out the coordination of metal. The thiol group and glutamic acid provide water solubility to PCS. Pcs is strongly induced by cd when compared to other metals mainly Pcs associated with homeostasis and exchange of essential metals and helps in detoxification of metals.

b. Metallothioneins (MTS): These are cysteine-rich proteins are induced by many abiotic stresses. It consists of mercaptide groups with molecule weight of 5-20 k Da, which leads to metal ions binding with this. Based on structure and cysteine content divided into MT_1 and MT_2 class. In class, 1 MTS cysteine aligns with MTS of mammalian but in class 2 MTS they do not align with class 1MTS. The above stated MTS shown several responses to Heavy metals. They have seen that participation in the protection of antioxidant and plasma membrane repair mechanism.

Other Alternatives:

Plants under heavy metal stress activate oxidative stress defence mechanism and them synthesis the reactive oxygen species, Stress-related proteins like (Heat shock proteins) and hormones.

Stress-Related Proteins:

When plants exposed to heavy metals it triggers the synthesis of these proteins, these are enriched in plasma membrane acts as a barrier for the inflow of metals leads to detoxifying and metal homeostasis. Heat shock proteins are common stress-related proteins, which act as molecular chaperones. Under stressful conditions, it helps in protection and protein repair.

ROS and Antioxidant Defence:

The initial response of plants increases the synthesis of ROS like O_2^-, OH^- and H_2O_2 to heavy metal stress. These are produced at low level continuously during normal metabolic processes.

In ROS, H_2O act as an important role for signalling the molecules to regulate defence mechanism. It poses a dual role, damages the cell at high concentration, and adapt to stress when they are present in moderate concentration and by inducing antioxidant response. At molecular and cellular levels complex of ROS scavenging mechanism decreases oxidative damage and increases the resistance to heavy metals.

Hormones

It plays an important role in the management of metal stress. They are involved in defence signalling pathways such as salicylic acid, jasmonic acid, ethylene, gibberellic acid. Glutathione is synthesised with jasmonic acid, which plays a key role in protecting plants from heavy metals. Salicylic acid activates genes related to H2O2-mediated signal transduction defence (or directly affects mechanisms related to metal detoxification). Heavy metal induces ethylene biosynthesis, acting as an endogenous signal, and activates the defence response of the plant.

Toxicity by Mineral Nutrition

The soil is composed of high amounts of minerals and organic materials. These materials have more importance for plant morphology and anatomical development. The fertility of the soil is an important factor in the supply of essential soil elements, which are absorbed by plants at an appropriate level for their metabolism. These nutrients are available in the form of ions. In plants, these mechanisms form a network with the root system. It is root interception, diffusion, mass flow. These essential elements are necessary for the metabolism and manufacture of simple and complex compounds for their development. They are classified minerals and non-minerals and subdivided into micronutrients and macronutrients. Hydrogen, oxygen and carbon are non-mineral nutrients used in tissue formation in plants responsible for protein, nucleic acids, carbohydrates, lipids and hormone composition. Essential elements classified as micronutrients and macronutrients and are essential for the completion of the cycle. Macronutrients are required in large quantities for plants and play an important role in the structure of the plant. It includes nitrogen, phosphorus, potassium, calcium, magnesium, sulphur. Micronutrients are absorbed in small quantities, which contribute to the regulation and activity of cell organelles. It is composed of zinc, boron, copper, iron, manganese, molybdenum and chlorine. Higher plants require micronutrients, but high concentrations of these nutrients usually lead to toxicity in plants and also stimulate negative effects such as the reduction of photosynthetic pigments and the absorption capacity of the membrane. The production and accumulation of ROS (Reactive Oxygen Species) are due to an increase in oxidative stress, which increases the activation of antioxidant enzymes and, in adverse conditions, leads to cell death. The excess supply of these nutrients in plants causes stress that increases their effects or the oxidant system, resulting in excess production of ROS such as H2O2 (Hydrogen Peroxide) and O2-(Superoxide Radical). High accumulation of reactive oxygen species causes oxidative damage and antioxidant enzymes such as catalase and glutathione peroxidase contribute to detoxification of the possible impact of ROS. Several mechanisms are proposed for the known tolerance of plant toxicity due to nutrients, including specific transporters, metal ion homeostasis (at high concentrations of nutrients (or by changing environments, these metal ions maintain internal steady-state balance within the cell) and compartmentalization of micronutrients in the vacuoles.

Toxicity by Salt Stress

Plants in soil with high concentrations of salt make it harder for the root to absorb water. Salts near the Rhizosphere have a direct effect on cell growth and metabolism. Sodium, calcium and magnesium chlorides are the various types of soluble salts, which are reduced by the weathering of parental rocks. Soil salinity is defined as a high concentration of EC soluble salts greater than 4 ds / m. In adverse conditions, salinity causes an increase in Na+ and Cl-ions. It causes critical conditions for the survival of plants and causes several physiological disorders in plants. It also affects root growth and physiology, as well as the uptake of nutrients by the roots.

The Response of Plants to Salt Stress

The response of plants to salinity depends on the genetic makeup, the stage of growth, their severity and the extent of stress. In general, the response of crops to salinity is a decrease in growth. Salinity causes osmotic stress, which leads to slow growth and developmental rate, absorption capacity, vegetative growth, leaf expansion, leaf area index (LAI) and a reduction in photosynthesis. Survival of these soil stress plants shows some adaptation features in their morphology, anatomy and physiological (or bio-chemical) under salt conditions. Plants under such conditions undergo a common mechanism viz. aggregation of certain consistent solutions such as sucrose, glycerol, trehalose, pinitol, proline and quaternary ammonium compounds such as Glycine betaine. These solutions protect plants from adverse stress conditions in a variety of ways, such as cytoplasm and chloroplast protection against salinity damage. In stressful conditions, plants survive by producing ROS (Reactive Oxygen Species) scavenge protein stabilisers and maintaining plant physiological stability. In plants, salinity leads to reduce the availability of CO_2 in the soil. This response leads to a reduction of oxidized $NADP^+$. The decrease in internal CO_2 concentration slow down reactions of the Calvin cycle, which compels photorespiration, results in the production of high H_2O_2 in the peroxisome. Salinity increases the production of ROS, which induces the phytotoxic reactions like lipid peroxidation, degradation of protein and mutations in DNA many reports showed the excess production of ROS in plants under salinity induces membrane damage, which mainly leads to cellular toxicity.

Non–Enzymatic based Defending Mechanism of Toxicity in Plants

Proline, ascorbic acid (Vit-C), Osmolytes (Glucose, tocopherols) comes under a non-enzymatic based defence mechanism in plants when exposed to toxicity (Fig. 2).

Fig. 2. Defense mechanism induced by plants in response to Abiotic stress

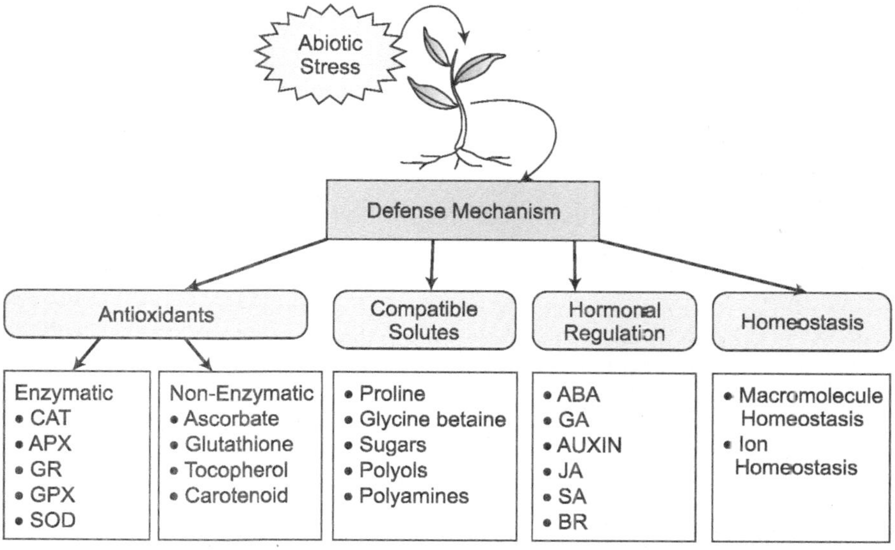

(Source: Sengupta et al., 2016, Springer.com)

Proline Accumulation and its Defensive Role under Diverse Stress Condition in Plants: an Overview

Adverse environmental conditions impose stress on plants, such as heat, cold, salinity, heavy metals, nutrients, drought stress; play a very important role in limiting plant growth and development. During exposure to these wide ranges of stresses, the plant itself changes or develops at the morphological, biochemical or molecular level. During morphological changes, the plant not only reduces the life span but also reduces plant body skeletons such as shortening of plant height, while ion transport, carbon metabolism and synthesis of osmoregulatory compounds or compatible molecules such as proline, glycine betaine, and sorbitol are adjusted at the molecular level. A variety of compatible solutions allows the plant to tolerate abiotic stress. Proline is one of the most effective compatible molecules that produce in a wide range of organisms, including plants, under adverse conditions. During osmotic stress, a transcriptional characteristic known as up-and down-regulation of gene expression may be observed. These changes can be made directly easier by different stress conditions. The up-regulation in plants results in proline accumulation under osmotic stress while the down-regulation process results in deterioration. Proline is a water-soluble amino acid and beneficial solute that accumulates in the plant under various kinds of stress such as drought, cold, heat, heavy metals, nutrients and salt stress. Initially, proline accumulation occurs in plant tissues in response to stress that reduces water levels, resulting in low water potentials, such as dryness, salinity and freezing, but does not reduce water levels. Response to heavy metal toxicity was also reported (Anaytullah, 2007; Verslues et al., 2006; Sharma and Dietz, 2009 and Siddique and Dubey, 2017). The capacity of proline accumulation, depending on the level of stress and may also vary from one species to another species, up to a high degree of folding control (Hayat et al., 2012 and Verbruggen and Hermans, 2008). It is now established that the accumulation of proline in plants under adverse or stressful conditions plays a significant role in the development of stress tolerance capacity and acts as an osmoregulatory compound (Hare and Cress, 1997). Heat stress, salt stress and heavy metal stress are another major abiotic stress (Anaytullah et al., 2012; Munns and Tester, 2008 and Siddique and Dubey, 2017) that inversely affect crop growth in terms of biomass and crop yield (Ahmad et al., 2012; Mantri et al., 2012 and Karur et al., 2012). In saline situations, osmotic regulation is an important mechanism at the cellular level because plants accumulate many compatible solutes inside cells, such as proline, polyols and betaine, which play an important role in alleviating the effects of salinity stress in barley plants (Hasegawa et al., 2000 and Chen et al., 2007). At the same time, heat stress also reduces crop growth and accelerates the accumulation of different organic base biochemical compounds, i.e. proline, betaine and sugar, by adopting specific strategies. However, the synthesis of proline under different types of stress results in the cell structure of the stabilization process (Anaytullah et al., 2012; Kaushal et al., 2011 and Ashraf and Foolad, 2007). Osmolyte synthesis (osmoregulatory compound), in particular proline, enables plants to develop osmotic adjustment through the osmoregulatory process under diverse environmental conditions. It is known that the osmoprotective compound can stabilize the effects of stress at the cellular level (Aziz and Khan, 2001).

Role of Proline in Heat Stress

Heat stress, in terms of temperature increase beyond the threshold, has a detrimental effect on crop growth and the development of crops around the world. The harmful effect of heat stress can be overcome by developing thermotolerance genotypes using modern approaches. There is, therefore, an urgent need to understand the plant's response to heat stress. In general, the term heat stress may be defined as an increase in temperature beyond the threshold for some time sufficient to cause permanent damage to plant growth and development. When the temperature rises more than 400C, plants suffer a higher amount of water loss due to stomata that reduce the chlorophyll content and ultimately the rate of photosynthesis and respiration and increase molecular membrane damage (Gosavi et al., 2014; Kumar et al., 2012 and Jagtap et al., 1998). A well-known concept of increased temperature. It creates an imbalance between the rate of photosynthesis and respiration. The rate of photosynthesis decreases due to a decrease in the activity of RUBISCO at the carboxylation site. The rate of photorespiration increases consistently as the temperature exceeds the threshold level (Fig.1) as a result of this, the growth and development of the plant are adversely affected; Genetic approaches may be one of the tools that help in the development of plant variety thermotolerance to mitigate the effects of heat stress (Wahid et al., 2007). Wide ranges of compounds are currently known to help protect the plant from heat stress injury within the plant by maintaining osmotic and turgor pressure that facilitates the movement of water within the plant cells. This type of compound is known as compatible solutes. Proline is one of the compounds that play a very important role in the mechanism of tolerance against various types of stress (Shevyakova et al., 2009 and Anaytullah et al., 2012). Proline not only acts as osmolytes but also plays a very fine role in the inner side of the plant, i.e. metal chelator, antioxidant defence and signalling molecule (Hayat et al., 2012). The effect of early-stage heat stress appears at the cellular level, in the particular plasma membrane and biochemical reactions (Sung et al., 2003 and Anaytullah et al., 2012). Increased plant proline content is osmoprotective for plant cells due to heat stress (Kumar et al., 2012). The most popular biosynthetic pathway of proline synthesis in the cytoplasm is Glutamate via a-Glutamyl phosphate and Glutamyl-a-semialdehyde, and the initial two steps of proline biosynthesis are initiated by a—1-pyrroline-5-carboxylate synthetase (Parida et al., 2008). Stress-related proline accumulation plays a multifunctional role within the plant, such as low molecular weight chaperons, stabilizing proteins and enzymes, and scavenging reactive oxygen species (Hameed et al., 2012, Anaytullah et al., 2012).

Role of Proline in Salinity Stress

Word "salinity" is a diverse form of soil-related problems when an excessive amount of salt is collected. This diverse form of salt collection within the soil can be divided into two types, particularly alkaline and saline soils. The main root cause of saline soil development is the mismanagement of soil irrigation water due to the scarcity of salt-leaching drainage facilities resulting from salty soil (IRRI, 2011; Iyenger and Reddy, 1994). As a result, soil salinity plays a very large role in improving crop growth and growth from the initial stage of seed germination to crop yield. The intensity of the side effect of salinity may vary depending on the amount of salt deposition in the soil and the tolerance level of the particular crop (Lauchli and Grattan, 2007 and Munnset al., 2002). Various attempts have been made by the worker to identify some important selection criteria for salinity tolerance during the early stage of sowing, especially up to the planting of seedlings, but initial screening or evaluation is of little importance in terms of yield under salinity stress conditions (Igartua et al., 1994 and Munns et al., 2002). Production of reactive oxygen species, i.e. H_2O_2,¬O_2-and OH^+ radicals, increased under various types of stress (e.g. salinity, heat, heavy metal, dry matter, high light intensity, etc.) in which the hydroxyl radical is one of the most reactive molecular compounds. These reactive oxygen species may, therefore, make a significant contribution as a harmful molecule to the adverse environment (Apel and Hirt, 2004; Das and Roychoudhury, 2014; Hasegawa et al., 2000; Prasad et al., 1994). The primary strategy to mitigate the effects of salinity stress is the accumulation of compatible plant solutions (Sakamoto and Murata, 2000 and Hasegawa et al., 2000). In the case of a salt-sensitive variety of barley, the accumulation of osmolytes proline, glycine betaine and polyols along with $NaCl_2$ induced K^+efflux was identified as an effective indicator of salt stress tolerance (Di Martino et al., 2003 and Chen et al., 2007). The study between the sensitive genotype and the tolerance genotype reported that the proline content increased under salt stress conditions. While the average increase in proline was fivefold in the root and twentyfold in the leaf (Chen et al., 2007. Munns, 2002) Instead of the conventional role of compatible solutes, the synthesizers play a multifunctional role in stabilizing the plan during salinity stress.

Role of Proline in Heavy Metal Stress

Industrialization, rock weathering and urbanization are the main source of soil and water pollution caused by heavy metals, which is one of the most important natural resources. Because of this activity, many harmful heavy metals such as Cu, Pb, Cd, As, Ni, Mn, Zn and others enter the soil, water and air, causing serious damage to plants and animals. These heavy metals are very useful for plant growth and development when they are within a limited range but will be very harmful beyond their limits (Hall, 1981). Various methods have been used to make plants defensive against heavy metal toxicity, including agronomic practices (Siripornadilsul et al., 2002). Many scientists believe that proline synthesis in the plant can be the result of any kind of injury from either heat, cold, salinity or heavy metal (Anaytullah et al., 2012; Anaytullah and Bose, 2007; Chen 2007and Lutts et al., 1996). Proline biosynthesis accelerated when plants are treated with toxic heavy metal content while the activity of SOD and membrane injury increased (Bose et al., 2008; Siddique and Dubey, 2017; Schat et al., 1997). Accumulation of highly induced toxicity within the plant (a type of stress) is therefore synthesized by proline and helps to protect the plant (Siddique and Dubey, 2017and Kavi Kishor et al., 1995). Various types of stress commonly induced oxidative stress, which leads to a process known as lipid peroxidation, thereby disrupting cellular levels, in the particular plasma membrane, resulting in potassium leakage from the plant cell, while research also shows that exogenous application of proline not only suppresses heavy metal-induced lipid peroxidation but also pot peroxidation (Table 1 & 2).

Table 1: Role of different enzymatic and non -enzymatic antioxidants: enzymatic antioxidants

TYPE ANTI-OXIDANT	LOCATION	ROLE
SOD	Chloroplasts, mitochondria, peroxisome, apoplast	It scavenges the O_2^- and transforms into hydrogen peroxide. Reduces the OH^- it is highly toxic
Ascorbate peroxidase	Chloroplast, cytosol, mitochondria, peroxisome, apoplast.	It scavenges the hydrogen peroxide
Catalase	Peroxisome, glyoxisome	It eliminates the H_2O_2.
GR	Chloroplast, mitochondria, cytosol	It reduces glutathione.
DHAR	Chloroplast, mitochondria, peroxisome	Restoration of ascorbate.
MDHAR	Chloroplast, mitochondria, cytosol	Formation of ascorbate by reducing of MDHA.

Non-Enzymatic Antioxidants

TYPE ANTIOXIDANT	LOCATION	ROLE
α -tocopherol	membranes	It removes OH, O_2^- and singlet oxygen.
carotenoids	chloroplast	It removes the singlet oxygen and protects from photo-oxidation.
Glutathione	Chloroplast, cytosol, mitochondria, apoplast.	It removes peroxides.
Ascorbic acid	Chloroplast,cytsol, mitochond ria,peroxisome,apoplast	It plays an important role in the removal of H_2O_2.

Reactive oxygen species and antioxidants like ascorbate, GR, Tocopherols (Vit-E) act as signal transmission in plants, which activates defence mechanisms against stress. Among them different non-enzymatic, ASA is an important component and plays a main role in the removal of Ascorbic acid has been involved in various roles in the plants and it protects from abiotic and biotic stress. It acts as an enzyme cofactor, involves in cell division, growth and expansion of the cell.

Ascorbic Acid as Antioxidant Molecules

In plants with the response to biotic stress and abiotic stress (salt, drought, heavy metals etc.,) it leads to an increase in the production of reactive oxygen species.

These radicals disturb the regular functions of the cell by oxidating the proteins, lipids, and DNA. The important role of ascorbic acid is to react with this Reactive oxygen species (ROS), it produces MDHA, and it is catalyzed by APX. Then two molecules of MDHA are transformed into DHA and ASA. DHA and MDHA must be transformed back to ascorbate to maintain the ascorbic pool or else it gets transformed to 2, 3-diketogulonic acid.

In plants, L-ascorbic acid is the most common water-soluble antioxidant. Under stress conditions, there is an increase in ROS and that orderly insist for antioxidant. Accordingly, during the crucial state, the ascorbic acid level and oxidoreduction condition of ASA is sustained via converting ASA-GSH cycle. These two molecules like GSH and ascorbic acid differs from others due to a few significant characters. They are: The specific compounds like ASA and GSH are associated with the metabolism of hydrogen peroxide. They can reclaim the reduced form through the enzymatic system via electron transport NADPH molecule.

Role of Ascorbic Acid as a Defense Mechanism Against Stress

When plants are exposed to adverse abiotic stress, affects the increased level of ROS production, which leads to oxidation of lipid and proteins, inhibits the activities of enzymes and finally causes the death of the cell. The compounds of enzymatic and non-enzymatic antioxidants support in-plant cell detoxification system. These antioxidants Help to maintain Reactive Oxygen species production to the optimum level.

During stresses conditions, compounds like ascorbate and glutathione perform an important role in defending the oxidoreduction signals in a plant cell. In the ascorbate-glutathione cycle, GSH acts as a reducing factor of DHA. In addition to this, it removes Reactive Oxygen Species. To maintain cellular equilibrium a redox couple (GSH/GSSG) is formed by combing the Reduced GSH and oxidized GSSG.

Under stress, conditions to conquer the negative effects of oxidative damage the accumulation of GSH is to be increased. Ascorbic acid is expressed to be a most significant antioxidant in plant cell and it found in many divisions pf plant cell like chloroplast, mitochondria, cytoplasm and apoplast during non-stress conditions ASA found (or) available in reduced forms. In many enzymatic and non- enzymatic reactions, ascorbic acid acts as an electron donor and it helps to protect the cell against oxidative damage.

Role of Ascorbic Acid against Toxicity Caused by Heavy Metals

Heavy metals aids to serious environmental pollution when they are available in excess quantities in soil and it leads to a decrease in plant growth and shows toxicity systems in plants. Moreover, it creates oxidative damage because of excess ROS synthesis. To defend against the oxidative damage plants develops antioxidant system consists of different enzymes and low molecular weight compounds. When plants are exposed to cadmium it shows that increase in the concentration of ascorbic acid and that indicates the association of this compound in maintaining the oxidoreduction equilibrium in a cell under stress conditions. If plants exposed to an excess concentration of cadmium i.e., 100 mM then ascorbic acid levels are decreased. Then an extrinsic application of ascorbate increases the levels of ascorbic acid, GSH and it acts against the cadmium stress in rice seedlings. When plants subjected to copper toxicity, it had shown the reduced activities of GSH and GP in the cell under short-term exposure of plants to Cu. The ratio of GSH/GSSG performed a major role under these conditions. The excess supply of copper contributes to increasing the content of ascorbic acid pool. In Arabidopsis when they are exposed to copper, GST and GPX help to govern the maintaining of redox balance in the cell. In potato, Al toxicity leads to inhibit the root growth. The increased activity of GST shows tolerance to Al toxicity. If Rice seedlings exposed to Al stress it showed the increased activity of enzymes like ascorbate-glutathione it helps plants to deal with oxidative damage.

In transgenic plants of tobacco, the overexpress of cytoplasmic MDHAR/ DHAR which can tolerant against theAl^{+3} stress because of the above enzymes. MDHAR and DHAR, which also helps in the restoration of, reduced ascorbic acid. Besides, found this DHAR in preference to MDHAR overexpress transgenic plants exhibited improved root growth than others. When plants exposed to Nickel stress, application of ascorbic acid shown better growth in plants because reduced the translocation of Ni from root to shoot when compared to plants without ascorbic acid. When soybean seedlings exposed to NaCl of 0.25 mM concentration for 5 days it induced the toxicity. Anyhow, these toxicity effects can be reversed by the external application of ascorbic acid with 1mM concentration. When leaves of *Arabidopsis thaliana,* exposed to Cadmium stress revealed that increased in the expression of gene related to the synthesis of GSH and Pcs, which are involved in different antioxidant enzymes and involved in different antioxidant enzymes and this help to scavenge the Reactive Oxygen species. However, genetically engineered crops can enhance the levels of endogenous ascorbic acid, which increases the tolerance to different abiotic stress.

Osmolytes

When plants are exposed to various abiotic stresses and it affects the plant growth by affecting the various physiological and biochemical mechanisms like photosynthesis, antioxidant etc. These stresses encourage the defence mechanisms in the plant, prevents injury, and raise their survival ability during adverse stress conditions. Due to stress conditions, plants adjust to their cellular and molecular action and this finally increases their growth and development. During abiotic stress, plants aggregate compatible solutes or osmolytes to protect the cell structure. The much known compatible solutes are GB (Glycine betaine), Sugars (mannitol, sorbitol & trehalose) polyamines and proline. These compatible solutes are obtained by accumulation during different abiotic stresses and it contributes to tolerance of cell without disturbing the cellular mechanisms of plants. During stress, a condition in plants accumulation of sugars like Trehalose, galactinol and mannitol and distinct genes performs a significant role in the synthesis of organic solutes, which helps to develop the tolerance in transgenic plants against abiotic stresses (Fig. 3).

Fig. 3. Schematic representation of the role of osmolytes during stress conditions

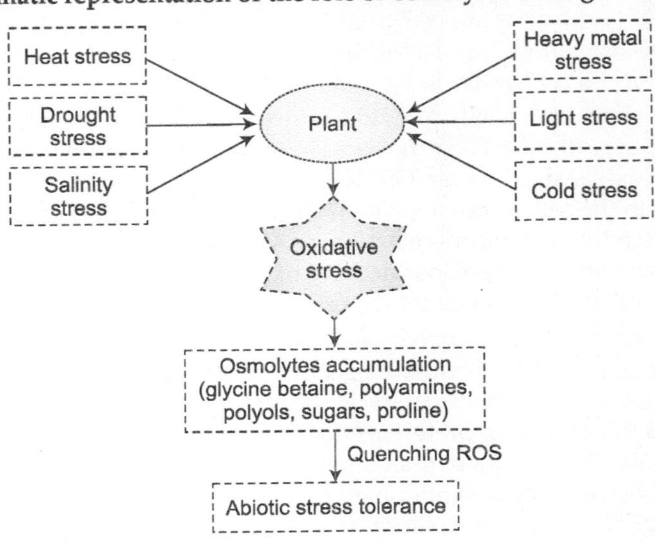

(Source: Sharma, 2019)

Osmolytes are consistent Osmoprotectant solutes that increase the cell capacity to maintain water level without inhibiting their functions. The main function of these compatible solutes is to adjust the osmotic regulation. These compatible solutes help to protect the plant from severe osmotic damage in their life cycle. These organic solutes balance the osmatic difference in surrounding cell and cytoplasm. Additionally, they protect from oxidative stress by constraining the production of reactive oxygen species, these are neutral compounds and protects the proteins and other membranes of the cell, cellular metabolism against different stresses.

Osmoprotectant is categorized into four groups. They are Amino acids, Quaternary ammonium compounds, sugar and sugar alcohols or polyols. Types of Osmolytes and their functions are mentioned in the below table.

Table 2: Group of Osmoprotectant, its types and functions

Group of Osmoprotectant	Types	Functions
Amino acid	Proline, analine, γ-aminobutyric acid etc.,	It acts as a precursor and protects from the toxicity of ions and membrane damage.
Quaternary ammonium compounds	GB, β-alanine betaine, proline betaine.	Prevents from membrane damage Detoxifying the ROS Activation of enzymes.
Sugars	Sucrose, trehalose, fructose, maltose	Osmotic balance and stabilize the membranes.
Sugar alcohols or polyols	Mannitol, sorbitol, pinitol	Maintains osmotic balance and acts as molecular signalling.

Role of Osmoprotectant as Antioxidants

The adverse effects of abiotic stress are prevented by the Osmolytes by the protection of membrane and inducing the ROS scavenger. These solutes also influence the activation of antioxidant enzymes. During stress conditions, accumulation of methylglyoxal takes place it is a cytotoxic compound. It can be detoxified by enzyme-like glyoxalase I & II and found that proline and GB also help in detoxification of methylglyoxal by increasing the activities of glyoxalase.

Amino acids

γ-aminobutyric acid: In plants, biosynthesis of γ-aminobutyric acid is from glutamic and an enzyme named glutamate decarboxylase catalyzes it. Normally, intrinsic accumulation of this solute is low in plants but during stress conditions, high levels of these solutes are produced to relieve the adverse stress conditions. However, these, γ-aminobutyric acid accumulated in high concentrations in tobacco leaves then the proline content because of high GADI gene expressions, which regulate the biosynthesis of r-aminobutyric acid under stress conditions. γ-aminobutyric acid has more ability to scavenge the H_2O_2 than proline and inhibits $^1O2$ production. The role of γ-aminobutyric acid acts as a stabilizer for oxidative stress. Besides, it reduced the H^+ and AL^{3+}, which influence the oxidative damage by raising the synthesis of antioxidant enzymes, which helps to scavenge the reactive oxygen species and decrease the peroxidation of lipids. Additionally, enhances the enzyme activation, which is involved in the AGH cycle. γ-aminobutyric acid is also used as priming of seedlings which reduce the accumulation of malondialdehyde it acts as a signal for oxidative damage and also shown early scavenging of ROS on the counter the seedlings are stressed by polyethene glycol are primed with γ-aminobutyric acid and seen that enhancing the activities of defence mechanisms of antioxidants. Proline is the most significant osmoprotectant or osmolytes because of its major role in stress tolerance. The extrinsic application of proline enhances the extrinsic contents in plant tissues when they are exposed to water stress conditions and it helps to maintain osmotic balance in plants. During stress, conditions in higher plants high amounts of proline accumulation take place and for maintaining the osmatic balance, membrane and proteins, scavenge the free radicals, Redox buffering during stress conditions. High accumulation of proline against stress and provide adequate reducing agents that maintain the oxidative phosphorylation in mitochondria and ATP produce to regain against and restore the damages caused by stress. This protects membranes against the extreme effect of the high range of inorganic ions and temperature. It also acts as protein compatible and scavenges the OH^- radical.

Quaternary ammonium compounds: In plants during adverse environmental conditions leads to accumulation of quaternary ammonium compounds like GB, β- adenine betaine, proline betaine.

Glycine betaine: In association with several quaternary ammonium compounds in plants, GB (Glycine-betaine) appears most generously in response to drought stress. It is luxuriant mostly in chloroplast where it performs an important role in adaptation and safeguard of the thylakoid membrane, so that maintains the photosynthetic ability. In plants, Glycine betaine is synthesized in chloroplast from serine through ethanolamine, choline and betaine aldehyde. The accumulation of two important osmoprotectants like GB and proline in various plant species due to abiotic stresses like drought, salt stress, UV radiation Heavy metal stress. These osmoprotectants have positive effects on enzyme and tissues besides adjust the osmotic balance in plants during stress conditions. When cells are exposed to stress involves alteration in cell cycle & cell division, cell wall structure, vacuoles all lead to an increase in the tolerance of cells against stress. Plants change mechanism in several ways to adopt the abiotic stress at biochemically by producing the osmoprotectant compounds like proline and GB. Some plants accumulate high amounts of GB when exposed to high salts stress, cold and drought. It also protects the protein and cell membranes during stress conditions. GB accumulates in main plants like sugar beet, spinach, barley, wheat and sorghum. Particularly tolerant varieties accumulate more Glycine-betaine than others do and they are variety specific. Not all plant species are equally capable of producing osmolytes when exposed to stress. It is very complex to complete plant the difficulty between the stress factors and various plant mechanisms affects the growth and development of the plant. The extrinsic application of proline as pre-sowing affects the K^+, Ca^{2+}, P and N levels of root and proline in shoot N levels.

Sugars: Among the Osmoprotectant, sugars like sucrose, Fructose, raffinose helps the cell to overcome the different stresses and regulates their normal plant metabolism. They act as an antioxidant in plants during stress conditions high amount of sugars are produced because of photosynthetic activity and it causes the high generation of cytoplasmic hydrogen peroxide and in converse, their sugars in pentose Phosphate pathway via oxidative stage induces NADPH which is required for synthesis glutathione, therefore it helps for scavenging of reactive oxygen species. Sucrose, in plants, it is abundant in nature. Additionally, acts as a substrate, molecular signalling and generation of osmotic pressure. It as positive function in scavenging of ROS in plants. It aids the ascorbic aids accumulation (an antioxidant), anthocyanin for quenching the various types of ROS and nitrogen species. The extrinsic supply of sucrose enhances the levels of ascorbic acid in plants. It activates the antioxidants like enzymatic and non-enzymatic systems in plants during the stress conditions and increases the tolerance of plants. Trehalose is an essential sugar soluble osmoprotectant commonly used by the cells to adopt osmatic pressure in the cell to prevent cellular damage due to oxidative stress. In the recent report shown that sugar signalling transmission plays an important role in hastening the photosynthetic operation of plants to its peak rate in relation with Trehalose. Due to these positive effects on gas exchange and osmoregulation and that influences on opening stomata. In wheat cultivars during water stress, condition application of trehalose increased the growth due to its role in osmatic balance. Various plant species behave dissimilarly on the extrinsic application of trehalose the development of a plant is restricted by the exogenous application of these compounds results in inhibition of growth or yield. In wheat cultivars, trehalose enhanced the biomass production in shoots and roots during water stress conditions. When plants exposed to stress conditions the extrinsic application of trehalose and proline enhances the internal levels of these solutes, commonly increases the plant growth and crop yield during stress conditions.

Sugar alcohols/ polyols: It is also known as polyols, consists of inositol, mannitol, sorbitol etc., It aids in scavenging of reactive oxygen species. OH^- is detoxified by mannitol, sorbitol and Myo-inositol are been reported earlier.

Myo-inositol: It is not directly involved in ROS scavenging. It acts as an initial substratum for the synthesis of ascorbic acid and indirectly scavenge the Reactive oxygen species. The overexpression of a gene, which helps in biosynthesis of ascorbic acid by an enzyme, called Myo-inositol oxygenase and it enhances the ascorbic acid levels two to three times. Hence, it is also called as a precursor for ascorbic acid. Also, it acts as a precursor for raffinose synthesis and if these solutes are accumulated in high levels acts as ROS scavenger. Present reports on osmoprotectants are based on the reduce the osmotic damage involves the usage of overexpressed transgenic plants, extrinsic application of osmoprotectants on plants.

Role of Osmolytes against Heavy Metal Stress

Heavy metal stress induces the oxidative damage and high production of reactive oxygen species in plants. It impairs the cellular biochemical mechanisms and plant physiology. When plants exposed to Pd it reduces the concentration of leaf pigment in *Corconopus didymus* and due to this reduced the synthesis of chlorophyll and degradation against the tolerance against the stress. Consequently, Plants are improved in different mechanisms for heavy metal detoxification and enhances the tolerance against stress. When plants are exposed to Cd, Ni, Cu, Zn, Mn, Cr. They observed an increase of various osmoprotectants like sucrose, mannitol and glycine-betaine. In wheat, plants during heavy metals stress the extrinsic supply of Glycine-betaine increased tolerance of plants and found that increase in root-shoot length, biomass, pigments and it supports to ROS scavenging and osmotic balance.

Role of Osmolytes against Salinity Stress

In plants salinity causes peroxidation of lipids, disturbs water balance, imbalance of nutrition, disturb the enzyme activity and production of reactive oxygen species that finally destroys the photosynthetic mechanisms and a reduction in growth due to initiation of ion toxicity and oxidative damage. Plants tolerant to salt stress by the accumulation of low molecules weight osmoprotectants like proline, Glycine-betaine and polyamines this helps to maintain the steadiness of membrane. This osmolyte increases the germination rate, growth and development by stimulating tolerance against the salinity.

In plants, proline accumulation is a defence mechanism against salinity. It enhances tolerance of plants against stress and activates of catalase and peroxidase enzymes that play an important role in antioxidant defence systems. GB is an organic water-soluble osmolyte protects during stress conditions by maintaining proteins (RUBISCO) and reduces the production of ROS and it enhanced the survival ability of plants against salinity.

Role of Osmolytes against Drought

When plants exposed to drought stress it effects on enzymes, which are related to the biosynthesis of chlorophyll. Therefore, that plant has adopted different mechanisms against drought stress. Among this accumulation of Osmoprotectant takes place they are sugars, GB, proline and these help in water balance during drought stress. Due to the accumulation of these solutes in the cell during stress, OP of the cell becomes negative which leads to endosmosis in the cell and sustain the TP of the cell. This osmotic balance is feasible due to the aggregation of Osmolytes. In that proline is most significant Osmoprotectant solute during drought stress. Glycine betaine also provides tolerance against stress in plants. Plant resist to drought stress is due to osmotic balance, antioxidant defence mechanisms that help in quenching of reactive oxygen species and maintains stability in membranes.

Tocopherol (Vitamin E)

Plants are exposed to various abiotic stress conditions like salinity, metal toxicity, drought, UV- radiation, ozone, temperature extremes. This is liable for inhibiting the plant physiology, growth and development and it produces the high ROS, which causes oxidative damage in plants. To overcome this condition plants possess defence mechanisms system like antioxidants. In that non-enzymatic antioxidants includes ASA, GSH, Tocopherols, amino acids, alkaloids etc., these are present in various cell organelles and they work concurrently to hold the ROS beneath the toxic level and prevents from oxidative stress.

Tocopherols are lipid-soluble antioxidants, which undergoes different functions in plants and animals. It occurs in four distinct forms they are α (Alpha), β (Beta), γ (Gama), δ (Delta). These are broadly synthesized in higher plants and algae. In plants synthesis of tocopherol are seen in the plastid. There are two compacting pathways for synthesis of tocopherol and various enzymes regulate it. However, tocopherol accumulation is varied among different plant species and various parts of the plant. Many studies stated that increased levels of tocopherol could provide tolerance to various abiotic stresses like salinity, metal toxicity, ozone, UV-radiation, extreme temperature. It plays an important role in maintaining oxidoreduction equilibrium. It acts as ROS scavenger mainly O_2 and OH^- reduces the lipid peroxidation.

It is clear that during stress conditions plant increase the synthesis of tocopherol and it contributes to protecting from oxidative damage by restricting the ROS production and they have proline role in the germination of seed and growth and development of plants. The roles are interlinked with phytohormones and take part in signal transmission during biotic and abiotic stresses to accelerate tolerance in plants. In plants, endogenous levels of tocopherols are regulated by the extent of the environmental stresses. Gene response and gene expression play an important role in increasing the tocopherol content in the plant for better tolerance. Additionally, the extrinsic supply of tocopherols also does success in protecting the plants during stress conditions.

Synthesis and Accumulation of Tocopherol in Plants

Tocopherols are broadly synthesized in photosynthetic organisms like higher plants and algae. In plants, synthesis of tocopherols takes place in plastid and enzymes are necessary for the synthesis of tocopherols is mostly related to the membrane of plastid. There are two compacting pathways for synthesis Toc. They are SHIKIMATE pathway and other MEP (methylerythritol phosphate pathway). These are confined to cytosol and plastid. These pathways caries the head group and side-chain block. Simultaneously, in detail, the shikimate pathway is important for making an aromatic ring at the same time MEP pathway forms PDP for the tail of tocopherol. In tocopherol, the hydroquinone ring is derived from the synthesis of aromatic amino acid of the shikimate pathway, it acts precursor, and it is constituted by oxygenation of HPP by the action of HPPD. This pathway consists of various steps; first Glycolysis (PEP) and another form the intermediate from pentose phosphate pathway and these are converts into chorismite. The extrinsic toxic substances (i.e., herbicides) can inhibit these steps. Glyphosate is herbicides it targets the EPSP, which inhibits the synthesis of various metabolites like tocopherols.

In tocopherols, the first intermediary compound is HPP is obtained from preheated by the act of PDH through arogenate and tyrosine. In higher plants, HPP is formed from tyrosine and it is catalyzed by TAT and that is affected by different factors (Jasmonic acid, caronatine etc.,) the extrinsic supply of these chemicals also increase the production of tocopherol by enhancing the activities of enzymes. Then HPP is later oxygenated to HGA by the activation of HPP dioxygenase. This product is combined with the MEP pathway products.

In MEP pathway has many steps for PDP formation. The product of PDP is obtained from precursor molecules like IPP and DMAPP. In higher plants, IPP is reported as an end product of the MEP pathway beside it forms C-20 intermediate, GGDP. It acts as a precursor for the synthesis of CAR, PDP, tocotrienols etc. This multi-step reaction takes place in the membrane of plastid. GGDP is transformed to PDP by GGR enzyme. In plants the reduced activity of GGR effect the level Chl and Toc. Due to geranylgeranylated chl.

In plants, PDP is formed by phytol through Chl catabolism by enzyme chlorophyllase. It is indirectly used as a precursor for the synthesis of tocopherols otherwise; it is phosphorylated into PP and PDP by the activity of PKK. The final step for synthesis of TOC is reduced of HGA with PDP by HPT enzyme and forms MPBQ is then methylated to DMPBQ by the action of MPBQ MT. gamma and delta-tocopherols are formed from MPBQ & DMPQ by the action of an enzyme Tocopherol cyclase. The synthesis of tocopherols is affected by various factors like the age of the plant, cell organelles, hormones etc. The synthesis of TOC is affected by plant hormones like abscisic acid, salicylic acid, JA. In barely and Arabidopsis, synthesis of tocopherol genes are controlled by jasmonic acid. In *Phillyea angustifolia* during drought stress. Endogenous salicylic acid enhanced the synthesis of α-Toc. Ascorbic acid also controls the genes, which are responsible for HPPD, which enhances the tocopherol synthesis. The Toc accumulation is most varied in plant species and various plant parts high levels of tocopherol are found in tissues of oilseed crops like soybean, mustard, coconut, sunflower etc., seed as total tocopherol content when compared to other plant parts. These tocopherols mostly occur in plastids but in few cases, it is seen in cytoplasmic lipids. Mostly in leaves α- tocopherols are seen but in seeds contain γ-TOC and other forms like β and δ are rarely seen in plants.

Role of Tocopherol as Antioxidant in Plants

It is a fat-soluble non-enzymatic antioxidant. It efficiently eradicates the ROS production especially O_2 and $OH^{\cdot}$. The synthesis of tocopherol during abiotic stress provides safeguard despite the oxidative damage by reducing the production of Reactive Oxygen Species. The high content of Tocopherols is present in the chloroplast thylakoid membrane and the main site for the production of ROS. Due to the same location for the synthesis of tocopherol and ROS, enhancement of α-tocopherol may provide the ability for ROS scavenging such as O_2 and peroxyl radicals. Protects the photosynthetic mechanisms from oxidative damage and peroxidation of lipid. The enzymes like MPBQ and DMPBQ exhibits in antioxidant activity by the offering of two electrons. Mostly the main role of Tocopherols is an antioxidant to balance the structure of the membrane, which are involved in reactions by polyunsaturated fatty acyl chains. (PUFA). In plants during oxidative damage causes lipid peroxidation it is an important parameter and it is started by formation of the alkyl radical by PUFA under oxidative damage. The main role of tocopherol is reacting with alkyl radical this is the significant result under peroxidation of lipid α- tocopherol has high antioxidant ability due to the high activity of vitamin E and fastly reacts with peroxyl radicals. Additionally, removes the properties of radical by oxidation of lipid.

Tocopherol is an antioxidant to scavenge the free radical is related with a capacity to donate of phenolic hydrogen to a free radical of lipid. Especially tocopherol as high ability to scavenge the O_2 and high amounts of ROS and reports stated that one molecule of α-tocopherols can scavenge O_2 up to 120. High synthesis of α-toc in transgenic. *A.thaliana* is due to overexpression of tocopherol synthesis gene and seen less production of O_2 and hydrogen peroxide when compared to non-transgenic plants. While the action of scavenging the O_2 irreplaceable of quinones and epoxides takes places, during the transformation of peroxyl radical to form tocopherol and it can be restored to α-tocopherol in the ascorbate-glutathione cycle. Therefore, tocopherol has become a correlation between the ASA-GSH cycles by maintaining the oxidative balance in the chloroplast. Tocopherol and plant hormones have a complex process, which is directly or indirectly involved in signalling the cell, regulation of gene against stress tolerance.

Role of Tocopherol in Metal Toxicity

Metal toxicity changes the physiology and biochemical aspects in plants. The long period causes of low concentration or short period of high concentrations of heavy metal are sufficient for consequent in death of the plant. In duckweed, when exposed to Zn and cadmium stress the enhanced content of tocopherol has provided tolerance. Similarly, results are seen when lupin is exposed to Pb stress.

In wheat, seedlings when exposed for nickel 50 and 100 mM for 9 days results are started that 38% and 50 % increase in the tocopherol content in shoots.

If *A. thaliana* is exposed to copper and cadmium ion and it has seen 6 to 5 times increase in the tocopherol level. Similarly, tocopherol deficient mutant showed suspectable to cadmium and copper than the wild type. In contrast, the level of tocopherol is decreased when exposed to stress. In transgenic *B. juncea* plants, overexpression of γ-TMT gene shown an increase in tolerance to cadmium stress. In Transgenic, plants have seen that increase in the action of SOD, CAT, GR and APX and reduction in the levels of H_2O_2, MDA and electrolyte leakage than wild type. Hence, the level of ASA and GSH are low in transgenic plants. In *B. juncea* exposed to $CdCl_2$ of 20mM shown 1.5-9.3 times enhanced levels of α and γ tocopherols.

Role of Tocopherols in Salt Stress

Salt stress is considered as important abiotic stress it causes disturbance in main physiological operations like photosynthesis, uptake of water and nutrients, respiration, toxic metabolites and generation of Reactive Oxygen species, peroxidation of fatty acids. All these are liable for obstructing growth and development of plants. To overcome these plants adopt several mechanisms and those are regulated by tocopherols are presented here: In L. *esculetum*, plants are exposed to NaCl stress of 50 and 150 mM it enhanced the levels of peroxidation of fatty acid, Chl degradation, engaged metabolism of antioxidant. During this increased tocopherol, level enhanced the GPV, GST, contents which are correlated with the antioxidant system and this tocopherol acts as signalling pathway during Stress conditions. In *B. juncea*, The overexpression of γ-TMT gene helps to regulate trans formation of γ-Toc to α-tocopherol which enhances the tolerance to salinity. Exogenous application of tocopherol in *T. aestivum* during salt stress considerably increases the activities of antioxidant enzyme and the accumulation of ASA, carotenoids, phenol takes places, which it helps to reduce the H_2O_2 and peroxidation of fatty acid and amended the permeability of membranes. By application of α-tocopherol reduced the accumulation of Na, K, Mg, Ca levels collectively seen in the delay of senescence and promotes tolerance in plants.

In transgenic plants tobacco plants with low levels of tocopherol are more susceptible to salinity contrarily, α-tocopherol accumulates has shown high resistance to oxidative damage when compared to γ-tocopherol against salt stress. It as functions in control the competence of PS-II, oxidative damage, Na^+/K^+ equilibrium, imbalance in hormones that are related to salinity. Mutants with high levels of α- and γ- tocopherols reduced the H_2O_2 and MDH levels in plants during salinity.

Role of Tocopherols in Drought Stress

In drought stress plants, tocopherol plays a significant role by depleting excess energy during oxidative damage and as an ability to scavenge the reactive oxygen species. Additionally, it regulates the levels of ROC, Peroxidation of fatty acid and also regulated the aggregation of PUFA, oxylipins and plays an important role in signal transmission during drought stress. Various plant studies stated that positive relationships between the synthesis of tocopherols and drought stress. Many reports stated that during drought conditions there is a prominent increase in levels of α-tocopherols in crops like pea, wheat, rosemary and cereals. In transgenic tobacco plants, the overexpression of gene Vte1 increased the synthesis of α-tocopherols, provides tolerance to drought conditions, and related with controlling the defence mechanisms of antioxidants like reducing peroxidation of fatty acids, leakage of an electrolyte, Hydrogen peroxide levels. In current studies showed that transgenic *B. juncea* plants have enhanced with α-tocopherol due to overexpression of γ-TMT gene during drought stress. These transgenic plants revealed the elevated the activities of SOD, CAT, APX, GR and reduced the contents of MDA, Hydrogen peroxide when compared to others. The increased level of α-tocopherol also increases the photosynthetic ability, reduces the lipid peroxidation, and maintains the osmotic balance in plants.

Conclusion and Future Perspectives

In worldwide agriculture the indisputable consequences of abiotic stress are exhibited by the facts of abiotic components is accumulated and constitute for serious prerequisites in worldwide crop production. Accordingly, advance measures to recognize the physiology and molecular mechanisms of abiotic stress resistance in plants are essential in agriculture. It is feasible to reduce the losses in production because of abiotic stress by a reasonable mix of knowledge in plant physiology and crop management methods. The generation, metabolism and detoxification of Reactive Oxygen species are necessary measures in the plant for their adoption, growth and existence. The production and scavenging of ROS are important in defence mechanisms of plants, controlled by overexpressing of genes for detoxification of ROS, and enhances resistance against abiotic stress. The ability of engineered plants, like overexpression of genes for defence mechanism system, contributes to the possibility to develop the plants with increased resistance to abiotic stress. Improving transgenic lines by changing the contents of antioxidants and enhancing resistance against oxidative damage in stress-induced plants as a significant development. Application of extrinsic or exogenous osmoprotectants like proline, GB, trehalose, Ascorbic acid is important for regulating the antioxidant defence and to mitigate the oxidative damage because of non-enzymatic antioxidants systems. Additionally, in transgenic plants, the overexpression of antioxidants enzymes has shown positive effects in stress resistance. Hence, enhanced prominence is being set on developing transgenic, overexpression of gene i.e., related with more than one antioxidant to attain tolerance for various stress.

References

Ahmad, P., Jackel, C., Azooz, M.M. and Nabi, G. (2009). Generation of ROS and non-enzymatic antioxidants during abiotic stress in plants. *Botany research international, 2*(1), 11-20.

Ahmad, P., Jaleel, C.A., Salem, M.A., Nabi, G. and Sharma, S. (2010). Roles of enzymatic and non-enzymatic antioxidants in plants during abiotic stress. *Crit. Rev. Biotechnol., 30*, 161-175.

Akram, A.N., Shafiq, F. and Ashraf, M. (2017). Ascorbic acid, a potential antioxidant scavenger and its role in plant development and abiotic stress tolerance. *Frontiers plant science, 8*, 613

Ashraf, M.A., Iqbal, M., Rasheed, R., Hussain, I., Perveen, S. and Saqib, M. (2018). Dynamic proline metabolism; importance and regulation in water limited environments. In: *Plant metabolites and regulation under environmental stress*, 323-336. Academic Press. http://dx.doi.org/10.1016/B978-0-12-812689-9.00016-9.

Aslam *et al.* (2017). Specific role of proline against heavy metals toxicity in plants. *Int. J. Pure App. Biosci., 5*(6), 27-34.

Ashraf, M., and Foolad, M.R. (2007). Roles of Glycine betaine and proline in improving plant abiotic stress resistance. *Environmental and experimental botany, 59*, 206-216.

Bhaskara, G.B., Yang, T.H. and Verslues, P.E. (2013). Dynamic proline metabolism: importance and regulation in water limited environments. *Front. Plant Sci., 6*(484).

Blokhina, O., and Fagerstedt, K. (2006). Oxidative stress and antioxidant defenses in plants. *Oxidative stress, disease and cancer*, 151 – 199.

Dalvi, A.A., and Bhalerao, S.A. (2013). Response of plants towards heavy metal toxicity: An overview of avoidance, tolerance and uptake mechanism. *Annals of plant sciences, 2*(9), 362-368.

Dar, M.I., Naikoo, M.I., Rehman, F., Naushin, F. and Khan, F.A. (2016). Proline accumulation in plants: roles in stress tolerance and plant development. In: Osmolytes and plant acclimation to changing environment: *Emerging osmics technologies, Springer*, 155-166.

da Silva Lobato, A. K., Lima, E. J. A., Lobato, E. M. S. G., Maciel, G. M., & Marques, D. J. (2016). Tolerance of plants to toxicity induced by micronutrients. In *Abiotic and Biotic Stress in Plants-Recent Advances and Future Perspectives*, 230-246, *IntechOpen*.

Delavoney, A.J., and Verma, D.P.S. (1993). Proline biosynthesis and osmoregulation in plants. *Plant. J., 4*, 1037-1047.

De Tullio, M. C. (2012). Beyond the antioxidant: the double life of vitamin C. *Water Soluble Vitamins*, 49-65.

Ejaz *et al.* (2020). Role of osmolytes in the mechanisms of antioxidant defense of plants. In: E. Lichtfouse (ed)., *Sustainable agriculture reviews, 39*, 95-117.

Emaverdian, A., Yerlong, D., Mokhbrerdoran, F. and Xie, Y. (2015). Heavy metal stress and some mechanisms of plant defense response. *The Scientific World Journal,2015* http://dx.doi.org/10.1155/2015/756120.

Foyer, C.H., and Noctor, G. (2005). Redox homeostasis and antioxidant signaling a metabolic interface between stress perception and physiological responses. *The Plant Cell, 17*(7), 1866-1875.

Freeman, B.C., and Beattie, G.A. (2008). An overview of plant defenses against pathogens and herbivores. *The Plant Health Instructer.* doi:10.1094/PH1-1-2008-0226-01.

Fryer, M.J. (1992). The antioxidant effects on thylakoid vitamin E (α-tocopherol). *Plant cell and environment, 15,* 381-392.

Gallie, D.R. (2012). The role of ascorbic acid recycling in responding to environmental stress and in promoting plant growth. *Journal.Exp.Bot ,64*(2), 433-443.

Gangwar, S., Singh, V.P., Tripathi, D.K., Chauhan, D.K., Prasad, S.M., & Maurya, J.N. (2014). Plant responses to metal stress: the emerging role of plant growth hormones in toxicity alleviation. In *Emerging technologies and management of crop stress tolerance,* 215-248, Academic Press.

Gill, S.S. and Tuteha, N. (2010) Reactive oxygen species and antioxidant machinery in abiotic stress tolerance in crop plants. *Plant physio Biochem, 48*(12), 909-930.

Giri, J. (2011). Glycine betaine and abiotic stress tolerance in plants. *Plant signaling & behavior, 6*(11), 1746-1751.

Gupta, B. and Huang, B. (2014). Mechanism of salinity tolerance in plants: physiological, biochemical and molecular characterization. *Internatinal journal of genomics.* http://dx.doi.org/10.1155/2014/701596.

Hasanuzzaman, M., Hossian, M.A., Jaime, A.S. and Fujita, M. (2012). Plant response and tolerance to abiotic oxidative stress: Antioxidant defense is a key factor. *Springer science,* 261-315.

Hasanuzzaman, M., Nahar, K., & Fujita, M. (2014). Role of tocopherol (vitamin E) in plants: abiotic stress tolerance and beyond. In *Emerging technologies and management of crop stress tolerance ,* 267-289, Academic Press.

Hasanuzzaman, M., Nahar, K., Gill, S.S., Fujita, M. (2014). Drought stress responses in plants, oxidative stress and antioxidant defense. In: Gill, S. S. Tuteja N. (Eds), *Climate change and plant abiotic stress tolerance,* 209-249.

Hayat, S., Qaiser, H., Alyemeni, M.A., Wani, S.A. and Ahmad, A. (2012). Role of proline under changing environments, *Plant signaling behaviour lands biosciences, 7*(11), 1456-1466.

Hare, P.D., Cress, W.A., Van Staden, J. (2001). The effects of exogenous proline and proline analogues on invitro shoot organogenesis in *Arabidopsis. Plant growth regulation, 34*(2): 203-207.

Hall, J.L. (2002). Cellular mechanisms for heavy metal detoxification and tolerances. *Journal of experimental botany, 53*(366), 1-11.

Heuer, B. (2010). Role of proline in plant response to drought and salinity. *Hand book of plant and crop stress, CRC Press, Boca Raton,* 213-238.

Hossain, M.A., Hoque, M.A., Burritt, D.J., & Fujita, M. (2014). Proline protects plants against abiotic oxidative stress: biochemical and molecular mechanisms. In: *Oxidative damage to plants*, 477-522, Academic press. http://dx.doi.org/10.1016/B978-0-12-799963-0.00016-2.

Hossain, M.A., Piyatida, P., da Silva, J.A.T., & Fujita, M. (2012). Molecular mechanism of heavy metal toxicity and tolerance in plants: central role of glutathione in detoxification of reactive oxygen species and methylglyoxal and in heavy metal chelation. *Journal of Botany, 2012*.

Hussain *et al.* (2013). Biosynthesis, structural and functional attributes of tocopherols in plant: past, present, future perspectives. *J. Agri.Food Chem, 61*, 6137-6149.

Iqbal, M.J. (2018). Roles of osmolytes and antioxidants enzymes for drought tolerance in wheat production global wheat production *Global Wheat Production, 51*. http://dx.doi.org/10.5772/intechopen.75926.

Ishikawa, T., Dowdle, J. and Simirnoff, N. (2006). Progress in manipulating ascorbic acid biosynthesis and accumulation in plants . *Plants. Physiol. Plant, 126*, 343-355.

Jovanovic, S.V., Vidovic, M. and Morina, F. (2017). Ascorbic acid in plant growth, development and stress tolerance, *Springer international publishing A.G.* http://doi.org/10.10071978-3-319.74057-73.

Kaur, R., & Nayyar, H. (2014). Ascorbic acid: a potent defender against environmental stresses. In *Oxidative damage to plants*, 235-287, Academic Press. http://dx.doi.org/10.1016/B978-0.12-799963-0.00008-3.

Khan, S.H, Ahmad, N, Ahmad, F. and Kumar, R. (2010). Naturally occurring organic osmolytes: from cell physiology to disease prevention. *IUBMB life, 62*(12), 891-895.

Khan, T.A., Mazid, M. and Mohammad, F. (2011). Review of ascorbic acid cellular redox imbalance. *Trends plant sci., 14*(1), 43-50.

Kumar, A., and Vara Prasad, M.N. (2018). Plant lead interactions: transport, toxicity, tolerance and detoxification mechanisms. *Ecotoxicology and environmental safety, 166*, 401-418.

Kumar, S., Singh, R. and Nayyar, H. (2013). Tocopherol application modulates the response of wheat (*Triticum aestivum. L,*) seedlings to elevated temperatures by mitigation of stress injury and enhancement of antioxidants. *J. plant growth regul., 32*, 307-314.

Maggio, A., Miyazaki, S., Veronese, P., Fujita, T. and Ibeas, J.I. (2002). Does proline accumulation play an active role in stress induced growth reduction. *Plant . journal, 31*, 699-712.

Mazid, M., Khan, T.A., Khan, Z.H., Quddusi, S., Mohammad, F. (2011). Occurrence, biosynthesis and potentialities of ascorbic acid in plants. *Int. J. plant, Animal and environmental sciences, 1*, 167-184.

Mattioli, R., Costantino, P., & Trovato, M. (2009). Proline accumulation in plants: not only stress. *Plant signaling & behavior*, 4(11), 1016-1018.

Mishra, P. K., and Prakash, V. (2010). Response of non-enzymatic antioxidants to zinc induced stress at different pH in *Glycine max*. L., Merrill, *Academic J. Of Plant Sciences, 3,* 1-10.

Munne-Bosh, S. (2005).The role of tocopherol in plant stress tolerance. *Journal Plant Physiology, 162,* 743-748.

Munns, R. and Tester, M. (2008). Mechanism of salinity tolerance, *Annu. Rev. Plant. Bio. 59,* 651-681.

Nahar, K., Hasanuzzaman, M. and Fujita, M. (2016). Roles of osmolytes in plant adaptation to drought and salinity. *Springer,* 37-68. doi 10.1007/978-81-322-2616-1-4.

Nayajyoti *et al.* (2010). Heavy metals, occurrence and toxicity for plants: a review. *Environmental chemistry letters, 8,* 199-216.

Nahar, K., Hasanuzzuman, M., Alam, M.M., M. Fujita (2015). Exogenous glutathione confers high temperature stress tolerance in mungbean (*Vigna radiate.* L.,*)* by modulating antioxidant defence and methyglyoxal detoxification system. *Environ. Exp. Bot, 112,* 44-54

Reddy, K.J. (2006). Nutrient stress. In: Physiology and molecular biology of stress tolerance in plants . K. V. Madhava rao. A.S. Raghavendra and R.K. Janardhan, (Eds)., 187-217, Springer, Dordrect, Netherlands.

Sadiq, M., Akram, N.A., Ashraf, M., Parqaiz Ahmad, F.Q. (2019). Alpha- tocopherol induced regulation of growth and metabolism in plants under non stress and stress conditions. *Journal of plant growth regulation.* http://doi.org /10.1007/ soo344-019-09936-7.

Sytar, O., Kumar, A., Latowski, D., Kuczynska, P., Strzałka, K., & Prasad, M.N.V. (2013). Heavy metal-induced oxidative damage, defense reactions, and detoxification mechanisms in plants. *Acta physiologiae plantarum, 35*(4), 985-999.

Smirnoff, N., Coklin, P.L., Loewus, F.A. (2001). Biosynthesis of ascorbic acid in plants: a renaissance . *Annu. Rev. plant Biology, 52*(1), 437-467.

Singh, M., Kumar, J., Singh, S., Singh, V.P., Prasad, S.M. (2015). Adaption strategies of plants against heavy metals toxicity: a short review. *Biochem pharmacol (Los. Angel*), *4,* 161.

Sharma *et al.* (2019). Phytohormones regulates the accumulation of osmolytes under abiotic stress. *Biomolecules, 9(7*), 285.

Sharma, P., Jha, A.B., Dubey, R.S., Pessarakhi, M. (2012). Reactive oxygen species, oxidative damage and antioxidant defense mechanism in plants under stressful conditions. *Journal of Bot,* 1-26.

Sharama, S.S. and Dietz, K.J. (2009). The relationship between metal toxicity and cellular redox imbalance. *Trends plant sci., 14*(1), 43-50.

Serraj, R. and Sinclair, T.R. (2002). Osmolyte accumulation can really help increase crop yield under drought conditions. *Plant cell environ., 25,* 333-341.

Mishra, S., and Dubey, R.S. (2006). Heavy metal uptake and detoxification mechanisms in plants. *International journal of agricultural research, 2,* 122-141.

Singh, M., Kumar, J., Singh, S., Singh, V.P. and Prasad, S.M. (2015). Roles of osmoprotectants in improving salinity and drought tolerance in plants: a review. *Review in environmental sciencebiotechnology, 14,* 407-426.

Siddiqui *et al.* (2018). Ascorbic acid application improves salinity stress tolerance in wheat. *Chiang mai. J. Sci., 45,* 1-11.

Suprasanna, P., Nikalije, G.C. and Rai, A.N. (2016). Osmolyte accumulation and implications in plant abiotic stress tolerance. *Springer,* 1-12.

Szabadas, L. and Savouse, A. (2010). proline a multifunctional amminoacid. *Trends Plant Sci., 15,* 89-97.

Szepesi, Á., and Szőllősi, R. (2018). Mechanism of proline biosynthesis and role of proline metabolism enzymes under environmental stress in plants. In: *Plant Metabolites and Regulation Under Environmental Stress,* 337-353, Academic Press.

Wingler, A. (2002). Function of trehalose biosynthesis in plants. *Phytochemistry. 60*(5), 437-440.

Yoshimura, K., & Ishikawa, T. (2017). Chemistry and metabolism of ascorbic acid in plants. In: *Ascorbic Acid in Plant Growth, Development and Stress Tolerance, Springer,* Cham., 1-23 http://doi.org/10.1007- 1978-3-319-74057-7-1.

Role of Transporter Proteins in Plants during Abiotic Stress

Khushbu Sharma
Department of Agronomy, School of Agriculture
Lovely Professional University, Jalandhar, Punjab, 144411, India
Email: sharma.khushbu2303@gmail.com

Abstract

Some of the heavy metals are under category essential micronutrients such as Fe, Zn, and Cu and some are not under the criteria of essentiality like Cadmium, mercury, lead, and aluminum. They have higher density and mobility. The presence of these heavy metals in the environment causes biotic stress, which leads to disturbance in metabolic activities of plants and causes a threat to crop production and productivity. Plants can respond under stress to continue their metabolic activities and minimize the losses that occur due to stress. During stress condition transport ions through the membrane is also suffered. There are various transporter proteins plays a significant and vital role in regulating the transportation of molecules through the plasma membrane to continues different metabolic processes of plant system such as osmotic regulation of ions, transduction of signals and induced metal ions. Aquaporins present in the cell membrane also plays a vital role in detoxification of heavy metals by inhibiting their entry within the cell through water and other molecules. Many transporters such as ZIP, HKT, SWEETs, ABC (families of the transporter), also carry out abiotic stress mitigating property. Over tuned study of these mechanisms adapted by the plant to adapt to biotic stress is necessary to maintain sustainable crop production. This review is focused on future advancement in research in constructing a proper approach to understand the role of this protein transporter in minimizing the effect of abiotic stresses in plant systems.

Introduction

Abiotic stress harshly affects the growth and development of plants with a decrease in quality production (Mickelbart et al., 2015, Challinor et al., 2016). The most common biotic stresses are salinity, heat stress, heavy metals stress, and drought stress (Thakur et al., 2010, Vishwakarma et al., 2017, Jaleel et al., 2009). The plant to grow under stress adopted several defense mechanisms. There are any changes adopted by the plant may be at the cellular or whole tissue system level. Examples of changes in a plant cell can be observed during stress such as maintaining the osmotic level of the cell, membrane stability, signaling and identifying the disturbance created by biotic stress. (Zhu et al., 2001, Amudha et al., 2011). In many studies, it was found the level of ABA and Ca is an increase in the membrane during stress. With injury found in the membrane and an increase in the secretion of the protein. Complete growth and development of plants get restricted during stress. The plant adopts static equilibrium after the detection of stress in the cell to detoxify the effect of heavy metals. There are some families of transmembrane proteins present in the cell to regulate signals, movement of anions across the membrane. They play a generous role in conciliating stress resistance, tolerance and avoidance by providing homeostasis to the cell under stress condition. Plant proteins regulate plant responses during stress. The nature of plants is static, not have the ability to escape unfavourable conditions. Transporter proteins commonly from a family of transmembrane proteins carry out the movement of selective molecules through the membrane. They provide sustainable strength to the cell in mitigation of metal ions, providing resistance, strong tolerance, and stress avoidance when the plant is under the situation of abiotic stress. Transportation of molecules, ions, minerals are necessary to maintain homeostasis of the cell. The role of ABA is also important in stress condition it is also known as the stress hormone.ABA plays a significant role in maintaining the activity of transporter proteins, stimulates solute concentration and transportation of molecules in a cell under stress. In the condition of heavy metal stress invasion of metal ions within the cell, protoplasm is regulated by the transporter and ABA (Smits et al., 2005). There are various transporters such as CDF (cation diffusion facilitators), ZIP(Zn –regulated iron-regulated transporter) all the above proteins can mediate heavy metal concentration in cell organelles such as vacuole (Smits et al., 2005; Bartels et al., 2009, Lunn et al., 2003). Sugars are also transported by specific sugar transporters example tonoplast monosaccharide transporters resist cell again chilling stress protects cells through the process of maintaining the subcellular amount of stress stimulators(Lunn et al., 2007). Additionally, Na and K exchange with high-affinity potassium transporters maintain homeostasis of the cell under extreme salinity conditions. Stomatal opening and closing are regulated by ABA under drought stress also maintain hydraulic conductivity of roots of plants, Aquaporins also plays a vital role in maintaining the ionic balance of cell in carrying of molecules. The role of transporters is complex and can be differentiated on the basis membrane, mode of action, species of plant and sensors too. This study provides a detailed expression of protein molecules, hormones released during stress, transporters and their role in mitigating abiotic stress obligatory under heavy metal stress, heat stress, salinity, and waterlogging. Moreover, it described the regulation of plasma membrane under stress, tonoplast, and changes adopted by these transporters proteins under stress. This review also described briefly the development of modified cultivars those shows tolerant behaviour under abiotic stress without compromising with yield and quality.

The Behaviour of the Membrane Transport System under Stress

Due to some of the natural and anthropogenic factors, there is a huge increase in heavy metal ion toxicity in water as well as soil which directly influences the life of plants and through enters in our food chain leads cause hazardable loss to mankind. Due to the continuous increase in the accumulation of heavy metals in soil and water, our agriculture worst affected. The presence of heavy metal ions in the plant system creates a disturbance in the physiological and biochemical activities of the plant, which adversely affect growth and development (Dalcorso et al., 2013; Hossain et al., 2013, Sharma et al., 2005). Further, it leads to toxicity and these ions interact with biomolecules of the plant cell, increases the production of ROS reactive oxygen species, etc. Again this imbalance creates by metal ions between antioxidant manage, also reduces the effect heavy metal these mechanisms are carried out by transporter proteins. Detoxification of heavy metal in cell organelles is carried out by transporters to maintain ion flux(Maestri et al., 2010; Hall et al., 2003). The role of these transporter proteins also varies with tissue and organ system.

Fig. 1. Diagrammatic representation of Heavy metal uptake by the plant during stress (a) Uptake of Heavy metal by ZIP protein in roots. (b)Movement of heavy metal ions from root to shoot (vacuole), (C) sequestration of heavy metal in a leaf

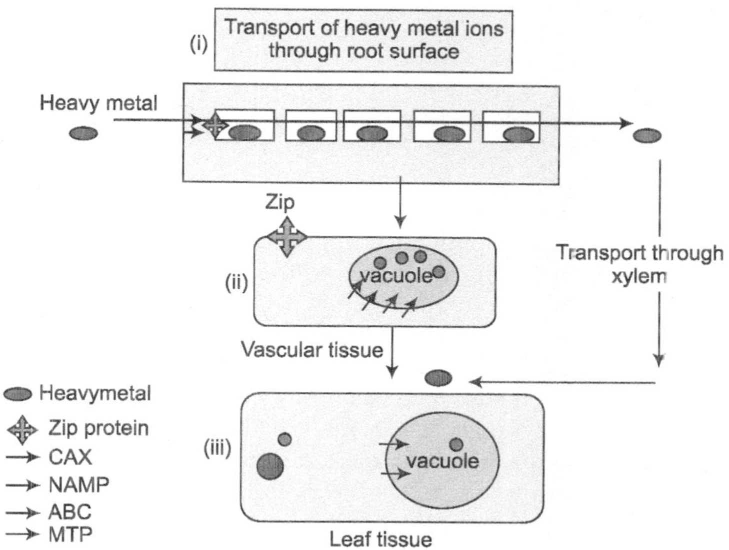

Movement of Heavy Metal Ions across Membrane

There are many transporter proteins involved in metal ion absorption. The families of transporter proteins are very specific with their role example members of ZIP family proteins are involved in the moment of metal ions through the plasma membrane, NARMP family members are identified as a transporter in the uptake of Fe and Cd during iron deficiency and maintain the metabolism of Fe.

Movement of Heavy Metal Ions among Cell Organelles

Metal transporters belonging from different families have a vital role in maintaining heavy metal ion concentration plant cell organelles. Although there are many transition metal transporters are as follows cation diffusion facilitator, ATP binding cassette, Heavy metal ATP as Zinc regulating transporter, etc (Hall et al., 2003). Along with heavy metal uptake of another trace element causes stress on the plant. Uptake of Boron with metal ions in the semi-arid region leads reduction of plant growth even it only nonmetal among micronutrients. Adaptation of wheat and maize under B accumulated soil is overcome by isolation of EMS3888 and 405, in maize tls1 or rte.

Cation Diffusion Facilitator Family (CDF) Transporters

CDF transporters work more significantly in Endoplasmic reticulum and cytosol to facilitate the movement of heavy metal ions such as Fe, Cu, Cd, Co, and Pb (Peiter et al., 2007). The CDF act as an antiporter for micronutrients such as Fe and Ni. In Arabidopsis CDF transporters and AtMTPI1 were showing tolerance behaviour with zinc and transport zinc from the cytoplasm of the cell (Zaal et al.,1999; Zhang et al., 2011)In the same way accumulation of zinc in roots of Arabidopsis is facilitated by another protein belongs to the same family, for example, AtMTP3 (Arrivault et al., 2006). Among all essential micronutrients, Fe and Zinc required in trace amounts for meeting the metabolic needs of the plant. Deficiency and toxicity of metal ions both are harmful to plant interveinal chlorosis in younger leaves is seen in case of deficiency of iron leads to stunted growth of plants (Hafeez et al., 2013, Fageria et al., 2002). Zinc is directly involved in protein synthesis in the case of Zinc deficiency synthesis of protein affected (Snowball et al., 1986, Fageria et al., 2004). These proteins ZIP (zinc–regulated transporter Protein) are commonly found in bacteria, animals, plants, and fungi. All ZIP proteins responsible for maintaining the transport of heavy metal ions within the membrane and also regulate homeostasis of cells in stress conditions. Binding of heavy metals is also associated with Histidine's present in proteins (Guerinot et al., 2000, Nishida et al., 2008). Recognition of these proteins involved in detoxification of heavy metals was done in yeast projected under iron stress. After analyzing the role of protein it was observed in the roots of Arabidopsis that protein plays a vital role in the efflux of iron during both the conditions either deficiency of iron or plant under stress (Vert et al., 2002; Barberon et al., 2011). Transportation of Zinc in Arabidopsis is carried out by a total of 8 proteins present of the ZIP family of transporter proteins named, AtIRT1, AtZIP1, AtZIP2, AtZIP3, AtpZIP7, AtpZIP10, AtpZIP11and AtpZIP12 (Grotz et al., 1998; Milner et al., 2013). Further, it was noticed different proteins out of these 8 are also involved in the uptake of Fe and Zn in field crops such as wheat, barley, rice and are responsible for transportation of Zinc, for Fe transport is shown in table 1.

Table 1: Responsible proteins for Metal translocation

S.No	Name of crop	The protein responsible for the transport of Fe	The protein responsible for the transport of Zn
1	Barley		HvIRT1, HvZIP3, HvZIP8, HvZIP5
2.	Rice	OSZIP6, OSZIP7, OSIRT 1, OSIRT 2	OSZIP3, OSZIP4, OSZIP5, OSZIP 8

Almost nine genes of ZIP are identified in maize are restrained to a small portion of the cell. Transporter protein plays a vital in detoxification, transpiration, uptake of ions with maintaining the equilibrium of cells. Encoding of the ZIP gene increased during heavy metal stress other hands lead to the accumulation of metal ions in roots (Talke et al., 2007). Few genes are found more expressive under very high threat of heavy metal stress in the soil such as asZIP4, 6, 9 and 10. TaZIP3, TaZIP6, 9 and 13 of wheat (Triticum aestivum L.)(Evens et al., 2017). Hypersensitivity of Zn in yeast even though they act as transporter protein of Wheat. The role of ZIP protein transporter is selective to ions and with crops, in some cases, it was found homologous to Barley genes but cannot sometimes translocate ions of iron in a plant cell. In Arabidopsis factors responsible for transcription such as AtbZIP19 and 13 involves in binding metal ions of zinc with the contribution of the isolated gene of ZIP. Transcription of ions in plants always varies with the plant-to-plant species is carried out by zinc transporter proteins and FbZIP (Assuncao et al., 2010) induces hyper-accumulation of zinc. In the case of Arabidopsis Iron zip proteins such as FbZIP with AtbZIP19 leads to regulate the uptake of Zinc ions during plant under deficiency of Zinc (Assuncao et al., 2010).Transporter proteins of Arabidopsis bzip19–4 bzip23-2aredouble null mutant unable to uptake Zn from the medium. (Li et al., 2013). It was found in research that there are almost nine ZIP genes in maize they are localized under cell. These genes within the cell detoxify divalent ions of metals. Accumulation of heavy metal ions is increased where these (ZIP) genes are expressed through encoding (Kim et al., 2007). In Arabidopsis hellers, ZIP4,6,9 and 10 were found to be highly articulated under the condition of soils that are drastically affected by heavy metals (Becher et al., 2004; Klein et al., 2006).

ABC Transporters for Heavy Metal Tolerance

There are many families of transporter proteins out of the ABC (ATP binding cassette) is considered as the largest family (Kang et al., 2011). These transporters realm in the cytosol where hydrolysis of ATP takes place and the domain of cell membrane to regulate the transport of ions, molecules, and signals (Higgins et al.,1992).These proteins also provide a channel for transportation with direct involvement in the catabolic activities of the cell (Verrier et al., 2008). Arabidopsis, some proteins have also been recognized to enhance the adaptability of the cell under heavy metal stress (Kim et al., 2007). Out of all AtABCC1 and 2 are commonly seen in the vacuole of the cell (Song et al., 2010). Phytochelatins present in the vacuole of cells minimizes the uptake of heavy metal ions through membrane also leads to optimize tolerance under stress because of unique behaviour they are also known as Multidrug Associated Proteins. MRPs is one of the largest group of transporter protein ABC. Frequent increase in transcription of MRPs specially AtMRPs was synchronized under cadmium toxicity in A.thaliana (Rea et al.,1998) Translocation of hormones along with ions is also carried out by ABA(Bovert et al., 2003).

ABC Transporters for Water Stress Tolerance

When plants subjected to stress there several mechanisms adapted by the plant to tolerate stress such as secretion of ABA, which is, also known as stress hormone particularly under water stress main function of the hormone is to regulate stomata opening and closing and prevent water losses from the cell. Water is the main structural component of plants required for meeting all metabolic needs, transportation of ions, molecules, and minerals. Formulation of ABA is taking place in roots and stems of plants under water stress (Finkelstein et al., 2008; Schachtman et al., 2008). A multiple-step process (Dong et al., 2010) follows synthesis, translocation in different parts of plants. Because of the lower acidity of ABA, it gets easily passes through the plasma membrane within the cell. Cell chloroplast is a specific organ for the absorption of ABA in alkaline form chloroplast also maintains the pH of the cell. stomatal opening and closing are regulated by gene transporters through apoplast of guard cells examples importer (AtABCG40) and exporter protein AtABCG25 they both proteins are also genetically involved in responding during stress and regulate stomata closing to minimize water losses(Kang et al., 2010, Kuromori et al., 2014). The rapid response of ABA in roots of plants has been leading to water stress and less subsequent response of stomata regulation by protein transporter (AtABCG40). It was found during studies transporter proteins treated through AA are found more sensitive to drought stress.

Fig. 2. Stress tolerance mechanism in Arabidopsis under osmotic stress

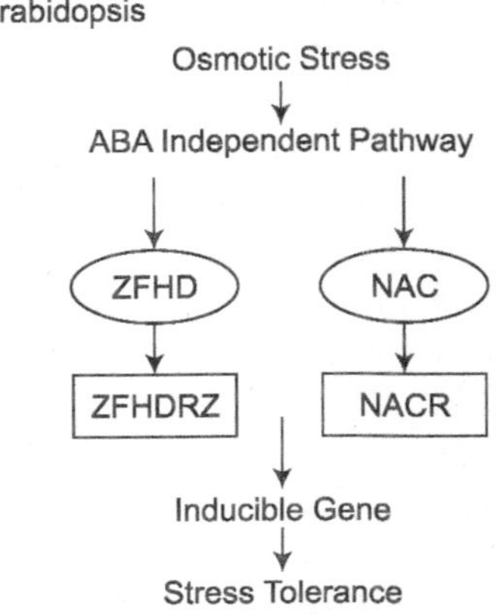

Low-temperature Stress

Rise and decline in temperature both the stresses are known as abiotic stress. The decline in temperature beyond the requirement leads to a great loss of productivity with the adverse effect of pests and diseases (Chinnusamy et al., 2007). Under the situation of cold stress, plant express many phenotype changes such as less development of new leaves, wilting, necrosis which further leads to the death of the cell (Milura et al., 2013). Reproductive phase of the plant is also get affected with cold stress in some cases dormancy appears and facilitation of genes and physiological process have been disturbed like transportation of molecules, ions, and minerals through the cell membrane (Hurry et al., 2002, Milura et al., 2013). The plant possesses its adaptability to tolerate cold stress it was found in different study sugars like maltose, fructose, etc are also involved in mediating stress. (Yuanyuan et al., 2009). These sugars interact with membrane to facilitate osmoregulation of cells (Garg et al., 2002). Stress response of plants is also affected by the amendment in the production of soluble sugars. Transportation of carbohydrates in the cell is regulated by specific cell organelles such as chloroplast and tonoplast there are mainly to transporters involved in regulation and transportation of carbohydrates such as SWEETs and TMTs provides tolerance under cold stress. (Klemens et al., 2013).

Sugars Will Eventually be Exported Transporters (SWEETs)

Plant survival can be determined by sugars. Sugars are the primary source of energy. Effect of stress also determined by bimolecular analysis of Translocation, transportation, and formation of sugars in different tissues of the plant. Sweet transporters act as catalase, which regulates the diffusion of sugars across the plasma membrane. Sweets transporters also facilitate the efflux of sugars and transport through phloem during biotic and abiotic stress (Patil et al., 2015, Streubel et al2013, Syed et al., 2015). In Arabidopsis, AtSWEET16 and AtSWEET17 are responsible for the augmentation of cold tolerance. Isolation of the AtSWEET16 gene has led to prominent baggage (cold) tolerance in Arabidopsis with vacuolar sugar transport (PA et al., 2013). AtSWEET17 is specifically known to transport fructose in leaf tissue and is thus considered as an indicator of total sugar (fructose) under chilling stress (Schulz et al., 2011). Recently, (Guo et al., 2004) recognized that the AtSWEET17 gene is the first transporter of fructose within the tonoplast where it translocates fructose from lumen to cytoplasm of the cell (Klemens et al., 2013)concluded that Transporter SWEETs 16 is efficiently used to transport of glucose, fructose and sucrose in (Xenopus oocyte) gene expression system in A.thaliana. The expression and role of these sugar transporters may be modulated by using different techniques such as by maintaining the equilibrium of sugars molecules in the cell, enhance homeostasis under stress response by improving redox potential to tolerate stress (Tarkowski et al., 2015).

Tonoplast Monosaccharide Transporters

The role of tonoplast transporters is limited to the membrane of tonoplast (Wormit et al., 2006) TNT Transporters also play a vital role in the transportation of glucose and fructose in the vacuole. Under cold stress, they provide tolerance to plant. It was observed in A. thaliana to the low capability of TNT in the uptake of sugars and even more minimize the number of sugar molecules in vacuoles. When two species of wild type and cultivar are grown in cold stress, it was observed lots of variation in influx/efflux of sugar molecules in the vacuole of the cell.

Salt Stress and Transport

Reported over 20% of cultivable land across the world is under salt stress. It occurs in areas where the accumulation of neutral soluble salts in soils increases the permissible limit if this condition continues then it leads to salt stress. Salt stress causes great damage to production and productivity in agriculture. Some plants that can grow under salt stress are known as halophytes. (Gupta et al., 2014) Among all halophytes, up to 1% of plant species cover monocots and dicots. The limit of halophytes to tolerate salt concentration is 500mM but some of the species show sensitivity at 50mM (Flower et al., 1986; Flower et al., 2005; Flower et al., 2008) Major Challenges faced by the plant during salt stress are tin uptake of water because of higher osmotic potential. For systematic regulation of transportation of ions needs a higher amount of K ions inside the cell and lees the amount of Na ions outside of the cell. Transporter carries out the whole process of regulation of Na and K ratio. Where NHX1/7 transporters present outside of the cell are responsible for minimizing the concentration of Na outside of the cell and lower Nap/Kp concentration in the vacuole. There are some genes of SOS family such as SOS1, 2 and 3 maintain adaptability of the cell during salt stress. Effect of osmotic stress is also observed in all physiological, structural and biochemical activities of plant-like integrity in the plasma membrane, production of ROS species, lower the rate of photosynthesis and disturb stomata opening and closing. (Janicka et al., 2015; Demidchik et al., 2014; Falhof et al.,2016). Conversely, Plant adopts different mechanism to tolerate osmotic stress such as absorption, production of antioxidants, secretion of hormones, production of NO and synthesis of polyamines. The normal ability of plants under salt stress is to accumulate salt ions in a particular tissue. Salt tolerance species accumulate Nap ions in leaves tissue and roots of plants to maintain the osmotic potential of plant and soil. The whole process of facilitation Na ions is carried out in the plasma membrane where the Hp-ATPase pump is created for the production of the electrochemical proton. It was noticed that Signal transduction in salt stress plays a vital role in regulating Na/K antiporters example MAPK (nitrogen-activated protein kinase), ZIP (zinc figure proteins), MYB, HSF families, etc. MAPK proteins usually generate responses toward ROS. Many transcription factors are activated by MAPK proteins gradient across the plasma membrane.

Na1/H1

There is exchangers (NHX) present inside the cell that maintains systematic control in decreasing concentration of Nap ions and regulation of higher concentration K ions. Two transporters one is active transporter maintain higher up/Na ratio and another is secondary transporter (Qiu et al., 2002). X1 and 7 are two antiporters that maintain a lower level of NaP/Hp in the vacuole or different cell organelles. The SOS transporter gene in the vacuole mediates the entry of Nap ions. Recent studies confirm the role of SOS1 in regulating the homeostasis of cells and increase the adaptability of cells in salinity. (Hasegawa et al., 2000; Sanders et al., 2000) SOS 2 and 3 regulate SOS 1 mediate intonation of Na/Hp transport system further transport of ions. Acclimatisation of salts has been increasing upregulation Nap/ Hp antiporter. SOS3 gene plays a vital role in maintaining Ca ions to provide tolerance under stress because they tend to bind Ca ions and the SOS2 gene improved its expression in the presence of a higher concentration of Ca ions (Shi et al., 2002; Ishitani et al., 2000). The presence of an excess amount of Ca ions in the cell protects the cell from Nap toxicity. An increase in Ca2p activity in the cytoplasm of root cells is due to the transduction of SOS signals specifically under drought stress. Primary calcium sensors increase Ca2p in cytosolic against response to Nap ions where SOS3 leads to binding with Ca2p improves the ability to interact after interaction it bonded with SOS 2 genes present in a cell. (Knight et al., 2000) The proteins of SOS3 and SOS2 family have an alternate controller like Calcium-binding protein. Within the Arabidopsis, the role of SOS 3 is more prominent in shoots and other species; it founds more active in roots. SOS 3 also regulates the phosphorylation process because of the association with kinase (cblscaBP) (Liu et al., 1998; Liu et al., 2000; Hrabak et al., 2003). Later interaction of SOS2 regulated in the plasma membrane of the cell to facilitate target SOS1. SOS1 found conserved usually in higher plant species. Efficient control of NHX is required to maintain a higher concentration of Kp inside the cell and a lower concentration of Nap outside of the cell to attain the adaptability of the plant in salt stress (Quan et al., 2007; Du et al., 2011)

K Transporter

K is the primary macronutrient required by plants always found in profuse inorganic form follows specific transporters, transport pathways, and channels. HKT high-affinity K uptake transporters maintain transportation of K in roots and shoots. There are two classes of HKT class I and class among which ClassI regulates higher accumulation of Na ions especially in leaves under salt stress. In Arabidopsis, (AtHKT1) leads to a decrease in the accumulation of Na icns in roots during salt stress. The K out channel promotes the efflux of Na ions into xylem and leaves.

Water Stress

Water stress is a situation when both the conditions are not suitable for plant growth either it may lack sufficient water or may excess amount of water. These stresses affect the physiological, metabolic activities of the plant. Plant face difficulty in respiration, photosynthesis, transpiration, reproduction including. Several defence mechanisms were adopted by the plant to tolerate stress such as stress adopting, stress escaping, stress tolerance or resistance commonly these phenomena are well studied in case of drought stress (Mittler et al., 2011; Ahuja et al., 2010). Under drought stress, CDPK (ca2p-dependent protein kinases) regulates stomatal opening and closing. Transporters CDPK carries out osmotic adjustment of the cell. (Osakabe et al., 2013). MSF family of membranous proteins regulates the transportation of monosaccharide and polysaccharides sugars there are two transporters identified one sucrose transporters (SUTs) and monosaccharide transporters (MST). The Expression of sugars found alters in the situation of abiotic stress. In Arabidopsis, the accumulation of monosaccharide sugars in Vacuole is increased due to the presence of TMT 1 and TMT 2 transporters. it was found in rice crops subjected to drought Sucrose transporter OoSUT2 regulates the activity of sugars specifically in leaves. In recent studies' role of NRT2 nitrate, transporters regulate accumulation nitrate in aerobic soils. In maintaining homeostasis, transportation and detection of signals in a cell is a complex process that involves the activation of stress responding genes. To understand, the process of plant response towards water stresses the necessary role of transporters, how they are involved in maintaining and their involvement in catalyzing the plant at the whole tissue level.

ABA-induced Transporters

ABA is phytohormone found very critical for plant growth. It integrates signals during stress to control the first step in stress response Phytohormone ABA plays a major role in response to drought stress by regulating inner water potential(Cutler et al., 2010, Fujii et al., 2012). Under conditions of water stress, ABA response more in guard cells for the closing of stomata by minimizing turgor pressure. ABA is transported from chloroplast to guard cells (Medici et al., 2008; Ibraheem et al., 2011). The most important transporters of ABA belong to the ABCG transporter family. Recent studies confirm, In Arabidopsis, there are three types of ABA transporters that have been recognized those are conscientious for stress tolerance (Kang et al., 2010, Kuromori et al.,2014; Kanno et al., 2012). The word transporter is k`+-now specifically for its unique ABA importing activity in Arabidopsis. Additionally, these three ABA transporters also form a signal pathway for improving stress tolerance. These signals are directly activated from phosphorylation.

Aquaporins

These are a group of channel proteins that play a major role in transportation and tolerance of biotic stress. Aquaporins are classified into seven major intrinsic protein families: nodulin 26, tonoplast, Glpf, plasma membrane intrinsic protein, hybrid intrinsic protein, small basic intrinsic protein (Henzler et al., 2014; Ahmad et al., 2016). PIP is a major family of aquaporins localized in the plasma membrane responsible for the transportation of water, solutes and improves tolerance against abiotic stress. It was reported that members of PIP families are engaged in the transportation of metalloids, boron, arsenic, and lead. Aquaporins are involved in maintaining hydraulic conductivity and 95% permeability of the plasma membrane. There are multiple isomeric forms of aquaporins are present in the plasma membrane to regulate trafficking, influx/efflux of guard cells. Generally, the activity of these channel proteins inhibited the condition of biotic stress.

When subjected to the water stress response of intrinsic proteins varies based on the extent of stress and plant species. The expression of PIP gene in cell reduce the loss of water by controlling reverse efflux of water in rhizosphere during water stress. Protein TIP is played a decisive role during water stress by maintaining equilibrium inside the cell by regulating the rapid fluctuation of water between cytosol and vacuole. Many research studies were made to understand the function of aquaporins more deeply there are two constrictions were identified at the centre of structure: Ar/s and NPA. The first constriction named Ar/s form four amino acids G-S-G-R. Second constriction form two extremely conserved NPA. (Deshmukh et al., 2015; Maurel et al., 2008; Mahdieh et al., 2008)

Conclusion

Understanding the mechanism of plant tolerance under (abiotic) stress is a major concern of agriculture. Plant stress response always depends upon the interaction between stimuli and signaling pathways to regulate gene expression, solute sequestration, and physiological responses. Transporter proteins or channel proteins carry out the whole process of signaling. Stress signaling pathways plays a vital role in maintaining the homeostasis of the cell. These pathways also found to be critical for the adaptation of new cellular responses and make a coordinated system for the transportation of sugars, hormones, osmolytes, and molecules. This study shows a practical review of the abiotic stress mediate transport system and signaling. The method of gene expression is more explored to increase the adaptability of the plant under stress. Recent studies show advancement in determining potential transporters with the help of phenotype (molecular, genomic, and physiological) which can be further used to generate new stress-tolerant varieties. Despite this progress remain to be explored more than a few transporters and the primary function of transportation during stress is known. Studies exploring the mechanism of transportation and signaling found a sustainable approach to minimize the threat of abiotic stress to agriculture.

References

Ahmad, P. (2010). Growth and antioxidant responses in mustard (Brassica juncea L.) plants subjected to combined effect of gibberellic acid and salinity. *Archives of Agronomy and Soil Science, 56*(5), 575-588.

Alejandro, S., Lee, Y., Tohge, T., Sudre, D., Osorio, S., Park, J., ...&Martinoia, E. (2012). AtABCG29 is a monolignol transporter involved in lignin biosynthesis. *Current Biology, 22*(13), 1207-1212.

Amudha, J., &Balasubramani, G. (2011). Recent molecular advances to combat abiotic stress tolerance in crop plants. *Biotechnology and Molecular Biology Reviews, 6*(2), 31-58.

Apel, K., &Hirt, H. (2004). Reactive oxygen species: metabolism, oxidative stress, and signal transduction. *Annu. Rev. Plant Biol., 55*, 373-399.

Arrivault, S., Senger, T., &Krämer, U. (2006). The Arabidopsis metal tolerance protein AtMTP3 maintains metal homeostasis by mediating Zn exclusion from the shoot under Fe deficiency and Zn oversupply. *The Plant Journal, 46*(5), 861-879.

Assaha, D. V., Ueda, A., Saneoka, H., Al-Yahyai, R., &Yaish, M. W. (2017). The role of Na+ and K+ transporters in salt stress adaptation in glycophytes. *Frontiers in physiology, 8*, 509.

Assunção, A. G. L., Martins, P. D. C., De Folter, S., Vooijs, R., Schat, H., &Aarts, M. G. M. (2001). Elevated expression of metal transporter genes in three accessions of the metal hyperaccumulatorThlaspicaerulescens. *Plant, Cell & Environment, 24*(2), 217-226.

Assunção, A. G., Herrero, E., Lin, Y. F., Huettel, B., Talukdar, S., Smaczniak, C., ...&Aarts, M. G. (2010). Arabidopsis thaliana transcription factors bZIP19 and bZIP23 regulate the adaptation to zinc deficiency. *Proceedings of the National Academy of Sciences, 107*(22), 10296-10301.

Barberon, M., Zelazny, E., Robert, S., Conéjéro, G., Curie, C., Friml, J., &Vert, G. (2011). Monoubiquitin-dependent endocytosis of the iron-regulated transporter 1 (IRT1) transporter controls iron uptake in plants. *Proceedings of the National Academy of Sciences, 108*(32), E450-E458.

Bartels, D., &Dinakar, C. (2013). Balancing salinity stress responses in halophytes and non-halophytes: a comparison between Thellungiella and Arabidopsis thaliana. *Functional Plant Biology, 40*(9), 819-831.

Bartels, D., &Sunkar, R. (2005). Drought and salt tolerance in plants. *Critical reviews in plant sciences, 24*(1), 23-58.

Becher, M., Talke, I. N., Krall, L., &Krämer, U. (2004). Cross-species microarray transcript profiling reveals high constitutive expression of metal homeostasis genes in shoots of the zinc hyperaccumulator Arabidopsis halleri. *The Plant Journal, 37*(2), 251-268.

Bonaventure, G., Beisson, F., Ohlrogge, J., & Pollard, M. (2004). Analysis of the aliphatic monomer composition of polyesters associated with Arabidopsis epidermis: occurrence of octadeca-cis-6, cis-9-diene-1, 18-dioate as the major component. *The Plant Journal, 40*(6), 920-930.

Bovet, L., Eggmann, T., Meylan-Bettex, M., Polier, J., Kammer, P., Marin, E., ...&Martinoia, E. (2003). Transcript levels of AtMRPs after cadmium treatment: induction of AtMRP3. *Plant, Cell & Environment, 26*(3), 371-381.

Bultreys, A., Trombik, T., Drozak, A., &Boutry, M. (2009). Nicotianaplumbaginifolia plants silenced for the ATP-binding cassette transporter gene NpPDR1 show increased susceptibility to a group of fungal and oomycete pathogens. *Molecular plant pathology, 10*(5), 651-663.

Campbell, E. J., Schenk, P. M., Kazan, K., Penninckx, I. A., Anderson, J. P., Maclean, D. J., ...& Manners, J. M. (2003). Pathogen-responsive expression of a putative ATP-binding cassette transporter gene conferring resistance to the diterpenoidsclareol is regulated by multiple defensesignaling pathways in Arabidopsis. *Plant physiology, 133*(3), 1272-1284.

Challinor, A. J., Koehler, A. K., Ramirez-Villegas, J., Whitfield, S., & Das, B. (2016). Current warming will reduce yields unless maize breeding and seed systems adapt immediately. *Nature Climate Change, 6*(10), 954-958.

Chardon, F., Bedu, M., Calenge, F., Klemens, P. A., Spinner, L., Clement, G., ...&Krapp, A. (2013). Leaf fructose content is controlled by the vacuolar transporter SWEET17 in Arabidopsis. *Current Biology, 23*(8), 697-702.

Chen, G., Komatsuda, T., Ma, J. F., Nawrath, C., Pourkheirandish, M., Tagiri, A., ...&Nevo, E. (2011). An ATP-binding cassette subfamily G full transporter is essential for the retention of leaf water in both wild barley and rice. *Proceedings of the National Academy of Sciences, 108*(30), 12354-12359.

Chinnusamy, V., Zhu, J., & Zhu, J. K. (2007). Cold stress regulation of gene expression in plants. *Trends in plant science, 12*(10), 444-451.

Choi, H., Ohyama, K., Kim, Y. Y., Jin, J. Y., Lee, S. B., Yamaoka, Y., ... & Lee, Y. (2014). The role of Arabidopsis ABCG9 and ABCG31 ATP binding cassette transporters in pollen fitness and the deposition of steryl glycosides on the pollen coat. *The Plant Cell, 26*(1), 310-324.

Conde, A., Chaves, M. M., &Gerós, H. (2011). Membrane transport, sensing and signaling in plant adaptation to environmental stress. *Plant and Cell Physiology, 52*(9), 1583-1602.

Crouzet, J., Roland, J., Peeters, E., Trombik, T., Ducos, E., Nader, J., &Boutry, M. (2013). NtPDR1, a plasma membrane ABC transporter from Nicotianatabacum, is involved in diterpene transport. *Plant molecular biology, 82*(1-2), 181-192.

DalCorso, G., Fasani, E., &Furini, A. (2013). Recent advances in the analysis of metal hyperaccumulation and hypertolerance in plants using proteomics. *Frontiers in plant science, 4*, 280.

De Luca, V., Salim, V., Thamm, A., Masada, S. A., & Yu, F. (2014). Making iridoids/secoiridoids and monoterpenoidindole alkaloids: progress on pathway elucidation. *Current opinion in plant biology, 19*, 35-42.

Delhaize, E., Gruber, B. D., Pittman, J. K., White, R. G., Leung, H., Miao, Y., ...& Richardson, A. E. (2007). A role for the AtMTP11 gene of Arabidopsis in manganese transport and tolerance. *The Plant Journal, 51*(2), 198-210.

Demidchik, V., Straltsova, D., Medvedev, S. S., Pozhvanov, G. A., Sokolik, A., &Yurin, V. (2014). Stress-induced electrolyte leakage: the role of K+-permeable channels and involvement in programmed cell death and metabolic adjustment. *Journal of experimental botany, 65*(5), 1259-1270.

Emamverdian, A., Ding, Y., Mokhberdoran, F., &Xie, Y. (2015). Heavy metal stress and some mechanisms of plant defense response. *The Scientific World Journal, 2015*.

Evens, N. P., Buchner, P., Williams, L. E., &Hawkesford, M. J. (2017). The role of ZIP transporters and group F bZIP transcription factors in the Zn-deficiency response of wheat (Triticumaestivum). *The Plant Journal, 92*(2), 291-304.

Fageria, N. K. (2002). Influence of micronutrients on dry matter yield and interaction with other nutrients in annual crops. *PesquisaAgropecuáriaBrasileira, 37*(12), 1765-1772.

Fageria, N. K. (2004). Dry matter yield and nutrient uptake by lowland rice at different growth stages. *Journal of plant nutrition, 27*(6), 947-958.

Falhof, J., Pedersen, J. T., Fuglsang, A. T., &Palmgren, M. (2016). Plasma membrane H+-ATPase regulation in the center of plant physiology. *Molecular Plant, 9*(3), 323-337.

Finkelstein, R., Reeves, W., Ariizumi, T., &Steber, C. (2008). Molecular aspects of seed dormancy. *Annu. Rev. Plant Biol., 59*, 387-415.

Flowers, T. J. (1985). Physiology of halophytes. In *Biosalinity in Action: Bioproduction with Saline Water* (pp. 41-56). Springer, Dordrecht.

Flowers, T. J. (2004). Improving crop salt tolerance. *Journal of Experimental botany, 55*(396), 307-319.

Flowers, T. J., &Colmer, T. D. (2008). Salinity tolerance in halophytes. *New Phytologist*, 945-963.

Flowers, T. J., Hajibagheri, M. A., &Clipson, N. J. W. (1986). Halophytes. *The quarterly review of biology, 61*(3), 313-337.

Flowers, T. J., Troke, P. F., & Yeo, A. R. (1977). The mechanism of salt tolerance in halophytes. *Annual review of plant physiology, 28*(1), 89-121.

Fourcroy, P., Sisó-Terraza, P., Sudre, D., Savirón, M., Reyt, G., Gaymard, F., ...&Briat, J. F. (2014). Involvement of the ABCG 37 transporter in secretion of scopoletin and derivatives by Arabidopsis roots in response to iron deficiency. *New Phytologist, 201*(1), 155-167.

Garg, A. K., Kim, J. K., Owens, T. G., Ranwala, A. P., Do Choi, Y., Kochian, L. V., & Wu, R. J. (2002). Trehalose accumulation in rice plants confers high tolerance levels to different abiotic stresses. *Proceedings of the National Academy of Sciences, 99*(25), 15898-15903.

Grotz, N., Fox, T., Connolly, E., Park, W., Guerinot, M. L., &Eide, D. (1998). Identification of a family of zinc transporter genes from Arabidopsis that respond to zinc deficiency. *Proceedings of the National Academy of Sciences, 95*(12), 7220-7224.

Guerinot,M.L.(2000).TheZIPfamilyofmetaltransporters.*BiochimicaetBiophysicaActa (BBA)-Biomembranes, 1465*(1-2), 190-198.

Guilfoyle, T., Hagen, G., Dong, T., Park, Y., & Hwang, I. (2015). Abscisic acid: biosynthesis, inactivation, homoeostasis and signalling. *Essays in biochemistry, 58*, 29-48.

Guo, W. J., Nagy, R., Chen, H. Y., Pfrunder, S., Yu, Y. C., Santelia, D., ...&Martinoia, E. (2014). SWEET17, a facilitative transporter, mediates fructose transport across the tonoplast of Arabidopsis roots and leaves. *Plant physiology, 164*(2), 777-789.

Gupta, B., & Huang, B. (2014). Mechanism of salinity tolerance in plants: physiological, biochemical, and molecular characterization. *International journal of genomics, 2014*.

Gustin, J. L., Loureiro, M. E., Kim, D., Na, G., Tikhonova, M., & Salt, D. E. (2009). MTP1-dependent Zn sequestration into shoot vacuoles suggests dual roles in Zn tolerance and accumulation in Zn-hyperaccumulating plants. *The Plant Journal, 57*(6), 1116-1127.

Hafeez, B., Khanif, Y. M., &Saleem, M. (2013). Role of zinc in plant nutrition-a review. *Journal of Experimental Agriculture International*, 374-391.

Hall, J. Á. (2002). Cellular mechanisms for heavy metal detoxification and tolerance. *Journal of experimental botany, 53*(366), 1-11.

Hammond, J. P., Bowen, H. C., White, P. J., Mills, V., Pyke, K. A., Baker, A. J., ... &Broadley, M. R. (2006). A comparison of the Thlaspicaerulescens and Thlaspiarvense shoot transcriptomes. *New Phytologist, 170*(2), 239-260.

Higgins, C. F. (1992). ABC transporters: from microorganisms to man. *Annual review of cell biology, 8*(1), 67-113.

Himabindu, Y., Chakradhar, T., Reddy, M. C., Kanygin, A., Redding, K. E., & Chandrasekhar, T. (2016). Salt-tolerant genes from halophytes are potential key players of salt tolerance in glycophytes. *Environmental and Experimental Botany, 124*, 39-63.

Hong, C. Y., Chao, Y. Y., Yang, M. Y., Cheng, S. Y., Cho, S. C., & Kao, C. H. (2009). NaCl-induced expression of glutathione reductase in roots of rice (Oryzasativa L.) seedlings is mediated through hydrogen peroxide but not abscisic acid. *Plant and soil, 320*(1), 103-115.

Ishimaru, Y., Suzuki, M., Kobayashi, T., Takahashi, M., Nakanishi, H., Mori, S., &Nishizawa, N. K. (2005). OsZIP4, a novel zinc-regulated zinc transporter in rice. *Journal of Experimental Botany, 56*(422), 3207-3214.

Jaleel, C. A., Manivannan, P. A. R. A. M. A. S. I. V. A. M., Wahid, A., Farooq, M., Al-Juburi, H. J., Somasundaram, R. A. M. A. M. U. R. T. H. Y., &Panneerselvam, R. (2009). Drought stress in plants: a review on morphological characteristics and pigments composition. *Int. J. Agric. Biol, 11*(1), 100-105.

James, R. A., Blake, C., Byrt, C. S., &Munns, R. (2011). Major genes for Na+ exclusion, Nax1 and Nax2 (wheat HKT1; 4 and HKT1; 5), decrease Na+ accumulation in bread wheat leaves under saline and waterlogged conditions. *Journal of experimental botany, 62*(8), 2939-2947.

Janicka-Russak, M., &Kabała, K. (2015). The role of plasma membrane H+-ATPase in salinity stress of plants. In *Progress in Botany* (pp. 77-92). Springer, Cham.

Jetter, R., Kunst, L., & Samuels, A. L. (2008). Composition of plant cuticular waxes. *Biol. Plant Cuticle, 23*, 145-181.

Kang, J., Hwang, J. U., Lee, M., Kim, Y. Y., Assmann, S. M., Martinoia, E., & Lee, Y. (2010). PDR-type ABC transporter mediates cellular uptake of the phytohormoneabscisic acid. *Proceedings of the National Academy of sciences, 107*(5), 2355-2360.

Kang, J., Park, J., Choi, H., Burla, B., Kretzschmar, T., Lee, Y., &Martinoia, E. (2011). Plant ABC transporters. *The Arabidopsis book/American Society of Plant Biologists, 9*.

Kang, J., Yim, S., Choi, H., Kim, A., Lee, K. P., Lopez-Molina, L., ...& Lee, Y. (2015). Abscisic acid transporters cooperate to control seed germination. *Nature communications, 6*(1), 1-10.

Kanno, Y., Hanada, A., Chiba, Y., Ichikawa, T., Nakazawa, M., Matsui, M., ...&Seo, M. (2012). Identification of an abscisic acid transporter by functional screening using the receptor complex as a sensor. *Proceedings of the National Academy of Sciences, 109*(24), 9653-9658.

Kavitha, P. G., Kuruvilla, S., & Mathew, M. K. (2015). Functional characterization of a transition metal ion transporter, OsZIP6 from rice (Oryzasativa L.). *Plant Physiology and Biochemistry, 97*, 165-174.

Kawachi, M., Kobae, Y., Mimura, T., &Maeshima, M. (2008). Deletion of a histidine-rich loop of AtMTP1, a vacuolar Zn2+/H+ antiporter of Arabidopsis thaliana, stimulates the transport activity. *Journal of Biological Chemistry, 283*(13), 8374-8383.

Kawachi, M., Kobae, Y., Mori, H., Tomioka, R., Lee, Y., &Maeshima, M. (2009). A mutant strain Arabidopsis thaliana that lacks vacuolar membrane zinc transporter MTP1 revealed the latent tolerance to excessive zinc. *Plant and Cell Physiology, 50*(6), 1156-1170.

Kim, D. Y., Bovet, L., Maeshima, M., Martinoia, E., & Lee, Y. (2007). The ABC transporter AtPDR8 is a cadmium extrusion pump conferring heavy metal resistance. *The Plant Journal, 50*(2), 207-218.

Klemens, P. A., Patzke, K., Deitmer, J., Spinner, L., Le Hir, R., Bellini, C., ...&Neuhaus, H. E. (2013). Overexpression of the vacuolar sugar carrier AtSWEET16 modifies germination, growth, and stress tolerance in Arabidopsis. *Plant Physiology, 163*(3), 1338-1352.

Korshunova, Y. O., Eide, D., Clark, W. G., Guerinot, M. L., &Pakrasi, H. B. (1999). The IRT1 protein from Arabidopsis thaliana is a metal transporter with a broad substrate range. *Plant molecular biology, 40*(1), 37-44.

Kramer, U. (2007). Talke IN Hanikenne M, Transitionmetal transport. *FEBS Lett, 581*, 2263-2272.

Krämer, U. (2010). Metal hyperaccumulation in plants. *Annual review of plant biology, 61*, 517-534.

Kuromori, T., Miyaji, T., Yabuuchi, H., Shimizu, H., Sugimoto, E., Kamiya, A., ...& Shinozaki, K. (2010). ABC transporter AtABCG25 is involved in abscisic acid transport and responses. *Proceedings of the National Academy of Sciences, 107*(5), 2361-2366.

Kuromori, T., Sugimoto, E., & Shinozaki, K. (2014). Intertissue signal transfer of abscisic acid from vascular cells to guard cells. *Plant physiology, 164*(4), 1587-1592.

Le Hir, R., Sorin, C., Chakraborti, D., Moritz, T., Schaller, H., Tellier, F., ...& Bellini, C. (2013). ABCG 9, ABCG 11 and ABCG 14 ABC transporters are required for vascular development in Arabidopsis. *The Plant Journal, 76*(5), 811-824.

Lee, M., Lee, K., Lee, J., Noh, E. W., & Lee, Y. (2005). AtPDR12 contributes to lead resistance in Arabidopsis. *Plant physiology, 138*(2), 827-836.

Lee, S., Jeong, H. J., Kim, S. A., Lee, J., Guerinot, M. L., & An, G. (2010). OsZIP5 is a plasma membrane zinc transporter in rice. *Plant molecular biology, 73*(4-5), 507-517.

Lee, S., Kim, S. A., Lee, J., Guerinot, M. L., &An, G. (2010). Zinc deficiency-inducible OsZIP8 encodes a plasma membrane-localized zinc transporter in rice. *Molecules and Cells, 29*(6), 551-558.

Li, S., Zhou, X., Huang, Y., Zhu, L., Zhang, S., Zhao, Y., ...& Chen, R. (2013). Identification and characterization of the zinc-regulated transporters, iron-regulated transporter-like protein (ZIP) gene family in maize. *BMC Plant Biology, 13*(1), 1-14.

Lunn, J. E. (2007). Compartmentation in plant metabolism. *Journal of experimental botany, 58*(1), 35-47.

Ma, L., Zhang, H., Sun, L., Jiao, Y., Zhang, G., Miao, C., &Hao, F. (2012). NADPH oxidase AtrbohD and AtrbohF function in ROS-dependent regulation of Na+/K+ homeostasis in Arabidopsis under salt stress. *Journal of Experimental Botany, 63*(1), 305-317.

Ma, X., Sela, H., Jiao, G., Li, C., Wang, A., Pourkheirandish, M., ...& Chen, G. (2012). Population-genetic analysis of HvABCG31 promoter sequence in wild barley (Hordeumvulgare ssp. spontaneum). *BMC evolutionary biology, 12*(1), 1-11.

Mach, J. (2014). Boron transport in maize.

Maestri, E., Marmiroli, M., Visioli, G., &Marmiroli, N. (2010). Metal tolerance and hyperaccumulation: costs and trade-offs between traits and environment. *Environmental and Experimental Botany, 68*(1), 1-13.

Mahajan, S., &Tuteja, N. (2005). Cold, salinity and drought stresses: an overview. *Archives of biochemistry and biophysics, 444*(2), 139-158.

Merilo, E., Jalakas, P., Kollist, H., &Brosché, M. (2015). The role of ABA recycling and transporter proteins in rapid stomatal responses to reduced air humidity, elevated CO2, and exogenous ABA. *Molecular Plant, 8*(4), 657-659.

Merilo, E., Jalakas, P., Kollist, H., &Brosché, M. (2015). The role of ABA recycling and transporter proteins in rapid stomatal responses to reduced air humidity, elevated CO2, and exogenous ABA. *Molecular Plant, 8*(4), 657-659.

Miao, Y., Laun, T., Zimmermann, P., &Zentgraf, U. (2004). Targets of the WRKY53 transcription factor and its role during leaf senescence in Arabidopsis. *Plant molecular biology, 55*(6), 853-867.

Mickelbart, M. V., Hasegawa, P. M., & Bailey-Serres, J. (2015). Genetic mechanisms of abiotic stress tolerance that translate to crop yield stability. *Nature Reviews Genetics, 16*(4), 237-251.

Milner, M. J., Seamon, J., Craft, E., &Kochian, L. V. (2013). Transport properties of members of the ZIP family in plants and their role in Zn and Mn homeostasis. *Journal of experimental botany, 64*(1), 369-381.

Miura, K., &Furumoto, T. (2013). Cold signaling and cold response in plants. *International journal of molecular sciences, 14*(3), 5312-5337.

Munns, R., & Tester, M. (2008). Mechanisms of salinity tolerance. *Annu. Rev. Plant Biol., 59*, 651-681.

Nazri, A. Z., Griffin, J. H., Peaston, K. A., Alexander-Webber, D. G., & Williams, L. E. (2017). F-group bZIPs in barley—a role in Zn deficiency. *Plant, cell & environment, 40*(11), 2754-2770.

Nishida, S., Mizuno, T., & Obata, H. (2008). Involvement of histidine-rich domain of ZIP family transporter TjZNT1 in metal ion specificity. *Plant Physiology and Biochemistry, 46*(5-6), 601-606.

Pallotta, M., Schnurbusch, T., Hayes, J., Hay, A., Baumann, U., Paull, J., ...& Sutton, T. (2014). Molecular basis of adaptation to high soil boron in wheat landraces and elite cultivars. *Nature, 514*(7520), 88-91.

Pang, C. H., & Wang, B. S. (2008). Oxidative stress and salt tolerance in plants. In *Progress in botany* (pp. 231-245). Springer, Berlin, Heidelberg.

Panikashvili, D., Shi, J. X., Schreiber, L., &Aharoni, A. (2011). The Arabidopsis ABCG13 transporter is required for flower cuticle secretion and patterning of the petal epidermis. *New Phytologist, 190*(1), 113-124.

Pasternak, T., Rudas, V., Potters, G., & Jansen, M. A. (2005). Morphogenic effects of abiotic stress: reorientation of growth in Arabidopsis thaliana seedlings. *Environmental and Experimental Botany, 53*(3), 299-314.

Patil, G., Do, T., Vuong, T. D., Valliyodan, B., Lee, J. D., Chaudhary, J., ...& Nguyen, H. T. (2016). Genomic-assisted haplotype analysis and the development of high-throughput SNP markers for salinity tolerance in soybean. *Scientific Reports, 6*(1), 1-13.

Patil, G., Valliyodan, B., Deshmukh, R., Prince, S., Nicander, B., Zhao, M., ...& Nguyen, H. T. (2015). Soybean (Glycine max) SWEET gene family: insights through comparative genomics, transcriptome profiling and whole genome re-sequence analysis. *BMC genomics, 16*(1), 1-16.

Pedas, P., Schjoerring, J. K., & Husted, S. (2009). Identification and characterization of zinc-starvation-induced ZIP transporters from barley roots. *Plant Physiology and Biochemistry, 47*(5), 377-383.

Pedas, P., Ytting, C. K., Fuglsang, A. T., Jahn, T. P., Schjoerring, J. K., & Husted, S. (2008). Manganese efficiency in barley: identification and characterization of the metal ion transporter HvIRT1. *Plant physiology, 148*(1), 455-466.

Peiter, E., Montanini, B., Gobert, A., Pedas, P., Husted, S., Maathuis, F. J., ...& Sanders, D. (2007). A secretory pathway-localized cation diffusion facilitator confers plant manganese tolerance. *Proceedings of the National Academy of Sciences, 104*(20), 8532-8537.

Petrussa, E., Braidot, E., Zancani, M., Peresson, C., Bertolini, A., Patui, S., &Vianello, A. (2013). Plant flavonoids—biosynthesis, transport and involvement in stress responses. *International journal of molecular sciences, 14*(7), 14950-14973.

Pighin, J. A., Zheng, H., Balakshin, L. J., Goodman, I. P., Western, T. L., Jetter, R., ...& Samuels, A. L. (2004). Plant cuticular lipid export requires an ABC transporter. *Science, 306*(5696), 702-704.

Pilon-Smits, E. (2005). Phytoremediation. *Annu. Rev. Plant Biol., 56*, 15-39.

Popescu, S. C., Popescu, G. V., Bachan, S., Zhang, Z., Gerstein, M., Snyder, M., & Dinesh-Kumar, S. P. (2009). MAPK target networks in Arabidopsis thaliana revealed using functional protein microarrays. *Genes & development, 23*(1), 80-92.

Rea, P. A., Li, Z. S., Lu, Y. P., Drozdowicz, Y. M., &Martinoia, E. (1998). From vacuolar GS-X pumps to multispecific ABC transporters. *Annual review of plant biology, 49*(1), 727-760.

Rout, G. R., &Sahoo, S. (2015). Role of iron in plant growth and metabolism. *Reviews in Agricultural Science, 3*, 1-24.

Schachtman, D. P., &Goodger, J. Q. (2008). Chemical root to shoot signaling under drought. *Trends in plant science, 13*(6), 281-287.

Schippers, J. H., Nguyen, H. M., Lu, D., Schmidt, R., & Mueller-Roeber, B. (2012). ROS homeostasis during development: an evolutionary conserved strategy. *Cellular and Molecular Life Sciences, 69*(19), 3245-3257.

Schulz, A., Beyhl, D., Marten, I., Wormit, A., Neuhaus, E., Poschet, G., ...&Hedrich, R. (2011). Proton-driven sucrose symport and antiport are provided by the vacuolar transporters SUC4 and TMT1/2. *The Plant Journal, 68*(1), 129-136.

Shabala, S., &Pottosin, I. (2014). Regulation of potassium transport in plants under hostile conditions: implications for abiotic and biotic stress tolerance. *Physiologiaplantarum, 151*(3), 257-279.

Shabala, S., Bose, J., Fuglsang, A. T., &Pottosin, I. (2016). On a quest for stress tolerance genes: membrane transporters in sensing and adapting to hostile soils. *Journal of Experimental Botany, 67*(4), 1015-1031.

Sharma, R. K., & Agrawal, M. (2005). Biological effects of heavy metals: an overview. *Journal of environmental Biology, 26*(2), 301-313.

Shi, J., Kim, K. N., Ritz, O., Albrecht, V., Gupta, R., Harter, K., ...&Kudla, J. (1999). Novel protein kinases associated with calcineurin B–like calcium sensors in Arabidopsis. *The Plant Cell, 11*(12), 2393-2405.

Shitan, N. (2016). Secondary metabolites in plants: transport and self-tolerance mechanisms. *Bioscience, biotechnology, and biochemistry, 80*(7), 1283-1293.

Shitan, N., Dalmas, F., Dan, K., Kato, N., Ueda, K., Sato, F., ...&Yazaki, K. (2013). Characterization of Coptis japonica CjABCB2, an ATP-binding cassette protein involved in alkaloid transport. *Phytochemistry, 91*, 109-116.

Skopelitis, D. S., Paranychianakis, N. V., Paschalidis, K. A., Pliakonis, E. D., Delis, I. D., Yakoumakis, D. I., ... &Roubelakis-Angelakis, K. A. (2006). Abiotic stress generates ROS that signal expression of anionic glutamate dehydrogenases to form glutamate for proline synthesis in tobacco and grapevine. *The Plant Cell, 18*(10), 2767-2781.

Snowball, K., & Robson, A. D. (1986). *Symptoms of nutrient deficiencies: lupins.* Department of Soil Science and Plant Nutrition, Institute of Agriculture, University of Western Australia.

Song, W. Y., Park, J., Mendoza-Cózatl, D. G., Suter-Grotemeyer, M., Shim, D., Hörtensteiner, S., ...&Martinoia, E. (2010). Arsenic tolerance in Arabidopsis is mediated by two ABCC-type phytochelatin transporters. *Proceedings of the National Academy of Sciences, 107*(49), 21187-21192.

Stitt, M., & Hurry, V. (2002). A plant for all seasons: alterations in photosynthetic carbon metabolism during cold acclimation in Arabidopsis. *Current opinion in plant biology, 5*(3), 199-206.

Streubel, J., Pesce, C., Hutin, M., Koebnik, R., Boch, J., &Szurek, B. (2013). Five phylogenetically close rice SWEET genes confer TAL effector-mediated susceptibility to Xanthomonasoryzaepv. oryzae. *New Phytologist, 200*(3), 808-819.

Sun, S. J., Guo, S. Q., Yang, X., Bao, Y. M., Tang, H. J., Sun, H., ...& Zhang, H. S. (2010). Functional analysis of a novel Cys2/His2-type zinc finger protein involved in salt tolerance in rice. *Journal of experimental botany, 61*(10), 2807-2818.

Sunitha, M. S., Prashant, S., Kumar, S. A., Rao, S. R. I. N. A. T. H., Narasu, M. L., &Kishor, P. K. (2013). Cellular and molecular mechanisms of heavy metal tolerance in plants: a brief overview of transgenic plants overexpressing phytochelatin synthase and metallothionein genes. *Plant cell biotechnology and molecular biology, 14*(1-2), 33-48.

Syed, N. H., Prince, S. J., Mutava, R. N., Patil, G., Li, S., Chen, W., ... & Nguyen, H. T. (2015). Core clock, SUB1, and ABAR genes mediate flooding and drought responses via alternative splicing in soybean. *Journal of experimental botany, 66*(22), 7129-7149.

Talke, I. N., Hanikenne, M., &Krämer, U. (2006). Zinc-dependent global transcriptional control, transcriptional deregulation, and higher gene copy number for genes in metal homeostasis of the hyperaccumulator Arabidopsis halleri. *Plant physiology, 142*(1), 148-167.

Tarkowski, Ł. P., & Van den Ende, W. (2015). Cold tolerance triggered by soluble sugars: a multifaceted countermeasure. *Frontiers in plant science, 6*, 203.

Thakur, P., Kumar, S., Malik, J. A., Berger, J. D., &Nayyar, H. (2010). Cold stress effects on reproductive development in grain crops: an overview. *Environmental and Experimental Botany, 67*(3), 429-443.

Tiong, J., McDonald, G. K., Genc, Y., Pedas, P., Hayes, J. E., Toubia, J., ...& Huang, C. Y. (2014). H v ZIP 7 mediates zinc accumulation in barley (H ordeumvulgare) at moderately high zinc supply. *New Phytologist, 201*(1), 131-143.

Tiong, J., McDonald, G., Genc, Y., Shirley, N., Langridge, P., & Huang, C. Y. (2015). Increased expression of six ZIP family genes by zinc (Zn) deficiency is associated with enhanced uptake and root-to-shoot translocation of Zn in barley (Hordeumvulgare). *New Phytologist, 207*(4), 1097-1109.

Van der Zaal, B. J., Neuteboom, L. W., Pinas, J. E., Chardonnens, A. N., Schat, H., Verkleij, J. A., &Hooykaas, P. J. (1999). Overexpression of a novel Arabidopsis gene related to putative zinc-transporter genes from animals can lead to enhanced zinc resistance and accumulation. *Plant physiology, 119*(3), 1047-1056.

Verma, P., Mathur, A. K., Srivastava, A., &Mathur, A. (2012). Emerging trends in research on spatial and temporal organization of terpenoidindole alkaloid pathway in Catharanthusroseus: a literature update. *Protoplasma, 249*(2), 255-268.

Verrier, P. J., Bird, D., Burla, B., Dassa, E., Forestier, C., Geisler, M., ...&Theodoulou, F. L. (2008). Plant ABC proteins–a unified nomenclature and updated inventory. *Trends in plant science, 13*(4), 151-159.

Vert, G., Grotz, N., Dédaldéchamp, F., Gaymard, F., Guerinot, M. L., Briat, J. F., & Curie, C. (2002). IRT1, an Arabidopsis transporter essential for iron uptake from the soil and for plant growth. *The Plant Cell, 14*(6), 1223-1233.

Vishwakarma, K., Upadhyay, N., Kumar, N., Yadav, G., Singh, J., Mishra, R. K., ...& Sharma, S. (2017). Abscisic acid signaling and abiotic stress tolerance in plants: a review on current knowledge and future prospects. *Frontiers in plant science, 8*, 161.

Wormit, A., Trentmann, O., Feifer, I., Lohr, C., Tjaden, J., Meyer, S., ...&Neuhaus, H. E. (2006). Molecular identification and physiological characterization of a novel monosaccharide transporter from Arabidopsis involved in vacuolar sugar transport. *The Plant Cell, 18*(12), 3476-3490.

Yadav, V., Molina, I., Ranathunge, K., Castillo, I. Q., Rothstein, S. J., & Reed, J. W. (2014). ABCG transporters are required for suberin and pollen wall extracellular barriers in Arabidopsis. *The Plant Cell, 26*(9), 3569-3588.

Yang, X., Huang, J., Jiang, Y., & Zhang, H. S. (2009). Cloning and functional identification of two members of the ZIP (Zrt, Irt-like protein) gene family in rice (Oryzasativa L.). *Molecular biology reports, 36*(2), 281-287.

Yazaki, K. (2005). Transporters of secondary metabolites. *Current opinion in plant biology, 8*(3), 301-307.

Yazaki, K., Shitan, N., Takamatsu, H., Ueda, K., & Sato, F. (2001). A novel Coptis japonica multidrug-resistant protein preferentially expressed in the alkaloid-accumulating rhizome. *Journal of experimental botany, 52*(357), 877-879.

Yu, F., & De Luca, V. (2013). ATP-binding cassette transporter controls leaf surface secretion of anticancer drug components in Catharanthusroseus. *Proceedings of the National Academy of Sciences, 110*(39), 15830-15835.

Yu, L., Nie, J., Cao, C., Jin, Y., Yan, M., Wang, F., ...& Zhang, W. (2010). Phosphatidic acid mediates salt stress response by regulation of MPK6 in Arabidopsis thaliana. *New Phytologist, 188*(3), 762-773.

Yuanyuan, M., Yali, Z., Jiang, L., &Hongbo, S. (2009). Roles of plant soluble sugars and their responses to plant cold stress. *African Journal of Biotechnology, 8*(10).

Yun, K. Y., Park, M. R., Mohanty, B., Herath, V., Xu, F., Mauleon, R., ...& de Los Reyes, B. G. (2010). Transcriptional regulatory network triggered by oxidative signals configures the early response mechanisms of japonica rice to chilling stress. *BMC plant biology, 10*(1), 1-29.

Zhang, H., Zhu, H., Pan, Y., Yu, Y., Luan, S., & Li, L. (2014). A DTX/MATE-type transporter facilitates abscisic acid efflux and modulates ABA sensitivity and drought tolerance in Arabidopsis. *Molecular plant, 7*(10), 1522-1532.

Zhang, M., Senoura, T., Yang, X., &Nishizawa, N. K. (2011). Functional analysis of metal tolerance proteins isolated from Zn/Cd hyperaccumulating ecotype and non-hyperaccumulating ecotype of Sedum alfrediiHance. *FEBS letters, 585*(16), 2604-2609.

Zhao, J., & Dixon, R. A. (2010). The 'ins' and 'outs' of flavonoid transport. *Trends in plant science, 15*(2), 72-80.

Zhu, J. K. (2001). Plant salt tolerance. *Trends in plant science, 6*(2), 66-71.

Influences and Future Approach of Rhizoremediation for Heavy Metal Stress

Khushbu Sharma

Department of Agronomy, School of Agriculture

Lovely Professional University, Jalandhar, Punjab, 144411, India

Email: sharma.khushbu2303@gmail.com

Abstract

The rapid increase in the concentration of heavy metals (HM) due to different anthropogenic activities is a major concern nowadays. Plants that are grown in (HM) contaminated areas lead to attaining stress and overall health of the plant is impaired. Several remediation strategies are adopted by scientists to mitigate the effect of HM stress in the plant such as microbe interaction with metals, phytoremediation, chelating agents, etc. Metal-microbe interaction is a widely adopted sustainable approach adopted for detoxification of metals but under specialized conditions. Microbes form, accumulation of HM and detoxification. Commonly suitable microbes are inoculated in the rhizosphere of plants and get nomenclature of rhizoremediation. Rhizoremediation approach also improves the health of the plant and physical or biological properties of soil with a qualitative increase in yield. In this chapter, the applied aspect of HM toxicity, causes, effects, plant adoption in stress tolerance or resistance provided by microbe's interaction (rhizoremediation and phytoremediation) in contaminated soils are discussed. The use of microbes (HMT-PGP) in contaminated soil for improving the growth of the plant is also discussed. Future aspects of (rhizoremediation or phytoremediation) techniques and its sustainability with low input costs are also described.

Introduction

The heavy metals (HM) are used particularly used for metals those are having higher density and have the potential to cause environmental toxicity with great damage to human health (Tchounwou et al., 2012). Although soil is considered as a primary or natural source of HM, some of the geological and anthropogenic activities lead to increasing their concentration at levels that are proved harmful for both plants and animals (Chibuike and Obiora, 2014). Mode of transportation of heavy metals also can be in gaseous form and as well as in particulate form (Adriano et al., 2005) which further leads to their accumulation in water, soil, and living systems. Among all essential micronutrients, almost all except Boron are metals required by the plant is very little concentration or in trace amount. Their excessive amount in plants after a certain limit causes toxicity. Although improper usage of agricultural goods and long-term use of industrial waste in agriculture, hazardous waste is the key cause of pollution of Arable land. In the end, arable lands, which are highly polluted by the impact of metal toxicity, pass them to the food chain (Zhuang et al., 2013, 2014). The plant absorbs HM ions from both from roots and shoots of the plant. (Patra et al., 2004). The direct influence of oxidative stress causes inhibition of cytoplasmic enzymes and destroys cell structures (Jadia and Fulekar, 2009). Abiotic stresses in plants lead to the production of Reactive oxygen species and disturb cell equilibrium (Hossain et al., 2012; Sytar et al., 2013). Many changes in the mechanism of the cell can be observed in plant cell including binding of protein sulfhydryl with metal ions and excess production of glutathione, antioxidative enzymes, and ROS-producing enzymes such as NADPH oxidases. (Valko et al., 2005 (Bielen et al., 2013). All metabolic activities of plants are directly or indirectly affected by HM. Although it has been reported that plants acquire several defence mechanisms to mitigate or tolerate HM toxicity to a certain extent but after a specific limit these mechanisms are not proved satisfactory (Clemens and Ma, 2016). So, it becomes very indispensable to eliminate accumulated HM ions for normal growth and development of plants and protect organisms that depend on particular plant species. The techniques being adopted for HM removal from the contaminated area through physical or chemical removal, in situ fixation, stabilization involves excavation. The physical process adopted for HM removal is not much efficient. (Schnoor, 1997). Along with effectiveness, there is a need of adopting certain kind of strategies for detoxification of heavy metal ions from contaminated site those could be implemented easily must be environment friendly and possess long-term durability. (Gavrilescu, 2004; Wuana and Okieimen, 2011). Phytoremediation comprises the use of growing plants to minimize the concentration of HM in the rhizosphere and rhizoremediation technique implies the use of soil microbes, which proves as an important alternative for detoxification of soil. It is reported that the use of rhizospheric microbes has abilities to prevent the plant from HM stress. Soil microbes also play a vital role in accumulating metal ions in the soil. Rhizospheric microbes made interaction with plant and metal ions to eliminate them from the soil surface. Microbes used for remediation have unique adaptability to perform under the very heavy concentration of HM ions.

HM and Rhizospheric Microbes

In an ecosystem, the soil is the major sink for metal toxification. (Gadd, 2010). The range of heavy metal varies from 1 to 100,000 mg/kg (Long et al., 2002) Metal ions interact with rhizosphere microbes and they are significantly transformed through geo-active action of soil microbes. Besides, this interaction of soil and microbes leads to colonization which results in detoxification of heavy metal. The population present in the rhizosphere directly influences HM elimination by the action of microbes. (Kuiper et al., 2004). This process of microbe interaction near the plant root zone increases the metabolic rate. According to Pires et al., (2017), the bacterial genus belongs to Bacillus and pseudomonas grows more rapidly in

HM contaminated sites. Among all plants, growth-promoting microbe's rhizobia and mycorrhiza are recommended best for metal detoxification. The activity of rhizobium was supposed to be sensitive for nitrogen fixation under HM stress but it founds during the experiment that rhizobia behave efficiently even under metal stress and it improves the quality of HM contaminated soils (Checcucci et al., 2017). Another microbe fungi, of the genus (Ascomycota and Basidiomycota), are most widely used for detoxification of HM from the soil (Narendrula-Kotha and Nkongolo, 2017). However, it has been observed that AM fungi can also make the colony even in HM contaminated soils with poor nutrient status. Metal-binding is linked to a number of intracellular AM fungal infection and certain soil microbes. Binding metal ions to the outside side of the cell surface or to bring inside the cell, are used to establish AM fungi and other rhizosphere microbes (Ehrlich, 1997). In soil, they can change the metal form; reduce toxic concentration, mobility, absorption in solution, and deterioration (Gadd, 2010). These rhizospheric microbes can translocate a large number of metals ions. This interaction of soil microbes with HMis depends upon the physical properties of soil, diversity of microbe and concentration of metal ions in the rhizosphere (Khan et al., (2005), Giller et al., (2009), Gadd (2010), and Kong and Glick (2017).

Fig. 1. Mechanisms involved in plant and soil for remediation of HM contamination

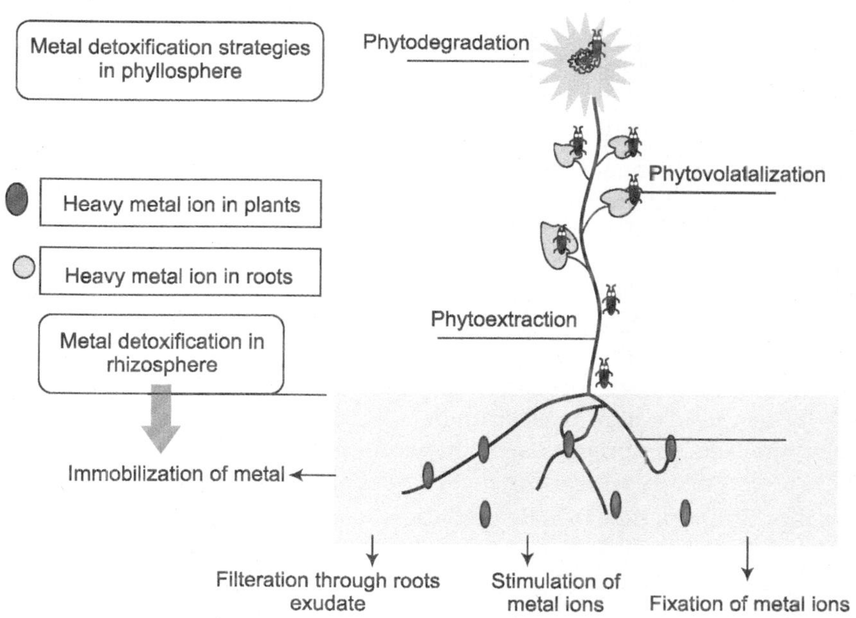

Table 1: Showing the result of phytoremediation strategies in several crops

S.No	Heavy metal contaminant	Crop	Observations	Reference
1.	Chromium	Mustard	Accumulation of Cr ions in the root zone	Diwan et al., 2010
2.	Lead	Zea maize	Significant increase in metal adsorption.	An et al., 2011
3.	Cadmium	Tomato+cabbage+ clover + Zea mays		
4.	Copper	Japan clover	Intercropping of tomato increase accumulation of metal ions	An et al., 2011

Interaction of Microbes with Metal Ions

Metal ions and plant growth-promoting microbes reduce metal availability through various processes such as complexation, chelation, precipitation, and reduction. Low pH favours the adsorption of heavy metal ions in the soil (Merdy et al., 2009). Microbial interaction of HM releases certain organic acids which lower the soil pH and confiscate metal ions. (Turnau and Kottke, 2005). Several experimental pieces of evidence propose that an extensive range of bacteria or fungi generate organic acids as usual chelating agents for HM (Seneviratne et al., 2017). There are some organic acids commonly suggested for heavy metal solubilization such as gluconic acid, oxalic acid, acetic acid, and mallic acids (Ullah et al., 2015; Gube, 2016). In a recent study, Fomina et al., (2005) reported that excess secretion of organic acids like oxalic and citric acid by HM tolerant species (Beauveria caledonica) solubilized many metal ions such as (Zn, Cd, Cu, and Pb). AM fungi also produced oxalate crystals that are known to detoxification or immobilization of HM (Gadd et al.,2014). Here, the filamentous structure of AM fungi penetrates soil aggregates for adsorption or chelation of HM ions. Kaewdoung et al., (2016).Fomitopsis cf. meliae or Ganoderma aff. Steyaertanum both species of fungus produce oxalate crystals (observed in scanning electron microscopy and X-rays powder diffraction highly specialized equipment used for photography of cell) are contributed in enhancing tolerance by converting metal ions to less toxic form.

Zinc Sulfate - Zinc oxalate dehydrate

Copper sulfate-Copper oxalate dehydrate

Lead nitrate - Lead oxalate

Cadmium sulfate- Cadmium oxalate trihydrate

Fig. 2. Showing the interaction of metal ions through microbes in oxidative and reduced form

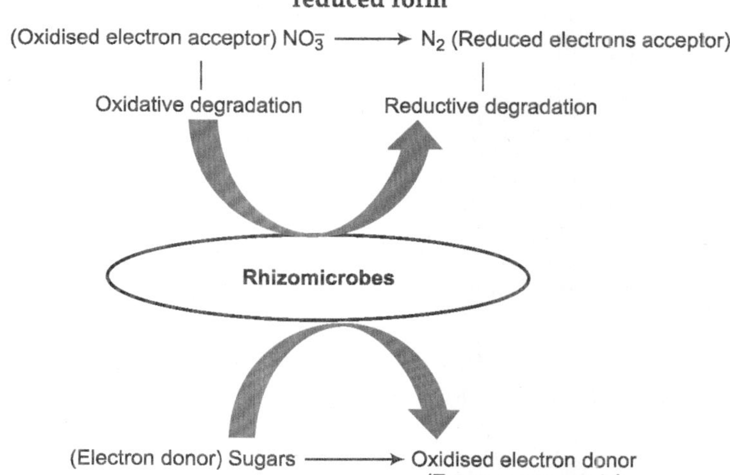

RootsExudates and their Role in Detoxification

Root exudates signify a suite of an organic compound with low molecular weight present in the rhizosphere that was secreted by roots of plants and microbial modified flora. It passively secretes root cell mucilage and also plays a very important role in converting metal bioavailability. And the release of specific organic compounds forms a metal complex which mobilizes metal ions and also applies nutrients to support plant growth. These compounds secreted by roots provide a source of energy to microbial biomass. Root exudates consist of amino acids, phytochelatins and organic acids which improve intra-cellular binding of heavy metal ions. These exudates substances also lead to reclaim alkaline soil by lowering pH due to the secretion of (H+ ions) with a certain group of enzymes (Ma et al., 2016). In the presence of root exudates compound the development of Biomarkers in rhizosphere occurred due to a decrease in the concentration of HM ions. A recent study,(gas chromatography-mass spectrometry) confirm that there are almost 15 compounds as a biomarker for Lead contamination are detected which can also change the form of exudates, Luo et al., (2017).

Fig. 3. Bioaccumulation of Heavy Metals

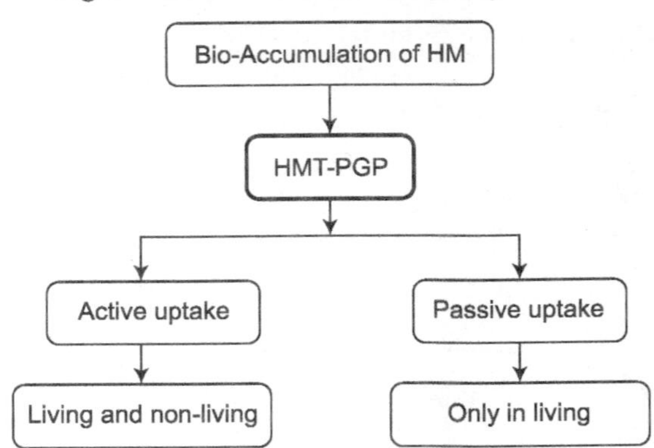

(Gutierrez-Corona et al., 2016).

Following process involved biosorption of heavy metals (Pokethitiyook and Poolpak, 2016)

- Complexation
- Coordination,
- Chelation,
- Ion exchange,
- Microprecipitation,
- Entrapment

Several Microbial communities present in the rhizosphere may also secrete several extracellular substances such as a variety of polysaccharide and lipopolysaccharide, glycoproteins and soluble peptides which acquire a considerable amount of anion groups those helps to reduce metals from the rhizosphere through the process of biosorption (Ayangbenro and Babalola, 2017). Production of EPS by different PGP microbes stimulate the formation of biofilm under abiotic stress response. The formation of Biofilm increases the tolerance of microbial cells through a protective sheath and also helps to transform metal from (toxic ions to non-toxic forms)after adsorption (Gupta and Diwan, 2017).

Plant Growth Promotion and HM Removal from Soil by HMT-PGP Microbes

The (HMT-PGP) microbes are contributing to growth and development of host plant However also accelerate the elimination of HM ion from soil contaminated with HM toxicity. This occurs due to enhancement in plant growth in HM stress or due to increasing the availability of metal ions for quick uptake by microbial cells and plants. In the rhizospheric- microbial population some of PGP traits release several extracellular enzymes, phytohormones, siderophores, solubilization of minerals from unusable to usable form especially P, K and Zn and biological fixation of nitrogen improves growth and concurrently minimize the harmful effect of HM on plant health. During the condition of abiotic stresses, plant cell induces the production of the stress hormone called ethylene in plants which leads to suppress the normal plant growth and decrease root proliferation. It was reported widely under stress there a huge number of enzymes produced such as 1-aminocyclopropane-1-carboxylate (ACC) deaminase. (Zhang et al., 2011; Han et al., 2015). Root growth of plants was directly influenced by the excess concentration of ethylene, concentration of ACC was regulated by metabolization of ethylene both are precursors under stress. Root elongation, development of tangential and adventitious roots to minimize HM concentration in the root zone is because of the application of HMT-PGP to increase the production of IAA and other phytohormones. The whole process of regulating the concentration of hormones in a balanced condition is responsible for better growth and development of the plant. It was reported that the strain of IAA producing B. subtilis SJ-101 regulates the growth of B.juncea in Ni-toxicity affected soil (Zaidi et al., 2006). Similarly, other strains of IAA production for micronutrients such as Zn, Ni, Cu, and Co regulates the growth of the plant in soil contaminated with Cd. (Belimov et al., 2005).Production of phytohormones, enzymes like IAA and ACC deaminase, PSB, siderophore and nitrogen-fixing(HMT-PGP) microbes also involved in plant growth and efficient root growth by enhancement in nutrient availability to plant as well as by transforming bioavailability of HM ions. (Wu et al., 2010; Gupta et al., 2014).

Significance of Study and Future Challenges

Advancement in the use of heavy metal tolerance and plant growth-promoting microbes in their host plants for mitigation of toxicity effect of HM is a sustainable and environmentally friendly approach. The commercialization of this technology requires more advancement to implement it in the future. It was found that the accumulation of metal ions at a harsh level in plant tissue downturns the mechanism of remediation (Ma et al., 2011). In some instances where the site is contaminated by many types of HM, the application of(HMT-PGP)microbes with additional nutrients is found to be more successful. Recently, (microcosm-scale phytoextraction experiments), Franchi et al., (2017) found that the addition of thiosulfate along with a specific range of HMT-PGP microbes improved mobilization and Hg uptake with As in Brassica. juncea and L. albus were grown in contaminated soil. The latest technique of GEM(genetically engineered microbes) is also well adapted for the efficient elimination of heavy metal from HM contaminated soil. (Das et al., 2016; Gupta and Singh, 2017). Biostimulation(the process of advancement of the environment to encourage the activity of existing bacteria) by the addition of limiting nutrients to remediate the toxicity of HM in soil. (Fulekar et al., 2012).. A recent study by Migahed et al., (2017) showed that the application of a mixture of (HMT) bacterial population and fungus species can efficiently remove metal ions of Fe and Chromium from a contaminated site near industry. (Gola et al., 2016). Transgenic and GEM plants use the expression of metal-binding proteins, metal detoxifiers, chelates, and enzymes for the development of heavy metal remediators. Recent research shows the exposure of plant microbial-biomass from the contaminated site can boost accessible phytoremediation technology as for the reclamation of HM ions (Thijs et al., 2017).

Conclusion

The main priority for sustainable agriculture is to develop certain kind techniques that eco-friendly, longer durability with low input cost to reclaim Arable soils affected with heavy metal toxicity. Heavy metal toxicity is a major threat to agriculture because it causes greater loss of quality yield and also leads to environmental pollution. The use of microbes to reclaim contaminated sites creates synergetic interaction between plant – microbes which lead to the elimination or transformation of HM ions from the rhizosphere. Application of rhizomicrobe for remediation is the most promising technique it encourages the use of HMT-PGP with immense multiple benefits such as detoxification of rhizosphere, better plant growth, improved physical-biological properties of soil. Conversely, further advancement in research is required to develop suitable biostimulants by using HMT-PGP microbes for reclamation and efficient utilization of contaminated soils.

References

Adriano, D. C., Bolan, N. S., Vangronsveld, J., & Wenzel, W. W. (2005). Heavy metals.

Ayangbenro, A. S., & Babalola, O. O. (2017). A new strategy for heavy metal polluted environments: a review of microbial biosorbents. *International journal of environmental research and public health, 14*(1), 94.

Belimov, A. A., Hontzeas, N., Safronova, V. I., Demchinskaya, S. V., Piluzza, G., Bullitta, S., & Glick, B. R. (2005). Cadmium-tolerant plant growth-promoting bacteria associated with the roots of Indian mustard (Brassica juncea L. Czern.). *Soil Biology and Biochemistry, 37*(2), 241-250.

Bielen, A., Remans, T., Vangronsveld, J., & Cuypers, A. (2013). The influence of metal stress on the availability and redox state of ascorbate, and possible interference with its cellular functions. *International Journal of Molecular Sciences, 14*(3), 6382-6413.

Checcucci, A., Bazzicalupo, M., & Mengoni, A. (2017). Exploiting nitrogen-fixing rhizobial symbionts genetic resources for improving phytoremediation of contaminated soils. In *Enhancing Cleanup of Environmental Pollutants* (pp. 275-288). Springer, Cham.

Checcucci, A., Bazzicalupo, M., & Mengoni, A. (2017). Exploiting nitrogen-fixing rhizobial symbionts genetic resources for improving phytoremediation of contaminated soils. In *Enhancing Cleanup of Environmental Pollutants* (pp. 275-288). Springer, Cham.

Clemens, S., & Ma, J. F. (2016). Toxic heavy metal and metalloid accumulation in crop plants and foods. *Annual review of plant biology, 67*, 489-512.

Blanco-Montenegro, I., De Ritis, R., & Chiappini, M. (2007). Imaging and modelling the subsurface structure of volcanic calderas with high-resolution aeromagnetic data at Vulcano (Aeolian Islands, Italy). *Bulletin of Volcanology, 69*(6), 643-659.

Ehrlich, H. L. (1997). Microbes and metals. *Applied microbiology and biotechnology, 48*(6), 687-692.

Fomina, M., Hillier, S., Charnock, J. M., Melville, K., Alexander, I. J., & Gadd, G. M. (2005). Role of oxalic acid overexcretion in transformations of toxic metal minerals by Beauveria caledonica. *Applied and Environmental Microbiology, 71*(1), 371-381.

Fulekar, M. H., Sharma, J., & Tendulkar, A. (2012). Bioremediation of heavy metals using biostimulation in laboratory bioreactor. *Environmental monitoring and assessment, 184*(12), 7299-7307.

Gadd, G. M. (2010). Metals, minerals and microbes: geomicrobiology and bioremediation. *Microbiology, 156*(3), 609-643.

Gadd, G. M., Bahri-Esfahani, J., Li, Q., Rhee, Y. J., Wei, Z., Fomina, M., & Liang, X. (2014). Oxalate production by fungi: significance in geomycology, biodeterioration and bioremediation. *Fungal Biology Reviews, 28*(2-3), 36-55.

Gavrilescu, M. (2004). Removal of heavy metals from the environment by biosorption. *Engineering in Life Sciences, 4*(3), 219-232.

Abdu, N., Abdullahi, A. A., & Abdulkadir, A. (2017). Heavy metals and soil microbes. *Environmental chemistry letters, 15*(1), 65-84.

Gola, D., Dey, P., Bhattacharya, A., Mishra, A., Malik, A., Namburath, M., & Ahammad, S. Z. (2016). Multiple heavy metal removal using an entomopathogenic fungi Beauveria bassiana. *Bioresource technology, 218*, 388-396.

Gupta, D. K., Chatterjee, S., Datta, S., Veer, V., & Walther, C. (2014). Role of phosphate fertilizers in heavy metal uptake and detoxification of toxic metals. *Chemosphere, 108*, 134-144.

Gupta, P., & Diwan, B. (2017). Bacterial exopolysaccharide mediated heavy metal removal: a review on biosynthesis, mechanism and remediation strategies. *Biotechnology Reports, 13*, 58-71.

Gupta, S., & Singh, D. (2017). Role of genetically modified microorganisms in heavy metal bioremediation. In *Advances in Environmental Biotechnology* (pp. 197-214). Springer, Singapore.

Gutiérrez-Corona, J. F., Romo-Rodríguez, P., Santos-Escobar, F., Espino-Saldaña, A. E., & Hernández-Escoto, H. (2016). Microbial interactions with chromium: basic biological processes and applications in environmental biotechnology. *World Journal of Microbiology and Biotechnology, 32*(12), 1-9.

Hao, X., Taghavi, S., Xie, P., Orbach, M. J., Alwathnani, H. A., Rensing, C., & Wei, G. (2014). Phytoremediation of heavy and transition metals aided by legume-rhizobia symbiosis. *International Journal of Phytoremediation, 16*(2), 179-202.

Hossain, M. A., Piyatida, P., da Silva, J. A. T., & Fujita, M. (2012). Molecular mechanism of heavy metal toxicity and tolerance in plants: central role of glutathione in detoxification of reactive oxygen species and methylglyoxal and in heavy metal chelation. *Journal of Botany, 2012*.

Jadia, C. D., & Fulekar, M. H. (2009). Phytoremediation of heavy metals: recent techniques. *African journal of biotechnology, 8*(6).

Jiang, J., Pan, C., Xiao, A., Yang, X., & Zhang, G. (2017). Isolation, identification, and environmental adaptability of heavy-metal-resistant bacteria from ramie rhizosphere soil around mine refinery. *3 Biotech, 7*(1), 5.

Kaewdoung, B., Sutjaritvorakul, T., Gadd, G. M., Whalley, A. J., & Sihanonth, P. (2016). Heavy metal tolerance and biotransformation of toxic metal compounds by new isolates of wood-rotting fungi from Thailand. *Geomicrobiology Journal, 33*(3-4), 283-288.

Khan, A. G. (2005). Role of soil microbes in the rhizospheres of plants growing on trace metal contaminated soils in phytoremediation. *Journal of trace elements in medicine and biology, 18*(4), 355-364.

Kong, Z., & Glick, B. R. (2017). The role of plant growth-promoting bacteria in metal phytoremediation. *Advances in microbial physiology, 71*, 97-132.

Kuiper, I., Lagendijk, E. L., & Bloemberg, G. V. (2004). Lugtenberg BJJ. *Rhizoremediation: A beneficial plant-microbe interaction bioremediation: a natural method, 17*, 6-15.

Frérot, H., Lefèbvre, C., Gruber, W., Collin, C., Dos Santos, A., & Escarré, J. (2006). Specific interactions between local metallicolous plants improve the phytostabilization of mine soils. *Plant and Soil, 282*(1), 53-65.

Luo, Q., Wang, S., Sun, L. N., & Wang, H. (2017). Metabolic profiling of root exudates from two ecotypes of Sedum alfredii treated with Pb based on GC-MS. *Scientific reports, 7*(1), 1-9.

Ma, Y., Oliveira, R. S., Freitas, H., & Zhang, C. (2016). Biochemical and molecular mechanisms of plant-microbe-metal interactions: relevance for phytoremediation. *Frontiers in plant science, 7*, 918.

Magoč, T., & Salzberg, S. L. (2011). FLASH: fast length adjustment of short reads to improve genome assemblies. *Bioinformatics, 27*(21), 2957-2963.

Merdy, P., Gharbi, L. T., & Lucas, Y. (2009). Pb, Cu and Cr interactions with soil: sorption experiments and modelling. *Colloids and Surfaces A: Physicochemical and Engineering Aspects, 347*(1-3), 192-199.

Migahed, F., Abdelrazak, A., & Fawzy, G. (2017). Batch and continuous removal of heavy metals from industrial effluents using microbial consortia. *International Journal of Environmental Science and Technology, 14*(6), 1169-1180.

Narendrula-Kotha, R., & Nkongolo, K. K. (2017). Microbial response to soil liming of damaged ecosystems revealed by pyrosequencing and phospholipid fatty acid analyses. *PloS one, 12*(1), e0168497.

de Azevedo Neto, A. D., Prisco, J. T., Enéas-Filho, J., de Abreu, C. E. B., & Gomes-Filho, E. (2006). Effect of salt stress on antioxidative enzymes and lipid peroxidation in leaves and roots of salt-tolerant and salt-sensitive maize genotypes. *Environmental and Experimental Botany, 56*(1), 87-94.

Pires, C., Franco, A. R., Pereira, S. I., Henriques, I., Correia, A., Magan, N., & Castro, P. M. (2017). Metal (loid)-contaminated soils as a source of culturable heterotrophic aerobic bacteria for remediation applications. *Geomicrobiology Journal, 34*(9), 760-768.

Pokethitiyook, P., and Poolpak, T. (2016). "Biosorption of heavy metal from aqueous solutions," in *Phytoremediation: Management of Environmental Contaminants*, Vol. 3, eds A. A. Ansari, S. S. Gill, R. Gill, G. R. Lanza, and L. Newman (Cham: Springer International Publishing), 113–114.

Mishra, J., Singh, R., & Arora, N. K. (2017). Alleviation of heavy metal stress in plants and remediation of soil by rhizosphere microorganisms. *Frontiers in microbiology, 8,* 1706.Schnoor, J. (1997). *Phytoremediation.* Technology Evaluation Report TE-98-01. Pittsburgh, PA: Groundwater Remediation Technologies Analysis Center.

Sytar, O., Kumar, A., Latowski, D., Kuczynska, P., Strzałka, K., & Prasad, M. N. V. (2013). Heavy metal-induced oxidative damage, defense reactions, and detoxification mechanisms in plants. *Acta physiologiae plantarum, 35*(4), 985-999.

Tchounwou, P. B., Yedjou, C. G., Patlolla, A. K., & Sutton, D. J. (2012). Environmental toxicology. *Molecular, clinical and environmental toxicology, Experientia Supplementum, 3,* 133-164.

Thijs, S., Sillen, W., Weyens, N., & Vangronsveld, J. (2017). Phytoremediation: state-of-the-art and a key role for the plant microbiome in future trends and research prospects. *International journal of phytoremediation, 19*(1), 23-38.

Turnau, K., & Kottke, I. (2005). Fungal activity as determined by microscale methods with special emphasis on interactions with heavy metals. *MYCOLOGY SERIES, 23,* 287.

Ullah, A., Heng, S., Munis, M. F. H., Fahad, S., & Yang, X. (2015). Phytoremediation of heavy metals assisted by plant growth promoting (PGP) bacteria: a review. *Environmental and Experimental Botany, 117,* 28-40.

Valko, M., Leibfritz, D., Moncol, J., Cronin, M. T., Mazur, M., & Telser, J. (2007). Free radicals and antioxidants in normal physiological functions and human disease. *The international journal of biochemistry & cell biology, 39*(1), 44-84.

Wu, G., Kang, H., Zhang, X., Shao, H., Chu, L., & Ruan, C. (2010). A critical review on the bio-removal of hazardous heavy metals from contaminated soils: issues, progress, eco-environmental concerns and opportunities. *Journal of Hazardous Materials, 174*(1-3), 1-8.

Hossni, Y. A. (1997). Safety of sewage water in the irrigation of plants. *Egypt. Horticulture, 24,* 261-270.

Harris, J., Mason, D. E., Li, J., Burdick, K. W., Backes, B. J., Chen, T., ... & Winssinger, N. (2004). Activity profile of dust mite allergen extract using substrate libraries and functional proteomic microarrays. *Chemistry & biology, 11*(10), 1361-1372.

Hanay, Ö., & Hasar, H. (2011). Effect of anions on removing Cu2+, Mn2+ and Zn2+ in electrocoagulation process using aluminum electrodes. *Journal of hazardous materials, 189*(1-2), 572-576.

Zhuang, P., Sun, S., Li, Y., Li, F., Zou, B., Li, Y., ... & Li, Z. (2018). Oral bioaccessibility and exposure risk of metal (loid) s in local residents near a mining-impacted area, Hunan, China. *International journal of environmental research and public health, 15*(8), 1

Zhuang, P., Li, Z. A., McBride, M. B., Zou, B., & Wang, G. (2013). Health risk assessment for consumption of fish originating from ponds near Dabaoshan mine, South China. *Environmental Science and Pollution Research, 20*(8), 5844-5854.

Xiao, X., Wu, Z. C., & Chou, K. C. (2011). A multi-label classifier for predicting the subcellular localization of gram-negative bacterial proteins with both single and multiple sites. *PloS one, 6*(6), e20592.

Phytoremediation and Its Possible Use For Agricultural Development

Poonam Kumari

Department of Agronomy, School of Agriculture

Lovely Professional University, Jalandhar, Punjab, India, 144411

Email: poonamnaik8988@gmail.com

Abstract

The growth of the plant mainly depends upon the mineral nutrition of the plants and the mineral nutrition is obtained mainly from the soil. Soil contains a lot of nutrients, metals and is absorbed by the plant's roots. The root hair accumulates the nutrients as well as the toxic elements which are then transported from below-ground parts of the plant to the above-ground parts of the plant. The toxic elements like Arsenic, Chromium, Cadmium, Nickel, Lead and Mercury etc. accumulate in the plants which hinder the growth and development of the plants. So, to correct the toxicity in plants Phytoremediation process is best. Phytoremediation process is used to accumulate the heavy metals or to degrade the toxic molecules of heavy metals into a non-toxic form. Various hyperaccumulator plants are used to extract the heavy metals because they do not have any toxic effects and can tolerate the toxic effect of the contaminants. The phytoremediation process is very important for agriculture because it is used for the degradation of toxic effluents which cleans the soil, water and plants so that the plants can easily grow. Along with this process Mycoremediation and Phytomining are also very beneficial because these processes also help in the extraction of heavy metals from the environment as well as funguses are used for their removal and the normal growth and development of the plants occurs which provides better yield.

Introduction

Phytoremediation helps in the mineralization of organic pollutants into non-toxic constituents like ammonia, carbon dioxide, chlorine and nitrate etc. In phytoremediation green plants are used to remove the pollutants from contaminated sites like mining areas, industrial area, dumping sites, near seas etc. In this technique, natural processes are used by the plants and the microbes degrade the organic and inorganic pollutants (Li et al., 2019). The inorganic pollutants are released by human activities like agrochemicals, mining areas, fertilizers, industrialisation etc. which causes harmful effects to the plants and human health as well. Various biological, chemical and physiological processes help in the biodegradation, storage, biotransformation and their detoxification also. The sequestration process helps in the mitigation of toxic level of contaminants (Saleem et al., 2020). Hyper accumulator plants are used for the extraction of metals and toxicants from soil and water. Many plants species of hyperaccumulator plants are there which are also known as Metallophytes. These plants extract the harmful contaminants through the soil by root hairs and then transport with the help of xylem into the upper parts of the plants like leaves and then the toxicant is stored in the ariel parts of the plant. After this process, the plant is harvested and burned and gone through with various smelting processes and at last, the metal is extracted and is used in various commercial purposes. Metals like Gold (Au), Silver (Ag), Platinum (Pt), Lead (Pb), Uranium (U), Arsenic (As), Mercury (Hg), Nickel (Ni) etc. are extracted with the process of Phytomining (Kumari et al., 2019). These processes are very easy and environmentally friendly. The techniques used in Phytoremediation are Phyto degradation, Phyto stabilization, Phytoextraction, Phyto volatilization, Phyto filtration and Rhizo degradation etc. These processes are very important in the process of phytoremediation, these processes enhance the process of degradation, their immobilisation, the extraction of metals etc. Without the use of this process, the phytoremediation does not complete. Enzymes like Catalase, Peroxidase, Laccase, Dehalogenase, Nitroreductase etc. are used to enhance the activity of the process. Mycoremediation is a process used as phytoremediation is. In this process, the fungus is used for the degradation of toxicants (Aken, 2010). The mycelium of the fungus helps in the degradation of the complex molecules, first, it degrades the complex compounds into simpler ones then after it the fungus uses the metal toxicant as feed material and after feeding on the toxicants it grows more but when the toxicant degrades completely then the fungus also dies because it does not get any feed material for its growth (Nwache et al., 2018). The leftover material is used as compost for the plant and that compost is applied to the plants for proper and development and provides nutrients to the plants. Many mushroom species are used for the mycoremediation of plants as well as food material for the humans. So, this process of phytoremediation is very important for agriculture as well as for commercial purposes because it cleans our environments and makes the soil and water toxicants free (Redfern et al., 2016).

Phytoremediation

Phytoremediation is a technique in which different plant species directly or indirectly used to remediate the pollutants of water and soil mainly heavy metals (DalCarso et al., 2019). Phytoremediation technique is a plant-based technology which uses plants for the extraction of heavy metals from contaminated zones. This technique is very cheap than other methods of remediation. As well as this also increases the aesthetic value, cleans the environment by not causing any pollution and harm to the soil, water and environment (Imadi et al., 2016, Peuke et al., 2005). Many other aspects are there as the use of electrokinetics, washing of soil and various physiochemical techniques are also used but these techniques are very costly, heavy machinery is required for the remediation process which directly leads to air and environmental pollution so, the alternative and easy aspects are to use the phytoremediation technique (McCutcheon et al., 2008).

Heavy Metal Toxicity

Heavy metals are the materials or metal compounds which are present in the earth crust and occurs mainly naturally as well as through human interventions also. Mainly the metals are originated in earth environment through volcanic eruptions, parent material of rocks metamorphically etc. or they may also originate from agricultural wastes like high and inadequate use of fertilizers, pesticides and fungicides etc. and high industrial activities (Image et al., 2016). The compounds of heavy metals enter through the water in the soil through which the soil, water and environment contaminated with these metal compounds. These compounds of heavy metals are toxic to human beings, animals, soils and environment. Their toxicity leads to reduced growth, poor yield and quality of crop as well as leads to the death of plants also. Heavy metals like Zinc, Cadmium, Chromium, Lead, Copper, Molybdenum, Gold, Mercury, Nickel, Arsenic and Iron etc. are mainly present in the earth and cause toxic effects when present in an excess amount by inhibiting the growth of plants (Kumari et al., 2019).

Hyperaccumulator Plants

Hyper accumulator plants are the plant's species which tend to tolerate the effect of heavy metals or the plant species which accumulates a large number of metals in it (Image et al., 2016). These plant species does not have any effect of heavy metal or stress conditions in it. These hyperaccumulator plant species are used in the phytoremediation of soils, water etc., mycoremediation and phytomining of various metals. These plant species can also accumulate more than one metal in it (Jan et al., 2016). Hyper accumulator plant species are mainly used in serpentine soils and can accumulate various heavy metal that is why these plant species are also known as Metallophytes (Pandey et al., 2019). Plant species belonging to family Brassicaceae, Euphorbiaceae, Asteraceae etc. are used for hyperaccumulation. Plants like Arabidopsis thaliana, *Berkheya coddii, Pteris vittata, Helianthus annus* are used for the extraction of metals from the soil, water and also for phytomining (Kumari et al., 2019). Along with this many fungi like mushroom species and algae species are also used for the extraction of heavy metals from the soil, water and environment etc. Basidiomycetes and Ascomycetes etc. are the groups which are used in the mycoremediation process mainly the species *Pleurotus spp., Agaricus spp., Lentinula spp.* etc. are used. Along with the extraction of heavy metals these species also help in the degradation of many compounds like plastic materials, PAHs etc. or also used in the preparation of organic compost from mushroom species (Weyens et al., 2015).

Techniques Used In Phytoremediation

There are various techniques which are used in phytoremediation based on the nature of pollutants. The growth of plants mainly depends upon the root system which is present in soil and the soil consists of various fungi, bacteria and actinomycetes etc. (Favas et al., 2014) which directly or indirectly helps in the growth and development of the plants and their association helps in the reclamation of soils also. Various techniques are used for the remediation of heavy metals from the contaminated soils (Dal Carso et al., 2019). There are six techniques which include i) Phyto degradation or Phyto transformation ii) Phytostabilization or Phyto immobilization or Rhizostabilization iii) Phytoextraction or Phyto sequestration or Phytoabsorption or Phytoaccumulation iv) Phytovolatization v) Phytofilteration or Rhizofilteratiation vi) Rhizodegradation etc. are used as listed in figure 1. These techniques are very useful in the process of phytoremediation (Pandey et al., 2019).

Fig. 1. Techniques of Phytoremediation

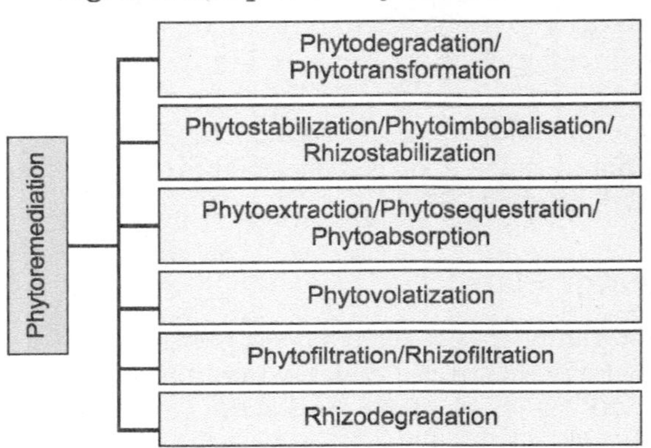

Phytodegradation/ Phytotransformation: This is a type of degradation process in which organic contaminants are metabolised or degraded with the help of some enzymes like dehalogenase, nitroreductase, nitrilase, oxidoreductase, phosphatase and laccase etc. In this degradation process complex compounds or contaminants are broken by these enzymes into simpler compounds which further become less harmful or toxic to plants (Das et al., 2011). This process is also known as Phyto transformation because the complex compounds are transformed into simpler ones. The plant's roots are associated with many enzymes and these enzymes are released when the contaminants are present in the rhizosphere. So, the plant should be able to translocate and uptake the contaminants for the transformation process (Wa et al., 2019). This process does not cause any harmful or lethal effects on the cells of plants. Contaminants are present in a wide range which may be munitions, organic or inorganic compounds, fertilizers, pesticides, herbicides, fungicides and chlorinated solvents etc (Kumar et al., 2018). Some enzymes are very good and show better results in low contaminated soil so the enzymes are used for the transformation of these contaminants. These types of enzymatic reactions mainly occur in the plant species of Populus spp. and Myriophyllium spectrum etc (Gil- Loaiza et al., 2018).

Phytostabilization/Phytoimmobalisation: In phytostabilization, the contaminants are immobilized so that there is no diffusion occurs in the soils that are why this process is also known as phytoimmobalisation. In this process, all the organic and inorganic contaminants are incorporated into the cell wall of lignin present in the cells of roots. The metals present in the soil are precipitated into the insoluble form by the exudates of roots which are in soil matrix (Schnoor et al., 1998). In this process vegetation is used to stop the mobility of contaminants in the soil, groundwater etc. by the help of various modifications like biological, physical and chemical etc. to stabilise the action of contaminants is known as phytostabilisation. Some selective plants are used and there is no environmental invasion (Naher et al., 2014). This technique is cheap and very simple than other ones. By increasing the organic matter in the soil or by doing amendments the cation exchange capacity also increases. This will prevent or protects the harmful contaminants from leaching to the groundwater thus the organic matter helps in the protection of groundwater from contaminants (Rosenkranz et al., 2019). This will also protect the soil from erosion and maintains the food chain through the process of phytostabilization. This process works on the changes that occur in the environment by stabilizing the solubility and mobility of metals like Mercury, Gold, Cadmium, Zinc, Chromium and Arsenic etc. The accumulations of heavy metals contaminants are important because they can also fix the low level of contaminated radioactive sites. In phytostabilization the reduction of metals occurs due to the precipitation process occurs in the rhizosphere and their sequestration, absorption also occurs within the tissues of roots as well as the adsorption into the cell wall of the roots (Sheoranz et al., 2009). Due to the addition of organic matter, the growth of plant increases which facilitates the soil health also in the metal-polluted areas. For the disposal of harmful biomass, the phytostabilization process is more beneficial than the phytoextraction process. The selection of plant species for a phytostabilization process is very important. The plant species used for this process may have heavy metal tolerance capacity. The roots of those plants play a very important role in stabilizing the soil structures, maintaining fertility, immobilizing the heavy metals and by preventing from soil erosion (Wiche et al., 2016). The plant's species which can accumulate a large amount of biomass and can grow faster are used for the phytostabilization of heavy metals in contaminated zones. The micro-organisms living inside the rhizosphere like bacteria, fungi and mycorrhiza can improve the biological properties, physiochemical properties, water holding capacity etc. The micro-organisms adsorb metals in their cell wall and then produce chelators which result in precipitation process (Spina et al., 2018). The mycorrhizal, fungi and bacteria also increase the root surface of plants and makes filtration barrier for heavy metals in the translocation of metals from the root portion of the plant to the shoot portion of the plant. Mainly the plant species of genus Alyssum, Ascolepsis, Haumaniastrum, Gladiolus and Eragrostis etc. are used for phytostabilization or Phytoimmobalisation (Wilson- Corral et al., 2012).

Phytoextraction Or Phytosequestration Or Phytoabsorption: Phytoextraction techniques include the adsorption of various contaminants with the help of roots and their accumulation and translocation in the remote parts of the plants. This technique is mainly applied to metals like Nickel, Zinc, Cadmium, Selenium, Copper, Arsenic and Lead etc. Hyper accumulator plant species are used to accumulate a large number of metals from the aerial parts of the plants (Zhang et al., 2014). This process of Phytoextraction is also known as Phytomining in which a large number of essential metals are extracted from the hyperaccumulator plant species. In phytomining process, the hyperaccumulator plant species are grown in metal contaminated areas. After the proper growth and development of those plant species, they are burned into ash and gone through smelting process from which metals can be extracted without causing any damage to the environment (Erakhrumen et al., 2017). After the incineration, the leftover ash is used in the filling of wastelands. This method of metal extraction is very easy, cheap, old and efficient through which the metals are translocated from below parts of the plant parts to the above plant parts. There are high translocation and high tolerance against the plant species that we are going to use in this phytomining process. They are easily harvested, there should be high production of biomass and bioconcentration factor should also be high which is good for the extraction of metals from the plant species (Erdei, 2005). This process is different from the phytostabilization because on that process the metals are not translocated and remain at the below-ground parts of the plants but in this phytoextraction process, the metals are translocated. This method is mainly used for commercial purposes because it has a permanent solution to get rid of heavy metals present in the soil. In phytoextraction, the sequestration of heavy metals occurs in the plant tissue that is why it is also known as photo sequestration (Kumari et al., 2019). The plant species that we are going to use in phytomining process should have a high tolerance towards heavy metals, can accumulate and translocate appropriate amount of metals, should be fast-growing and have high production of biomass, the root and shoot system should be extensive, can adopt a wide variety of environmental conditions and can also grow in poor soils, may have resistance towards various insects pests as well as herbivorous animals. These plant species can accumulate a large number of heavy metals without showing any toxicity symptoms (Bani et al., 2015). If the root to shoot ratio of heavy metals is more than one then they can easily and better translocate the metals from below-ground parts to the above-ground parts of the plants. The shoot to soil ratio is also greater than one only then they can take up heavy metals from the soil (Tognacchini et al., 2020). The plant species of hyperaccumulator mainly belongs to the family Euphorbiaceae, Brassicaceae, Scrophulariaceae, Asteraceae and Lamiaceae etc. Mostly the plants' species used for the phytoextraction process for the accumulation of various metals like Copper, Cadmium, Arsenic, Zinc, Nickel etc. are *Thlaspi caerulescens, Elcholtzia splendens, Pteris vittata, Alyssum bertolonni* and *Arabidopsis thialiana* etc. (Kumari et al., 2019).

Phytovolatization: Phyto volatilization is a process in which the plants can absorb the metals and then violate them. Firstly the metals are absorbed through roots and then converted into some toxic forms and then released to the environment with the help of leaves. This method is mainly used for the organic pollutants, Selenium, Mercury and Arsenic etc. (Keeling et al., 2003). The mechanism of evapotranspiration is used in this process. Phyto volatilization process mainly depends on the climatic conditions, conditions of contaminated sites, translocation, absorption, accumulation and chemical nature of the metals present in the contaminated sites. This method is much different from phytoextraction because phytoextraction provides complete removal of the contaminants but this process Phyto volatilization does not provides complete removal of the contaminants (Ha et al., 2019). It just accumulates the metals from the soil, groundwater etc. and releases it in the atmosphere. The released contaminants in this process are less toxic because they are present in low concentration. From this process, the pollutants can enter from atmosphere to soil and then contaminates the soil and water again (Chaney et al., 2007). The plant species of Brassicaceae family is best suited for this process. The plant species *Stanleya pinnata, Brassica napus, Arabidopsis thaliana, Liriodendron tulipifera, Astragalus bisulcatus* and *Nicotiana tabacum* etc. are mainly used for phytoremediation (Kumari et al., 2019).

Phytofiltration Or Rhizofiltration: In this process, the plant roots can absorb the heavy metals and the radioactive elements and then the contaminants are precipitated from the aqueous medium with the help of plant roots (Favas et al., 2014). Aquatic hyperaccumulator plants are used for this process because they can accumulate more contaminants and radioactive substances, have high root biomass and absorption is also good. In this, the plants are kept in any hydroponic medium from where the contaminants are passed and filtered through the roots. This practice involves the filtration with the roots that is why it is also known as Rhizo filtration. The contaminants are absorbed with the help of roots until the saturation point occurs. After that, the plants are harvested along with the root portion (Dal Carso et al., 2019). In this process, translocation does not occur and the contaminants remain at the root portion that is why this process does not cause any atmospheric pollution or contamination (Harrier et al., 2004). The Phytofiltration is used for the remediation of metals like Molybdenum, Cadmium, Gold, Mercury, Nickel, Arsenic, Chromium, Cobalt, Copper, Lead, Manganese, Selenium and Zinc etc. This process is also used for the filtration of radioactive substances like 90So, 234U, 137Cs and 239Pu etc. The plants with fibrous roots mainly the terrestrial plants are used because they are fast-growing. Along with roots, this technique also involves the use of shoots that are known as Caulofiltration and the use of seedlings is known as Blastofiltration etc. (Redfern et al., 2016). Plant species used in this technique are Azolla, Cattail, Duckweed, Populus, Hyacinth etc. for aquatic plants and terrestrial plants the *Helianthus annus, Brassica juncea, Nicotiana tobacco, Spinacia oleracea, Fontinalis antipyretic, Salix sp., Callithricha spp., Phragmites australis* and *Lemma spp.* etc. are used.

Rhizodegradation or Phytostimulation:- In this process, the roots of the plants promote the proliferation of micro-organisms. The carbon is used as the energy source for plant metabolites. The micro-organisms used in this process possess heterogeneous nature. The species used for this process is Pseudomonas. This process is also known as Phytostimulation (Yan et al., 2020).

Chemical and Physiological Aspects of Phytoremediation

The heavy metals and contaminants are present nearby the plants which are then accumulated by the roots and then translocated to the different parts of the plants. The metals and contaminants are also get trapped in the leaves of the plants through the spray of chemicals by some agricultural operations. The contaminants get attached to the root portion through the diffusion process (Silambarasan et al., 2013). The transport rate of xenobiotics depends upon the physical and chemical nature of the pollutants. If the process of transportation is passive then the rate of transportation is evaluated and transportation occurs by different modes.

Agricultural activities enhance the rate of transportation, activities like irrigation, application of fertilizers etc. Plants having a strong root system can uptake good amount of contaminant (Gil- Loaiza et al., 2018). When the metals are entered into the xylem various reactions like hydrolytic and oxidoreduction reactions takes place, and conjugates with the metabolites like sugar, glutathione and organic acids etc. The plant species Salix and Populus are mainly used because they have good root structure and transpiration activity. Binding of xylem with metabolites increases the solubility (Imadi et al., 2016). The bound contaminant is accumulated in vacuoles and for their biological degradation, many metabolic reactions take place. The compartmentalizations of cells are important for the detoxification of metals. For the degradation of organic contaminants various enzymes like Dehalogenase, Carboxylase, Nitroreductase, Phosphatase, Dioxygenase, Peroxidase, Nitrilase, Hydrolase and Monooxygenase etc. are used. The biocatalysts help in the easy and early degradation of contaminants. The hyperaccumulator plant species like *Arabidopsis thaliana* secrets the enzyme Laccase which helps in the early degradation of heavy metals (Keeling et al., 2003). Some enzymes like CytochromeP450 mono-oxygenase gave better results in the degradation of enzymes. This enzyme increases the hydroxylation process. Due to this enzyme, the plants can survive under high toxicity conditions (Singh et al., 2005). For the degradation of herbicides, the plant's *Helianthus tuberosus* C4P76B1 and the biocatalyst Glycine max C4P71A1D are best. Many genetically engineered plants like *Arabidopsis thaliana* and *Nicotiana tobacco* have a gene P450 shows tolerance towards herbicides (Tripathi et al., 2020).

Detoxification Process of Phytoremediation

Many transporters, ligands and chaperons help in the transportation inside the cells. Proteins are used to transfer metals between the membrane of cells. They form influx- efflux pathway for the translocation of metals from the root portion to the shoot portion of the plant (Dal Carso et al., 2019). Those transporters help in various mechanisms occurs in between the chloroplast, plasma membrane and tonoplast etc. The mechanisms like exportation, trafficking, import and sequestration occur in between the membranes. Proteins like ZIP family that have ZRT IRT like proteins and NRAMPs that is naturally resistant associated macrophage proteins are very useful in the detoxification process (Bani et al., 2015). The ZRT IRT like proteins helps in maintaining the homeostasis during the translocation of roots to the shoots and the naturally resistant associated macrophage proteins like NRAMP3 and NRAMP4 helps in the transport of iron within the plastid and uptake of Manganese from the soil. Proteins like heavy metals P1B type ATPase helps in the pumping of cation from the cytoplasm. HMA1 and HMA3 help in the detoxification of Zinc, Cadmium, Lead and Cobalt etc. The protein HMA4 plays a role of the transporter in the plasma membrane. Some proteins also provide tolerance against metals like Zinc, the proteins used for that purpose are Metal tolerance proteins (MTPs) (Weyens et al., 2015). Detoxification of heavy metals is an important process in the phytoremediation process. Various plant strategies are used to tolerate the effect of heavy metals (Okparanma et al.,

2011). The main two defence mechanisms are Avoidance and Tolerance. With these two mechanisms, the plants can maintain the concentration of heavy metals below the threshold level.

1. **Avoidance:** The mechanism of detoxification blocks the movement of heavy metals into the plant tissues. This mechanism works on an extracellular level and forms a line of defence. This mechanism includes various activities like precipitation of metal ions root sorption and exclusion of metals. In the detoxification process, the plant first goes through the process of root sorption. Various root exudates like amino acids and organic acids are used to tolerate the effect of heavy metals in the rhizosphere (Yan et al., 2020). The metal ligands present in root exudates changes the pH of soil and precipitation of heavy metal occurs so that the bioavailability and toxicity of metals stop. The metal exclusion system present in plants makes a barrier between the root portion and the shoot portion of the plants due to which the transportation of heavy metals stops from below-ground parts to the above-ground parts of the plants. Some micro-organisms like Vascular Arbuscular Mycorrhizae (VAM) also acts as a barrier between the transportation of metals in the plants to the rhizosphere. The cell wall also acts as a barrier between the transportation of metals because it contains carboxyl group that is negatively charged and can bind the heavy metals and also acts as an exchange of cations, therefore, it restricts the heavy metal entry into the plant cell (Erdei L., 2005).
2. **Tolerance:** Tolerance is said to be the second line of defence mechanism after the avoidance. Intolerance various other mechanisms work like compartmentalization of metals, inactivation of heavy metals and chelation etc. The tolerance strategy is available to the plant only when the heavy metals entered into the cytosol. Plants can only detoxify the metals to get rid of toxic effects and only when the heavy metals are present in a huge amount in the cytosol (Spina et al., 2018). Chelation process is used for the detoxification of contaminants with the help of various ligands. Chelation process reduces the concentration of heavy metals. Various organic and inorganic ligands are present in the cytoplasm of the cell which promotes the chelation process. Inorganic compounds, the organic acids are present which reduces the penetration of metal ions in the plant cell (Thijs et al., 2016). Many Phytochelatins, Amino acids, Metallothioneins, Phenols, Pectins and Proteins are present which reduces the penetration also. After the complete process of chelation, the ligands are transported from cytosol to the inactive compartments. It protects in removing the toxic effects of metals. This process will also reduce the damage caused to the cells. In an environment when the heavy metals accumulate so that the avoidance and tolerance strategies becomes inadequate for detoxification so that the production of Reactive Oxygen Species (ROS) increases and the oxidative stress also increases that disturbs the oxidation of proteins, damage to DNA, disturbs the homeostasis of a cell etc. To get rid of all these oxidative stress, the anti-oxidative defence mechanisms should be activated (Yan et al., 2020).

Strategies Used In Phytoremediation

The pollutants present in the environment are manmade or either organic. Even the organic pollutants are carcinogenic for humans and toxic in nature. The organic pollutants arise in the environment from the oil spills, chemicals industries, chemical fertilizers like herbicides, pesticides and fungicides etc., explosions, weapons and urbanisation etc (Tripathi et al., 2020). The organic pollutants like polycyclic aromatic hydrocarbon, linear halogenated hydrocarbons and polychlorinated biphenyl etc. are used for the phytoremediation process. These pollutants can be remediated directly by the direct uptake and the process used in this is called direct phytoremediation. The direct uptake of contaminants by the plants hinders the transport mechanism of plants (Favas et al., 2014). The transport of any contaminant depends on the physiochemical nature of the plant like the concentration of metals present, toxicity and acidity etc. The herbicides and pesticides transported through the soluble portion of the soil into the plant. Different plant species have different absorption power and different transportation system. There are changes in the rate of transpiration and uptake of contaminants also. The solubility of contaminants in the soil is very important for the process of phytoremediation in the removal of contaminants from the environment, soil and water also. Various surfactants are used to increase the solubility of contaminants so that they are easily absorbed and translocate in the plant system. The surfactant also helps in the degradation of complex molecules, catalytic reactions and solubility of xenobiotics etc. For the phytoremediation process, first, the information about the parent material should be there according to that surfactant is used and based on physical and chemical properties of contaminants, the accumulation of toxicants in roots and the shoot portion is determined (Sheron et al., 2009). The second strategy to remediate the pollutants is through phytoremediation explants. In this, the metabolites of roots are exploited which increases the multiplication of microbes and facilitates the metabolic process also in the rhizosphere area for the remediation of toxicants. Roots secrets the organic substances like organic acids, proteins, phenols and alcohols etc. and these are the good source of nitrogen and carbon which helps in the multiplication of micro-organisms used for the catalysis of pollutants. The rate of secretion occurs by the root directly or indirectly depends upon the type of plants species which is different species. So, the plant species that we are going to use in this process must have good and fast root secretion and is enriched with phenols which help in the biodegradation of PCB. The amount of phenols present in plant species differs from species to species. With this, the volatilization process occurs with the degradation of PAHs, TCE and TNT also and results in the synthesis of humans (Wiche et al., 2016). Many large trees having a deep root system can help in the good amount of transpiration by removing a large quantity of water from soil and plants. Plant tissues are sequestered and the enzyme helps in the metabolism of contaminants. The vegetation cover decreases the infiltration of rainwater and spread the pollutants. This will also increase the biodegradation process, transpiration process, evaporation and maintains the aeration in the soil. Organic soil, aquatic plants, micro-organisms, bacteria and fungi work together for the removal of toxicants and help in the exchange of ions, precipitation increases infiltration and adsorption (Redfern et al., 2016). The treatment of wastewater is also used in the phytoremediation. The construction of wetlands is a cheap, efficient method for the treatment of industrial waste, agriculture and also in drainage. There is one technique in which halophytes are used to remove a high amount of salts present in saline soil through the process called Phyto desalination (Zhang et al., 2014). The plant species like *Sesuvium portulacastrum* and *Suaeda maritime* are used for the removal of sodium chloride from the soils. Various strategies are used for the degradation of contaminants and prevention from plant pathogens, these strategies are a selection of plants for the interaction of microbes and changes in the diet of microbes, competition between microbes and environment, host and microbes and microbes and microbes and increasing the immigration ratio etc. (Schnoor et al., 1998). In phytoremediation plants with high biomass are adopted for the accumulation of contaminants. The transgenic crops have less microbiome in total heterotrophic bacteria and active heterotrophic bacteria. The selection of plant strategy is very important for the high biomass yields and phytoremediation. The cultivars of transgenic Nicotiana tobacco are used for the improved detoxification of contaminants and the bacteria Nitroreductase present here also increases the positive effects and even doubled it. So, the plants should be selected based on mycorrhizal habitats which have no antagonistic effects. The second strategy used here is the change that occurs in the diet mainly in the exudates of roots (Wa et al., 2019). Here the process of Rhizo engineering is adopted. In this strategy, some changes are formed in the quality and the number of root exudates present in the rhizosphere for the proper and good growth of plants as well as degradation. The transgenic plants like Arabidopsis thaliana in transgenic form exuded some xenotropic substances like octopene which will increase the ratio of degradation. Proper selection of plants should be there for the interaction with the microbiome in the rhizosphere. The seedling of maize has a high amount of DIMBOA which is transformed by some anti-microbial activities in the soil rhizosphere (Saleem et al., 2020). The compound DIMBOA increases the catabolic reactions which are beneficial as well as the species *Pseudomonas putida* also helps in the biodegradation as well as beneficial for rhizome engineering. The metal strategy used in the process of phytoremediation is feeding the supply lines, first degradative strains of culture are isolated. Various culture-based techniques are used for improving the culture media and for this different technical and intelligent devices like i-Chip are used, which helps in the cultivation of the culture that is very difficult for culturing purposes of micro-organisms (Tripathi et al., 2020). The selection of strains for the media and its inoculation is also very important, the strains mainly depend upon the genotypes and they remain consistent for different development stages and the different soil types, this will facilitate the enrichment of microbes. The proper time of inoculation is a must. The colonization of the leaves of plants having some useful bacteria reduces the reproduction ability of secondary colonizers, this will directly influence on phytoremediation. The last strategy in the phytoremediation process is the modification of some driving forces. The driving forces are enhanced by increasing the interactions between the plant and microbiome (Yan et al., 2020). The antibiotic hinders the gastrointestinal structures of microbiota and the addition of antibiotics into the soil changes the microbiome in the soil which ultimately increases the biodegradable rates of hydrocarbons. So, the reduction in competition enhances catabolic activity. The microbe and microbe interaction, as well as the microbe and the environment interaction, should be there; these interactions will increase the rate of phytoremediation in the plants, soils as well as in the environment. So, these strategies should be adopted for proper, easy and fast effects of phytoremediation. This will also enhance the activities of microbes (Thijs et al., 2016).

Role of Enzymes In The Process Phytoremediation

An enzyme plays an important role in the detoxification and transformation of various organic pollutants and helps in the restoration of heavy metals polluted environment. Enzymes are very useful in the phytoremediation process. Enzyme increases the microbial activity and helps in the degradation of contaminants. The contaminants are of very small size that's why they are easily transportable (Image et al., 2016). Microbial enzymes are used for the degradation of contaminants and are environmentally friendly also. The action of enzymes occurs both inside and outside the cell and are heterogeneous or homogeneous in nature. The terrestrial and aquatic plants also secrets some biocatalysts in the environment which leads to an increase in the multiplication of microbial activities occurs in the rhizosphere (Aken 2010). Various enzymes are secreted by different types of plants, enzymes like peroxidases, laccase, nitroreductase, dehalogenase and nitrilases etc. are secreted and are most common. The enzymes like laccase and nitroreductase secreted by plants and helps in the degradation of organic pollutants and various fertilizers and chemicals which contains nitrogen. Nitroreductase enzyme also mineralizes different contaminants of nitroaromatic compounds. The enzyme halogenate works in the mineralization of compounds of hexachloroethane whereas the enzyme nitrilase catabolises the reaction of 4 chloro benzonitrile (Bani et al., 2015). The organic pollutants are transformed with the help of enzymes in lab conditions also. The enzymatic reactions of enzymes like proteases, esterases and amidases degrade the bonds of peptides, ester and amide at the time of formation of by-products, along with that no toxicity symptoms appears. The bacterial species like Pseudomonas, Nocardia, Achromobacter, Flavobacterium and Bacillus etc. gives better results. The enzymes can participate in the mineralization of pollutants when secreted from plant system to the roots and ultimately helps in the degradation of harmful chemical compounds which may be man-made and organic ones (Erdei, 2005).

How Genetic Modified Organisms are Used In Phytoremediation

New achievements in phytoremediation are maintained through the use of genetically engineered host plants and microbes. Genetically engineered microbes are used in the host plant to increase the tendency of the phytoremediation process. The process of genetic engineering in phytoremediation increases the species of microbes associated with the plant system and the isolation of the bacteria, followed by the transfer of genes into the enzyme that contributes to the degradation of contaminants (Redfern et al., 2016). Genetically engineered micro-organisms used in field conditions do not grow much faster and faster than in laboratory conditions, thus affecting the degradation process. The mineralization process is affected by the metabolic activities of the enzymes and it is not possible to introduce all beneficial genes into a single organism for the proper degradation of the contaminants (Yan et al., 2020).

Phytoremediation of PCBs

Phytoremediation helps in the remediation of many toxic substances such as heavy metals, aromatic compounds and polychlorinated biphenyls. Many transgenic plants, bacteria and genetically modified organisms (GMOs) are used for soil and water treatment. Polychlorinated biphenyl (PCB) compounds are present in soil and water, which are also very harmful to soil, water and the environment (Erakhrumen et al., 2017). For the treatment of this polychlorinated bi-phenyl transgenic plant, bacteria are used which are environmentally friendly and cost-effective. Phytoremediation is a very good and easy technique used for the degradation of polychlorinated biphenyls. In this treatment, the phytoextraction process helps in the extraction of harmful contaminants of polychlorinated biphenyl from the soil, and the phytotransformation process helps in the transformation of polychlorinated biphenyl molecules during this process by releasing various metabolites that help to improve microbial activity (Sheoron et al., 2009). The process of remediation of rhizomes is a process that can degrade the pollutants and remediate the root zone. Degradation of polychlorinated biphenyl is a very slow process, but it produces better results, and the bacteria, as well as the transgenic plants, also help in this treatment (Aken et al., 2010).

Phytomining

Phytomining process is a method in which metals are extracted from the hyperaccumulator plants for economic purposes. Some metals like Gold, Arsenic, Chromium, Uranium, Cadmium and Platinum etc. are very important in day to day life (Kumari et al., 2019). These metals are present in the soil at a very good amount and can be extracted with the help of phytomining. Many other processes are also used but they are costly, cause pollution and their maintenance are also very high, therefore the environmentally friendly, easy and low-cost method is used that is phytomining. The heavy metals present in soil and from there they are extracted. In phytomining process the hyperaccumulator plants like *Arabidopsis thaliana, Berkheya coddii, Streptantus polygaloides, Alyssum bertolonni, Thlaspi caerulescens, Serbertia accuminata, Alyssum murale* and *Bornmullerea tymphaea* etc. are used for the extraction of metals (Favas et al., 2014). The plants are grown and harvested after their proper growth and development. With the help of xylem and phloem, the metals are transported from below-ground parts to the above-ground parts of the plants. After the harvesting of the plants, they are burned and gone through with the smelting process after that the metal is extracted. This process helps in the phytoremediation because it cleans the environment from the harmful contaminants of heavy metals and other toxicants. So, the phytomining process can be used for commercial purposes as well as it helps in the cleaning of the environment, soil and water etc. (Kumari et al., 2019).

Mycoremediation

Mycoremediation is a process used as phytoremediation in which various types of fungus species are used to clean the environment, contaminated soil, water bodies etc. The contaminants like dyes, PAHs, xenobiotics, fertilizers, pesticides, heavy metals, aromatic compounds etc. The fungal mycelium is used for the remediation of toxic compounds (Li et al., 2009). The fungal mycelium changes the toxic effluents into non-toxic effluents. The mycelium of fungus eats or feeds on the pollutants or any contaminant and after that, the fungal mycelia disappear and do not leave any toxic material after degradation. The mycelium used those harmful toxicants as food material and digest them as a source of nutrient that is why mycoremediation process is used for the removal of toxicants (Wu et al., 2019). Many fungal species are used which may belong to the phylum Zygomycota, Ascomycota, Deuteromycota and Basidiomycota etc. The phylum Zygomycota contains a round shape fungus which is mostly present on the bread and other eatable products. The fungus Rhizopus, Albugo and Mucor are used. In the phylum Ascomycota contains the fungi having sac-like structure. The fungus Claviceps, Neurospora and Aspergillus etc. belongs to this phylum. The fungus phylum Deuteromycota have Thallus and is found almost everywhere nearby water, soil, animals etc. The phylum Basidiomycota contains the fungi having club shape and the body is known as Basidium (Jan et al., 2016). This fungus can reproduce sexually and asexually. It contains the fungus Ustilago, Agaricus and Puccinia etc. Fungi can tolerate various environmental conditions and can be found everywhere. The fungal species like *Dichomiyus squalens, Planerochaete sp., Ganoderma applantatum, Pleurotus spp., Irpexlactans sp., Lentinula edodes, Trametes versicola, Bjerkandera adusta, Phlebia tremeblosa spp., Inonotus hispidus., Hispchiopomus larincinnus., Arbascular mycorrhizae, Mucor racemesus, Sordaria fimicola, Pennicilium raistrickii* and *Flavadon flavus* etc. are used (Kumari et al., 2019). As mushrooms are very beneficial for human health like that they are also helpful for the environment. Mushroom helps in the mycoremediation of pollutants. Three processes are there through which mushrooms clean the harmful contaminants. The three processes are Biodegradation, Biosorption and Bioconversion. In the biodegradation process, the complex molecules of toxicants are recycled. Many enzymes like Xylanase, Oxidase, Pectinase, Cellulose, Ligninase and Peroxidase are used for the degradation of complex molecules like PAHs, dyes, nitroreductase etc. (Gil- Loaiza et al., 2018). The second process used here is the biosorption process. This process is mostly used for the remediation of toxicants released from industries. The sorption of xenobiotics occurs due to which formation compost occurs (Rosenkranz et al., 2019). In this process, mushrooms species absorb the pollutants and the dead biomass is used for the preparation of compost. The third process used here is Bioconversion. The mushroom-like Pleurotus, Rusulla, Polypomus, Agaricus, Boletus, Termitomyces, Volvariella, *Flammulina velutipes, Calocype indica, Fomes fasciatus* etc. are used for the conversion of mushrooms to the compost after the process of mycoremediation (Kumari et al., 2019). The mycoremediation process is environmentally friendly because it does not cause any pollution or harmful effects to the environment. This technique does not use any harmful chemicals for the degradation and no maintenance is required as well as it provides nutrients to the plant growth. Fast results are there and the cost of applying this treatment is very low than other techniques on which large machinery is required. The contaminants are used as food material for the fungal mycelium for its proper growth and development (Chaney et al., 2007).

Uses and Advantages of Phytoremediation

The phytoremediation process has many advantages because phytoremediation technique helps in the washing of soil, provides stabilization, soil excavation etc. Phytoremediation is based upon the in-site strategy in which soils are treated and agronomical methods are applied which manages the availability of pollutants (Tripathi et al., 2020). The technique is natural and mostly solar energy is used due to which is cheap. With phytoremediation the organic matter, oxygen and nutrients are applied and the soil which makes the soil more fertile and improves the structure. Phytoremediation is used for checking the soil erosion, water erosion and wind erosion by planting the hyperaccumulator plants, this will helps in the stabilization of soil by plant roots (Kumari et al., 2019). This treatment is used for the phyto treatment of wastewater, industrial water etc. The phytoremediation technique is an in-situ and passive technique. It is of low cost because solar energy is used. This provides habitats for micro-organisms, birds and animals and reduces the runoff of surface water and is highly accepted by the public because of low cost and easy process (Favas et al., 2014). It also reduces the dust particles and harmful contaminants from the air. This technique also helps in the improvement of landscapes as well as the losses due to which leaching also reduces and contaminants can mobilize easily in the soil. The harvesting process is also very simple and the biomass is economically more valuable (Aken 2010). Phytoremediation is used for the remediation of contaminated sites like mining sites, dumping sites, oil refineries etc. Many companies are working on the phytoremediation process which can remodify the land and provides revitalized land for the urban programs. The remedied lands are used for recreation purposes like for public parks, gardens, stadiums like hockey, cricket and football (Pandey et al., 2019). Phytoremediation is mainly based on the plant which helps in the sequestration of carbon. The remediation strategy mainly depends upon the soil, plant and their physical, chemical and biological properties. This technique increases the fertility of the soil by the detoxification process and the quality of soil is increased by the sequestration of carbon (Gatliff et al., 2016). Various other benefits are there like phytoremediation enhances the aesthetic value and also provides fresh air on the polluted land. Many ornamental crops like flower crops are used for aesthetic as well as aromatic crops, pulp crops, energy crops, paper crops, grasses, trees and for timber purposes etc. and have much economic value. Some plants are also grown for seed purposes, plant species like Pongamia, Jatropha and Ricinus etc. are used for the extraction of seed and plant species like Switchgrass, Populus spp., Miscanthus spp., Canary grass and Salix sp. are used as biomass energy crops (Pandey et al., 2019). These crops also provide aroma in the environment. Many agronomic practices, soil contaminants etc. are used to enhance the quality of soil, plants and environment etc. Agronomic practices like irrigation, crop rotation, pest control, tillage, harvesting, planting etc. affects the pollution level by decreasing it and increasing the quality and quantity of soil and plant yield (Schnoor et al., 1998). Along with this the conditions of flora and fauna also increases. The invertebrates like spiders, bees, beetles, centipedes, scorpions, mites, cricket, ants etc. help in the dispersal of seeds of hyperaccumulator plants, pollination and the formation of humus layer etc. The presence of hyperaccumulator plants like many herbs, shrubs and trees etc. also maintains the phyto diversity. The biomass produced in the process of phytoremediation helps in the generation of income. The hyperaccumulator plants like *Tectona grandis, Salix spp., Eucalyptus spp.,* are used for the preparation of furniture. Like these many other applications of phytoremediation are there which makes this process more valuable, efficient and better for the remediation of heavy metals (Wa et al., 2019).

Limitations and Disadvantages of Phytoremediation

Like advantages, there are very limitations and disadvantages also of phytoremediation. In this process of phytoremediation, the stress conditions occur mostly in this process which creates a reduction in the growth of a plant and proper nutrients are also unavailable to the plants. Proper farming techniques are not applied in phytoremediation (Erdei, 2005). Plants are not able to use the food and for the fodder, because most of the hyperaccumulator plants are used for the fodder purpose and these plants are cultivated in high salt conditions due to which many toxic substances accumulate in the plants which are dangerous for the health of the plant and are not edible and not good for fodder purpose. In this process some plant species like Brassica spp. which is used for oil and food purposes for humans also are not edible due to the excess amount of toxicants in it. In this technique, plant growth is a very slow process. Leaching of toxicants is also the main problem which makes this technique less efficient. The fertility of soil also reduces due to the high amount of salt and the presence of heavy metals in the soil (Wiche et al., 2016). Various mitigation techniques are used to reduce the limitation occurs due to phytoremediation. So, for the phytoremediation process right plant should be selected and correct treatments should be given. If high salt levels are there then the use of halophytes should be used which maintains the levels of salts. Pre-treatment system should be there and used so the level of remediation is maintained (Tripathi et al., 2020).

Conclusion

The pollution caused by heavy metals affects the environment, soil and plants as well. The toxicity caused by heavy metals hinders the growth of the plants. Phytoremediation technique helps for the removal of toxicants as well as the degradation of the harmful toxic compounds into non-toxic compounds. Along with this funguses and mushrooms are also used for the degradation of toxic compounds and the leftover material from mycoremediation process is used as compost which increases the soil structure, soil fertility and is very useful for plant growth. In this review, we concluded that with the help of various improved techniques like the use of genetic engineering, transgenic crops in phytoremediation the harm and effect of pollution can be reduced. Phytoremediation technique is very important on the point of view of agriculture because due to this the farmers get a quality product and good yield also. Phytomining is also with this technique due to which the economy of crop increases and is beneficial for commercial purposes. In this technique, various heavy metals like Cadmium, Mercury, Lead, Arsenic, Chromium, Nickel, Uranium, Platinum and Gold are extracted very easily. This process does not cause any harm to the environment, soil and water and is easy, cheap than other methods. Phytoremediation method is very useful for the environment as well as in agriculture for enhancing productivity. The combined use of all the techniques like mycoremediation, phytomining, transgenic crops, genetic engineering and use of chelates are very effective and appreciatable approach.

References

Aken, B. V., Correa, P. A., & Schnoor, J. L. (2010). Phytoremediation of polychlorinated biphenyls: new trends and promises. Environmental science & technology, 44(8), 2767-2776.

Bani, A., Echevarria, G., Sulçe, S., & Morel, J. L. (2015). Improving the agronomy of Alyssum murale for extensive phytomining: a five-year field study. International Journal of Phytoremediation, 17(2), 117-127.

Chaney, R. L., Angle, J. S., Broadhurst, C. L., Peters, C. A., Tappero, R. V., & Sparks, D. L. (2007). Improved understanding of hyperaccumulation yields commercial phytoextraction and phytomining technologies. Journal of Environmental Quality, 36(5), 1429-1443.

DalCorso, G., Fasani, E., Manara, A., Visioli, G., & Furini, A. (2019). Heavy metal pollutions: State of the art and innovation in phytoremediation. International journal of molecular sciences, 20(14), 3412.

Das, N., & Chandran, P. (2011). Microbial degradation of petroleum hydrocarbon contaminants: an overview. Biotechnology research international, 2011.

Erakhrumen, A. A., & Agbontalor, A. (2007). Phytoremediation: an environmentally sound technology for pollution prevention, control and remediation in developing countries. Educational Research and Review, 2(7), 151-156.

Erdei, L. (2005). Phytoremediation as a program for decontamination of heavy-metal polluted environment. Acta Biologica Szegediensis, 49(1-2), 75-76.

Favas, P. J., Pratas, J., Varun, M., D'Souza, R., & Paul, M. S. (2014). Phytoremediation of soils contaminated with metals and metalloids at mining areas: potential of native flora. Environmental risk assessment of soil contamination, 3, 485-516.

Gatliff, E., Linton, P. J., Riddle, D. J., & Thomas, P. R. (2016). Phytoremediation of soil and groundwater: economic benefits over traditional methodologies. In Bioremediation and Bioeconomy (pp. 589-608). Elsevier.

Gil-Loaiza, J., Field, J. P., White, S. A., Csavina, J., Felix, O., Betterton, E. A., ... & Maier, R. M. (2018). Phytoremediation reduces dust emissions from metal (loid)-contaminated mine tailings. Environmental science & technology, 52(10), 5851-5858.

Ha, N. T. H., Sakakibara, M., & Sano, S. (2011). Accumulation of Indium and other heavy metals by Eleocharis acicularis: an option for phytoremediation and phytomining. Bioresource technology, 102(3), 2228-2234.

Harrier, L. A., & Watson, C. A. (2004). The potential role of arbuscular mycorrhizal (AM) fungi in the bioprotection of plants against soil-borne pathogens in organic and/or other sustainable farming systems. Pest Management Science: formerly Pesticide Science, 60(2), 149-157.

Imadi, S. R., Shah, S. W., Kazi, A. G., Azooz, M. M., & Ahmad, P. (2016). Phytoremediation of saline soils for sustainable agricultural productivity. In Plant Metal Interaction (pp. 455-468). Elsevier.

Jan, S., Rashid, B., Azooz, M. M., Hossain, M. A., & Ahmad, P. (2016). Genetic Strategies for Advancing Phytoremediation Potential in Plants: A Recent Update. In Plant Metal Interaction (pp. 431-454). Elsevier.

Keeling, S. M., Stewart, R. B., Anderson, C. W. N., & Robinson, B. H. (2003). Nickel and cobalt phytoextraction by the hyperaccumulator Berkheya coddii: implications for polymetallic phytomining and phytoremediation. International Journal of Phytoremediation, 5(3), 235-244.

Kumar, P, Pathak, S, Kumar, M and Dwivedi, P. 2018cd. Role of secondary metabolites for the mitigation of cadmium toxicity in sorghum grown under mycorrhizal inoculated hazardous waste site. In: Biotechnological Approaches for Medicinal and Aromatic Plants. Springer, Singapore, pp. 199-212.

Kumar, R., & Kaur, A. (2018). Oil Spill Removal by Mycoremediation. In Microbial Action on Hydrocarbons (pp. 505-526). Springer, Singapore.

Kumari, P., & Kumar, P. (2020). Trichoderma fungus in mitigation of rhizosphere arsenic: with special reference to biochemical changes. *Plant Archives, 20*(2), 3512-3517.

Li, C., Ji, X., & Luo, X. (2019). Phytoremediation of heavy metal pollution: a bibliometric and scientometric analysis from 1989 to 2018. International Journal of Environmental Research and Public Health, 16(23), 4755.

Naher, L., Yusuf, U. K., Ismail, A., & Hossain, K. (2014). Trichoderma spp.: a biocontrol agent for sustainable management of plant diseases. Pak. J. Bot, 46(4), 1489-1493.

Nwauche, K. T., Anacletus, F. C., & Berezi, E. P. (2018). Influence of Mycoremediation and Selected Surfactants on Growth Performance of Fluted Pumpkin In Crude Oil Impacted Soil and the Nephrotoxic Effects of Leaves Aqueous Extract on Wistar Rats. International Journal of Biochemistry Research and Review, 23(2), 1-12.

Okparanma, R. N., Ayotamuno, J. M., Davis, D. D., & Allagoa, M. (2011). Mycoremediation of polycyclic aromatic hydrocarbons (PAH)-contaminated oil-based drill-cuttings. African Journal of Biotechnology, 10(26), 5149-5156.

Pandey, V. C., & Bajpai, O. (2019). Phytoremediation: from theory toward practice. In Phytomanagement of Polluted Sites (pp. 1-49). Elsevier.

Peuke, A. D., & Rennenberg, H. (2005). Phytoremediation: molecular biology, requirements for application, environmental protection, public attention and feasibility. EMBO reports, 6(6), 497-501.

Redfern, L. K., & Gunsch, C. K. (2016). Endophytic phytoaugmentation: treating wastewater and runoff through augmented phytoremediation. Industrial biotechnology, 12(2), 83-90.

Rosenkranz, T., Hipfinger, C., Ridard, C., & Puschenreiter, M. (2019). A nickel phytomining field trial using Odontarrhena chalcidica and Noccaea goesingensis on an Austrian serpentine soil. Journal of environmental management, 242, 522-528.

S.C. Mc Cutcheon., S.E. Jorgenson. (2008). Phytoremediation. Ecological Engineering, 2751-2766.

Saleem, M. H., Ali, S., Hussain, S., Kamran, M., Chattha, M. S., Ahmad, S., ... & Abdel-Daim, M. M. (2020). Flax (Linum usitatissimum L.): A Potential Candidate for Phytoremediation? Biological and Economical Points of View. Plants, 9(4), 496.

Schnoor, J. L., Light, L. A., McCutcheon, S. C., Wolfe, N. L., & Carreia, L. H. (1995). Phytoremediation of organic and nutrient contaminants. Environmental science & technology, 29(7), 318A-323A.

Sheoran, V., Sheoran, A. S., & Poonia, P. (2009). Phytomining: a review. Minerals Engineering, 22(12), 1007-1019.

Silambarasan, S., & Abraham, J. (2013). Mycoremediation of endosulfan and its metabolites in aqueous medium and soil by Botryosphaeria laricina JAS6 and Aspergillus tamarii JAS9. PloS one, 8(10).

Singh, M., Srivastava, P. K., Verma, P. C., Kharwar, R. N., Singh, N., & Tripathi, R. D. (2015). Soil fungi for mycoremediation of arsenic pollution in agriculture soils. Journal of applied microbiology, 119(5), 1278-1290.

Spina, F., Cecchi, G., Landinez-Torres, A., Pecoraro, L., Russo, F., Wu, B., ... & Zotti, M. (2018). Fungi as a toolbox for sustainable bioremediation of pesticides in soil and water. Plant Biosystems-An International Journal Dealing with all Aspects of Plant Biology, 152(3), 474-488.

Thijs, S., Sillen, W., Rineau, F., Weyens, N., & Vangronsveld, J. (2016). Towards an enhanced understanding of plant–microbiome interactions to improve phytoremediation: engineering the metaorganism. Frontiers in microbiology, 7, 341.

Tognacchini, A., Rosenkranz, T., van der Ent, A., Machinet, G. E., Echevarria, G., & Puschenreiter, M. (2020). Nickel phytomining from industrial wastes: Growing nickel hyperaccumulator plants on galvanic sludges. Journal of environmental management, 254, 109798.

Tripathi, S., Singh, V. K., Srivastava, P., Singh, R., Devi, R. S., Kumar, A., & Bhadouria, R. (2020). Phytoremediation of organic pollutants: current status and future directions. In Abatement of Environmental Pollutants (pp. 81-105). Elsevier.

Weyens, N., Thijs, S., Popek, R., Witters, N., Przybysz, A., Espenshade, J., ... & Gawronski, S. W. (2015). The role of plant–microbe interactions and their exploitation for phytoremediation of air pollutants. International journal of molecular sciences, 16(10), 25576-25604.

Wiche, O., Székely, B., Kummer, N. A., Moschner, C., & Heilmeier, H. (2016). Effects of intercropping of oat (Avena sativa L.) with white lupin (Lupinus albus L.) on the mobility of target elements for phytoremediation and phytomining in soil solution. International journal of phytoremediation, 18(9), 900-907.

Wilson-Corral, V., Anderson, C. W., & Rodriguez-Lopez, M. (2012). Gold phytomining. A review of the relevance of this technology to mineral extraction in the 21st century. Journal of Environmental Management, 111, 249-257.

Wu, T., Xu, J., Liu, J., Guo, W. H., Li, X. B., Xia, J. B., ... & Wang, R. Q. (2019). Characterization and initial application of endophytic Bacillus safensis strain ZY16 for improving phytoremediation of oil-contaminated saline soils. Frontiers in microbiology, 10, 991.

Yan, A., Wang, Y., Tan, S. N., Yusof, M. L. M., Ghosh, S., & Chen, Z. (2020). Phytoremediation: a promising approach for revegetation of heavy metal-polluted land. Frontiers in Plant Science, 11.

Zhang, X., Houzelot, V., Bani, A., Morel, J. L., Echevarria, G., & Simonnot, M. O. (2014). Selection and combustion of Ni-hyperaccumulators for the phytomining process. International Journal of Phytoremediation, 16(10), 1058-1072.

Plant Metabolites under Heavy Metal Research

Monika Sharma

Department of Agronomy, School of Agriculture

Lovely Professional University, Jalandhar, Punjab, India, 144411

Email: ms9375033@gmail.com

Abstract

Heavy metals are naturally occurring compounds having 5 times greater density than that of water. Heavy metals include metalloids, Arsenic (As), Cadmium (Cd) etc. which causes toxicity at a lower amount of exposure. With the increasing activities of humans, there is an increase in metal exposure in the environment. Various agricultural, industrial and technological activities lead to the exposure of heavy metals in the atmosphere. Accumulation of plant metabolites in plant system is important to boost the functioning of the plant life processes. Plant primary and secondary metabolites play a major role in the growth and functioning of the plant and help in protecting the plant from a different type of biotic and abiotic stresses. Plant secondary metabolites alkaloids, terpenes, phenols are not synthesized in the plant system, but they play a major role in the functioning of the plant. Various environmental factors like temperature, humidity, wind velocity, and light influence the synthesis of primary and secondary metabolites. Heavy metal toxicity in plants is a major concern nowadays. Metal toxicity contaminates the plant and environment and thus causes many health hazards to men and animals. Heavy metals heavily pollute the soil, food and water and decrease the quality of the product which ultimately causes dangerous diseases to the population. In heavy metal research, these are found very toxic to the plants which affect the growth and functioning of the plant and decrease its quality. Plant metabolites play a crucial role in protecting the plants against heavy metal stress and increase the productivity of the produce.

Introduction

Heavy metals are the natural elements which are present in the earth crust and expose to the environment by the various anthropogenic and industrial activities. These are the metals having density 5 times greater than that of water. The common features of these heavy metals about biological life are that they are toxic in excess amount and cause the death of the living organisms. They are poisonous to the plant and animals and cause the pollution of soil, water and air when exposed in the environment in a large amount. Some heavy metals are toxic even at low amounts and pollute the quality of air and water. The heavy metal toxicity contaminates all the natural resources and creates an issue for fresh air and water required by the population. Due to anthropogenic activities, there is continuous stress on the biotic and abiotic factors which adversely affect the land and water resources. Soil is mostly affected by the heavy metal toxicity as it reacts with the soil particles and decreases the fertility of the soil by decreasing the physical and chemical properties of the soil. Soil pollution can be defined as the addition of any toxic compound to the soil which affects its quality and functioning which is responsible for the poor growth and less productivity of the plants on its surface. It is estimated that about 70% of the heavy metal ions are introduced in the human body due to means of food (Jaishankar et al., 2014). The concentration of heavy metals affects all the processes of plant life and decreases its quality which leads to the various health hazards to the animals and human population. Pollution of soil due to heavy metal toxicity is of great concern to the scientists of different countries of the world. It is a serious problem which leads to less production and various health hazards to the population. Plants grown on heavy metal toxic land are unfit for human consumption and can lead to the poor condition of the population. Continuous consumption of toxic products regularly can ultimately cause the death of humans and animals. Various diseases are found in human and animals due to the regular consumption of heavy metal toxic components. The toxicity of heavy metals, Arsenic, Nickel, Chromium, Cadmium, Lead etc. are of great concern. These all are toxic to the plant and animals. Various biotic and abiotic stress affect the plant growth and development, but heavy metal stress is one of the most important abiotic stresses which is of major concern in terms of its toxicity and production of quality products of the plants. Drought, salinity, freezing is the stresses which affect the growth of the plant and productivity of the produce, but heavy metal stress affects the quality of the product and causes a great threat to the population. The release of various components and chemicals in the soil can affect the soil which is the mother of infinite life. Various studies of heavy metal research show its adverse effects on the land, water and environment. The contaminated land and atmosphere reduce the capacity of the plant and animals to fight against stress conditions. Heavy metal toxicity in metal research and advances shows its adverse effect on human and other animals. Scientists across the world conduct various studies on heavy metal research to know about its toxicity for the plants at various levels. The heavy metals toxicity causes a reduction in the productivity and quality of the produce. These are toxic at the lower and higher level and concentration at different stages of plant growth and development. Heavy metals toxicity is one of the most challenging concerns to our ecosystem. Heavy metals are non-biodegradable and remain at its place for a long time to infect it. Bioaccumulation of heavy metals in the environment is of great concern (Gautam, 2015). Different microbes are used for Bioremediation of heavy metal toxicity and reduce the effect of heavy metal toxicity in the environment. Bioremediation is employed to transform heavy metals toxicity into a less harmful state using microbes (Ndeddy Aka & Babalola, 2016). It is also employed to facilitate enzymes to clean-up polluted environment. Bioremediation is a cost-effective technique to reduce the harmful effect of toxic heavy metals (Hrynkiewicz & Baum, 2014). The technique has various advantages and disadvantages also but the results show a positive interest in decreasing the effect of heavy metal toxicity.

The study of plant metabolites helps mitigate the toxicity of heavy metal and can decrease the stress caused by heavy metal stress. Plant primary and secondary metabolites play a major role in tolerating the various biotic and abiotic stresses. Primary metabolites are mainly synthesized in the plant system. The primary metabolites include carbohydrate, proteins, fats and hormones which are important for the growth and development of the plants. Besides primary metabolites, various organic compounds are required for the plant development and protection which are not synthesized in the plant system. They are required to apply externally on the plant surface to facilitate the functioning of plant processes. Heavy metal toxicity is of great concern nowadays worldwide, therefore, the role of plant primary and secondary metabolites is helpful to mitigate the stress or toxicity caused by heavy metals in the plants and to reduce the effect of heavy metal ions inside the plant system to formulate all the activities and functioning of the plant system. Plant metabolites play a major role in transforming the adverse effect of toxic heavy metals to influence the productivity of the plants. These organic compounds function to enhance the yield and production of the plant and are required in the plant system to carry out different functions of regulating heavy metal toxicity.

Plant metabolites are crucial for plant life to grow better and to produce good production. Secondary metabolites do not possess the main function of plant life processes but are important for the plant to deal with the environment or other factors especially in case of any type of stresses. Plant secondary metabolites are generally not synthesized in the plant system and also are of great importance in protecting the plants from various kinds of biotic and abiotic stresses about the environment. Secondary metabolites regulate the defence system of the plant and enhance the adaptation in all type of environment. Plant primary and secondary metabolites are mainly important for facilitating the growth and quality yield of the plant. In some plants, primary metabolites may lead to the production of secondary metabolites and other organic compounds which are important for the plant development. The secondary metabolites synthesized from the primary metabolites include carbohydrates, lipids and amino acids which are needed for special functions performed in the plant system to regulate its growth and development. Secondary metabolites are required to influence the defence system of the plant and provide protection against various microbes and pathogens (Seigler, 1998). Plant secondary metabolites also impart colour and odour and taste in the plants which are an important component of the plant life processes (Bennett, 1994). These are the important constituent of pharmaceuticals and industrial products. Primary and Secondary metabolites in the plant system regulate the synthesis of plant growth hormones like Abscisic acid (ABA, Salicylic acid (SA) and Jasmonic acid (JA) etc. Plant metabolite accumulation in the plants influences the signal molecules and protects the stressed environment. Plant secondary metabolites under heavy metal stress influences signal molecules to promote the plant growth under stressed conditions. Plant metabolites facilitate the physiological and biochemical processes of the plant and increase the phenomenon of yield and productivity. The special increase in the photosynthesis rate, transpiration rate and other plant life processes can further increase both quality and quantity of the product which is of great importance to the economy of a nation.

Toxicity of Metals, ROS and Plant Metabolite Production

Heavy metals are taken up by the plant roots and are transported from one place to another within the plant system by the means of cells or tissues. Heavy metal toxicity interacts with the physiological aspects of the plant growth and development which ultimately inhibits the germination and photosynthesis (Latef, 2018). The metal toxicity decreases the proteins and carbohydrate content in the plants and reduces nutrient uptake or absorption in the plant system. The metals like Copper (Cu), Zinc (Zn), Manganese (Mn), Nickel (Ni) and Iron (Fe) are the main components which are required by the plant to carry out various functions of plant life. They play a significant role in the plant life in the development of various plant cells and synthesis of different enzymes which are very useful the plant growth and development. The excess concentration of heavy metals proves to cause great damage to the plant and its processes. The accumulation of toxicity of heavy metals in the plants can be mitigated by the action of plant cells and concerning plant primary and secondary metabolites. Due to this, it has become productive to mitigate or reduce the heavy metal stress on plants up to some extent. Plant secondary metabolites are organic compounds well known for their ability of chelation, which is necessary to remove the heavy metal stress from the plant species. Many scientists across the world had been studied the function of plant molecules, vitamins, polyphenols and metabolites on the plants to remove the stress caused by heavy metals (Asgari Lajayer et al., 2017; Kısa et al., 2016). Others show the physiological and biochemical response of plant species under heavy metal stress. With the increased concentration of heavy metals in the plant there is a malfunctioning of various processes and decrease in the level of flavonoids in roots and shoot tissues. Direct interaction between the accumulation of heavy metal and genetic expression of plant shows the relation of plant mechanism concerning its life processes. The heavy metal research also shows the results of toxicity to the agricultural and environmental components. The interaction of heavy metal and gene expression leads to the production of phenylpropanoids provide information about the production of phenolics acids in response to metal stress. Different impact of heavy metals on secondary metabolite production can serve as a key factor in mitigation of heavy metal stress. Plants produce phenolics acids and other metabolites to relieve the stressed condition and to save its energy it does not react with expressions of genes. Phenols are synthesized to protect against heavy metal stress to the plants. These are the compounds which are main chelators of metal ions (Cd) and its production in plant roots (Kovácik and Klejdus, 2008). Some studies show the activation of genes involved in plant metabolite pathway to synthesize the production of lignin. In heavy metal stress, there is a decrease in flavonoids content and defence mechanism influence the release of some enzymes and plant metabolites to protect the plant under heavy metal stress. The gene expression of some genes involved in the biosynthesis of flavonoids and to regulate the plant metabolism under metal stress. Heavy metal stress is a major threat affecting plants and secondary metabolite production to access malfunctioning of all plant processes. It has been shown that heavy metals cadmium and copper at different concentration enhance the production of secondary metabolites in the plants. In some plants, it also leads to a decrease in the synthesis of plant metabolites (Ibrahim et al., 2017).

Fig. 1. Heavy metal toxicity in plants

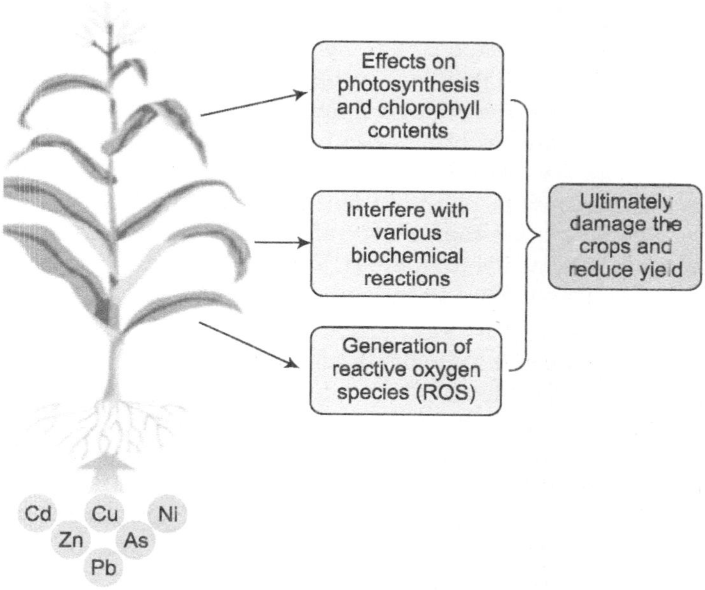

Effects on photosynthesis and chlorophyll contents

Interfere with various biochemical reactions

Generation of reactive oxygen species (ROS)

Ultimately damage the crops and reduce yield

Cd Cu Ni Zn As Pb

Source:https://www.google.com/search?q

Plant Defense Mechanisms under Heavy Metal Toxicity

To tolerate the toxicity of heavy metal stress, plant influences the network of defence mechanism or strategies to overcome the stress. A plant defence mechanism is the main component of the plant system which is very essential to protect the plant from any external damage or hard conditions. Under heavy metal stress, at very first plant physical components come in contact with it which primarily act as a barrier to the heavy metal stress. These physical or morphological characters include cuticle layer, cell wall, tissues, trichomes and hydropotes which act as a very first barrier to the heavy metal stress (Hall, 2002 and Harada, 2010). Vacuole, Trichome and hydropotes serve as an accumulator of heavy metals in one place either in vacuoles or in shoot part of the plant. This store the heavy metal ions at one place to detoxify the heavy metal stress and to synthesize the production of secondary metabolites within the plant itself (Hauser, 2014). After overcoming the first barrier the heavy metal ions interact with the cellular defence mechanism to reduce the effect of heavy metal stress. To tolerate heavy metal toxicity, the biosynthesis of cellular molecules takes place which is a way to protect the plants from metal ion toxicity. Biosynthesis of the cellular mechanism includes the production of low molecular weight proteins, organic acids, flavonoids, plant chelators, amino acids, plant growth regulators etc. which plays a significant role in protecting the toxicity under heavy metal stress (Viehweger, 2014). The increased production of Reactive Oxygen Species (ROS) is possible after synthesis of above low molecular weight compounds (Mourato et a., 2012). The plant has developed an antioxidant defence mechanism to decrease the harmful effect of heavy metal toxicity. The antioxidant defence mechanism of plants includes several compounds namely: Superoxide dismutase (SOD), Catalase (CAT), Glutathione reductase (GR), Ascorbate (AsA), phenols and lignin mitigate the heavy metal stress in plants. These act as the scavengers of free radicals in the plant system to decreases the heavy metal stress. Many studies indicate that the biological molecules are involved in detoxification of heavy metal stress and perform various functions in regulating the functions of the plant system. The mitigation of heavy metal stress and regulation of plant mechanism varies with plant to plant and species to species. Their level of tolerance also depends upon the concentration of metal ions and the capacity of the plant species to tolerate the toxic effect. Various compounds include in defence mechanism against heavy metal stress are:

1. **Phytochelatins (PCs):** Phytochelatin is found in microorganisms yeast, bacteria, fungi to plants and animals. It acts as a source of mitigating the metal ion toxicity and plays a role in plant defence mechanism. It is responsible to protect the plants from various biotic and abiotic stresses. Abiotic stresses like drought, salinity, heavy metal stress, the high and low temperature can cause serious damage to the plant root and shoot part and the further plant may die due to malfunctioning of all the processes of plant life. Plant chelators can be used further to know about the heavy metal stress in the plant system. Plant chelators are produced in cytosol and are of low molecular weight compound but after interaction with heavy metals, they become of high molecular weight compounds which accumulate metal ions and transport it to the trichomes and vacuoles where it detoxifies into another compound (SONG et al., 2014). Forms of ATP are used as a binding agent or transporters of toxic metal ions from plant to vacuoles (Sytar, 2013).

2. **Metallothionein (MTs):** These have different characters and functions in response to metal stress. The structure and functioning of metallothionein (MT) vary with plant species to species. The properties of metallothionein alter within the plant system in response to heavy metal stress. The role of metallothionein increases as the concentration of heavy metals in the plant increases. They have more wide functions in the plant as compare to other living organisms and protect the plant in metal stressed conditions. They play a major role in sequestration of heavy metals in the plants and accumulate the metal ions in vacuoles and trichomes which is a site for detoxification of heavy metals. Metallothionein is a key component to mitigate the heavy metal stress by the means of metal transport in the cellular level, accumulation of metal ions in the plant tissue and synthesis of various enzymes in response to heavy metal stress. Rather their role in detoxification of heavy metals, these are also important for ROS scavenging, maintaining the cellular mechanism, enzymatic activities, maintaining plasma

membrane, growth and repair of cells and gene expressions (Wong et al. 2004). There is a need to study more in this field to properly know the structure and functioning of the metallothionein in protecting the plant against heavy metal stress. The behaviour of plant under heavy metal stress will show the need for these components in response to mitigate or reduce the toxicity of heavy metals which is a function of plant metabolite. Various experiments were done with plant metallothionein in expressing gene expressions and tolerating heavy metal toxicity so that a combination of components will be formed to overcome the stressful conditions. The study of metallothionein also provides an opportunity to know about phytoremediation and other strategies to control or tolerate heavy metal stress in the plants.

3. Proline(Pro): Heavy metals cause pollution of the soil, water and environment which are the source of life for all living organisms. Weathering of rocks and minerals, smelting, industrialization and other anthropogenic activities are the main source of releasing heavy metal ions in the atmosphere to cause its pollution. The heavy metals like Cu, Cd, As Zn enters in the soil and atmosphere by the various man-made and industrial activities and cause toxicity to pollute our environment. The heavy metal toxicity causes serious damage to the plant and animals when applied in excess amount. A lower dose of metal ions are required by the plant to carry on its growth processes and better production but its excess concentration proves to be harmful to both plants as well as animals (Hall, 1981). Different techniques are used to tolerate the toxicity of heavy metals in the plants and to detoxify the toxicity of heavy metal ions. Proline synthesis takes place in case of any injury occurs to the plant and is believed by many scientists that the injury to the plant leads to the synthesis of proline and other compounds in plant system to protect them from heavy metal toxicity (Anaytullah et al., 2012). Under heavy metal stress, the injury to the plasma membrane increased and there is a level up in SOD content in the plants which further leads to the production of proline in the plants to overcome this heavy metal stress (Siddique and Dubey, 2017). The higher amount of metal ions cause serious damage to the plant but biosynthesis of proline helps in protecting the plant against heavy metal stress. Heavy metal stress damages the different processes of plant life and causes a reduction in the quality and yield potential of the plant. The production of proline and other plant metabolites help in achieving the better quality and production of the plant yield by mitigating heavy metal ions from the plants. Proline production interacts with heavy metal ions at the cellular level and transports the metal ions to the vacuoles and trichomes where it detoxifies into valuable compounds and helps in getting good productivity and quality of the produce. Proline plays a significant role in achieving a better quality of the produce by influencing various physiological processes of the plant. Thus it is a factor that enhances the yield under heavy metal stress conditions which is a need to feed the population of human and animals. Proline is a proteinous metabolite which helps in the synthesis of other enzymes also in protecting the plants from heavy metal toxicity and to produce a better yield and quality of the produce. Proline is a constituent of the cell wall, antioxidant and ROS scavengers which stabilizes the micro and macromolecules in the plant. It helps in the growth and development of seed, its germination and other physical characters of the plant from the vegetative stage to the reproductive stage. Under abiotic stresses like drought, salinity, heavy metal stress, the proline production takes place to control the activities of the plant and enhances the functioning of physiological functions to facilitate a better production. It shows a typical response in response to heavy metal ions which serve as a role to conduct the plant physiological and biochemical activities. Under heavy metal stress in the plants, proline performs multi-functions to facilitate the plant growth and development characteristics. Various studies have been proposed that increased level of proline in the plant help them to tolerate heavy metal stress and influence many enzymes to serve as a resistant mechanism against heavy metal stress. Nowadays spraying of proline is also done on the surface of the plant which is effective against toxicity of heavy metal ions and detoxifies the concentration of heavy metal ions under less time. Many poisonous metals can be detoxified with the use of proline in the plants and activate the functioning of molecular and chemical compounds. In chickpea, there is an increase in antioxidant enzymes and photosynthetic parameters due to the increase in the proline content in the plants which further increases the tolerance capacity of the plants against heavy metal (Cadmium) stress (Hayat et al.2015).

4. Arbuscular Mycorrhizal (AM): These are symbiotic fungi that help in protecting plants from abiotic stresses, especially heavy metal stress. They influence photosynthetic rate and transpiration rate which helps in the growth and development of the plant. It accumulates the heavy metal ions in the plant roots and facilitates its transport from root to shoot within the cells and tissues. It improves the absorption of micro and macronutrients from the soil and plays a significant role in smoothly expressing the function of photosynthesis, transpiration and evapotranspiration which influence the yield and productivity of the plant. Arbuscular Mycorrhizal fungi form a symbiotic bond between plant roots and vascular tissues that help in the transport of heavy metal ions to the shoot tip to detoxify its toxic effect. It enhances the tolerance of plants towards heavy metal stress and another type of stresses. These fungi also improve the uptake of mineral nutrition in the plants to enhance its productivity (Johansson, J. F., Paul, L. R., & Finlay, R. D. (2004). Plant fungi use necessary measures to protect the plants from heavy metal stress. The symbiotic mechanism helps plant to synthesize various compounds in the plant system which helps to reduce the toxicity of heavy metal ions in the plant. It increases the absorption capacity of the plant from the plant roots and makes a path to heavy metal stress for its transportation to the shoot tissues. The mycorrhizal symbiosis between plant roots and heavy metal ions leads to bioremediation of the plant from heavy metal stress. Under heavy metal stress, plant decreases its capacity to hold nutrients from the soil, Arbuscular Mycorrhizal fungi takes part in symbiotic fixation of plant nutrients from the soil to the root zone of the plant and further its movement from one cell to another. It plays a major part in a defence mechanism to protect the plant from heavy metal ions for its quality yield and better production. The main function of Mycorrhizal fungi is to form a barrier to the deposition of the heavy metals in the plants and facilitates its transport from root to shoot part for detoxification. It helps in sequestration of heavy metal ions to increases the production. It also leads to the synthesis of various proteins and compounds which reduce the strength of heavy metal ions and cause less damage to the plant. these fungi enhance the root and shoot growth of the plant system by providing mineral nutrition to the plants. In the rhizosphere, the Mycorrhizal fungi adopt various remedial measures to protect the plant from stresses condition.

Fig. 2. Plant defence mechanism under heavy metal stress

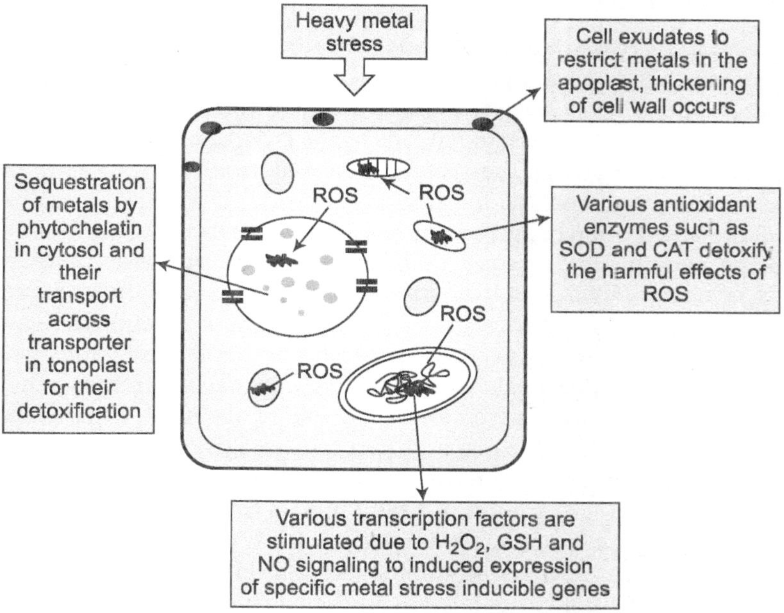

Source:https://www.google.com/search?

The entry of heavy metal ions in the atmosphere contaminate the soil, water and other natural resources which cause damage to the plant as well as the population of animals and humans. It decreases the quality of food grains and leads to various health hazard which can cause the death of the various living organisms. Infection to plants can cause serious damage to the population because every living being depends upon food and water to live. The human population is responsible for this condition of plant and food resources. Various anthropogenic factors involved in the release of heavy metal ions in the environment which affect the quality of the product and decrease its production and yield which is a great issue of concern. Man inductive activities disturb the biological status and interaction between plant animal and environment. Nowadays plants are exposed to heavy metal toxicity which is not noticed as much in our history. The release of heavy metals I the soil and water decreases the fertility of the soil and exert its toxicity in the growing plants which are a major issue of food safety. The excess use of industrialization and chemicals in agriculture causes a release of heavy metal ions in the place where it is placed. They are non-biodegradable and remain at their place for a long time to cause its toxicity. The necessities of the population create a major issue to expose the heavy metal ions in the plants and decrease its functioning and development.

Various mechanisms of the plant help to decrease the metal toxicity and increase the plant yield. Plant primary and secondary metabolites function in the process of detoxification of heavy metal ions within the plant system. It influences the defence mechanism of the plant to facilitate physiological activities and to reduce heavy metal toxicity.

Remediation of Soils by Hyper Accumulators

With day by day increase in the exposure of heavy metals in the environment, the pollution of soil, water and atmosphere also increases may affect the quality of food products and is a challenge to food security. Various anthropogenic activities involved in the release of heavy metal ions to the environment are the main cause of decreasing productivity of the crop and creates problem to the population. It affects the environment worldwide and leads to various health hazards (Zheljazkov et al., 2008a). Remediation of soil containing heavy metals using various strategies facilitates the growth and development of the plant. The processes involved in the separation of heavy metal ions from metal contaminated soil include landfilling, chemical leaching, electrokinetics etc. (Lasat, 2000). The separation of heavy metal ions from the soil is quite an expensive method and is labour intensive also, so phytoextraction of heavy metals is an effective approach to remove the ions of heavy metal from the plant as well as soil (Gupta et al., 2013). Phytoremediation is a technique that involves the use of different plants to remove the heavy metal ions from the soil. This process interacts with various microorganisms to remove or degrade the toxic concentration of heavy metals from the soil (Ouyang, 2002). Lower concentration does not affect as much as it's higher concentration damage the plant development and contaminates the soil. Hyper accumulators are the plants accumulate the heavy metal ions in their leaves, shoot portion roots etc. and remove the concentration of metal ions from the soil. Hyper accumulator plants can accumulate a large concentration of heavy metals and detoxification (Chaiyarat et al., 2011). These transfer the heavy metal ions from root to shoot portion to detoxify the metal stress and regulation of other processes of plant life. Hyperaccumulators are the plants accumulate the heavy metal ions in their roots and transfer to the aerial parts for further detoxification (Baker et al., 2000; Lasat, 2002). Heavy metal toxicity is of great concern nowadays affected the growth and quality of the plants and their productivity. Different medicinal, aromatic and other woody species are used for phytoremediation of the plant and soil. These plants function as an accumulator of heavy metal ions and enhance various physiological and biochemical functions of the plant to facilitate its growth and production. Generally aromatic or other plants are used to accumulate heavy metal ions rather than edible plants. Edible plants are not used for phytoremediation of heavy metal ions because heavy metal ions can enter in the food chain and may affect human and animal health which has become a great issue (Gupta et al., 2013). Some aromatic plants do not accumulate heavy metal ions because they cannot accumulate a higher concentration of heavy metals. Various other hyperaccumulator plants are used to accumulate heavy metal ion in the aerial parts of the plant. Non-edible plants used for phytoextraction of heavy metal ions is advantageous because there will be no harm to the food safety and population. Aromatic plants produce essential oils that are not affected by the accumulation of heavy metal ions and can successfully remove the metal toxicity fro plants and soil. Heavy metals remain attached in the tissues of the plant during the process of oil extraction and detect the adequate quantity of heavy metals in oil production (Zheljazkov et al., 2008b; Khajanchi et al., 2013). The proper disposal of these hyperaccumulator plants can lead to the removal of heavy metal ion from the soil. In this way, a proper amount of heavy metal ions can be removed from the soil and it is safe for the cultivation of other food crops. The disposal of plant residue became the land metal-free and safe production and extraction of oils and other products. Like food crops, aromatic and medicinal plants are also very important to develop our economy so phytoextraction of heavy metal ions is of the great need for future development. Phytoremediation is a very effective technique that we can conduct to disinfect the land and water resources from the toxicity of heavy metals. This phytoextraction promotes various cellular mechanisms and other enzymes to influence the growth of the plant and to facilitate the production of quality yield. The removal of heavy metal toxicity from plants and soil will improve the conditions of plants growing in the soil and promotes physiological functions of the plant. The medicinal plants are rarely used for the process of phytoremediation because it can affect the quality of the plant and is not further used for its processing and so on. Mostly non-edible crops are used for phytoremediation to accumulate the heavy metal ions and disinfect plant tissues. Complete uprooting of hyperaccumulator plants is done for complete removal of heavy metals from the plant tissues and cells. This uprooting of hyperaccumulator plant from the soil remove the maximum concentration and amount of heavy metal from the land and make it suitable for the production of plants.

Production of Plant Secondary Metabolites under Heavy Metal Treated Soil

Heavy metal toxicity affects various plant tissues including agronomical, horticultural and forest plants. Along with agronomical crops, medicinal and aromatic plants are equally affected by heavy metal toxicity. Plant primary metabolites which are synthesized in the plant system itself let further to the production of plant secondary metabolites. Plant primary metabolites include amino acids, carbohydrate, lipids and proteins are used for the production of secondary metabolites. These are generally used by the higher plants and the process proceeds to produce secondary metabolites to protect the plants from heavy metal toxicity (Hatami and Ghorbanpour, 2016). Plant secondary metabolites do not have any direct significant role in the plants but they play a major role in the interaction of the plant with the environment and their adaptability under stressed conditions. Plant secondary metabolism also plays an important role in plant defence mechanism to serve as a backbone of plant system under stresses conditions (Ramakrishna and Ravishankar, 2011). Secondary metabolites are of low molecular weight can produce less than 1% of dry matter weight. These perform other functions of imparting colour, odour, taste, texture also in the plants which are very important to represent the quality of produce (Mazid et al., 2011). They enhance the quality of the product by adding flavour to them. Plant secondary metabolites can further be divided into three groups such as phenols, terpenes and nitrogen-containing compounds based on their biosynthetic pathway (Bourgaud et al., 2001). In this the terpenoid/terpenes group has various compounds namely; isoprene, monoterpenes, sesquiterpenes having several carbons 5, 10 and 15 respectively, are volatile in nature (Nagegowda, 2010). Secondary metabolites containing nitrogen compounds are alkaloids, glycosides and phenolics compounds such as phenolic acid, anthocyanin and lignin (Taiz and Zeiger, 2006). Secondary metabolites are produced in the plants under a different type of biotic and abiotic stresses. When the plant faces any stressed condition, naturally there is a production of plant metabolites to enhance the defence mechanism of the plant to overcome that stress. They help in regulating the various factors of plant life processes which include physiological and biochemical processes. In case of medicinal and aromatic plants, they are generally affected by abiotic stresses in which heavy metal stress is one and main which cause great damage to the plant yield and quality (Asgari Lajayer et al., 2017). Heavy metal stress affects the quantity and quality of the crop and leads to the synthesis of primary and secondary metabolites in the plants. The new studies of heavy metal research show various exposure to heavy metal toxicity in the environment and measures to overcome it. Biosynthesis of heavy metal induces the production of primary and secondary metabolites to influence the defence mechanism to tolerate any type of stresses and enhance the production and quality of produce.

Plant Secondary Metabolites

Secondary metabolites are organic compounds synthesized after the production of primary metabolites in the plants. These are natural products which are not directly essential in the processes of plant life. Secondary metabolites usually not directly involved in the physiological and biochemical functions of the plant. In all the processes of plant life from growth, development, reproduction to maturity, secondary metabolites play a secondary role and help in plant development. These are the most important component of the plant defence mechanism as it protects the plant from various kinds of biotic and abiotic stresses. Secondary metabolites are very important for plants to cope with the stresses conditions. Plant processes like respiration, synthesis of protein, nutrient absorption, photosynthesis etc. are regulated by plant primary metabolites and other compounds, and leads to the synthesis of secondary metabolites. Under heavy metal stress, the absence and non-synthesis of secondary metabolites can lead to the death of the plant tissues. Absence of secondary metabolites decreases the functioning of the plant system and nutrient absorption which can cause the immediate death of the plant. Secondary metabolites are complex compounds and are involved in the defence mechanism to stimulate its functioning. These plant metabolites involved in the chemical defence of the plant system against heavy metal stress held at the tissue and cellular level. Plant primary metabolites are present in almost all cells and perform its function in cell division, elongation and maturing. On the other hand, secondary metabolites do not directly involve in all these processes. They play a secondary and major role in the plant to protect the plant from various kinds of stresses. Theses metabolites are widely present in the plants and play a significant role in the plant system to carry out its different functions. Secondary metabolites are present at a lower concentration in the plants and perform multiple tasks in the plant system. They play a significant role in plant stress conditions and make preferable functions to mitigate the stress from the plant to enhance its production. These are also involved in the production of enzymes and components for good growth of the plants. Secondary metabolites function in the production of medicines, cosmetics, nutrition and play a significant role in physiological processes of the plant.

Under the heavy metal, stress plants tend to decrease their immunity, increase in the production of secondary metabolites influence defence mechanism of the plant to overcome stresses environment (Edreva and others 2008). There is a balance between the advantages and limitations of plant metabolites under the conditions of heavy metal stress. The level of relative oxygen species (ROS) is brought lower under normal conditions and secondary metabolites stimulate antioxidant mechanism and other enzymes to control stressful environment. Sequestration of heavy metals involves a great role of secondary metabolites in the accumulation of heavy metals in the plant from root to shoot part. Various researches on heavy metals show the role of flavonoids, antioxidants and phenolics acids to overcome heavy metal stress (Khan and others 2010).

Secondary metabolites are grouped into 3 categories (Michalak, 2006):

1. Terpenes:- Terpenes include terpenoids or isoprenoids
2. Phenolics: - phenolics compounds include flavonoids, tannins, lignins, and so on.
3. Nitrogen-containing secondary metabolites: - these include alkaloids such as cocaine, caffeine, and morphine; cyanogenic glycosides and glucosinolates, and so on.

Characters and Functions of Secondary Metabolites

Under biotic and abiotic stress, secondary metabolites perform specific functions include antioxidant and antimicrobial activities to protect the plants from the stressed external environment. Various physical and chemical properties of plant secondary metabolites perform functions to mitigate the stressed conditions. The chemical properties of plant secondary metabolites perform various functions in the plant system to protect them from heavy metal stress. Generally, secondary metabolites are produced in the plants from the synthesis of primary metabolites and under the condition of any stress to the plant. Various abiotic stresses lead to the synthesis of secondary metabolites in the plants to remove the stress from the plant species. Phenyl prostanoids and flavonoids are major compounds or secondary metabolites used for the removal of heavy metal toxicity from the plants. These also perform acclimatory responses to protect plants from heavy metal stresses. Heavy metal research has been shown that proteins, lipids, alkaloids involve in the functioning of all plant processes. Under heavy metal stress, secondary metabolites lead to the production of different components and enzymes to protect the plants. Various proteins and carotenoids perform a great role in plant cellular mechanism to protect the plants (Havaux and Gruszecki 1993; Havaux, 1998).

Fig. 3. Role of Primary and secondary metabolites under heavy metal stress

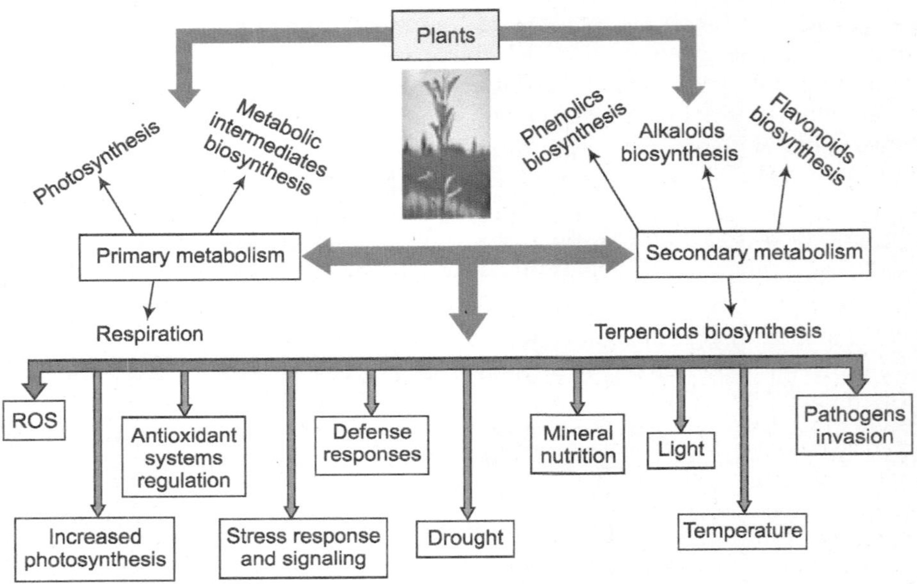

Source:https://www.google.com/search?q=secondary+metabolites+in

Phenolics compounds include flavonoids, anthocyanin; lignin etc. plays an important role in plants and is the main components of plant secondary metabolites. These components accumulate under stressed conditions and perform the functions to mitigate the stress conditions to the plant so that they can achieve a better production and quality yield. These decreases peroxidase polyphenyl lyase and modulate the level of flavonoids under stress condition and facilitate the plant growth (Sgherri et al., 2004; Taiz and Zeiger, 2006). In aromatic plants, the high temperature decreases the production of anthocyanin in the plants. In plants, some volatile terpenoids help the plant to mitigate heavy metal stress. another component of terpenes i.e. isoprene helps plant by protecting and regulating photosystem II from increasing effects of relative oxygen species (ROS) (Taiz and Zeiger, 2006). These terpenoids form a double bond with oxygen molecules to protect the biological functioning of the plant. Plant secondary metabolites involve in osmoregulation, water solubility and protect cellular mechanism under plant stress. They protect plant membranes and stimulate the production of enzymes to tolerate the stressed environment. Plant primary and secondary metabolites also play a crucial role as signalling molecules in the plants to detect the concentration and damage by heavy metals. Various nitrogen-containing secondary metabolites also function in the process of removal of heavy metal toxicity. These also involve in other functions such as imparting colour, flavour, taste and odour to the plants which enhance the quality of the plant produce. Plant primary and secondary metabolites are crucial for the plant to stimulate various functions in the plants to stimulate their quality and production for the benefit of the population of human and animals.

Terpenes

A large group of plant secondary metabolites contains terpenes which represent a big role and characters in the functioning of plant secondary metabolites against heavy metal stress. Several terpenes structure isolated and purified to perform the functions against plant stress. The structure of terpenes present in almost all plant species to express its activities (Obst, 1998). These are the structures obtained from different types of hydrocarbons. Terpenes as secondary metabolites perform different functions in the plant system. It influences the defence mechanism of the plant and regulates plant physiological and biochemical activities. They are also involved in the functioning of plant hormones and other enzymes to enhance the development of the plant and increase its yield. Terpenes perform a significant role in the defence mechanism of the plant and synthesis of other compounds to facilitate plant growth. Some studies show that the plant secondary metabolites terpenes increase the mechanism of photosynthesis and tolerance to heat and heavy metal stress (Sharkey and Singsaas, 1995; Loreto et al., 1998). These perform specific functions in reducing or removal of stressed conditions and to influence the quality of the crop produce. Terpenes help tolerate various abiotic stresses like heat, temperature and heavy metal stress. It induces various qualities of the plant and to recover the damage to the plant tissues. It is also useful for the interaction of the insect, plant and environment to check out the influence of insect pest on the plant tissues. Some of the types of terpenes perform a direct specific role in the plant system which is of great advantage to the plants (Khan and Mohammad, 2011). The various structure of terpenes are present in the plant cell wall and performs a variety of functions to regulate or to stabilize the interactions with proteins and phospholipids to influence its development. These protect the process of photosynthesis from oxidation and induce colour to the plant. Some plant growth hormones like abscisic also function as terpenes to influence the physiological processes of the plant (Taiz and Zeiger, 2006). Secondary metabolites serve as a key component to protect the plant system from heavy metal and other stresses and to influence its growth and development. In the process of photosynthesis, terpenes play a major role to avoid photo-oxidation and to impart colour to the plant. They can absorb the somewhat large wavelength of light and pass it to the chlorophyll to impart colour to the plant. The presence of terpenes protects the process of photosynthesis from the harmful rays of sunlight and help in promoting all the physiological processes of the plant. Terpenes as secondary metabolites perform a great role directly or mostly indirectly in the plant development to enhance its quality and yield.

Phenols

Secondary metabolites phenols include phenylpropanoids, flavonoids, tannins and lignins. Under stress conditions, they detoxify the level of hydrogen peroxide in the plants and remove the concentration of heavy metal ions in the plant's system. The phenolics compounds reduce the metallic potential of the plants and stable the pH and solubility of the molecules (Decker 1997). Different phenolic compounds synthesized in plant species in lower concentration like flavonoids and functions in the plant processes. Generally, the production of plant secondary metabolites takes place by the influence of biotic and abiotic stresses on the plants (Yamasaki et a., 1995). These stresses include drought, salinity, heat, temperature etc. which damages the plant adversely and reduces the functioning of the plant system. These flavonoids are responsible for scavenging of hydrogen peroxide and to function in the defense system. Plant phenolic compounds like lignin and tannins also involved in plant defence mechanism and protect plant from various stressed conditions. Various phenolic compounds are synthesized in trees after a wound or some damage. When the plant is get affected by any external damage, the oxidation of phenolic compounds going to occur in the plant tissues. The plant growth hormones secreted by the plants also functions as a protection against this damage. Phenolic compounds act as a strong antioxidant agent to protect the plant from external damage. Tannin and lignin also function in the biosynthesis of antioxidants. Phenolic compounds consist of 6 carbon atoms with a hydroxyl group attached to it. They are the main components of plant secondary metabolites to induce the protection against plant stresses conditions. These groups of secondary metabolites attach metals like iron and copper at lower concentrations to meet the need of the plant (Jung et al., 2003). Plant phenols involve chelating of metal ions is an important source of reactive oxygen species (ROS). Many studies have shown the synthesis of phenolic compounds in wheat and maize plant under heavy metal stress of Ni and Al respectively (Winkel-Shirley, 2002). Phenolic compounds are supposed to be synthesized under heavy metal stresses to protect the plant from these stresses and induce various functions of plant growth and development. In pulse crop, biosynthesis of phenols takes place under cadmium toxicity and stimulates production. Phenolic compounds induce anti stressed environment to promote plant growth under natural conditions. Accumulation of phenolic compounds under heavy metal stress expresses all its functions in response to plant growth and to avoid heavy metal stress to facilitate its growth and development. Under stressed conditions, all the activities of the plant system influenced by phenolic compounds to represent its role and functions. Under heavy metal stress conditions, the roots of some plants promote the biosynthesis of phenolic compounds in an excess amount to reduce the damage caused by heavy metal stress (Winkel-Shirley 2002). Tannin is the component of plant secondary metabolites reduces the manganese accumulation and chelation in the plants. Phenolic compounds generally have the ability of chelating; they act as a chelating agent in some plants to mitigate stress. These phenol compounds inhibit lipid peroxidation in the plants and facilitate plant production. The activity of phenolic compounds is different to the different plant molecules but is an important component to tolerate stressed conditions. Phenolic compound flavonoids involve in the lipid peroxidation and stimulate the activity of plant growth. These are important components of the plant system which decreases the plant cell wall from any damage and harmful environmental conditions. Flavonoids and lignins play a role to protect the signalling of molecules and stimulate the gene expression. Scavenging of relative oxygen species (ROS) is also facilitating by flavonoids. The phenolic compounds act as antioxidant agents and are capable to perform many functions in the plant system in response to plant protection. Plant phenolic compounds are the main component of the plant system which functions as the plant metabolites and reduce the stress caused by heavy metal to the plants. Polyphenols and flavonoids are observed to be an act in recovering of plant cell wall and tissues under heavy metal stress conditions and influence the growth and quality yield production.

Nitrogen-Containing Secondary Metabolites

Increasing heavy metal exposure to the environment causes a great reduction in quality produce and pollutes the natural resources and atmosphere unknowingly. The plant primary and secondary metabolites serve as a key role in plant system to protect from these stresses. Nitrogen-containing secondary metabolites also perform work in the process of protecting plants from heavy metal stress. These in contact with alkaloids inhibit single oxygen molecule and forms various types of structures to conduct plant protection. They influence the physiological behaviour of the plant and enhance production. These alkaloids are not damaged by the process of quenching (Larson, 1988). On a molecule of alkaloid, they inhibit many molecules of single oxygen. The alkaloids are synthesized by polyamines and perform different physiological roles in plant metabolism (Ali, 2000). They also function in the lipid peroxidation and other plant processes to stimulate plant production. Nitrogen-containing secondary metabolites include alkaloids such as cocaine, caffeine, and morphine; cyanogenic glycosides and glucosinolates, and so on. These conduct various properties of antioxidant to tolerate the stressed conditions. Phytochemical functions also influence by these compounds to carry out the plant growth and production of quality yield. These are the compounds which also functions in filtering the UV rays to protect the plant species. These alkaloids commonly accumulate in the plant tissues or cells after when plant faced any type of stressed conditions. Nitrogen-containing secondary metabolites include alkaloids, glycosides, glucosinolates etc. to remove the heavy metal stress from the plant species. Some of these components are volatile, toxic and quickly acts to serve its functions. Cyanogenic glycosides are the compounds of secondary metabolites which release poisonous gas i.e. Hydrogen cyanide when it starts its functioning. This will help the plant to remove insect-pest and their species from the plants. Generally, it is well known to resist the pest from the plant and make plant able to resist the pest species. As known earlier these compounds also involves taking part to impart smell, colour, odour etc. in many fruit and vegetable crops. The defence mechanism is widely influenced by compound glucosinolates so that it can grow well till harvesting. Various other products of glucosinolates function in the defence mechanism of the plant system to facilitate its production and quality yield (Taiz and Zeiger, 2006).

Conclusion

The increasing population today is the main cause of exposing heavy metals to the environment using industrialization and other factors. Heavy metals are poisonous to the plant and animals and cause the pollution of soil, water and air when exposed in the environment in the large amount. Some heavy metals are toxic even at low amounts and pollute the quality of air and water. The heavy metal toxicity contaminates all the natural resources and creates an issue for fresh air and water required by the population. Plants perform a variety of functions in response to heavy metal stress and to facilitate its growth and yield. Various primary and secondary metabolites functions to cope with heavy metal stress and to stimulate quality yield production. Plant metabolites represent a biological molecule which performs functions to regulate plant activities and to enhance quality yield production. Decreasing day by day agricultural fields due to increase in population, there is a need to grow more and more food with quality produce. In this, it is suitable or important to focus on the role of plant metabolites to enhance the quality and production of the plant species under heavy metal stress. Secondary metabolites show a variety of structures and compounds to facilitate functions of protection against heavy metal stress and to enhance the yield. To know the role of plant metabolites in response to quality crop production and to tolerate heavy metal stress is of great importance and help to enhance our food security. There is a need to conduct more research on this aspect and to increase the process of improving the quality and yield of the plants. The improved use of techniques and practices to regulate the function of plant metabolites provides a better understanding to increase the yield and quality of the crop plants.

References

Akula, R., & Ravishankar, G. A. (2011). Influence of abiotic stress signals on secondary metabolites in plants. *Plant signaling & behavior*, *6*(11), 1720-1731.

Ali, R. M. (2000). Role of putrescine in salt tolerance of Atropa belladonna plant. *Plant Science, 152*(2), 173-179.

Anaytullah, S. A., & Bose, B. (2012). Impact of seed hardening treatment with nitrate salts on nitrogen and anti oxidant defense metabolisms in Triticum aestivum L under different sowing conditions. *Vegetos, 25*(1), 292-299.

Baker, A. J. M. (2000). Metal hyperaccumulator plants: a review of the ecology and physiology of a biological resource for phytoremediation of metal-polluted soils. *Phytoremediation of contaminated soil and water.*

Bennett, R. N., & Wallsgrove, R. M. (1994). Secondary metabolites in plant defence mechanisms. *New phytologist, 127*(4), 617-633.

Bourgaud, F., Gravot, A., Milesi, S., & Gontier, E. (2001). Production of plant secondary metabolites: a historical perspective. *Plant science, 161*(5), 839-851.

Chaiyarat, R., Suebsima, R., Putwattana, N., Kruatrachue, M., & Pokethitiyook, P. (2011). Effects of soil amendments on growth and metal uptake by Ocimum gratissimum grown in Cd/Zn-contaminated soil. *Water, Air, & Soil Pollution, 214*(1-4), 383-392.

Edreva, A., Velikova, V., Tsonev, T., Dagnon, S., Gürel, A., Aktaş, L., & Gesheva, E. (2008). Stress-protective role of secondary metabolites: diversity of functions and mechanisms. *Gen Appl Plant Physiol*, *34*(1-2), 67-78.

Gautam, R. K., Soni, S., & Chattopadhyaya, M. C. (2015). Functionalized magnetic nanoparticles for environmental remediation. In *Handbook of research on diverse applications of nanotechnology in biomedicine, chemistry, and engineering* (pp. 518-551). IGI Global.

Gupta, A. K., Verma, S. K., Khan, K., & Verma, R. K. (2013). Phytoremediation using aromatic plants: a sustainable approach for remediation of heavy metals polluted sites.

Gupta, A. K., Verma, S. K., Khan, K., & Verma, R. K. (2013). Phytoremediation using aromatic plants: a sustainable approach for remediation of heavy metals polluted sites.

Hall, J. Á. (2002). Cellular mechanisms for heavy metal detoxification and tolerance. *Journal of experimental botany*, *53*(366), 1-11.

Harada, E., Kim, J. A., Meyer, A. J., Hell, R., Clemens, S., & Choi, Y. E. (2010). Expression profiling of tobacco leaf trichomes identifies genes for biotic and abiotic stresses. *Plant and cell physiology*, *51*(10), 1627-1637.

Hatami, M., & Ghorbanpour, M. (2016). Retracted Chapter: Changes in Phytochemicals in Response to Rhizospheric Microorganism Infection. In *Microbial-mediated Induced Systemic Resistance in Plants* (pp. 1-14). Springer, Singapore.

Hauser, M. T. (2014). Molecular basis of natural variation and environmental control of trichome patterning. *Frontiers in Plant Science*, *5*, 320.

Havaux, M., & Gruszecki, W. I. (1993). Heat-and light-induced chlorophyll a fluorescence changes in potato leaves containing high or low levels of the carotenoid zeaxanthin: Indications of a regulatory effect of zeaxanthin on thylakoid membrane fluidity. *Photochemistry and photobiology*, *58*(4), 607-614.

Hrynkiewicz, K., & Baum, C. (2014). Application of microorganisms in bioremediation of environment from heavy metals. In *Environmental deterioration and human health* (pp. 215-227). Springer, Dordrecht.

Huang, W., Shao, H., Zhou, S., Zhou, Q., Li, W., & Xing, W. (2017). Modulation of cadmium-induced phytotoxicity in Cabomba caroliniana by urea involves photosynthetic metabolism and antioxidant status. *Ecotoxicology and Environmental Safety*, *144*, 88-96.

Ibrahim, M. H., Chee Kong, Y., & Mohd Zain, N. A. (2017). Effect of cadmium and copper exposure on growth, secondary metabolites and antioxidant activity in the medicinal plant Sambung Nyawa (Gynura procumbens (Lour.) Merr). *Molecules*, *22*(10), 1623.

Jaishankar, M., Tseten, T., Anbalagan, N., Mathew, B. B., & Beeregowda, K. N. (2014). Toxicity, mechanism and health effects of some heavy metals. *Interdisciplinary toxicology*, *7*(2), 60.

Johansson, J. F., Paul, L. R., & Finlay, R. D. (2004). Microbial interactions in the mycorrhizosphere and their significance for sustainable agriculture. *FEMS microbiology ecology*, *48*(1), 1-13.

Jung, C., Maeder, V., Funk, F., Frey, B., Sticher, H., & Frossard, E. (2003). Release of phenols from Lupinus albus L. roots exposed to Cu and their possible role in Cu detoxification. *Plant and Soil, 252*(2), 301-312.

Khan, A. L., Hussain, J., Hamayun, M., Gilani, S. A., Ahmad, S., Rehman, G., ... & Lee, I. J. (2010). Secondary metabolites from Inula britannica L. and their biological activities. *Molecules, 15*(3), 1562-1577.

Kısa, D., Elmastaş, M., Öztürk, L., & Kayır, Ö. (2016). Responses of the phenolic compounds of Zea mays under heavy metal stress. *Applied Biological Chemistry, 59*(6), 813-820.

Kováčik, J., & Klejdus, B. (2008). Dynamics of phenolic acids and lignin accumulation in metal-treated Matricaria chamomilla roots. *Plant cell reports, 27*(3), 605-615.

Lajayer, H. A., Savaghebi, G., Hadian, J., Hatami, M., & Pezhmanmehr, M. (2017). Comparison of copper and zinc effects on growth, micro-and macronutrients status and essential oil constituents in pennyroyal (Mentha pulegium L.). *Brazilian Journal of Botany, 40*(2), 379-388.

Lal, K., Yadav, R. K., Kaur, R., Bundela, D. S., Khan, M. I., Chaudhary, M., ... & Singh, G. (2013). Productivity, essential oil yield, and heavy metal accumulation in lemon grass (Cymbopogon flexuosus) under varied wastewater–groundwater irrigation regimes. *Industrial Crops and Products, 45*, 270-278.

Lasat, M. M. (1999). Phytoextraction of metals from contaminated soil: a review of plant/soil/metal interaction and assessment of pertinent agronomic issues. *Journal of Hazardous Substance Research, 2*(1), 5.

Latef, A. A. (2018). Growth and some physiological activities of pepper (Capsicum annuum L.) in response to cadmium stress and mycorrhizal symbiosis.

Lee, S., Moon, J. S., Domier, L. L., & Korban, S. S. (2002). Molecular characterization of phytochelatin synthase expression in transgenic Arabidopsis. *Plant Physiology and Biochemistry, 40*(9), 727-733.

Leung, H. M., Ye, Z. H., & Wong, M. H. (2007). Survival strategies of plants associated with arbuscular mycorrhizal fungi on toxic mine tailings. *Chemosphere, 66*(5), 905-915.

Loreto, F., Förster, A., Dürr, M., Csiky, O., & Seufert, G. (1998). On the monoterpene emission under heat stress and on the increased thermotolerance of leaves of Quercus ilex L. fumigated with selected monoterpenes. *Plant, Cell & Environment, 21*(1), 101-107.

Mazid, M., Khan, T. A., & Mohammad, F. (2011). Role of secondary metabolites in defense mechanisms of plants. *Biology and medicine, 3*(2), 232-249.

Mazid, M., Khan, T. A., & Mohammad, F. (2011). Role of secondary metabolites in defense mechanisms of plants. *Biology and medicine, 3*(2), 232-249.

Michalak, A. (2006). Phenolic compounds and their antioxidant activity in plants growing under heavy metal stress. *Polish Journal of Environmental Studies, 15*(4).

Mourato, M., Reis, R., & Martins, L. L. (2012). Characterization of plant antioxidative system in response to abiotic stresses: a focus on heavy metal toxicity. *Advances in selected plant physiology aspects, 12*, 1-17.

Nagegowda, D. A. (2010). Plant volatile terpenoid metabolism: biosynthetic genes, transcriptional regulation and subcellular compartmentation. *FEBS letters, 584*(14), 2965-2973.

Ndeddy Aka, R. J., & Babalola, O. O. (2016). Effect of bacterial inoculation of strains of Pseudomonas aeruginosa, Alcaligenes feacalis and Bacillus subtilis on germination, growth and heavy metal (Cd, Cr, and Ni) uptake of Brassica juncea. *International journal of phytoremediation, 18*(2), 200-209.

Obst, J. R. (1998). Special (secondary) metabolites from wood. *Forest Products Biotechnology*, 151-165.

Ouyang, Y. (2002). Phytoremediation: modeling plant uptake and contaminant transport in the soil–plant–atmosphere continuum. *Journal of Hydrology, 266*(1-2), 66-82.

Seigler, D. S. (2012). *Plant secondary metabolism*. Springer Science & Business Media.

Sgherri, C., Stevanovic, B., & Navari, F. (2000). Role of phenolic acid during dehydration and rehydration of Ramonda serbica.

Sharkey, T. D., & Singsaas, E. L. (1995). Why plants emit isoprene. *Nature, 374* (6525), 769-769.

Sharma, S. S., & Dietz, K. J. (2006). The significance of amino acids and amino acid-derived molecules in plant responses and adaptation to heavy metal stress. *Journal of experimental botany, 57*(4), 711-726.

SONG, W. Y., MENDOZA-CÓZATL, D. G., Lee, Y., Schroeder, J. I., AHN, S. N., LEE, H. S., & Martinoia, E. (2014). Phytochelatin–metal (loid) transport into vacuoles shows different substrate preferences in barley and A rabidopsis. *Plant, cell & environment, 37*(5), 1192-1201.

Sytar, O., Kumar, A., Latowski, D., Kuczynska, P., Strzałka, K., & Prasad, M. N. V. (2013). Heavy metal-induced oxidative damage, defense reactions, and detoxification mechanisms in plants. *Acta physiologiae plantarum, 35*(4), 985-999.

Viehweger, K. (2014). How plants cope with heavy metals. *Botanical Studies, 55*(1), 1-12.

Winkel-Shirley, B. (2002). Biosynthesis of flavonoids and effects of stress. *Current opinion in plant biology, 5*(3), 218-223.

Wong, H. L., Sakamoto, T., Kawasaki, T., Umemura, K., & Shimamoto, K. (2004). Down-regulation of metallothionein, a reactive oxygen scavenger, by the small GTPase OsRac1 in rice. *Plant physiology, 135*(3), 1447-1456.

Wong, H. L., Sakamoto, T., Kawasaki, T., Umemura, K., & Shimamoto, K. (2004). Down-regulation of metallothionein, a reactive oxygen scavenger, by the small GTPase OsRac1 in rice. *Plant physiology, 135*(3), 1447-1456.

Yamasaki, H., Heshiki, R., & Ikehara, N. (1995). Leaf-goldenning induced by high light in Ficus microcarpa L. f., a tropical fig. *Journal of Plant Research, 108*(2), 171-180.

Zheljazkov, V. D., Jeliazkova, E. A., Kovacheva, N., & Dzhurmanski, A. (2008). Metal uptake by medicinal plant species grown in soils contaminated by a smelter. *Environmental and experimental botany, 64*(3), 207-216.

Role of Enzymatic Antioxidants for the Amelioration of Toxicity in Plants

D. Vasundhara*, M Lalitha, E Lokesh Goud
Department of Agronomy, School of Agriculture
Lovely Professional University, Jalandhar, Punjab, India,144411
**Email: vasudevagari07@gmail.com*

Abstract

The yield of the crop mainly depends on the capacity of the plant to adjust to various adverse conditions of the environment. Stress interferes with the normal functioning of the plant which stops them from showing their complete potential for the growth, development and the reproduction which finally decreases yields by reducing the productivity of the plants. The stresses in the environment are of biotic and abiotic types. Here, the question is how the plants handle these stresses of the environment. As plants are immobile, they cannot migrate to an optimal environment from the adverse environment. Then the only option for the plants is to deal with various stresses in the environment through adaptive processes like tolerance and the resistance. So, it is crucial to gain enough knowledge about the various processes which helps to expel the damages caused by stress. Various stresses in the environment exert their impact by triggering reactive oxygen species which causes oxidative damage and finally leads to the death of the cell. Hence, the role of enzymatic antioxidants in protecting and detoxifying the plants from various stress-induced toxins is gaining significant interest. In the process of evolution, plants have formed an intricated antioxidant system for maintaining balance and for improving the functions of cells by ameliorating the generated toxins. To increase the tolerance and resistance in plants for various stresses induced toxin, transgenic plants have been developed with changing levels of the antioxidant genes. This increased tolerance by antioxidant genes is characterized by many antioxidant enzyme activities. The antioxidant system constitutes of enzymatic and non- enzymatic antioxidants, this chapter deals with enzymatic antioxidants and their role in detoxification.

Introduction

Climate change is a major concern for human beings. Climate change severely affects the livelihood of farmers by causing droughts, floods, cold and heatwaves. In India, it is more severe because here the economy depends more on the agriculture sector. It causes crop losses by reducing the yield and quality of the crop. To handle these problems there is a need for developing resistance in plants. Different biochemical and the physiological reactions and various levels of tolerances observed in different plant species. When the plants are exposed to various stresses in the environment like drought, extreme temperatures, salinity, mineral deficiencies, there is a disturbance in balancing the production of ROS and the extinguishing activities of the antioxidant causing the oxidative damage. There are various sources of reactive oxygen species in the plants. ROS are produced during photosynthesis and respiration in plants with normal metabolism. Reactive oxygen species also produced during various pathways of the biotic and abiotic stresses like the reaction of photorespiration where the glycolate oxidase in the peroxisome triggers the production of superoxidase. Cell wall-bound peroxidase, Amine oxidase and NADPH oxidase are some new sources of reactive oxygen species. These reactive oxygen species cause cellular destruction. Reactive oxygen species like superoxide, Hydroxyl radical and Hydrogen peroxide cause oxidative destruction to nucleic acids, Proteins and lipids which results in disruption of normal metabolism activities in plants. Plants have got numerous antioxidant enzymes which saves them from those cytotoxic effects. Plants which posses more antioxidants are more resistant to the oxidative damages. Antioxidant enzymes like (SOD) Superoxide dismutase, (GR) Glutathione reductase and (APX) Ascorbate peroxidase search for the H_2O_2 present in mitochondria and chloroplast. Other antioxidant enzymes like (POD) Peroxidase and (CAT) Catalase removes $H_20_2$and neutralizes various free radicals. Superoxide dismutase, Peroxidase, Catalase, Ascorbate peroxidase and some other enzymes are the principal enzymes which are involved in ROS detoxification. During stress situations, antioxidants augment the activities of various enzymes. Oxidative destruction induced ROS like H_2O_2 is changed into water, molecular oxygen by catalase enzyme. This reaction is catalyzed by tetrameric haeme groups. Some other enzymes like Ascorbate peroxidase and Glutathione reductase which were part of defence mechanism against drought, chilling and salinity stresses are also enhanced H_2O_2 to water which results in detoxification of H_2O_2. Ascorbate peroxidase enzyme uses ascorbate for the reduction of H_2O_2. The tolerance levels of plants to many stresses vary with different physiological growth and development stages and size of the plant. Efficient use of antioxidant enzyme helps in detoxification of reactive oxygen species and finally decides the ability of the plants to withstand various stresses.

Antioxidants

Antioxidants are the molecules present in the plants in lesser concentrations which have the capability of inhibition of oxidation of another molecule.

Plants produce antioxidants to guard themselves against sun's UV radiations and Reactive Oxidant Species occurred during stress which causes oxidative damage. The antioxidant defence system is universal and varies from the species to species.

1. Enzymatic antioxidants
2. Non-Enzymatic antioxidants

This chapter discusses the role of enzymatic antioxidants for detoxification in the plants.

Enzymatic Antioxidants

Enzymatic antioxidants are the proteins which are involved in the catalytic transfiguration of the Reactive Oxygen Species and they're by-products to stable non-toxic compounds. Hence, Antioxidant enzymes form the major relevant defence system by protecting the plants from various oxidative stress triggered cell damages. Enzymatic antioxidant defence systems include the enzymes which can neutralize and scavenge the oxy-intermediates and free radicals. In the absence of this system, the plant would not convert efficiently the light energy into the chemical energy.

Fig. 1. Plant under Oxidative Stress

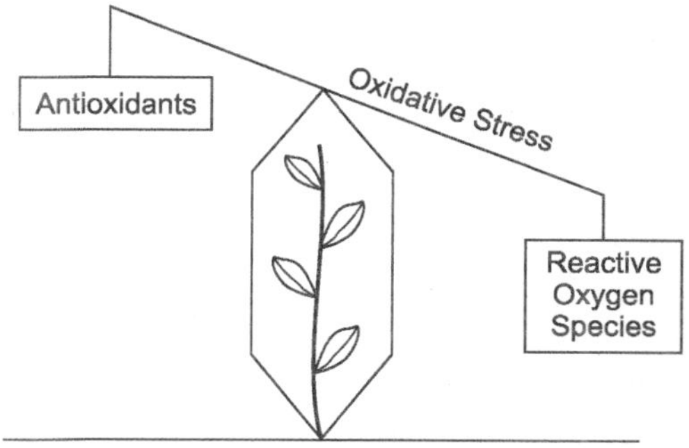

Reactive Oxygen Species and Responses of Enzymatic Antioxidants

Plants are exposed to various stresses in the environment leading to Reactive Oxygen Species production which causes oxidative damage to cells. Incomplete reduction of oxygen leads to the production of ROS. Initially, reactive oxygen species were recognized as the toxic intermediates of aerobic metabolism. In recent years, It was being seen that reactive oxygen species have significant importance as they control growth and development in plants and responses to the abiotic and biotic environment's stimuli. Free radicals such as $O_2^{\cdot-}$ and $OH^{\cdot}$ and non-radicals such as H_2O_2, O_2 are the major reactive oxygen species. Primary sites of reactive oxygen species production are chloroplast, mitochondria, peroxisomes. Secondary sites of ROS production are endoplasmic reticulum, apoplast, cell wall and the cell membrane. There is the double role of ROS, they perform like secondary messengers in different physiological mechanisms and also they triggers oxidative damage during various stresses of the environment such as drought, salinity, heavy metal's toxicity, UV Radiation and extreme temperatures, when the balance is disturbed needed for usual cellular homeostasis reactive oxygen species production and removal.ROS produced constantly at the basal levels during favourable conditions but they were not able to trigger damages since they are scavenged by various antioxidant defence phenomenons. So, the damage is observed only when there is a disturbance in their generation and removal. The damages in the cellular level constitute a reduction of the biomolecules such as proteins, carbohydrates,

DNA, lipids and pigments which finally merge in the cellular death. Under such conditions, plants survival depends on various factors such as changes in growth conditions, duration and severity of the stress conditions and plant's ability to adapt to the changes. The estimates show that only about 1-2 per cent of the oxygen consumption by the plants triggers ROS formation. The reactive oxygen species are more destructive and induce damages to lipids, proteins, and DNA affecting the usual functions of the cell. Redox balance is maintained in plants by antioxidants defence mechanisms. Enzymatic antioxidants constitute SOD Superoxide dismutase, APX Ascorbate peroxidase, GPX Guaiacol peroxidase, GST Glutathione-s-transferase, CAT catalase.

CLASSIFICATION OF ROS:

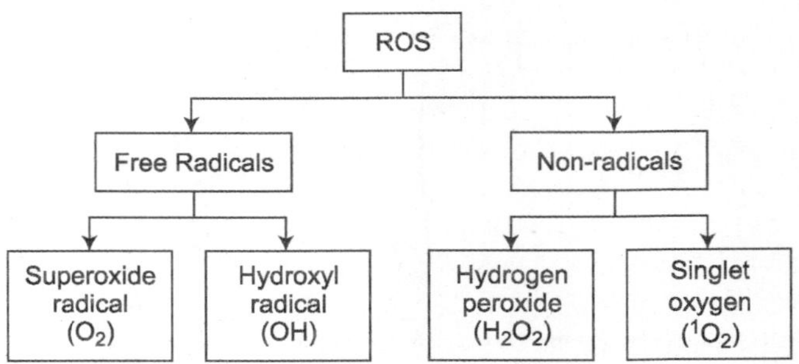

1. **Superoxide radical:** Reactive oxygen species constantly produced in chloroplast because of incomplete O_2 reduction or as outcomes of energy transfer to oxygen. Superoxide radical produced principally in thylakoid located photosystem I during the non-cyclic electron transport chain and also another cell compartments. Commonly, H_2O is produced during interactions of cytochrome c oxidase with oxygen. O_2 now and then react with various components of the electron transport chain to produce the superoxide radical. This is commonly the Ist reactive oxygen species to be produced. Superoxide radical produces other components of reactive oxygen species family by undergoing some more additional reactions. Superoxide radical has short (2 to 4 μs) half-life. They are medium in reactions so do not leads to damages by itself. Alternatively, it triggers lipid peroxidation by undergoing transfigurations into the highly reactive toxic hydroxyl ions and singlets oxygen.
2. **Hydroxyl radical:** Among all the ROS components, hydroxyl radical is highly reactive and also highly toxic reactive oxygen species known. OH is produced at natural ranges of pH by Fenton reaction in between hydrogen peroxide and superoxide radicals which is catalyzed with a transition metal such as Fe. Hydroxyl radical got the capacity to damage various components of cells by membrane degradation, damaging proteins and by lipid peroxidation. When there is no enzymatic defence system to remove toxic hydroxyl radical, more hydroxyl radical accumulations lead to the death of the cell.

3. **Hydrogen peroxide:** H_2O_2 is medium reactive and produced when superoxide further undergoes univalent reduction and protonation. Hydrogen peroxide generated by the reaction which catalyzing with superoxide dismutase or by non enzymatically by being caused to Hydrogen peroxide during lower pH situations. In plants, Hydrogen peroxide is generated during normal situations and also during oxidative stress occurred due to factors such as UV radiation, drought, intense light, infection by a pathogen, chilling and wounding. Because of the closure of stomata and less CO_2 availability and its less fixation, oxygenation of Ribulose 1-5 bisphosphate is occurred which leads to enhanced photorespiration. This constitutes about more than 70 per cent of hydrogen peroxide generated as the outcome of drought stresses. In plants, the main sources of hydrogen peroxide generation include electron transport chain in the chloroplast, endoplasmic reticulum, mitochondria, photorespiration, cell membranes and beta-oxidation of the fatty acids. Some other sources constitute of the various reactions including photo-oxidation by nicotinamide adenine dinucleotide phosphate oxidase and the xanthine oxidase.

   In plants hydrogen peroxide has a parallel role, its behaviour looks like the double edge sword. At fewer concentrations H_2O_2 is beneficial, but at high concentrations H_2O_2 damages the cells. Hydrogen peroxide behaves like signal-regulating tool for indispensable physiological phenomenons such as photorespiration, senescence, growth, development, photosynthesis, stomatal moments and cell cycle. As a result of hydrogen peroxide's longer half-life of the 1 ms, in comparison to another Reactive Oxygen species components, it could traverse long-distance and crosses cell membranes of the plants. H_2O_2 crosses the cell membranes through aquaporins and covers significant lengths inside cells which then leads to oxidative damage. Hydrogen peroxide in higher concentrations in the cells oxidizes cysteine and methionine remnants and by oxidizing thiol group, H_2O_2 inactivates copper/zinc superoxide dismutase, Fe-superoxide dismutase and Calvin cycle's enzymes. H_2O_2 leads to about 50 per cent losses in the activities of various enzymes such as sedoheptulose 1,7-bisphosphatase, phosphor ribulokinase and fructose 1,6-biphosphate and finally causes programmed cell's death at high concentrations. Hydrogen peroxide is mediumly reactive like superoxide radicals, hence hydrogen peroxide's damage completely shown when only it is transformed into the higher reactive oxygen species.

4. **Singlet oxygen:** $^1O_2$ is unusual reactive oxygen species component which is produced by the reaction of the chlorophyll triplet condition in antenna's system by O_2. Various stresses in the environment such as drought, heavy metal's toxicity and salinity lead to the closure of stomata, causing insufficient intracellular carbon dioxide concentration. This promotes the generation of singlet oxygen. $^1O_2$ can induce more damages to photosystem I and photosystem II and harms whole photosynthetic mechanisms of the plants. Although singlet oxygen has a shorter half-life of 3μs, it could handle to spread a few 100 nm and induces damages to the wider ranges of the targets. These targets constitute molecules such as pigments, lipids, nucleic acids

and proteins. Singlet oxygen is the main reactive oxygen species accountable for the light instigated losses of photosystem II activities, provoking the death of the cell. Plants efficiently handle scavenging of singlet oxygen with the aid of plastoquinone and could react with PS II's DI proteins. Also, $^1O_2$ has a major role in directing genes which were answerable for preparing protection against the photo-oxidative damage.

Fig. 2. Different factors responsible for the production of Reactive Oxygen Species

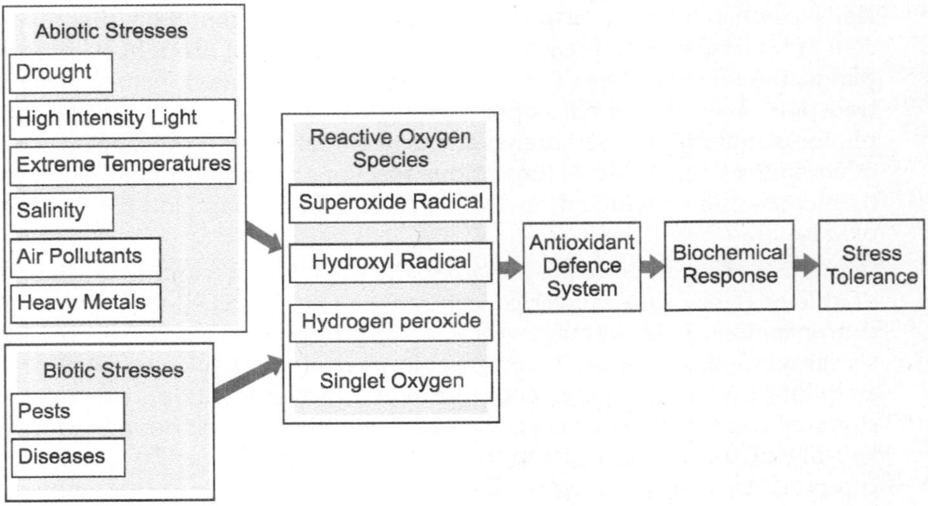

Locations of Reactive Oxygen Species (ROS) Generation in the Plant Cells

ROS is produced In both the normal and stress conditions at different sites in mitochondria, chloroplast, peroxisomes, plasma membrane, apoplast, cell wall and endoplasmic reticulum. Chloroplasts and the peroxisomes are main producers of reactive oxygen species in the presence of sunlight. In dark, mitochondria are the main source of reactive oxygen species production.

Mitochondria

Mitochondria are one of the locations of production of damaging ROS such as hydrogen peroxide and superoxide radicals in smaller amounts. Plant's mitochondria vary from the animal's mitochondria by involving in photorespiration by having oxygen and carbohydrates rich surroundings. Mitochondrial electron transport chain is the main thing of fault because it accommodates electrons which are adequately energized to reduce oxygen to produce ROS. Complex I and the complex III are the 2 main factors of mitochondrial electron transport chain accountable for the generation of ROS. Complex I or NADH dehydrogenases reduces directly oxygen to superoxide radical under its flavoprotein area. At the complex-I, ROS generation is additionally augmented during reversed electrons flow from III complexes to I complex because of deficiency of the NAD^+ connected substrates. ATP hydrolysis controls this reversed flow of electron. Also, there are some other additional metabolic mechanisms such as flavin oxidases pathways, Beta oxidation of the fatty acids and redox reaction of superoxide radical for the peroxisomal production of ROS.

Chloroplasts

Chloroplast constitutes a highly commanded system of the thylakoids which accommodates sunlight capturing photosynthetic mechanisms and also anatomical necessities for enhanced sunlight harvesting. In thylakoids, photosystem I and photosystem II which forms a central part of sunlight harvesting mechanism are the main procedures of ROS. Environmental stresses such as salinity, drought, extreme temperatures which induce water stresses and reduces CO_2 levels, combined with intense sunlight, triggers the superoxide radical formation at the photosystem. Accordingly, the membrane-bound copper/zinc superoxide dismutase at the photosystem I transform superoxide radical into hydrogen peroxide.

The two groups of 2 iron -2 sulphur and 4 iron-4sulphur are the other promoters of electron leakages from electron transport chain of photosystem I. Leakage of the electrons in photosystem II occurs through Q^A, Q^B electron acceptor and these are accountable for the production of superoxide radical. This superoxide radical converts itself into highly toxic reactive oxygen species such as hydroxyl radical through hydrogen peroxide mediated by Fenton reactions at centres of iron-sulphur clusters. The photosystem II is too responsible for the production of singlet oxygen and this takes place in 2 ways. First, the surrounding stresses of the environment disturb the balance between sunlight harvesting and energy utilization, which causes the generation of ^{3}chl* which then reacts with $^3O_2$ and finally produces $^1O_2$. Second, sunlight harvesting complex of photosystem II produces singlet oxygen when the electron transport chain is more reduced. These accumulations of singlet oxygen in chloroplast lead to peroxidation of the lipids membranes and PUFA and then damages protein membranes. It can directly lead to the death of the cell. Therefore, chloroplasts are the main source of the ROS generation in the plants. To make sure the survival of the plants in stress conditions, handling and removing ROS in the chloroplast is crucial which can be done by enzymatic antioxidants like superoxide dismutase.

Peroxisomes

Peroxisomes are spherical single membrane-bounded microbodies, acting as a major location of the intracellular hydrogen peroxide generation because of their immanent oxidative mechanism. Like the mitochondria and chloroplast during different metabolic phenomenon, peroxisomes also generate superoxide radicals. Here in 2 different sites, $O_2^{\cdot -}$ is produced. Xanthine oxidase in the matrix of peroxisomes metabolizes xanthine and the hypoxanthine to uric acid and produces superoxide radical as the byproduct. The second way is NADPH dependent smaller electron transport chain, comprised of NADH, cytochrome b situated in the membrane of peroxisomes which uses oxygen as an electron acceptor and then releases superoxide radicals into the cytosol. Besides, PMPs with molecular masses of 18, 29, 132 kDa are 3 immanent proteins accountable for superoxide radical generation.

Under environmental stress conditions, during the scarcity of water and stomata closure situations, the ratio of carbon dioxide to oxygen decreases which leads to increase of photorespiration causing glycolate generation. This formed glycolate oxidized in the peroxisome by glycolate oxidases to produce hydrogen peroxide. Therefore this makes it a leading generator of hydrogen peroxide during the photorespiration. Additionally, other metabolic systems such as flavin oxidases pathways, Beta oxidation of the fatty acids and redox reactions of superoxide radicals also cause ROS generation in peroxisomes.

Plasma Membrane

Plasmalemma surrounds whole-cell and plays a major role during interacting with environmental situations and stipulates necessary information for continuing plant cell's survival. NADPH oxidase located in plasmalemma is important because of gene expressions and availability of various homologs in stress situations. NADPH oxidases generate superoxide radical from transferring the electrons in cytosolic NADPH into oxygen which dismutases into hydrogen peroxide or undergoes catalyzation with superoxide dismutase. The supported fact is NADPH oxidase has major significance in the plant's defence system against the pathogens and environmental stresses.

Apoplast

Apoplast, forespread area around the cell membrane is accountable to convert carbon dioxide to soluble and diffusible forms, which then moves to the cytosol for undergoing photosynthesis. During adverse conditions of the environment, signals of stress combine with ABA and make apoplast as the eminent location for hydrogen peroxide generation. Besides, some more supplemental ROS producing enzyme are peroxidase, germin oxalate oxidase, polyamine oxidase and cell wall connected oxidase.

Cell Wall

At times of stresses in the environment, cell wall located lipoxygenase leads to PVFA's hydroperoxidation by making it as a powerful site for generation of ROS members like superoxide radicals, singlet oxygen, hydrogen peroxide and hydroxyl radical.

Endoplasmic Reticulum

NADPH intermediated electron transports including cytochrome P450, located in endoplasmic reticulum produces superoxide radical.

Damages of Reactive Oxygen Species

ROS generated due to the normal metabolism activities and stressful conditions injure biomolecules like lipids, DNA, and proteins and trigger damages of the membranes, destruction of normal functions of the cell, loss of efficiency in metabolic activities, mutation, leakages of electrolytes, breaks of chromatids, decreased carbon fixation, all of these finally leading to the death of the cell.

Damages to Lipids

Lipids form the main parts of plasmalemma which covers cells and assists it for adjusting to altering environments. Under environmental stress conditions, during the raising of ROS levels above threshold levels, LPO acts as more injuring and it is regarded as a single variable to measure the destruction of lipids. It starts the chain reaction then additionally aggravates oxidative damage by forming lipid radical then damaging DNA and proteins. Two major goals of ROS in the membrane phospholipid are ester linkages between fatty acids and glycerol and double bonds between the carbon atoms. PUFA are major substances of plasmalemma which serves as hotspots to ROS injury. The PUFA's such as linolenic and linoleic acids are distinctively prone for damage by the ROS such as singlet oxygen and hydroxyl radical. OH is a major attacking member of ROS by having the capability to induce the cyclic chain reactions further causing another PUFA's peroxidation.

The overall mechanism of the LPO split into 3 clear phases. They are the Initiation phase, Progression phase and the Termination phase. Initiation phase includes invigorating the oxygen to produce radicals such as superoxide radicals and hydroxyl radicals. This $O_2^{\cdot-}$ and OH· reacts with methylene group of PUFA, generating hydroperoxide conjugated dienes and lipid peroxyl radicals. Polyunsaturated fatty acid peroxyl radicals once generated, they inhold the capability for further reproduce LPO by removing one hydrogen atom from adjacent side chains of the PUFA. Lipids hydroperoxide go through a splitting after reaction by reduced metals like iron. It also goes through decomposition producing various reactive species like aldehydes, lipid alkoxyl, alcohols lipid epoxides and alkanes. LPO breaks off by the production of various lipid dimers formed by various radicals which are lipid-derived. Entirely, LPO raises fluidity of membranes including membranes to occur leaky for compounds, otherwise which go into cells through specialized channels damages membrane protein, prevents the action of membrane receptor, membrane located ion channels and enzymes.

Damages To DNA

Since nuclear DNA of plants is accurately kept safe by the connected proteins and histones, each of the two chloroplastic and mitochondrial DNA tolerates various attacks by ROS because of the absence of histones and the ROS generating mechanism. Oxidative injury to DNA as the effect of ROS arises at the various levels including the oxidation of residues of deoxyribose sugars, alteration of nucleotide bases, separation of nucleotide, cross-linkages of proteins and DNA and breaks in the strands of DNA. OH· Injures backbone of deoxyribose sugar by removing hydrogen atom and react with the double bond of pyrimidine and purine bases. ROS separates deoxyribose sugar's carbon-4-hydrogen atom and produces the deoxyribose radicals which further react causing breaks in a single strand of DNA. Injured components as the outcome of the base oxidation include more common 8 hydroxyquinoline and the less usual compounds such as thymine, glycol, thymine and opened ring of adenine. Hydroxyl radical is notorious to DNA protein cross-links when OH reacts either with connected proteins or DNA. These cross-links are irreparable further becomes lethal, if they are not rectified within time before the origin of critical phenomenons such as replication and transcription.

Damages to Proteins

ROS generated under stressful situations induces oxidation of the proteins. Proteins go through various alterations which might be either indirect or direct. At the time of direct alterations, protein's activity varies as the outcome of various chemical alterations like glutathionylation, carboxylation and nitrosylation. Carboxylation of protein is usually utilized as the marker to evaluate the oxidation of proteins. Indirect alteration of protein occurs as the outcome of interaction by LPO products. ROS levels beyond crossing threshold levels induce location-specific as changes of the amino acids such as proline, tryptophan, lysine, arginine and threonine and raised vulnerability for proteolytic destructions. Amino acids vary in vulnerability for ROS injuries. Amino acids with sulphur and thiol clusters are more susceptible. Cysteine and methionine both are susceptible to injuries by singlet oxygen and hydroxyl radical. Enzymes with Fe-s clusters are inactivated irreversibly after oxidized with superoxide radical. Oxidized proteins, therefore, become targets to proteolytic degradation.

Role Of Enzymatic Antioxidants In Detoxifying the Generated ROS

Generally, plants are sufficiently equipped for keeping ROS under the limit which is produced under normal metabolic activities of the cell. However, during stressful environments, ROS production exceeds the whole cell's antioxidative ability causing the stress triggered effects on the plant's growth. A stable balance is necessary for protecting plants against oxidative damage. As a result, the plant has enzymatic antioxidant defence machinery which protects them from the above-mentioned damages. This enzymatic antioxidant machinery works together to handle the series of uncontrolled oxidative damages and protects plants by removing ROS. In all the cellular compartments, this defence system is present. This defence system not limited to intracellular compartments, but also visible in apoplast in limited ranges.

Role of Superoxide Dismutase In ROS Detoxification

During environmental stressful conditions, superoxide dismutase acts as front line defence versus ROS triggered injuries. Superoxide dismutase catalyzes scavenging of superoxide radical by disproportionating it to oxygen and hydrogen peroxide.

This abstracts the chances of hydroxyl radical generation by Haber Weiss reactions. Based on metal ions which bind to it, superoxide dismutase is divided into 3 isozymes. They are copper /zinc-superoxide dismutase located in chloroplasts, cytosol and peroxisomes, Iron- superoxide dismutase located in the chloroplasts and Manganese-superoxide dismutase located in the mitochondria. The augmented actions of superoxide dismutase reduce oxidative damages therefore SOD has major importance in plant's adaptation to adverse environmental conditions.

Role of Ascorbate Peroxidase In ROS Detoxification

Ascorbate peroxidase is an immanent part of ascorbate glutathione cycle. Catalase primarily removes hydrogen peroxide in peroxisomes, while Ascorbate peroxidase executes the same action in chloroplast and cytosols. Ascorbate peroxidase uses ascorbic acid as a reductant agent during the reduction of hydrogen peroxide into water and DHA.

The family of Ascorbate peroxidase constitutes 5 isoforms depending on various locations and amino acids namely mitochondrial, chloroplastid, cytosolic and peroxisomal. Ascorbate peroxidase acts as the most crucial remover of hydrogen peroxide during the stressful conditions because it is distributed widely and also has got more affinity to hydrogen peroxide than catalase. Therefore, Ascorbate peroxidase plays the main role in removing ROS and thus protects the plant cells. Some studies revealed that zinc proteins ZAT12 essential for expression of the cytosol localized APX1 at times of oxidative damages induced by H_2O_2, woundings in the arabidopsis and heat. Furthermore, ZAT12 increases oxidases of NADPH expression, which shows nicotinamide adenine dinucleotide phosphate oxidase using ROS generation also involved in handling plants responses to oxidative damages (Rizhsky et al., 2004)

Role of Glutathione Reductase In ROS Detoxification

Glutathione reductase is the flavoprotein oxidoreductase.GR reduces GSSG into GSH by utilizing NADPH reducing agent. Reduced GSH is useful for regenerating ascorbic acid from DHA and MDHA. As the outcome is converting into its (GSSG) oxidized forms.

Glutathione reductase is the crucial enzyme of the cycle "ASC-GSH" and catalyses the production of disulfide bonds in glutathione disulfides for maintaining the higher GSH to GSSG cellular ratios. It primarily visible in the chloroplast and also occurs in smaller amount in the cytosols and mitochondria. Reduced Glutathione is the compound of lower molecular weight and behaves as the major reducing agent for preventing the thiol clusters from being oxidized and then reacts with harmful members of ROS such as singlet oxygen and hydroxyl radical. Glutathione reductase has a major role in maintaining tolerance of plants to different stresses and can change the redox position of different components of the electron transport chain. Many studies revealed GR activities increases in plants during different environmental problems. This increase might be because of raised synthesis of a protein or altered kinetic attributes of enzyme causing the GR isoform generation which then leads to detoxifying hydrogen peroxide formed during the normal and stressful situations.

Role Of Guaiacol Peroxidase in ROS Detoxification

Guaiacol peroxidase is the enzyme-containing heme group and it constitutes 40 to 50 kda monomers which used to remove exceeded amounts of hydrogen peroxide during both the stress and also normal conditions. Guaiacol peroxidase has a significant position in biosynthesizing the lignins and also in defending the environmental biotic stresses by destructing IAA and using hydrogen peroxide in this mechanism. Guaiacol peroxidase forechooses aromatic components such as pyrogallol and guaiacol as the donors of electrons. In the cells, it is observed in vacuoles and cytosols. In cells walls and also extracellularly, GPX is regarded as crucial enzymes for scavenging the hydrogen peroxide.

Role of Catalase in ROS Detoxification

Catalase is the tetrameralheme comprising enzyme accountable to catalyse the disproportionation of hydrogen peroxide into water and oxygen. CAT has more affinity to hydrogen peroxide and less specificity to organic peroxide. CAT has higher rates of H_2O_2 dismutation viz. 6×10^6 hydrogen peroxide molecules into water and oxygen per minute. As a result of beta-oxidation of the fatty acids, oxidative damages, catabolism of purines and photorespiration, peroxisomes acts as a site for the high activity of hydrogen peroxide generation. Recent findings mention that catalase also saw in the other cellular components like chloroplast, mitochondria and cytosol. Catalase shows dual activities depending on hydrogen peroxide levels. During lower levels that is less than 10 to 6M of hydrogen peroxide, catalase behaves peroxidatically where many donors of hydrogen such as ascorbic acid and ethanol could oxidize. Whereas during a higher level of the substrate, it degrades hydrogen peroxide at higher rates by utilizing catalytic reactions where hydrogen peroxide behaves like both the acceptor and the donor of hydrogen. Some spectrophotometric observations mention that catalases utilize 2 steps in peroxidative and catalytical reactions. Iron peroxide rich in oxygen was formed by the interaction of heme iron with hydrogen peroxide during the 1st step. This iron peroxide considered as a compound –I which could be noticed in both in-vitro and the invivo as it changes spectrophotometric attributes of heme catalases. In 2nd step, 2nd hydrogen peroxide acts as 2 electron donor and reduces compound -I and generates ferric catalase, H_2O and O_2. During lower hydrogen peroxide levels, catalase thus shows peroxidative activities. With 2 electron-donating substrates of peroxide, this compound-I reduced to the ferric catalase. With 1 donor of electrons such as superoxide radical or phenol, compound-II is generated, which is inactive and could not react with hydrogen peroxide. This inhibition could be altered with second-1 electron transfers from peroxidase substrates leading to the reformation of the ferricatalse. During higher hydrogen peroxide levels, compound-II converts into the compound-III which also denoted as the oxycatalase. This compound-III which is inactive and unstable could also generate from ferri catalase reacting with superoxide radical. The extended assemblage of the compound-III lead to the catalase inhibition. This finally induces catalase inactivation in presence of higher levels of hydrogen peroxide or superoxide radicals with the help of nicotinamide adenine dinucleotide phosphate against to hydrogen peroxide or superoxide radical inactivation in the mammalian catalase. NADPH reduces the generation of compound-II and increases its scavenging. CAT is unique among hydrogen peroxide degrading enzymatic antioxidants for not needing the reducing agent. Thus it enables cells with an effective system to scavenge H_2O_2stress environments insist more energy production and cost of the cell. This was accomplished by incremented catabolism rates which produce hydrogen peroxide. Catalase efficiently scavenges that accumulated H_2O_2. These outcomes a gain of more reducing agent and levels of cellular energies. It proves the unique model of catalase in handling the H_2O_2 balance in cells.

Role of Monodehydroascorbate Reductase in ROS Detoxification

Monodehydroascorbate reductase is accountable for the production of ascorbic acid from MDHA which has a short life span, by utilizing NADPH reductant, finally filling up the cell's ascorbic acid pools. As MDHAR produces ascorbic acid, it is located together with ascorbic peroxidase in mitochondria and peroxisomes, where ascorbic peroxidase removes hydrogen peroxide and oxidizes ascorbic acid in the mechanism. Monohydroascorbate reductase has various isozymes that are within the mitochondria, chloroplast, cytosol, glyoxysomes and peroxisomes.

Role of Dehydroascorbate Reductase In ROS Detoxification

Dehydroascorbate reductase reduces DHA into ascorbic acid by utilizing GSH as the donor for electrons. This made DHAR as another agent besides monodehydroascorbate reductase, which produces the cellular pools of ascorbic acid. DHAR is crucial in handling the size of the ascorbic acid pool both in apoplast and symplast, hence regulating the redox conditions of the cell. Dehydroascorbate reductase is located in roots, etiolated and green shoots and also in seeds.

Types of Antioxidant Enzymes

Superoxide Dismutase

Superoxide dismutases first were isolated from the blood of bovine as the copper proteins which believed to perform the role of storage of copper. At first SOD's authorized as the indophenol oxidases, tetrazolium oxidases and erythrocupreins. McCord and Fridovich in 1969 discovered the catalytic role of SOD. Superoxide dismutase belongs to metalloprotein's family. It is universally present in all the living organisms and also in some anaerobes. Copper/zinc SOD and mostly prokaryotic manganese SOD, Iron SOD are dimers and Manganese SOD of mitochondria and thermophile bacteria are tetramers. To perform catalyzing the disproportionation reactions SOD mechanism uses alternate reductions or oxidations of corresponding associated metal with the enzyme. 3 noticeable SOD types based on metal ions situated on their active site are observed. They are Cu/ZnSOD which holds copper, zinc, MnSOD holding Manganese and FeSOD holding an iron. As their date of origin is not clear, some sequential data describes 3 SOD types falls into 2 phylogeny families, copper/zinc SOD and Iron /Manganese SOD. The Iron and manganese SOD's are related closely as they similarly share a high degree of amino acid sequences and structural homology which are different from copper and zinc SOD's (Stalling et al,1984). Presently available data shows that these 2 superoxide dismutase families might have independently evolved and selected to responses for an environmental stressful situation and biosphere oxygenations by the photosynthetic organisms. Copper/Zinc SOD's are sensitive to cyanide and Iron/Manganese SODs remains unaffected from cyanide so it is easy to distinguish Cu/ZnSOD's from Fe/MnSOD's. Iron SODs are inactivated irreversibly by hydrogen peroxide but manganese SODs remains unaffected which distinguishes them from the Fe/SOD. Plants have various isozyme forms of SOD which are not seen in many organisms. Existence of these isozymes of SOD first observed in the maize crop by Baum and Scandalios in 1979 and 1982. Biological functions and role of superoxide dismutase as the protective enzyme versus toxicity of O_2 were visible in various studies with the prokaryotes and the lower, higher eukaryotes by including the higher plants. Analyzing the genes which were responsible to code the different types of SOD in the various organisms is possible with the approach of recombinant DNA technologies. Cloning, sequencing of superoxide dismutase genes provides information and knowledge of grasping the various structures and regulations of these genes. Besides, clones of SOD's furnish the chances of progress for generating larger active enzyme quantities for the detailed evolutionary and structural resemblances, enzymatic analysis and functional domain analysis. Further, these clones used as the probes to isolated genes which are homologous from different species. Genome clone isolation furnishes an opportunity for determining intron, Extron boundary, stop and start locations of the transcription, cis regulations and identification of signal responding compounds such as the sequences responsible for established oxidative damage signals which may describe how these genes are useful in responding and protecting cells against the oxidative damages.

Superoxide Dismutase Role In Alleviating Stress-Induced Toxins

SOD Role During Waterlogging

Waterlogging establishes oxygen scarcities on plant's organs which are submerged. Plants show responses to this stress by modifying protein generation patterns. Expressions of proteins which are already existing were rejected and synthesis of new proteins sets like anaerobic polypeptides occurs. Various plants capable of surviving the anoxia stage only for dying upon exposures to air, revealing oxidative injuries might occur also under-recovery phases. More lipid peroxidation observed in rhizomes of waterlogging sensitive *I. germanica* than intolerant one *I. pseudacorus* rhizomes after exposure to the air. Subsequent raise occurs in activities of SOD under the anoxic stage in resistant variety and not insensitive one. This raise is because of copper or zinc SOD but the location of this enzyme in the cell was not described. It was mentioned that this raised activities of SODare important for saving plants from oxidative injuries produced on exposure to the air. SOD application before oxygenation sometimes could save plants from further damages.

SOD Role During Drought

Responses of plants for drought are a complex process which includes accumulations of polyamines, new protein sets whose functionings were not known. ABA is the main response as it induces closure of guard cells of stomata which decreases loss of water. This mechanism also decreases CO_2 availability leading to ROS generation which causes misdirection of electrons in photosystems. Thus, a process which decreases oxidative damages has the main role under tolerances to drought. Cytosol situated copper or zinc SOD triggered more by drought in tomato and chloroplast situated copper or zinc SOD was not affected. In 2 types of mosses, the comparison was done for drought triggered alterations on lipid peroxidations as well as SOD and catalase activities. After exposing to drought, twisted moss showed lesser lipid peroxidation rates with raise in those enzymes and C.filicinum shown more lipid peroxidation rates and decreased enzyme activities. Drought tolerant and sensitive inbreeds of maize were studied by Malan et al. They found resistant plant correlated with copper/zinc SOD as well as GR activities.

Catalase

In 1818, the action of catalase was observed by Thenard in the tissues of plants and animals. He observed that those tissues were easily degraded H_2O_2. Catalase first was purified and crystallized out of beef's liver and its identification distinctly made in 1937 by Sumner and Dounce. Catalase was discovered in almost all the plants and was observed genetically, molecularly and biologically in the major agronomic crop, maize in 1990 by Scandalios. Further in 1965 and 1968 Scandalios discovered that catalase has existence in manifold molecular types or the isozymes and encoded with the several genes.

Catalase which is an enzyme comprising of tetrameric heme observed in all living organisms. Catalase plays a major role in hydrogen peroxide degradation. It acts as a highly active catalyst generated by nature.

Kinds of Catalase

3 main kinds of catalase which are not related phylogenetically are

1. Monofunctional catalase
2. Bifunctional catalases-peroxidase
3. Non-heme Manganese comprising catalase.

The active sites of non-heme catalase comprise reaction centre which is rich in manganese. Another 2 types of catalase comprise heme. Catalase-peroxidase which are dimers seen in fungi, archaebacteria, and cyanobacteria. Monofunctional catalase type is prevalent and more widely observed. Based on evolutionary changes, this monofunctional catalase differentiated into 3 groups. They are clade-I which comprises smaller subunits from the plants and bacterial catalase like CatF from Pseudomonas species, clade -II which comprises larger subunits from the fungi and 2nd bacterial groups, clade-III which comprises smaller subunits from 3rd true bacterial group, archaebacteria, fungi and animals.

Occurrence, Location of the Catalase

Catalase in plants mainly situated in the peroxisomes. Catalase activities are absent in various types of algae. Catalase not observed in prasinophytes, diatoms and euglena. Algae comprise glycolate dehydrogenases situated in the mitochondria. Heterokonts comprise catalase and peroxisomal type of glycolate oxidases but leaf type peroxisomes are not observed. Sometimes the catalses are seen in sites other than peroxisomes. Catalases are situated in mitochondria in Chlamydomonas, yeast Claviceps fungi release catalase into extracellular spots, the released catalase scavenges the hydrogen peroxide and protects the fungi during oxidative stress in the host plant.

Characteristics of the Catalase

Catalases, the antioxidant enzymes are the proteins comprising tetrameric heme with 240 KDa molecular wt. Unusual and green colour catalases of spinach leaf with 125 KDa molecular wt. shows that catalases are dimers. Other unusual types of catalases were found in thylakoid parts of the chloroplast of spinach and mitochondria of Chlamydomonas. Catalase is inhibited by some inhibitors like Azide, Cyanide, Salicylic type acid and amino triazole. The catalase enzymes interact by $^1O_2$. $^1O_2$ changes heme and generate electroporated movements in catalases, hence expressing isozyme behaviour. Manifold isozymes of catalases are observed in Maize, Tobacco, Sunflower, Cotton, Tomato and pumpkin. These manifold isozymes show manifold functionings of catalases in many plants during many growth and development phases and stress conditions. At present, interest in catalase antioxidant enzymes is increasing due to its importance in signals identification, metabolism and defence. The catalase genes were thoroughly described in maize crop which are CAT-1, CAT-2, CAT-3 genes which encode CAT-1,2,3 isozymes. In endosperm, matured pollens, scutellum at kernel formation stages and the aleurone, only isozyme of catalase observed is CAT-1. In blanched leaf and maize seedling's coleoptile, CAT-1 and 3 isozymes of catalases were seen. There is more CAT 2 isozyme assemblage in the leaves which are exposed to sunlight because of the increase in transcript assemblage and CAT 2 gene message translations. CAT 2 isozyme seen in peroxisomes and CAT 1 and CAT 3 isozymes seen in mesophyll tissues in the maize leaf. Biochemically CAT 3 catalase isozyme is somewhat distinct from CAT 1 and CAT 2. Degree of resistance to inhibitors like azide, cyanide, amino triazole and salicylic type acid varies in these 3 catalase forms. CAT 3 type isozyme shows some specific functions of metabolism which are different from the other 2 isozymes. Though these 3 isozymes are biochemically differentiated, these can interact invitro to generate heterotetramers. CAT-1 and 2 type isozymes of catalase when together present in within the living organisms, forms heterotetramers.CAT 3 type forms homo tetramers within the living organisms. This shows this CAT 3 type isozyme fractionated compartmentally from other 2 isozyme types of catalase. Target specificity of CAT 3 type is observed. CAT-1 and 2 types are seen in cytosol, peroxisomes and CAT 3 type seen in mitochondria.

Role of Catalase in Alleviating Stress-induced Toxins

In drylands and arid areas, 2 major abiotic stresses are Drought and Salinity. Degradation of plants growth and loss of productivity are causes of drought and salinity and then leads to the generation of some complicated responses at various levels. Responses generated because of the intensity of photons which oversteps the ability of stressed crop for absorbing it. Photorespiration removes all toxic intermediates like phosphoglycolate and makes oxygen-related photosynthesis process possible, but also causes CO_2 losses in plants. As catalase has a major role in photorespiration, authors concentrated on catalase roles in catalysis processes during drought and salinity conditions. Catalase activities regulation in plants under drought allows photorespiratory generated H_2O_2 removal during plants affected by water deficits of salinity stress. Therefore photorespiration and catalase processes are concluded as a waste process but they are important as the main elements of photosynthesis and main elements of responses for stresses for removing ROS assemblages. Extreme drought and salinity stress influences photoinhibition, Causing photosystem-II inactivations which last after watering. Photosynthesis is the main mechanism affected by the scarcity of water and extreme salt levels. The effect of these prescriptions varies with stress intensities and species. ROS generation increases under stress situations by electron transfer process degradation and by an oxidizing metabolic process in mitochondria and chloroplast.

ROS under non-stress situations removed by antioxidant enzymes, while under drought and salinity situations, ROS generation oversteps antioxidants enzyme capacities for removing them and leading to oxidative damage. Isoforms of catalase which are Fe porphyrin enzymes plays the main role as enhanced ROS removing systems by avoiding oxidative damages occurred due to these stresses.

Catalase Role During Drought

Recent concepts focused on raise in the activities of catalase generally correlated positively with the level of droughts experienced by the plants. During drought situations in little millets length of root gradually raised during all the growth stages but there was a decrease in height and pigments of chlorophyll. Moreover, stress applications triggered raise in enzymatic antioxidant activities such as SOD, peroxidase and catalase which allowed this plant to show higher levels of tolerance attributes to the drought. In a physic nut, plant which is drought tolerant, carbon dioxide assimilation levels in leaf and efficient attributes of carboxylation reduced with the increase of water shortage. H_2O_2 levels of the leaf, lipid peroxidation in this plant were highly and inversely related with activities of catalase, showing drought triggered stopping of the enzyme may allow oxidative stress. Drought-induced a raise in catalase levels and activities in 2 species cleome spinosa and cleome gynandra . Outcomes showed that antioxidant enzyme defence process in spinosa species was capable of handling ROS generation during drought. Fujita and Hasanuzzaman observed selenium role about AsA and GSH changes and antioxidant enzyme activities in rapeseed by exposing them to 2 level drought situations. Catalase activities reduced and APX remained unaffected during drought. While selenium pre-treated rapeseed seedlings under drought shown increase in APX activities and also in catalase activities when compared with plants without selenium treatments. Outcomes showed that selenium external applications raised tolerance in plants for drought triggered oxidative injuries by an enhanced enzymatic antioxidant defence mechanism.

Catalase Role During Salinity

Worldwide in agriculture lands salinity is the main problem showing immediate restrictions on the growth of crop and productivity in various areas. This proliferated salinity in cultivable lands expected to have increased global impacts. Although plants differ for their sensitiveness for salt accumulations, severe salinity triggers water deficiency and ion toxicities in various plants. Studies related to transgenic plants presented that consistent solutes like trehalose, glycine betaine and proline are assembled during salinity at millimolar levels in plants having an osmo scavenging roles in physiologically related maintenances and enables plants to tolerate salinity stresses. Furthermore, GB external applications in lower levels or formed by the transgenes to consistent solute biosynthesis may trigger stress-related genes expressions by including these to enzymes which removes ROS. Moreover many efforts made for inspecting responses of genes under salinity in many plants by utilizing proteomics and many other approaches. Impacts of NaCl on hydrogen peroxide levels and catalase activities were discussed in the diversified plant groups like chlorella, annual seabilte and rice field water nymph, these all show more levels of NaCl tolerances. H_2O_2 assemblages observed in all these plants on NaCl exposures and also catalase activities raised gradually under responses to treatments of NaCl. Authors also identified plants growing under severe salinity levels also outcomes in new catalase isoform synthesis. Some studies conducted in leaves, roots of tomato shown peroxisome located catalase activities were more in the plants under various salinity degrees when compared with controlled ones while in other plants like pea, catalase activities in peroxisomes were not raised in salinity. Another major gene which has importance in salinity is AtWNK8. Catalase activities were more in AtWNK8 mutants during stress than in the wild plants. Authors observed tolerance in AtWNK8 mutants is due to increased catalase and Glutathione peroxidase inside activities, with increased proline productions and assemblages. Further, it was identified that salinity triggered many modifications in activities of catalase in wild species of almonds. Gondium et al., in other studies revealed the impacts of pre-treatments of H_2O_2 sprayings on the growth of the plants. This study showed sprayings of H_2O_2 raised the enzymatic antioxidant activities and catalase showed more responses to sprayings of H_2O_2. Increased activities of catalase correlated with regulations of expressions of genes and lesser oxidative injuries were observed in the plants with more catalase activities, taking into account of the protecting functionings of catalase.

Ascorbate Peroxidase

Ascorbate peroxidase is the proteins which comprise heme type prosthetic clusters in which Fe has major significance in catalytic location. This is proved by providing

Fe deficiency situation in the sugar beet which caused a doubling of ascorbate levels and having of cytosolic ascorbate peroxidase in the roots which reflects the exhaustion of Fe in roots (Abadia and Zaharieva, 2003)

Ascorbate peroxidases are highly sensitive for ascorbate levels. Ascorbate is a compound employed in H_2O_2 reduction and acts as a donor of electron for which enzymes have more affinity. If ascorbate levels are lesser then twenty μM, APX activities fall down and also it loses its stability. This reveals the obstacles that occurred in the trials of isolating enzymes from the plants (Grazyna et al., 2007). For the first time, Zielinskas and Mittler isolated and purified cytosol situated ascorbate peroxidase from the pea in 1991. APX mainly seen in plants, cyanobacteria and algae and also seen in some insects like Trypanosoma and also found in eyes of bovine and in ascorbate enriched tissues.

Isoforms and their locations in APX

APX isoforms in the plants vary in pH, molecular weight, the specificity of substrates and stability. These ascorbate peroxidases in plants seen in various cell compartments.

Cytosol Situated APX

Cytosol isoforms of APX are homodimers. The residues of amino acids which participate in the interaction of electrostatics between the subunits are more conserved.

Concerning their localization, ascorbate peroxidase 1 transcript observed in leaf, weakly observed in the stem but absent in roots of sweet potato. (Grazynab et al., 2007) Their appearance is more leaves which are mature when compared with the immature leaves. Ascorbate peroxidase 2 expressed in the leaf in bundle sheaths, when exposed to more sunlight.

Peroxisome Situated APX

The mechanisms such as beta-oxidation of the fatty acids, metabolism of ureide and photorespiration also release ROS as the byproducts. Peroxisome situated APX has major importance in removing H_2O_2. Peroxisome located ascorbate peroxidase saw more in spongy parenchymas and vascular tissues when compared with palisade parenchymas.

Peroxisomal located ascorbate peroxidase activities decrease in the tomato after the pathogen inoculations. This may be due to the reduction of peroxisome located ascorbate reaction to the pathogen infection also from NO inhibition.

Chloroplast Situated APX

Thylakoid and stromal located forms of the ascorbate peroxidase involved in the water cycle in which water is formed after O_2^- reduction in 2 steps which is catalysed by SOD and ascorbate peroxidase. Thylakoid bound type ascorbate peroxidases are less responsive for external influences, causing oxidative damages in plants. This clarifies that this thylakoid type APX is responsible only for direct H_2O_2 detoxification but not for scavenging against generated ROS in stress situations. Over occurrence of cytosol type ascorbate peroxidase in the chloroplast in tobacco generated enough resistance to the oxidative damages caused by salinity and paraquat which finally shows the thylakoid type APX functionings as the immediate ROS scavenger. NO along with hydrogen peroxide acts as partner type molecules, handles hyper activities sensitiveness responses versus attacks by pathogens, inhibits assemblages of transcript thylakoid type ascorbate peroxidase and reduces enzymatic activities of tAPX. Ascorbate peroxidase subcellular locations are described by the occurrence of target peptides and domains in proteins. All the chloroplastidial APX types have peptides which are hydroxylated near terminus N which is refined in the matured type protein. Thylakoid bound forms of APX presents transmembrane domain near terminus COOH.

Genes of Ascorbate Peroxidase

Various trials and studies conducted on many genes of APX, the structure of gene which encodes APX 1 is well identified. Ascorbate peroxidase 1 encoded with the single gene. It has a normal structure and has various regulation elements such as HSE and ARE. HSE also has seen in APX 2. APX 3 and 5 comprises the same heat shock elemental sequences.

Ascorbate peroxidase comprises nine introns with more A, T nucleotide contents. 1st intron situated in the uncommon site in mRNA(5[1]UTR site). Chloroplastidial ascorbate peroxidase comprises of twelve introns in the species where stromal thylakoid type acerbate peroxidases produced from 1 gene alternate splicing. This is observed in spinach. Such type of genes has splicing regulation cis-elements to handle alternate splicings. In plants which have separate specific genes for the chloroplastidial types, these elements were absent. Recent studies show interest in hormonal activities of ascorbate peroxidase. some studies reveal the increased APX 1 transcription on ABA applications in Ipomea batatas, whereas there is no effect of ethylene application on ascorbate levels, expression of the involved genes in the ascorbate metabolisms.

Characteristics of Ascorbate Peroxidase

APX has prokaryotic type origin. 2 histidine remnants such as His 42, His 163 observed in APX. These remnants enhance catalytical activities in APX. His 42 situated at heme molecule's distal position. His 163 creates heme ligands interconnected with heme molecules in its proximal position

APX binds to single potassium ion at the proximally positioned domain. This binding location of potassium ion remains constant. K^+ is needed for the activity of APX. Besides, ascorbate peroxidase has purported metal-bound location in a distal position.

Unlike the III-class peroxidases, APX has tryptophan remnants in place of the phenylalanine at the 41 positions.

Role of Ascorbate Peroxidase in Alleviating Stress-induced Toxins

Good knowledge of the impacts of water scarcity and excess salts on the plant's biochemistry has a significant role in the enhanced management practices, improved programmes of breeding and also for foretelling growth of the plant. During this consideration, main H_2O_2 detoxification process in plants at abiotic stress is ASA-GSH cycle. Genes of APX in the mutant plants shown modifications in physiology, growth, antioxidant system uncovering these enzymes inclusions in the normal development of plants. APX gene expressions maintained in reply to the drought, salinity and also underdevelopment of the plant. APX encoded genes important for handling the balance between ASA and GSH for cell's redox potentials and redox signals and also regulates directly and indirectly more rates of photosynthesis during adverse situations in plants. In a study, it was identified that small molecules of RNA were responsible for handling various pathways related to drought, salinity like stress situations. APX gene coding expressions were up-regulated under drought and salinity like stress situations, showing the main significance of molecular handling phenomenon triggered by these stresses.

APX Role During Drought

Redox positions of mitochondria and chloroplast are regulated by homeostasis between generation and consumption of energy and then affected by the necessity of avoiding raised ROS generation. These needs are very critical under stress exposures. During drought ROS produced in mitochondria, electron transfers are needed for regulating normal metabolism, carrying out various functions of cells, regulating the plant's whole signalling systems and activating responses by retrograde signals. In complex regulation system networks of antioxidant enzyme defence, APX is a major enzymatic antioxidant which has the main role under drought and handles recoveries from this stress. In soybeans, Kausar et al. identified an increase of APX rates and also activities with increased water scarcity. Zarei et al observed APX levels in transgenic plant tobacco and found increased APX activities during drought revealing P5CS is the triggering gene and triggering of activities of APS included in drought stress tolerance process. Cruz et al found in crabwood that external glycine betaine applications might weaken the impacts of gentle water deficiencies in gas exchanges of leaf and the lipid peroxidation. They identified these glycine betaine applications increased activities of APX under drought in plants. Therefore glycine betaine weakened lipid peroxidations in plants under drought by positive modification of the APX activities.

Transgenic plants like tobacco with the over expressions of cytosol situated APX were generated and examined for tolerating gentle water scarcity to study the significance of cytosol situated enzymatic antioxidants during drought protection. Outcomes shown over expressions of these APX genes in tobacco relieves damages generated during drought to some extent. Oxidative damages were more in non-transgenic plants when compared with transgenic plants advising that cytosol situated APX gene over expressions saves membranes of tobacco from drought-induced toxin. Zhang et al found OsAPX2 gene over expressions in rice raised activities of APX under drought and protects plants from drought triggered toxins. Singh et al found SbpAPX gene over expressions in transgenic species has shown significant drought tolerance and protected the plant from drought-induced toxins.

APX Role During Salinity

APX is a major antioxidant enzyme of ROS removing system which has importance in uplifting salinity tolerances, detoxifying hydrogen peroxide in various cells of the plants, involved in maintaining balances of AsA. Many studies have shown mutants which are deficient in cytosol situated APX were not tolerant to oxidative injuries triggered by salinity. Diaz.Vivancos et al. generated transgenic plant plum encoding cytosol situated APX genes to strengthen the enzymatic antioxidant ability in plum. These transgenic plum showed tolerance to salinity. Guan et al. examined the relation of PutAPX gene of the puccinellia, tenuiflora (wild species) grown under salinity stress. PutAPX over expressions in transgenic arabidopsis raised tolerance of the plants pretreated with NaCl, reduced lipid peroxidations. Transgenic species showed lesser H_2O_2 amounts and more chlorophyll rates than wild species during normal and NaCl applied stress situations. Further, expressions of ascorbate peroxidase proteins and its activities in transgenic species raised to rates which were more than in wild species under NaCl exposures.

While plants silenced with APX could capable of up-regulating other enzymes like peroxidases enabling mutants to handle salinity like non-transgenic species. Bonifacio et described APX 1/2s mutant in rice have a modified redox balance, as expressed by raised H_2O_2 levels and AsA-GSH redox positions but antioxidant restitution process expressed by mutants together associated with raised OsGpx expression of genes which outcomes high GPX activities in chloroplast and cytosol. Hence cytosol situated APX deficiencies restituted effectively by upregulation of another peroxidase. Salicylic acid affects photosynthesis during normal situations and triggers tolerances for various stress in plants by enhancing physiological systems. Nazar et al examined and compared mungbean cultivars with varying tolerance degrees to salinity. They investigated how these systems were triggered by salicylic acid applications to ameliorate reduction of photosynthesis during salinity. SA applications raised APX activities. Hence there was a raise in photosynthesis during salinity situations, ameliorated the reduction of photosynthesis during stress, maintained water levels and also ameliorated salinity induced toxins.

Glutathione Peroxidase

Glutathione peroxidase seen in all the eukaryotes and intensively studied in mammals but less studied enzymatic antioxidant in plants. GPX utilizes glutathione for catalyzing H_2O_2 reduction into water or alcohols. Some complex evolutionary related studies reveal that GPX in all the kingdoms might originate from transfer, loss or duplication of genes among vertebrates and invertebrates or the plants. Glutathione peroxidase is a common name of a family of the manifold isozymes which catalyzes hydrogen peroxide reduction into water or alcohol.

In this process, glutathione acts as a donor for electrons. In the metazoa, certain glutathione peroxidases have activities depending on selenium. In mammals 4 main isozymes of GPx which are selenium-dependent are GPx1 also called as classical glutathione peroxidase, GPx2 also known as gastrointestinal located glutathione peroxidase, GPx3 called as plasma glutathione peroxidase,GPx4 also known as phospholipid glutathione peroxidase and other 2 forms are GPx5 and 6 which related closely to the GPx3. Another new form identified in mammals is PHGPx which includes cysteine in place of selenocysteine. This PHGPx acts as 7thgroup. Plant glutathione peroxidase divided as 5th group of the peroxiredoxins since they are capable of efficient peroxide reduction with thiroredoxine system(Trx) but not with GSH. Trx forms of GPx are recently observed in a common fruit fly. The genes of plant's GPx closely similar to the animal's PHGPx, their proteins form 3 widely conserved remnants of cysteine, which essential for enzymatic catalytic reactions. 8 genes of GPx were identified in arabidopsis. Based on virtual reality analysis, GPx estimated in mitochondria, chloroplast, ER and cytosol locations and have similarities with cysteine. In plants, glutathione peroxidase has cysteine remnant in active locations, functioning in GSH and classes of thiol peroxidase of non-heme family.GPx shows responses in stress conditions. Studies revealed that during abiotic and biotic stresses like oxidative damage, drought, salinity and attacks by pathogens, there is a rise in the mRNA levels of the glutathione peroxidase. In rice, more GPx gene regulations were observed during increased cold stress and H_2O_2 levels. Many transgenic experiments also support GPx functions such as GPx upregulations in transgenic plant tomato outcomes more abiotic stress tolerances. GPX besides to stress responses also handles cell's redox balances by changing thiol-disulfide homeostasis. GPx regulated redox balances during oxidative damage in mustard. The expression of GPx is associates more with the cell's metabolic position.

Glutathione Peroxidase Role in Alleviating Stress-Induced Toxins

Role of Glutathione Peroxidase during Stress

It is accepted widely that severe stresses in the environment like higher salt, metal concentrations, extremes of temperature, drought triggers oxidative damages. Various species showed responses to these stresses with raising GPx expressions which would save plants from oxidative injuries. Sakamoto and Sugimoto in the arabidopsis shown NaCl, Fe or Al applications to seedlings of arabidopsis raised the levels of AtGPx6 mRNA. Salinity caused raise in AtGPx 1,2,5,6 levels and AtGPx 3 remained unaffected. This raise is observed within 3 hours of applications and maximum increased rates observed after 12 hours of application osmotic stresses induced by the mannitol outcomes the raise of these genes. The AtGPx gene responses to toxic levels of iron were more for AtGPx 2,5 and 6 but levels of responses of AtGPx 1 and 3 were slow. AtGPX 6 presented strong responses to toxins of iron. Copper exposures outcomes in raised AtGPx 2,5 and 6. These outcomes show AtGPx member genes correspond with wider levels of various stress and during this stresses, AtGPx up-regulates and protects the plants.

Guaiacol Peroxidase

Guaiacol Peroxidase is the heme comprising enzyme which is involved in various physiological phenomena. The pattern of this antioxidant enzyme's expression is organ-specific and maintained developmentally. Peroxidase which oxidizes the guaiacol, as a usually utilized reduced agent is known as the guaiacol peroxidase. It is situated in the vacuole, cytosol, extracellular locations, apoplast and cell walls but absent in cell organelles. This enzyme participates in various growth and developmental processes of plants. This enzyme is studied extensively and involves in various physiological mechanisms but their physiological and biochemistry related roles in plants are not completely clarified. Cyanide and azide are 2 inhibitors of peroxidases. Isoforms of guaiacol peroxidases are sensitive to azide and cyanide. When it is studied under 2 cultivars of safflower revealed that azide efficiently degrades guaiacol peroxidase activities than cyanide.

Guaiacol Peroxidase role in Alleviating Stress-Induced Toxins

Mokoena et conducted a study to observe and identify the roles of ascorbate and guaiacol peroxidase involvements during drought stress at various growth stages of soybean. They observed activities of guaiacol peroxidases were triggered for high levels in resistant types than the susceptible types only at a flowering stage but no increase at the development of pod stages showing the enzyme's protective implications versus drought. Tolerance to drought correlates with scavenging ability of ROS. Extent up to which the enzyme inhibited or weakened during drought is highly different among the species and also varies from cultivars to cultivars belonging to same plant species. Lesser activities of POD during flowering than at the pod development stages shown generation of enzymatic antioxidants raised with the maturity in soybean. More increases of activities of POD after the drought generation at the flowering additionally shown that enzyme was component of drought tolerance process during triggering of stress at flowering stage. They concluded by suggesting POD upregulation activity intolerant cultivar of soybean shown its implication in drought resistance.

PEROXIREDOXINS

Peroxiredoxins are family of the peroxidases which depends on thiol group. It is observed in all the kingdoms. 5 groups of peroxiredoxins are identified in plants. These groups catalyze peroxide decomposition since they do not comprise prosthetic class, this catalytic cycles outcome the formation of inactive types of the peroxiredoxins. Prx depends on the outside donors of electrons like thioredoxins, cyclophilins, NTRC and glutaredoxins. Besides Prx significance in antioxidant defence systems, it also identified in showing importance in growth and developmental activities, tolerances to desiccation in dormant stages of seed, photosynthesis protection, defence versus redox signals and pathogens. Peroxiredoxins also forms water to water cycles for H_2O_2 detoxification. There are various isozymes of peroxiredoxins in organelles of the cells. Peroxiredoxins classified into 2 groups. They are 1-cys, 2-cys Peroxiredoxins. This classification is dependent on several remnants of cysteine which involved in the catalytic phenomenon. Further 2-cys peroxiredoxins divided into atypical and typical groups.

Peroxiredoxins role in Alleviating Stress-Induced Toxins

Biotic as well as abiotic stresses commonly involve with redox positions of cell and trigger ROS, RNS productions. Peroxiredoxin transcripts maintenance has observed and shown regulatory differences depending on the group of peroxidase, plant type, the intensity of the stress and development phase. Transcripts of peroxiredoxins respond for all stress types like heavy metals, drought, salinity, chemicals and extreme ranges of temperature. There was an increase in PrxQ1 and 2 transcript rates under exposure to nutrient deficiencies, more H_2O_2 levels and under higher intensities of light. AtPrxQ levels were raised during the oxidative damages. 1cysPrx is present in dehydration resistant *X. viscosa* but absent in completely hydrated cells. There are assemblages of the transcript in cells during stress like heat, under ABA application, higher intensities of light and dehydration. This shows the major role of 1cysPrx in drought resistances like in again raising plants and shown also that uncommon peroxidase expression regulations would trigger tolerance to the stress.

Monodehydroascorbate Reductase

MDAR in plants is an enzyme component of glutathione ascorbate cycle which is one main antioxidant enzyme systems of plants for scavenging damages generated by ROS. Activities of MDHAR in the cytosol, chloroplast, glyoxysomes, peroxisomes in leaf and mitochondria. In the ascorbate glutathione (AsA-GSH) cycle APX reduces H_2O_2 into the water by utilizing AsA as a donor for electron which leads to the generation of MDHA. This MDHA reduces again back into ascorbic acid by MDAR or it may undergo redox reactions and forms ascorbic acid and DHA. MDAR is the efficient mediator in photo reductions of the dioxygen in $O_2^{\cdot-}$ in the chloroplast. It is a FAD type enzyme and it was first to utilize the substrates organic radicals. MDAR catalyze MDA reduction into ascorbate by utilizing NADPH as a donor for electron. In MDAR reaction phenomenon, Its FAD reduced by the NADPH with the constant rates of the 1.8×10^8(rate)$M^{-1}S^{-1}$. FAD which is reduced form of MDAR is then oxidized by the MDA with 2.6×10^8(stable rates) $M^{-1}S^{-1}$. Therefore, MDAR and substrate interactions proceed at the limited rates of diffusion. The MDAR in the chloroplast is attached with membranes of thylakoids. Studies reveal 2 physiological related significances of MDAR situated in the chloroplast. They are ascorbate reformation from the MDA and dioxygen photo reductions into $O_2^{\cdot-}$ during the absence of MDA substrate

MDAR Role In Alleviating Stress-Induced Toxins

Role of MDAR During Salinity

Activation of MDAR has identified in various plants on exposure to several stresses. Further overexpression of MDAR in the transgenic species shows a high level of tolerances to stress. Eltayeb et al., In 2006 identified that transgenic plant tobacco overexpressions of cytosol situated AtMDAR1 genes shown lesser levels of H_2O_2, improved ascorbate levels and high rates of photosynthesis during salinity when compared by wild plant types.

MDAR involvements in salinity tolerance are described by mangrove species cytosol situated MDAR's overexpression in the rice crop. Transgenic rice showed good tolerances for salinity at germination, seedling phases. Further, there was a rise in several tillers and weight of grain which finally causes an increase in yields. Sudan et al., In 2015 suggested in the millets, there is upregulation of expressions of MDAR but not its activities. Here it is unknown about the relationship between the expression of mRNA and activities of MDAR because of not identified genes. Therefore consensus is there that the MDAR improves ascorbate levels and saves plants from salinity and also from the salt stress-induced toxical compounds.

Role of MDAR During Freezing Stress

In arabidopsis, tolerance for freezing stress is done by overexpression of mustard's MDAR. The impact is more when the BrDHAR expressed together with MDAR, glutathione content is improved and the redox levels are weakened (Shin et al., 2013)

Li et al., In 2010 identified chloroplast situated tomato isoforms overexpressed in the arabidopsis cause tolerances for high as well as low temperatures. There was a raise in MDAR activities in chilled seedlings of rice and also increase in protein levels. Various studies suggest MDAR involvement in heat and cold stresses but it is difficult to describe the significance of various isoforms due to insufficient data. Activities of MDAR also improves under other abiotic stresses like drought, severe light, UV radiations. (Sudan et al.,2015)

Dehydroascorbate Reductase

Dehydroascorbate reductase reduces dehydroascorbate into reduced ascorbic acid by utilizing GSH. The outcome of dehydroascorbate reductase reaction is GSSG which is recycled by GR. Nicotinamide dinucleotide phosphate acts as a donor of electron for the activities of GR and MDAR. This cycle observed in chloroplast, cytosol, peroxisomes and mitochondria and involves in regulation of cell's redox positions. Hence, recycling of ascorbate through MDAR and DHAR is essential for the regulation of cells redox position, responses of plants to stresses and photosynthesis.

Physiology related significance of DHAR observed in the tobacco by utilizing reverse genetical attempts. DHAR in tobacco raises the levels of ascorbate and redox state. Suppression of cytosol situated DHAR by introducing cDNA of DHAR have the opposite impact. This process also suppresses chloroplast situated DHAR and cytosol situated DHAR protein expressions. Hence, recycling of ascorbate through cytosol situated DHAR is limiting process handling the accumulation of ascorbate and redox positions in the tobacco. Besides, utilizing this transgenic tobacco it is identified DHAR is needed for the growth and developmental activities, photosynthesis and tolerance to oxidative damages. Also, DHAR shows some negative impacts on the closure of stomata by regulating levels of H_2O_2 in tobacco's guard cells. Therefore its reduction raises tolerance to drought in the tobacco. This study shows the physiological related significance of the DHAR during abiotic stresses in plants. Positive impacts of DHAR overexpression on redox positions and ascorbate levels and also on the levels of tolerances to several forms of stresses identified in various plants. Although, only some studies undersee eminent physiological related significances like offering to recycle of ascorbate, responses to drought of DHAR isozymes in other various plants by including the arabidopsis utilizing the reverse genetical attempts.

DHAR Role in Alleviating Stress-Induced Toxins

Role of DHAR During Ozone Stress

There was an increase of levels of AsA in DHAR transgenic species combined by enhanced redox potential of AsA would be because of improved cytosol situated DHAR activities which decrease oxidized DHA spread from chloroplast, apoplast into the cytosol. DHAR transgenic species shown greater resistances to ozone like stresses. Sensitivity related to ozone is mainly correlated by AsA levels in leaf (Barth and Conklin, 2004.), studies show high activities of scavenger enzymatic antioxidant would save from oxidative damages. Improved ozone stress tolerance in transgenic tobacco would be because of raised AsA amounts, which is due to improved DHAR activities. Moreover, pools of AsA inside the cell as well as outside of the cell handled by cytosol situated DHAR (Gallie and Chen, 2005). With exchange for AsA, apoplast situated DHA is transferred with carriers which are specific in the cytosol. As apoplast situated AsA forms front line system of defence versus damaging oxidants like ozone, DHAR over expressions in the cytosol may regulate exchanges of AsA and DHA(oxidized form) and presence good protection to the cell from ozone.

Role of DHAR During Drought And Salinity

DHAR transgenic species recorded more rates of photosynthesis during drought and water stresses triggered by PEG. This is due to more AsA rates in these species. Adriano et al., in 2005 identified AsA levels would inhibit damages triggered by the AOS is major character-related for tolerance to drought and DHAR activities raised in correlation to extreme drought situations in 4 prunus interspecific hybrids. More H_2O_2 assemblages cause damages during salinity and drought situations. For H_2O_2 detoxification and regulation AsA activities, its reformations and biosynthesis are necessary. Hence, good DHAR transgenic species performances during drought and salinity would be because of fastest H_2O_2 scavengings outcomes from raised AsA levels. Results show raised AsA rates via DHAR overexpression in the cytosol could lead to the enhanced tolerance of plants to oxidative stresses.

Glutathione Reductase

GR regulates the homeostasis in between the GSH and ascorbate collections, additionally, it also regulates the cell's redox phase. GR is the flavoprotein oxidoreductase seen in prokaryotes as well as eukaryotes. GR first was described in the eukaryotes, yeast and also in the plants. It is purified from cyanobacteria, yeast, fungi, maize, pea, tobacco, spinach and wheat. Though the main site of GR is chloroplast which is almost 70 to 80 per cent, it also is seen in the cytosol, peroxisomes, mitochondria and also even in the tissues of non-photosynthetic and in organelles. GR observed from various organisms shown optimum activities at various levels of pH. At pH value 8, the chloroplast situated GR of wheat shown optimal activities and chloroplast situated GR in Pea, the pH value for optimal activities is 7.6 and pH for human's erythrocyte is 6.8. In the single plant, there are various values of optimal activities of pH for various isoforms. GR is inhibited with low NADPH concentrations in the liver of Mouse, Escherisia coli and yeast. GR activities are not fully inactivated in Maize, anabena and Spinach. In 1997 Creissen and Mullineaux reported that GR in plants altered their activities for scavenging themselves from NADPH levels produced in the photosynthesis process. Glutathione reductase comprises various domain to bind with FAD, NADPH and interface domains to join 2 subunits. Active location is in between 2 subunits where the GSSG binds. GR mainly exists in homodimer form. It may also assemble to tetramer and to even more aggressive state which is observed in spirulina. High ordered state types depends on temperature and pH and independent of their products or substrates. This might be one process of management of glutathione reductase activities.

GR'S Catalytic Phenomenon

Glutathione reductase goes through redox conversions which depend on the availability of substrates. Oxidized GR form is highly stable than reduced GR form. GR's oxidized form has tolerance for Cu^{+2}, Zn^{+2} and Fe^{+2} metal ions. In the process of GSSG conversion into GSH, GR splits 2 electrons furnished by the nicotinamide adenine dinucleotide phosphate and donate them to each GSSG's sulphur atoms. GR comprises FAD and redox disulphide enzyme. For the GSSG reductions depending on NADPH, GR contains residues of cysteine in active locations for catalyzation. Sequences of reactions are:-

- In nicotinamide dinucleotide adenine phosphate domain of binding, GR comprises GXGXXA fingerprint pattern. With the help of residues of arginine, NADPH binds these domains. These remnants of arginine are conserved greatly.
- Conversion of FAD into $FADH_2$ occurs. Then the reaction of $FADH_2$ takes place by disulphide.
- The reaction occurs between GSSG and reduced enzyme forms.
- Half glutathione portion forms bonds with histidine group (-SH) and also forms a bond with GSS enzyme.
- The bond between glutathione and histidine group produces GSH molecule and after that SH-FAD enzyme complex reacts with GSS enzyme to generate 2nd GSH which finally outcomes in production of oxidized enzyme form.

Genes of Glutathione Reductase

GR in higher plants encoded with more than 1 gene. In pea, tobacco and *A. thaliana*, 2 genes are reported. In *P. trichocarpa* and rice, 3 genes were reported. Native plant GR molecular mass range is 60 to 190 kd. Molecular GR mass of Chlamydomonas is 56kd. Rouhier et al., in 2006 divided GR into 2 classes depending on N terminal extensions. They are GR-1 and GR-2

GR ISOFORMS

Recent studies show the gaining of interest to grasp modifications in glutathione reductase isoform levels than the modification of the total activities of GR in plants which are under stress. Activities of GR increased during ozone-related stress conditions in pea but there was no change in protein and mRNA contents. Isoforms of GR identified in spinach, tobacco, mustard, *Chlamydomonas reinharditi*, wheat and pea. By utilizing gels, chromatography ion exchanges and focused isoelectric, GR isoforms separated. In red spruce during freezing conditions, 2 isoforms were identified, revealing proofs that there is the existence of more than 1 gene. Although, all isoforms are not encoded by the various genes. In pea, 2 genes encode 8 isoforms. In plants, the sequences of GR comprises an extension of c-terminal which might be accountable for various isoform appearances of glutathione reductase in plants during stress and other 2 reasons may be post translation and post transcription changes.

GR Role in Alleviating Stress-Induced Toxins

GR Role During Drought

Under drought, stomata close then CO_2 availability for assimilation of carbon reduced then outcomes in the unavailability of acceptors of electrons causing free radical formation. Further in enzymatic antioxidants already mentioned a raise in activities of GR under the closure of stomata in responsive to drought has identified. Thus, drought stress-tolerant species would not only capable to keep adequate water but it in holds also an effective antioxidant enzyme system which saves from oxidative injuries and stress-induced toxins. As the component of this mechanism, GR has a major role and shows an increase in its activities under the stress. This increase in activities of GR was identified in many plants such as Maize, Barley, Wheat, Pea, Tobacco and Rice.

GR Role During Salinity

Salinity is the main abiotic factor of stress which severely affects the productivity of the crop. Salinity outcomes osmotic restrictions, ion toxicities which shows the impact on biochemical as well as physiological functioning of the plant. Salinity shows various symptoms which were similar to symptoms of drought. Raise of activities of GR identified in cantaloupe, pea, citrus, wheat, tomato, chickpea, soybean, green gram, sunflower and rice.

GR Role During Chilling Stress

Chilling damage is physiological entropy usually observed in the crops belonging to subtropical areas as well as temperate areas. Chilling stress decreases activities of the enzymes involved in the calvin cycle by interrupting groups of sulphhydryl thus decreasing usage of energy absorbed for assimilation of carbon dioxide (Leegood 1985). Chilling stresses limits carbon dioxide metabolisms which cause insufficient supplies of acceptors of electrons leading to more decrease of PS reaction centre causing finally raise in ROS generation. Adaptation for chilling stress causes a rise in GSH levels as well as activities of GR. (Anderson et al.,1992). A positive significant correlation exists between the chilling triggered photoinhibition resistances and higher activities of GR observed in seedlings of the french bean, tomato, maize and rice.

GR Role During Heavy Metal Stress

Minings, industries, burning of fossil fuel has infected many land areas in the developed countries by more heavy metal amounts leading to pollution. Unlike the organic pollutants, heavy metals are not degraded chemically and are not bio disrupted with microorganisms also. Phytoremediation is the only best and cheap approach to the problem. In small and normal amounts some heavy metals might essential for the growth of the plant but excess levels cause toxic impacts on the growth of plant finally leads to plant's death. Many crops are affected by toxicities of Cr, Nickel, Zinc and some other. By regulating higher phytochelatin or GSH levels, plants can survive these toxicities of heavy metals. Heavy metals trigger free radicals and cause oxidative injuries. GR has a significant role in Cadmium triggered ROS detoxification through AsA-GSH cycle. Cadmium stops activities of various group's enzymes included in Calvin cycle, sugar and nitrogen metabolisms. Raise in activities of GR in roots under cadmium exposures was identified in Potato, Soybean, Sugarcane and Radish. Raise of activities of GR also observed in plants which exposed to another HM such as copper in mulberry and iron in rice GSH induces phytochelatin generation. Phytochelatin observed in the plants and mainly activates from HM stresses. Phytochelatins reacts with HM and finally causes the death of plants. Hence decreased GSH is needed for every reaction to segregate HM's and this decreased GSH contents regulated with GR which describes pertinent roles of GR for detoxification of heavy metals.

GR Role During Biotic Stress

Though the interactions of plants and pathogens were clearly described, negligible information available about AsA-GSH cycle's involvement in defence system versus biotic stresses in plants (Baker and Orlandi, 1995). Though pathogen triggered stresses occur outside the cell but including organelles such as mitochondria, chloroplast, peroxisomes and their enzymatic antioxidant systems would be targeted under pathogen production. Increase in activities of GR was found in tomato during stress which shows the significance of GR under biotic stress situations and detoxifications of GR. In 1998 Vanacker et al., reported enzymes of AsA-GSH cycle in the apoplast of leaf in barley and significance of GR in leaves of barley under biotic stress is more.

Conclusion

Various factors of environment like salt, drought, pests, temperatures affect growth and developmental activities in plants. When there is an excess of any one of those factors will cause stress in plants. Stress then triggers ROS generations. These generated ROS are toxic compounds. ROS damages lipids, DNA, proteins in the cells. ROS also leads to PUFAs peroxidations in membranes. The toxic impacts of ROS will be counteracted with antioxidant enzyme mechanisms like Catalase, Superoxide dismutase, Glutathione reductase, APX, GPX, Prxs, DHAR and MDAR. Generally, every cell's compartments possess these antioxidant enzymes which play a major role in detoxification of generated ROS. Plants which overexpress genes of enzymatic antioxidants were genetically engineered to achieve the increase of tolerance to stress by absolutely changing expressions of these antioxidant enzymes which scavenge generated ROS. Analytical studies could be done to know responses of plants to various environmental stresses and could be expressed together with the usage of mutants which has ROS removing capabilities. As scavenging systems were somewhat complicated, transgenic species usages explores the role of enzymatic antioxidants and indicated as a major significant opportunity. Along with improving the activities of enzymatic antioxidants, enhancing of metabolical activities included by ROS removing essential to be described. Further enzymatic antioxidant processes castellated by altering the enzymes. Various signal transfer processes behave independent and have important talks between them. Multiple types of genes which were influenced during various stresses shows there would not the single form of markers for tolerances against stress. Usage of external applications of chemicals such as glycine betaine and selenium and also applications of NO like molecules indicates enhanced upregulations in enzymatic antioxidant defence process and ameliorates stress-induced toxins. Overexpressions of enzymatic antioxidant combinations in the transgenic species shown synergistic impacts on plants against tolerance to the various stresses. Hence, there is a rise in attention to generating transgenic species which overexpresses genes linked with more than one enzymatic antioxidants for accomplishing tolerances for manifold stresses in the environment.

References

Adriano, S., Antonio, S., Maria, N. and Antonella, V. (2015). Ascorbate peroxidase and catalase activities and their genetic regulation in plants subjected to drought and salinity stresses. *Int.J .Mol.Sci., 16,* 13561-13578.

Ahmad, P., Abdul, C.J., Mohamed, A.S., Gowher, N. and Sharma, S. (2010). Roles of enzymatic and non-enzymatic antioxidants in plants during abiotic stress. *Biotechnology, 30*(3), 161-175.

Ahmed *et al.* (2006). Glutathione peroxidase like protein of synechocystis PCC 6803 confers tolerance to oxidative and environmental stresses in transgenic *Arabiodopsis. Physiology plantarum, 128,* 251-262.

Anderson, J.V. and Davis, D.G. (2004). Abiotic stress alters transcript profiles and activity of glutathione S-transferase, glutathione peroxidase, and glutathione reductase in euphorbia esula. *Physiol Plant, 120,* 421-433.

Apel, K., and Hirt, H. (2004). Reactive oxygen species: metabolism, oxidative stress, and signal transduction. *Annu. Rev. Plant Biol., 55,* 373-399.

Asada, K. (2006). Production and scavenging of reactive oxygen species in chloroplasts and their functions. *Plant Physiol., 141,* 391-396. DOI:10.1104/pp.106.082040.

Asada, K., Urano, M. and Takahashi, M. (1973). Subcellular location of superoxide dismutase in spinach leaves and preparation and properties of crystalline spinach superoxide dismutase. *Eur. J. Biochem., 36,* 257-266.

Bhatt, I., and Tripathi, B. N. (2011). Plant peroxiredoxins: catalytic mechanisms, functional significance and future perspectives. *Biotechnology advances, 29*(6), 850-859.

Bowler, C., Montagu, M. V., & Inze, D. (1992). Superoxide dismutase and stress tolerance. *Annual review of plant biology, 43*(1), 83-116.

Bridges, S. M., and Salin, M. L. (1981). Distribution of iron-containing superoxide dismutase in vascular plants. *Plant Physiology, 68*(2), 275-278.

Brigelius-Flohé, R. (1999). Tissue-specific functions of individual glutathione peroxidases. *Free Radical Biology and Medicine, 27*(9-10), 951-965.

Caverzan, A., Casassola, A., and Brammer, S.P. (2016). Reactive oxygen species and antioxidant enzymes involved in plant tolerance to stress. Shanker, A.K. & Shanker, C. *Abiotic and biotic stress in plants-Recent advances and future perspectives. Publisher InTech,* 463-480.

Dąbrowska, G., Kata, A., Goc, A., Szechyńska-Hebda, M., and Skrzypek, E. (2007). Characteristics of the plant ascorbate peroxidase family. *Acta Biologica Cracoviensia Series Botanica, 49*(1), 7-17.

Dalal, M., and Khanna-Chopra, R. (2001). Differential response of antioxidant enzymes in leaves of necrotic wheat hybrids and their parents. *Physiologia Plantarum, 111*(3), 297-304.

Das, K., and Roychoudhury, A. (2014). Reactive oxygen species (ROS) and response of antioxidants as ROS-scavengers during environmental stress in plants. *Frontiers in environmental science,* 2, 53.

Dietz, K. J. (2007). The dual function of plant peroxiredoxins in antioxidant defence and redox signaling. *Peroxiredoxin Systems,* 267-294

Dietz, K. J. (2011). Peroxiredoxins in plants and cyanobacteria. *Antioxidants & redox signaling, 15*(4), 1129-1159.

Foyer, C. H., and Shigeoka, S. (2011). Understanding oxidative stress and antioxidant functions to enhance photosynthesis. *Plant physiology, 155*(1), 93-100.

Ghamsari, L., Keyhani, E., and Golkhou, S. (2007). Kinetics properties of guaiacol peroxidase activity in Crocus sativus L. corm during rooting.

Gill, S. S., and Tuteja, N. (2010). Reactive oxygen species and antioxidant machinery in abiotic stress tolerance in crop plants. *Plant physiology and biochemistry, 48*(12), 909-930.

Gull, A., Lone, A. A., and Wani, N. U. I. (2019). Biotic and abiotic stresses in plants. *Abiotic and biotic stress in plants,* 1-19.

Gupta, A. S., Heinen, J. L., Holaday, A. S., Burke, J. J., and Allen, R. D. (1993). Increased resistance to oxidative stress in transgenic plants that overexpress chloroplastic Cu/Zn superoxide dismutase. *Proceedings of the National Academy of Sciences, 90*(4), 1629-1633.

Gupta, D. K., Palma, J. M., and Corpas, F. J. (Eds.). (2018). *Antioxidants and antioxidant enzymes in higher plants.* Berlin: Springer International Publishing.

Herbette, S., Brunel, N., Prensier, G., Julien, J. L., Drevet, J. R., and Roeckel-Drevet, P. (2004). Immunolocalization of a plant glutathione peroxidase-like protein. *Planta, 219*(5), 784-789.

Jung, B. G., Lee, K. O., Lee, S. S., Chi, Y. H., Jang, H. H., Kang, S. S., ... and Lee, S. Y. (2002). A Chinese cabbage cDNA with high sequence identity to phospholipid hydroperoxide glutathione peroxidases encodes a novel isoform of thioredoxin-dependent peroxidase. *Journal of Biological Chemistry, 277*(15), 12572-12578..

Kumar, A. A., Mishra, P., Kumari, K., and Panigrahi, K. C. (2012). *Environmental stress influencing plant development and flowering. Front Biosci, 4,* 1315-1324.

Margis, R., Dunand, C., Teixeira, F. K., and Margis-Pinheiro, M. (2008). Glutathione peroxidase family–an evolutionary overview. *The FEBS journal, 275*(15), 3959-3970.

Mittler, R., and Zilinskas, B. A. (1994). Regulation of pea cytosolic ascorbate peroxidase and other antioxidant enzymes during the progression of drought stress and following recovery from drought. *The Plant Journal, 5*(3), 397-405. http://dx.doi.org/10.1111/j.165-313x.1994.00397.x

Milla, M. A. R., Maurer, A., Huete, A. R., and Gustafson, J. P. (2003). Glutathione peroxidase genes in Arabidopsis are ubiquitous and regulated by abiotic stresses through diverse signaling pathways. *The Plant Journal, 36*(5), 602-615.

Hasanuzzaman, M., Hossain, M. A., da Silva, J. A. T., and Fujita, M. (2012). Plant response and tolerance to abiotic oxidative stress: antioxidant defense is a key factor. In *Crop stress and its management: perspectives and strategies* (pp. 261-315). Springer, Dordrecht.

Mittova, V., Volokita, M., Guy, M., and Tal, M. (2000). Activities of SOD and the ascorbate-glutathione cycle enzymes in subcellular compartments in leaves and roots of the cultivated tomato and its wild salt-tolerant relative Lycopersicon pennellii. *Physiologia plantarum, 110*(1), 42-51.

Miyake, C., Schreiber, U., Hormann, H., Sano, S., and Kozi, A. (1998). The FAD-enzyme monodehydroascorbate radical reductase mediates photoproduction of superoxide radicals in spinach thylakoid membranes. *Plant and cell physiology, 39*(8), 821-829.

Moloi, M. J., Mwenye, O. J., and Van der Merwe, R. (2016). Differential involvement of ascorbate and guaiacol peroxidases in soybean drought resistance. *South African Journal of Science, 112*(9-10), 1-4.

Noshi, M., Yamada, H., Hatanaka, R., Tanabe, N., Tamoi, M., and Shigeoka, S. (2017). Arabidopsis dehydroascorbate reductase 1 and 2 modulate redox states of ascorbate-glutathione cycle in the cytosol in response to photooxidative stress. *Bioscience, biotechnology, and biochemistry, 81*(3), 523-533.

Örvar, B. L., and Ellis, B. E. (1997). Transgenic tobacco plants expressing antisense RNA for cytosolic ascorbate peroxidase show increased susceptibility to ozone injury. *The Plant Journal, 11*(6), 1297-1305.

Ohya, T., Morimura, Y., Saji, H., Mihara, T., and Ikawa, T. (1997). Purification and characterization of ascorbate peroxidase in roots of Japanese radish. *Plant Science, 125*(2), 137-145.

Ozyigit, I. I., Filiz, E., Vatansever, R., Kurtoglu, K. Y., Koc, I., Öztürk, M. X., and Anjum, N. A. (2016). Identification and comparative analysis of H2O2-scavenging enzymes (ascorbate peroxidase and glutathione peroxidase) in selected plants employing bioinformatics approaches. *Frontiers in plant science, 7*, 301.

Passardi, F., Longet, D., Penel, C., and Dunand, C. (2004). The class III peroxidase multigenic family in rice and its evolution in land plants. *Phytochemistry, 65*(13), 1879-1893.

Scandalios, J. G. (1993). Oxygen stress and superoxide dismutases. *Plant physiology, 101*(1), 7.

Scandalios, J. G., Guan, L., and Polidoros, A.N. (1997). Catalases in plants: gene structure, properties, regulation, and expression. *Cold Spring Harbor Monograph Series, 34*, 343-406.

Schultes, N. P., Zelitch, I., McGonigle, B., and Nelson, T. (1994). The primary leaf catalase gene from Nicotiana tabacum and Nicotiana sylvestris. *Plant physiology, 106*(1), 399.

Shilpi, G., Shilpi, S., and Sunita, S. (2015). Tolerance against heavy metal toxicity in cyanobacteria: role of antioxidant defense system. *International journal of pharmacy and pharmaceutical sciences, 7*(2), 1-8.

Suzuki, N., and Mittler, R. (2006). Reactive oxygen species and temperature stresses: a delicate balance between signaling and destruction. *Physiologia plantarum, 126*(1), 45-51.

Teixeira, F. K., Menezes-Benavente, L., Margis, R., and Margis-Pinheiro, M. (2004). Analysis of the molecular evolutionary history of the ascorbate peroxidase gene family: inferences from the rice genome. *Journal of Molecular Evolution, 59*(6), 761-770.

Tsang, E. W., Bowler, C., Hérouart, D., Van Camp, W., Villarroel, R., Genetello, C., ... and Inzé, D. (1991). Differential regulation of superoxide dismutases in plants exposed to environmental stress. *The Plant Cell, 3*(8), 783-792.

Vianello, A., Zancani, M., Nagy, G., and Macrì, F. (1997). Guaiacol peroxidase associated to soybean root plasma membranes oxidizes ascorbate. *Journal of Plant Physiology, 150*(5), 573-577.

Wood, Z. A., Poole, L. B., and Karplus, P. A. (2003). Peroxiredoxin evolution and the regulation of hydrogen peroxide signaling. *Science, 300*(5619), 650-653.

Young, L.C.T., and Conn, E.E. (1956). The reduction and oxidation of glutathione by plant mitochondria. *Plant physiology, 31*(3), 205.

Zhang, X., Zhang, L., Dong, F., Gao, J., Galbraith, D. W., and Song, C. P. (2001). Hydrogen peroxide is involved in abscisic acid-induced stomatal closure in *Vicia faba*. *Plant physiology, 126*(4), 1438-1448.

Boron Based Plant Responses Cultivated under Acidic Soil

Pankaj Kumar Mishra[1], Surendra Singh[2]

[1]*Incharge, District Soil Testing Laboratory, Krishi Bhavan, Mangurahi, Roberts Ganj Sonebhadra, U.P., 231216, India*

[2]*Department of Soil Science and Agril. Chemistry Institute of Agricultural Sciences Banaras Hindu University, Varanasi, 221005*

**Email: MISHRAIAS@gmail.com*

Abstract

The new emphasis on current research is to fill real and pressing needs for answers to grand challenges in food, health, energy, climate, and national security. *Food is the basic need of human life. A human or any living thing can live without any luxury of life but cannot survive without food. Hence,* there is widespread consensus that going forward, farmers must produce more food per unit of land, water, and agrochemicals. To do so, however, they simply cannot continue producing in the same way. They will have to do this while facing climate change, volatility, shifting nutrition needs, and the increasing scarcity of most of the physical factors of production. Agriculture is at the threshold of a necessary paradigm shift.

Introduction

Micronutrients deficiencies have emerged as one of the major constraints in successful crop production in Indian soils. During the last three decades, micronutrient deficiencies have grown in both magnitude and extent due to increased use of high analysis fertilizers, high yielding crop varieties and increase in cropping intensity. This situation has aggravated a major threads in the reduction of productivity of cereal, oilseeds and pulses. Thus, there is an urgent need for correction of individual nutrient deficiency for arresting its further spread. Availability of micronutrients is closely related to Physico-chemical nature of different soils type and plays an important role in sustaining higher crop productivity and soil health. Boron is one of the essential micronutrients for plant growth. Its deficiency is much more common in crops that are grown in soils containing a higher amount of free carbonates, low organic matter, and high pH (Lindsay, 1991 and Rashid, 1996). Besides, abundant soil moisture also causes

Bleaching, beyond the root zone (Keren and Bingham, 1985). On the other hand, the dry surfaces of soil layers also inhibit root absorption of boron (Shorrocks, 1997). Fertilizer use in the country predominantly pertains to Nitrogen (N) and Phosphorus (P). Potassium (K) use is confined to crops and Boron use is negligible. Many introduced crop varieties are more susceptible to boron deficiency (Chaudhry *et al.*, 1977; Rashid and Din, 1992). Hence, soil conditions and agronomic practices are conducive to the incidence of Boron deficiency in plants.

Field research has found that suitable balancing uses of Boron on several crops, such as wheat, rice, maize, groundnuts, tobacco, and potato enhance the yield (Shaphiq, 2008). In all the mineral nutrients, the role of boron in plant nutrition is still the least understood. The studies have limited to explore the effect of resupplied of boron after its deficiency in crops. This poor knowledge is surprising, because on a molar basis the requirement for boron, at-least for dicotyledons, is higher than that for any other micronutrient (Shelp, 1992). In several parts of the world, cereals such as maize, sorghum, wheat, rice and barley, which are less sensitive to B deficiency than dicotyledons, are found to be strongly affected by Boron deficiency (Rashid *et al.*, 2004). Boron neither is an enzyme constituent nor is their convincing evidence that it directly affects enzyme activities. There is a long list of postulated roles of boron (Parr and Loughman, 1983) such as: (a) sugar transport; (b) cell wall synthesis; (c) lignification; (d) cell wall structure; (e) carbohydrate metabolism; (f) RNA metabolism; (g) respiration; (h) indole acetic acid (IAA); metabolism; (i) phenol metabolism; (j) membranes.

Tourmaline, complex borosilicate, is the main B-containing mineral found in most of the soils. Release of boron from this mineral is quite slow. The concentration of Boron varies in soils of arid climates, where borates of alkali and alkaline earth elements predominate. The total concentration of Boron in soils ranges is found to vary from 20-200 mg kg^{-1} (Mengal and Kirby, 1987). Generally, the available Boron (hot water-soluble) fraction of soil ranged from 0.4-5.0 mg kg^{-1} (Gupta, 1979). Quantity of Boron decreases with increasing the pH of the soil, thus it is often inadequate available in calcareous soils. It is absorbed by plants and correlates well with the level of hot water soluble and HCl extractable soil of Boron (Rashid *et al.* 1994). Boron fertilization methods and rates should be carefully defined because the small amounts are needed for its correction to avoid over-application and possible toxicity in the soil. Wide application of Boron should be followed by incorporation to the full cultivation depth before seedbed preparation. Surface broadcast application on hay crops is an effective application technique (Will Peacock and Peter Christensen, 2005).

The adequate management of Boron in the soil-plant system is usually difficult because the range between B deficiency and toxicity is relatively narrow. Thus, knowledge of the chemical behaviour of B in the soil is particularly important. To recommend, the correct amount of Boron, it present in the soil and needs to the plants should be determined. The availability of it depends upon adsorption-desorption processes, which are influenced by various physicochemical properties of soils (Arora and Chahal, 2005). The extent of it adsorption in soils depends on solution pH, soil texture and mineral composition. The soil pH has been reported as the main factor affecting the B adsorption in the soil (Saltali *et al.*, 2005; Soares *et al.*,

2008). It is found that Boron adsorption increases with increasing pH and reaches a maximum around pH 9.0 and decreases with further increase in pH (Goldberg, 1997). The maximum development of adsorption sites has found to occur at a pH equivalent to the dissociation constant (pKa) of boric acid, approximately 9.2. At below pH 7.0, $B(OH)_3^0$ predominated, the amount of adsorption is small because the affinity of the clay from this species is relatively low. As the pH increased, the $B(OH)_4^-$ concentration increased rapidly. The amount of adsorbed B increased rapidly because of the relatively strong affinity of the clays. Further, increase in pH resulted in an enhanced OH- concentration relative $B(OH)_4^-$, and B adsorption decreased rapidly due to the competition of OH^- for the adsorption sites (Goldberg *et al.*, 2005).

The range of soil solution B concentration between plant deficiency and toxicity is narrow. Plant B deficiency is observed in areas of plentiful rainfall, especially on sandy soils. Boron toxicity symptoms are often observed in arid areas due to high levels of B in soil solution and use of irrigation waters has high B content. Both B deficiency and toxicity conditions can lead to marked yield reductions of crop plants and resulting economic losses. Soil solution B concentrations equilibrate with B adsorbed into various organic and mineral surfaces (Goldberg, 1993). Adsorption sites on organic matter, oxide minerals, clay minerals and carbonates act as sources and sink for B. Adsorbed B is neither directly available nor toxic to plants (Keren *et al.*, 1985). Thus, the adsorption complex plays a critical role in controlling B availability in soils. Despite being categorized as less sensitive to B deficiency (Rerkasem and Jamjod, 1997), both rice and wheat crops are suffering from this nutritional disorder in the worldwide (Chaudhry *et al.*, 1977; Rashid *et al.*, 2002, 2004, Shorrocks, 1997; Rerkasem and Jamjod, 2004). Although B fertilization is a simple and cost-effective solution to the problem, however, practical constraints are prohibitive in its adoption by farmers cultivating in low B soil status. Therefore, B fertilizer has used in cereals is in infancy, especially in developing countries. Consequently, these staple food crops for the majority of the world population are suffering yield losses when grown in low B soils.

Pulses are the second most important group of crops after cereals. In 2009, the global pulses production was 61.5 million tons from an area of 70.6 million ha with an average yield of 871 kg ha^{-1}. Dry beans contributed about 32% to global pulses production followed by peas (17%), chickpea (15.9%), broad beans (7.5%), lentils (5.7%), cowpeas (6%) and pigeon pea (4.0%). The latest triennium (2007-09) average of 61.2 million tons of production shows a positive annual growth of 0.7% per annum over 55.03 million tons recorded in 1997. Comparative data for the eighties reveal that phenomenal annual growth of 2.85% mainly attributed to positive growth of 0.87% in area and 1.83% in productivity. Developing countries contribute about 74% to the global pulses production and the remaining comes from developed countries. India, China, Brazil, Canada, Myanmar and Australia are the major pulse producing countries with a relative share of 25, 10, 5, 5 and 4 per cent, respectively. Countries recording annual production growth of more than 4% are Myanmar (11.48%), Canada (10.80%), Germany (8.27%), Sudan (8.08%), Spain (7.37%), Ethiopia (4.92%), China (4.67%) and Syria (4.12%). The major producers of pulses in the country are Madhya Pradesh (24%), Uttar Pradesh

(16%), Maharashtra (14%), Rajasthan (6%), Andhra Pradesh (10%) followed by Karnataka (7%) which together share about 77% of total pulse production while remaining 23% is contributed by Gujarat, Chhattisgarh, Bihar, Orissa and Jharkhand (Anonymous, 2011).

The importance of pulses are very much relevant for food and improving the farm family income to ensure food security, nutritional security and economic security. According to an estimate by the Indian Institute of Pulse Research, Kanpur the pulse requirement for the year 2030 is 32.00 million tons will be needed with an anticipated growth rate of 4.2% whereas the current production remained 14.66 million tones and leaving a huge gap of 17.34 million tones. In India, pulses occupy a prominent position and accounted for 35.2% of world pulses areal production. The neglect of pulses in the cropping system has not only lead to soil sickness and break down in the sustainability of production system but also lead to the soil to progressive decline in the per capita availability of pulses from 60.0g per person in 1951 to 34 g today (Nandakumar *et al.*, 2010). Pulses are the primary and cheapest source of vegetable proteins for human and animal nutrition. A major part of pulse cultivation is concentrated in the semi-arid region of the country.

Pulse crops meet a major portion of their nitrogen requirement through biological fixation of atmospheric nitrogen. Seed treatment with a crop-specific strain of *Rhizobium* bacteria promotes N-fixation and improves crop yields by 10 to 20 %. Use of *Rhizobium* inoculants in pulse crops is an integral part of an improved package of practices for pulses. It is also important for the sustainability of our agricultural production system through biological nitrogen fixation process in soil and nutritional security of the weaker section of the society, especially the children and lactating woman. Among the pulses, mung bean is an important grain legume with its peculiarity to growth at all the three seasons in a year. It contains 11.0 per cent water, 25.0 per cent protein, 1.3 per cent fat, 60.4 per cent carbohydrate, 0.145 per cent, Ca, 0.078 per cent Fe. Mung bean enrich soils in terms of enhancing of nitrogen & organic matter content through biological means of Nitrogen fixation and addition of crop residue, respectively and ultimately this lead to maintain higher soil fertility land. Mung bean (*Vigna radiata* L) is an important legume of Asian origin, now widely cultivated throughout Asia, Australia, New Zealand and Africa. It is a valuable green manure crop for improving soil fertility, and its sprouts and starch are also popular health foods.

The total area under pulse cultivation is 23.63 million tones with an annual production of 14.76 million tones in the country. Mung bean occupies 3.44 million hectares and contributed to1.4 million tones in pulses production (Anonymous, 2011). The important mug bean growing states are Orissa, Maharastra, Andhra Pradesh, Tamil Nadu, Utter Pradesh, Madhya Pradesh, Rajasthan and Bihar. In eastern Uttar Pradesh, the productivity of pulses in Sonabhadra and Mirzapur districts is very low due to poor farming of upland soils, and deficiencies of nutrient in soils. Pulse yield is strongly influenced by the application of the nutrients. The magnitude of yield losses is due to nutrient deficiency and varies among the nutrients (Ali *et al.*, 2002). Since, micronutrient availability for the plant depends, among other factors, soil texture, organic matter and soil pH. The response to added fertilizers application varied with soil type, soil moisture and management practices, besides the seasons in different areas particularly in rainfed agro-ecosystem. The major causes of Boron deficiency of in Indian soils are coarse-textured, low pH and low organic matter content. In Boron deficient soil, both yield and quality of pulses are poor. The inclusion of this nutrient is essential in the fertilization scheduled. Thus Boron has become a key nutrient to increase productivity in B deficient light-textured soil. Neglecting the B nutrient, will lead to a low yield, inferior crop quality and reduce the efficiency of plant nutrient. The boron requirement of the pulse is fairly high in comparison to oilseeds, cereal and millet crops (Gupta, 1985).

Red & lateritic soils comprise an area of about 70 million hectares covering around 28% of the total land area of the country. *Vindhyan* region is located in eastern Uttar Pradesh and spread over Mirzapur, Sonbhadra, and some parts of Allahabad. These soils are accounted for 5.1% of the total area of Uttar Pradesh. The climate of *Vindhyan* hills is subtropical. Mean annual rainfall of this region varies between 1000 to 1200 mm, most of which is received during the monsoon (July to September). These soils are predominantly low crop productive mainly owing to non-availability of almost all major nutrient and few micronutrients (B and Mo). Boron deficiency in *Vindhyan* soils is being increased due to light textured soil, low organic matter content and continuous nutrient mining by pulses. The entire area of *Vindhyan* region is rainfed and about more than 50 per cent, the land is suitable for oilseeds and pulses cultivation in upland topography. Generally, farmers do not apply fertilizers in pulses due to lack of knowledge, as result productivity of the pulses in this region is quite low. At present, little scientific information is available on the distribution of boron in surface and subsurface reds soils of *Vindhyan* region. The response of boron application to pulses in red soils of Mirzapur region is also lacking. Boron is involved in numerous important processes, including protein synthesis, transport of sugars, respiration, RNA and carbohydrate metabolism, and the metabolism of plant hormones (Indole Acetic Acid). Moreover, functions of boron are related to cell wall synthesis, lignifications, and cell wall structure by cross-linking of cell wall polysaccharides as well as the structural integrity of biomembranes. It increases the transport of chlorine and phosphorus as a result of plasmalemma ATPase induction. Other investigations have shown that boron can stimulate proton pumping that causes hyperpolarization of the membrane potential. More than 90% of the boron in plants is found in cell walls, and rhamnogalacturonan II was shown to bind boron. Because the wall-associated kinase in the plasma membrane has an extracellular matrix connection with the pectin molecule, the membrane cell wall connection is finally also boron-dependent. Boron was found to promote the structural integrity of biomembranes and the formation of lipid rafts. Since all these functions are fundamental to meristematic tissues, boron deficiency is predominantly amazing actively growing organs such as shoot and root tips so that the whole plant may be stunted (rosetting). Flower retention, pollen formation, pollen tube growth or germination, nitrogen fixation, and nitrate assimilation are also affected by boron.

Boron (B) deficiency is a widespread problem for field crop production where large losses of yield occur annually both quantitatively and qualitatively (Dell and Malajczuk, 1994; Hopmans and Flinn, 1984; Ram *et al.*, 1989; Shorrocks, 1997; Wei *et al.*, 1998). Although reports from agricultural practice have well established that adequate B supply is imperative for obtaining high yields and good quality and increasing evidence suggests a metabolic function or at least beneficial effects of B in animal metabolism (Hunt, 2003), knowledge about metabolic functions of B in plants remains incomplete. Recent research findings have greatly improved our understanding for B uptake and transport processes (Brown and Shelp, 1997; Brown *et al.*, 2002; Takano *et al.*, 2002, 2005a, b, 2006), and roles of B in cell-wall formation (Matoh, 1997; O'Neill *et al.*, 2004), cellular membrane functions (Goldbach *et al.*, 2001), and anti-oxidative defence systems have been suggested (Cakmak and Römheld, 1997). Boron (B) is often considered the most difficult essential nutrient to manage in traditional agronomic situations because of the narrow range between deficient and toxic conditions in soils (Ponnamperuma, *et al.*,1981). On B deficient soils, positive responses to B fertilizer applications have been reported in more than 80 countries on 132 crops over the past 70 years (Shorrocks 1997). The literature about the present investigation has been reviewed and presented under the following heads:

- Soil available status of Boron
- Availability of Boron about soil properties.
- Boron adsorption in soils
- Factor & affecting boron adsorption in Soils
- The response of boron application to cereals
- The response of boron application to vegetable
- The response of boron application to oilseed
- The response of boron application and *Rhizobium* inoculation to legume/ pulses
- Soil fertility after post-harvest pulses

Status of Soil Available Status of Boron

Numerous methods have been developed to determine available boron in soils. Berger and Truog (1939) reported that extraction of the soil available B with boiling water for 5 minutes was formed to be the best method of estimation of available B in soils. Despite its capacity to estimate plant-available B, several workers have noted that 5 minutes hot water extraction procedure was not convenient for routine analysis and have suggested various adoptions (Gupta, 1967; John, 1973). Boron deficiency has emerged as an important micronutrient problem in Indian soils and crops next to zinc. Survey studies revealed that deficiency of boron ranges from 1-84% with a mean of 33% in different states of India (Singh, 2006). Several workers have estimated the hot water extractable B in Indian soils (Kanwar and Randhwa, 1974) but practically little information is available on the content of extractable B determined by other chemical extractants. The extent of B deficiency found in soils of eastern Uttar Pradesh (Annual Report, 2012). A survey was conducted by Srinivasarao *et al.* (2008), in Madhya Pradesh for the characterization of soil resources in the benchmark watersheds (Semi and Shyampura Watersheds, Dewas district and Madhusudangadh watershed, Guna district in Madhya Pradesh and

Bundi Watershed, Bundi district in Rajasthan) as an entry point activity. Stratified sampling of fields on a toposequence in the watersheds revealed a widespread deficiency of multiple nutrients particularly Zn (67-100 %), B (72-100 %) with an exception of Guna and S (72-100 %) along with N and P which could be limiting the yield potential in the watersheds. Fanners' participatory research trials with different crops showed tremendous yield advantages of up to 79 % over the farmer's practice.

Bendale *et al.* (1951) reported that the normal cotton soils of old Bombay contained 0.10 to 0.60 mg kg^{-1} hot water extractable B. Gandhi and Mehta(1958) found that normal soils of Gujarat contained 0.20 to 1.50 mg kg^{-1} hot water extractable B. Grewal *et al.* (1969) collected soil samples from different agro-climatic zones of Punjab, Haryana and Himachal Pradesh and found that hot water extractable B in these soils varied from 0.56 to 5.30 mg kg^{-1}. Nathani *et al.* (1970) reported that hot water extractable B in some soils of Rajasthan ranged from 0.42 to 8.2 mg kg^{-1}, with an average value of 3.22 mg kg^{-1}. Balasundram *et al.* (1972) analysed different types of soils from Tamil Nadu and reported that hot water extractable B ranged from 0.61 to 2.51 mg kg^{-1} in these soils with mean values of 0.64, 1.03, 0.5 and 0.61 mg kg^{-1}, respectively in alluvial, black, laterite and red soils. Anitha *et al.* (2013) reported found that in different fractions of boron, residual boron contributed around 98 per cent and ranged from 125 in the central dry zone to 567 mg kg^{-1} in northern dry zone followed by oxide bound boron that ranged from 2.15 in the coastal zone to 8.21 mg kg^{-1}in eastern dry zone and the least amount was readily soluble boron that ranged from 0.03 mg kg^{-1} in the southern transition zone to 0.78 mg kg^{-1} in the northern dry zone. Total boron was in the highest amount in the northern dry zone (582.3 mg kg^{-1}). Residual boron has minimum effect on available boron, whereas readily soluble boron, specifically adsorbed boron, oxide bound boron and organically bound boron contributes towards boron nutrition in all the agro-climatic zones. Relationships of individual B fractions with physical and chemical properties of the experimental soils confirmed the general validity of the proposed fractionation scheme, which is satisfactory in terms of sensitivity and position for soil boron.

The hot water extractable B in some soil of Delhi was found (Gajbhiye *et al.*,1980) to vary from 0.43 to 1.29 mg kg^{-1} in non-saline soils. Chakraborty *et al.* (1982) reported that hot water extractable B varied from 0.02 to 1.43 mg kg^{-1} in some acid soils of Assam. In some of the apple orchards of Himachal Pradesh hot water extractable B was found (Bhandari and Randhawa, 1985) to range from 0.1 to 1.0 mg kg^{-1} with a mean value of 0.5 mg kg^{-1}. Datta and Rathore (1992) reported that hot water extractable B content of some acid soils of Madhya Pradesh varied from 0.06 to 3.47 mg kg^{-1} with an average value of 0.47 mg kg^{-1}. There are few studies on available B status of soils of West Bengal. Hot water extractable B in some soil of Murshidabad (WB) was reported that about 80 per cent of the soil sample collected from the Northern West Bengal were deficient in hot calcium chloride extractable B. In some acid soils of Northern West Bengal tartaric acid, ammonium acetate, mannitol - calcium chloride and hot extractable B were found to vary from 0.15 to 1.34, 0.07 to 0.47, 0.15 to 0.49 and 0.07 to 0.50 mg kg^{-1}, respectively with the corresponding mean values of 0.35, 0.31, 0.28 and 0.27 mg kg^{-1} (Mandal and De, 1993).

Gestring and Soltanpour (1984) reported that extractable B varied from 0.03 to 0.31, 0.02 and 0.22 and 0.08 to 0.47 mg kg^{-1} when extracted by ammonium bicarbonate – DTPA, saturation and hot water, respectively in some Colorado soils. Carwright *et al.* (1983) reported that the mean values of extractable B were 1.28, 0.58, 0.47, 0.45, 0.85, 0.15 and 0.56 mg kg^{-1} with hot water, 0.01 M calcium chloride – 0.05 M mannitol (pH 8.5), 0.01 M calcium chloride – 0.01 M mannitol, 0.01 calcium chloride (pH 8.5), 0.5 M sodium bicarbonate (pH 4.8), respectively. Aitken *et al.* (1987) found that B extracted by hot water, hot 0.01 M tartaric acid and 5 percent glycerol – 0.01 M calcium chloride, 0.05 M mannitol (buffered at pH 7.5), 1 M ammonium acetate (pH 4.8) tartaric acid and 5 percent glycerol – 0.1 M calcium chloride, varied from 0.12 to 0.93, 0.12 to 0.09 mg kg^{-1}, respectively in some acid soils. Similarly, the amount of B extracted from some acid of New Zealand by hot water, hot 0.02 M calcium chloride,0.01 M calcium chloride- 0.05 M mannitol and B-specificc resin, were found (Adam *et al.*, 1991) to vary from 0.32 to 2.67, 0.30 0.16 to 0.52 and 0.32 to 0.64 mg kg^{-1}, respectively.

According to Chude (1988), the profile distribution of total and extractable B was determined in 16 Nigerian cacao-growing soil profiles formed from different parent materials. Total B for all soils ranged from 8 to 54 $\mu g g^{-1}$ with a mean of 24 $\mu g g^{-1}$. The soils formed from sandstones in the rainforest zone contained higher amounts of total B than soils derived from basement complex.

Boron Availability of Boron about Soil Properties

Several soil properties such as texture, soil pH, Organic matter; cation exchange capacity and oxides of iron and aluminium are known to alter the availability of B, particularly in acid soils. Soil texture appears to be the most important factor, governing the B availability in soils. In general, sandy soils contain less available B than heavy soils (Kubota *et al.*, 1948). It is reported that available B content in soils increased by 20 to 25 per cent with the increase in fine particles content from 4 to37 per cent (Kluge and Podlesak, 1981). The positive relationship of hot water extractable B with clay plus silt content was also obtained by Singh and Sinha (1987) but they failed to establish a relationship with clay content alone. Whereas, Datta and Rathore (1992) reported of both of these soil. Properties (clay and silt) on hot water extractable B in some acid soils. On the contrary, Chakraborty *et al.*, (1982) did not find any correlation between hot water extractable B and soil texture. The influence of soil pH on the availability of B was studied by Truog (1948) and found that the availability of B increased up to pH 5.0 and it remained constant up to pH 7.0; above this pH range, it became less available. In some acid of Armorican Massif, hot water extractable B was found to be positively correlated with soil pH (Maurice, 1977). However, in some investigations, a significant relationship between hot water extractable B and soil pH was not obtained Chakraborty *et al.*, 1982; Shuman *et al.*, 1992). Soil organic matter plays an important role in B availability; it can minimize Bleaching from acid soils and maintain B in a relatively available for (Page and Paden, 1954; Okazaki and Chao, 1968). It was found that the availability of B in acid soils increased with an increase in organic matter content (Berger and Troug, 1945; Ghani and Haque, 1945; Bhandari and Randhwa, 1985).

On the other hand, several researchers in organic matter content relationship between organic matter and hot water extractable B (Datta and Rathore, 1992). A negative relationship between hot water extractable B and organic matter has also been reported by kushwaha and Singh (1981). Apart from these soil properties, in some acid soils, iron oxide was found to be negatively correlated with native extractable soil B (Mandal and De, 1993). However, aluminium hydroxide was not related significantly to the content of extractable B in soils

Boron Adsorption in Soils

When B is released by minerals or by organic matter during its mineralization, or when it is added to the soil by fertilization, a large part of it remains in solution, while another part is absorbed by soil particles (Gupta *et al.*, 1985). Boron can be absorbed by iron and aluminium oxides and their hydroxides or by several clay minerals (Keren & Gast, 1983; Goldberg & Glaubig, 1985). Bingham *et al.* (1971) found a positive correlation between the content of Al_2O_3 and the adsorption of B in four Mexican soils and six Hawaiian soils. The greater the amounts of amorphous oxides (extracted with ammonium oxalate), the greater the capacity for B adsorption (Sims & Bingham, 1968a; Goldberg & Glaubig, 1985). Soil organic constituents also play an important role in soil B adsorption. Available B is found associated mainly with organic matter and this explains the high available B contents in surface horizons (Evans & Sparks, 1983). Boron adsorption on the soil is affected by many factors.

Dey *et al.* (2013) Studies on adsorption of boron (B) undertaken in 12 soils representing different agro-ecological regions of India revealed that the adsorption of added B was in the order: Vertisols > Inceptisols > Alfisols. Higher the concentration of added B solution, greater was the variability in B adsorption. The soil within each order also differed in B adsorption. Of the three adsorption equations namely, Freundlich, Langmuir and Temkin fitted to the data, Freundlich and Langmuir equations explained the adsorption behaviour of the soils, yet based on prediction coefficients (R^2) Freundlich appeared the best fit for Vertisols and Inceptisols, and Langmuir for Alfisols which showed lowest B adsorption. Freundlich constant 'a' and Langmuir constant 'b' were positively correlated with free $CaCO_3$, CEC and clay content. Desorption of added B was in the order: Inceptisols > Vertisols > Alfisols. Desorption of B increased with an increase in soil pH, whereas it decreased with an increase in organic C, clay and CEC

Soil Factor & Affecting Boron Adsorption in Soils

Soil pH

Boron adsorption is highly influenced by pH, increases as the pH increases, attaining a maximum in the alkaline range and diminishes abruptly at high pH (Keren & Gast, 1983; Goldberg & Glaubig, 1985). The maximum development of adsorption sites occurs at a pH equivalent to the pK of boric acid, approximately 9.1 (Schalcha *et al.*, 1973). The agricultural practice that is most often used to raise pH is liming. It is expected that after liming, there will be an increase in soil B adsorption, especially in soils that contain high amounts of iron and aluminium oxides. Hatcher *et al.* (1967) showed that the adsorption of B increased two to five times after liming in various acidic North American soils, accompanied by a pH increase. In a Brazilian medium-textured Oxisol (Cruz *et al.*, 1987), lime application to increase the pH from 4.2 to 5.6 significantly increased (around 14%) the amount of adsorbed boron.

Minerals Oxide

The adsorption behaviour of B on Al and Fe oxide minerals (Sims and Bingham, 1968; McPhail *et al.*, 1972; Choi and Chen, 1979; Goldberg and Glaubig, 1985) as well as on layer silicate clay minerals (Sims and Bingham, 1967; Keren and Mezuman, 1981; Mattigod *et al.*, 1985; Goldberg and Glaubig, 1986) has been investigated previously. Boron adsorption on both types of minerals exhibited strong pH dependence. Adsorption peaks for the oxide minerals occurred in the pH range 7 to 9 (Goldberg and Glaubig, 1985). For the clay minerals, the range for the adsorption peaks was shifted upward to pH 8 to 10 (Goldberg and Glaubig, 1986). Boron adsorption on both crystalline and amorphous aluminium and iron oxides increased with increasing pH up to an adsorption maximum at pH 6 to 8 for aluminium oxides and pH 7 to 9 for iron oxide (Bloesch *et al.*, 1987; Choi and Chen, 1979; Goldberg and Glaubig, 1985; McPhail *et al.*, 1972; Metwally *et al.*,1974; Scharrer *et al.*, 1956; Sims and Bingham, 1968; Su and Suarez, 1995). Above the maximum, B adsorption decreased with increasing pH. Boron adsorption was greatest on freshly precipitated solids and decreased with ageing due to increasing crystallinity (Hatcher *et al.*, 1967; Sims and Bingham, 1968). Boron adsorption per gram was greater for aluminium than iron oxides (Goldberg and Glaubig, 1985; Sims and Bingham, 1968). This is likely due to the higher surface area of aluminium oxides since adsorption per square meter was similar in magnitude for aluminium and iron oxides (Goldberg and Glaubig, 1985). Boron adsorption on oxide minerals occurred rapidly, being virtually complete after one day of reaction time (Choi and Chen, 1979). Mechanism of B adsorption on aluminium and iron oxide minerals is considered to be ligand exchange with reactive surface hydroxyl groups (Goldberg *et al.*, 1993b; McPhail *et al.*, 1972). Ligand exchange with surface hydroxyl groups is a mechanism whereby anions become specifically adsorbed on mineral surfaces. This specific adsorption produces a shift in the point of zero charges (PZC) of the mineral to a more acid pH value. Boron adsorption occurs specifically since it produces a shift in PZC of aluminium and iron oxides (Alwitt, 1972; Beyrouty *et al.*, 1984; Blesa *et al.*, 1984; Fricke and Leonhardt, 1950; Goldberg *et al.*, 1993b; Su and Suarez, 1995). Specifically adsorbed ions are held in inner-sphere surface complexes that contain no water between the adsorbing ion and the surface functional group. Kinetic experiments using pressure jump relaxation indicated that B adsorbs as an inner-sphere surface complex on aluminium oxide via ligand exchange of borate with surface hydroxyl groups (Toner and Sparks, 1995). Fourier transform infrared spectroscopic analyses have shown that B is adsorbed via ligand exchange as both $B(OH)_3$ and $B(OH)_4$ species on amorphous aluminium and iron oxide (Su and Suarez, 1995).

Soil Texture

Adsorbed B is dependent on soil texture and increases with increasing clay content (Bhatnager *et al.*, 1979; Elrashidi and O'Connor, 1982; Mezuman and Keren, 1981; Wild and Mazaheri, 1979). Boron adsorption maxima increased with increasing clay content (Biggar and Fireman, 1960; Goldberg and Glaubig, 1986a; Nicholaichuk *et al.*, 1988; Singh, 1964). A highly significant correlation was found between soil content of the clay minerals kaolinite, montmorillonite, and chlorite and the B adsorption maximum. Hingston (1964), Keren *et al.* (1981) showed that the rate of B adsorption on clay minerals consisted of a fast adsorption reaction and a slow fixation reaction. Short term experiments have shown that B adsorption reaches equilibrium in less than one day. Long term experiments indicated that fixation of B increased even after six months of reaction time (Harder, 1961; Jasmund and Lindner, 1973). Boron adsorption by clay minerals is considered to be a two-step process. Initially, B adsorbs onto the particle edges, subsequently migrates, and incorporates structurally into tetrahedral sites replacing structural silicon and aluminium (Couch and Grim, 1968; Fleet, 1965; Harder, 1961). The mechanism of the B adsorption step is considered to be ligand exchange with surface hydroxyl groups on the clay particle edges (Couch and Grim, 1968; Goldberg *et al.*, 1993b; Keren and Sparks, 1994; Keren and Talpaz, 1984; Keren *et al.*, 1994). Boron deficiency may occur in heavy-textured soils with high pH because under these conditions boron is readily adsorbed. Boron toxicity may also occur in boron-rich soils or in soils that have been exposed to boron-contaminated irrigation waters or the excess application of boron rich fertilizers, sewage sludge, and fly ashes (Nable *et al.*, 1990).

Soil Moisture

Adsorbed B was independent of variations in soil moisture content from 50 to 100 per cent of field capacity in one study (Gupta 1968) and increased with decreasing soil water content in another (Mezuman and Keren, 1981). Wetting and drying cycles increased the amount of B fixation (Biggar and Fireman, 1960). The effect of drying became more pronounced with increasing additions of B (Biggar and Fireman, 1960). It is usually accepted that boron availability is decreased under dry soil conditions. Thus, boron deficiency is often associated with dry weather and low soil moisture conditions. This behaviour may be related to the restricted release of boron from organic complexes which ultimately impaired ability of plants to extract B from the soil due to lack of moisture in the rhizosphere. Even if B levels in the soil is high, then also low soil moisture impairs transport of B to absorbing root surfaces (Das 2007). The rainy season also affected the availability of Boron. Besides its effect on Soil B leaching rate, low-pressure deficits in the atmosphere during the rainy season slow down the transpiration rate this may inhibit uptake and translocation of B in the plant (Rawson,1996a,b).

Soil Temperature

Boron adsorption increases with increasing soil temperature. However, this may be due to an interactive effect of soil temperature with soil moisture since B deficiency is associated with dry summer conditions. Boron adsorption decreased as a function of temperature in the range of 10–40°C on soils dominant in crystalline minerals (Biggar and Fireman, 1960; Goldberg *et al.*, 1993a). In contrast, B adsorption increased slightly as the temperature of an amorphous soil was increased from 10 to 40°C (Bingham *et al.*, 1971). The effect of temperature on B adsorption by clays has been investigated (Couch and Grim, 1968; Goldberg *et al.*, 1993a; Harder, 1961; Jasmund and Lindner, 1973; Singh, 1971).In these clays (crystalline clay minerals) for a very short reaction time of two hours, B adsorption in the pH range of 5.5 – 9.5 decreased with increasing temperature (Goldberg *et al.*, 1993a). Also in clays B adsorption for longer reaction times of twelve hours to sixty days increased with increasing temperature (Couch and Grim, 1968; Jasmund and Lindner, 1973; Singh, 1971). These results suggest initial adsorption of B on crystalline clay minerals is exothermic while the subsequent B fixation reaction is endothermic.

Competing Ions

Competing ions such as silicate, sulfate, phosphate, and oxalate decreased the magnitude of B adsorption on oxides (Bloesch *et al.*, 1987; Choi & Chen, 1979). The effect of competing ions on B adsorption can be slight, in the case of sulfate, or substantial in the case of phosphate (Bloesch *et al.*, 1987; Metwally *et al.*, 1974). The ability to compete for anions to leach adsorbed B from oxides increased in the order: chloride < sulfate arsenate < phosphate (Metwally *et al.*, 1974).Significant silicate adsorption produced only a slight decrease in B adsorption by aluminium oxide suggesting that some B sorption sites show B preference (Goldberg and Glaubig, 1988).

Calcium Carbonates

Application of lime increases B fixation by soils because it raises the soil solution pH (Elseewi, 1974; Elseewi and Elmalky, 1979). In addition to its effect on soil pH, calcium carbonate also acts as an important B adsorbing surface in calcareous soils (Elseewi, 1974; Elseewi and Elmalky, 1979; Goldberg and Forster, 1991). Boron adsorption was greater on soils having higher calcium carbonate content (Elrashidi and O'Connor, 1982; Elseewi, 1974). Acid soluble B was highly significantly correlated with a calcium carbonate content of soils (Elseewi and Elmalky, 1979). Retention of B on calcium carbonate occurs via an adsorption mechanism. The mechanism could be the exchange with carbonate groups. The magnitudes of the B adsorption maxima for soil samples treated to remove calcium carbonate were statistically significantly lower than those for untreated soil samples indicating that calcium carbonate acts as an important sink for B adsorption in calcareous soils (Goldberg and Forster, 1991). Boron adsorption on reference calcites increased with increasing solution pH from 6 to 9, exhibited a maximum at pH 9.5, and decreased with increasing solution pH from 10 to 11 (Goldberg and Forster, 1993). The magnitude of B adsorption on clay minerals is affected by the exchangeable cations (Keren and Gast, 1981; Keren and Mezuman, 1981 and Sarkar, *et al.*, 2013). Calcium clays adsorbed more B than sodium and potassium clays (Keren and Gast,

1981; Keren and O'Connor, 1982; Mattigod *et al.*, 1985). An explanation for this result is that calcium 2:1 clays occur as tactoids consisting of several clay particles while the sodium forms exist in solution as single particles (Keren and Gast, 1981). In tactoids, the diffuse double layer and negative electric field from the planar surfaces are less extensive, making the edge sites of calcium clays more accessible to adsorbing borate anions than those of sodium clays (Keren and Gast, 1981). An alternative explanation for increased B adsorption in the presence of calcium is the formation and adsorption of the calcium borate ion pair (Mattigod *et al.*, 1985).

Soil Humus

Humus extracted from a soil retained significant amounts of B and was considered to play an important role in B adsorption (Parks and White, 1952). Soil organic matter adsorbs more B than mineral soil constituents on a weight basis (Gu and Lowe, 1990; Yermiyaho *et al.*,1988)). Boron adsorption on a mineral soil increased with increasing additions of composted organic matter (Yermiyahu *et al.*, 1995). The presence of organic material can also occlude B reactive adsorption sites on clays (Gu and Lowe, 1992) and soils (Harada and Tamai, 1968; Marzadori *et al.*, 1991). Boron adsorption on organic soil and composted organic matter increased with increasing pH (Yermiyaho *et al.*, 1988). Adsorption on a soil humic acid increased with increasing pH up to a maximum near pH 9 and decreased with increasing pH above 9 (Gu and Lowe, 1990). Boron adsorption on composted organic matter occurred rapidly, reaching equilibrium after three hours and increased with increasing solution ionic strength (Yermiyaho *et al.*, 1988).

The Response of Boron Application to Cereals

Although cereals are generally considered insensitive to boron (B) deficiency, incidences of B deficiency have been reported from many of the world's wheat-growing countries since the 1960s. The most extensive region of B deficiency in wheat so far reported in the adjoining area of eastern Nepal, northeastern India and northwestern Bangladesh, through to southwestern China (Ganguly, B., 1979, Rerkasem and Jamjod, 2004). Among the micronutrients, deficiency of B is the most frequently encountered in the field (Gupta, 1979). Grain set failure associated with male sterility was observed in wheat in Brazil in 1962 (da Silva and de Andrade, 1980). Diagnosis of B deficiency was confirmed by positive responses to B application. The same happened in Nepal in 1964, among introduced high yielding Mexican and Indian wheat germplasm (Misra *et al.*, 1992, Singh *et al.*, 1976; Ganguly, 1979; Sarkar and Chakraborty, 1980; Mandal and Das, 1988; Dwivedi *et al.*, 1990). Most of these were from the northeastern states of Bihar, Uttar Pradesh West Bengal, Orissa, Meghalaya and Assam. Boron deficiency was found to be the cause of almost complete crop failure in some 40,000 ha of wheat in Heilongjian Province in the north of China in 1972 and 1973 (Li *et al.*, 1978). According to Aref, 2011, the effect of Zn and B fertilizers on the concentration of Zn and B in the corn leaves two field experiments were conducted in a calcareous soil at Fars Province of Iran during 2009 to 2010. Treatments included five levels of Zn (0, 8, 16 and 24 kg haG1 Zn added to the soils and Zn foliar spray with a

0.5 per cent concentration) and four levels of B (0, 3 and 6 kg haG1 B and B foliar spray with a 0.3 per cent concentration) in a completely randomized block design. Results of this study clearly show that there was an antagonism between the Zn and B. No B content in the soil (zero and B spraying levels) helped increasing leaf Zn content, but the presence of B in the soils prevented from an increase of leaf Zn content. In rice, Chaudhry et al. (1976) initially observed positive crop responses to B application in cvs. Basmati-370 and IR-6, grown in major rice-growing areas of Punjab, with a mean paddy yield increase of 14%. Recently, paddy yield increases of 14-25% over control have been observed in cvs. Super Basmati, Basmati-385, and in KS-282 grown in Sindh province (Rashid *et al.*, 2004). Yield increases with B accrued because of reduced panicle sterility (on the lower portion of the ear) and increased productive tillers per hill. Post-harvest grain shedding is also reduced with improved B nutrition. Also, B deficiency in rice causes uneven crop maturity. Optimum B dose for effective management of B deficiency in rice is 0.75 kg ha^{-1} (Rashid *et al.*, 2004). These paddy yield results are in agreement with those of Rashid *et al.* (2002), Ali *et al.* (1996), Rakshit *et al.* (2002), Tandon (1999) and PARC (2002). Contrary to the perception that B deficiency hampers grain set more than vegetative growth (Rerkasem and Jamjod, 1997), the mean yield reduction with B deficiency was slightly greater in straw (i.e.20%) than paddy (*i.e.*18%).

As fertilizer boron dosage for correcting the deficiency is very small (*i.e.*, only 0.75kg B ha^{-1}) and crop yield increases with B applications are appreciable, its use is highly cost-effective. Wheat yield increased by 13% over control with the application of one kg B ha^{-1} (Hussain & Yasin, 2004). Similarly, 16% yield increase of paddy over control was observed with the same level of boron. Rashid *et al.* (2002) reported an increase of 5 to 26% in rice yield with B application. Ali *et al.* (1996) reported a maximum increase in paddy grain yield due to B application, which was 34.6 and 19% at 2.0kg B ha^{-1}, on Miranpur and Satgara soils, respectively. Hussain & Yasin, (2004) conducted field experiments to study the residual effect of Zn and B in rice-wheat system. Wheat grain yield ranged from 3.45 to 3.53 t ha^{-1}. The highest yield was produced from 5kg Zn+2kg B ha^{-1}whereas, the lowest yield was produced from control. Cumulative application of boron gave an increase of 10% over control; direct application gave an increase of 9% over control. However, the residual effect of B increased paddy by 4% over control. B deficiency symptoms of open spikelets in wheat were observed in specific cultivars and supposedly related to low boron soils and differential B requirement among cultivars (Furlani *et al.* 2003).

The effects of boron on the yield, and concentrations and uptake of N, P and K by wheat in red and laterite soils of West Bengal, India, were studied during the winter seasons of 2000-01 and 2001-02 (Ghatak *et al.* 2006). The application of boron significantly increased plant height, number of effective tillers, panicle length, number of grains per panicle, test weight and grain and straw yields of wheat, but had no significant effects on N, P and K concentrations and uptake. The application of 15 and 20 kg borax ha^{-1} resulted in higher values of yield attributes and yield, and the increase in grain yield over the control ranged from 4.5 to 7.7%. The optimum rate of borax was 14.0 kg ha^{-1} in the first year and 10.4 kg ha^{-1} in the second year. Thus, 10 and 15 kg borax ha^{-1} may be beneficial for the increased production of wheat in this region.

Wheat sterility (the term commonly associated with the depressed percentage of florets with the grain) as a result of B deficiency, low-temperature stress during reproductive development (Subedi,1992), waterlogging at flowering, low soil nitrogen and hot dry wind (Misra *et al.*, 1992) has been reported from many parts of Nepal. Wheat sterility is a widespread problem in rice-wheat farming systems of Asian countries, such as Bangladesh, Thailand, Nepal, China and India, that has lead to significant losses of grain yield (Rerkasem, 1996a). Many factors may contribute to the formation of sterile florets of wheat, including B deficiency (Rerkasem, 1996a, b), water deficit (Saini and Aspinall, 1981), high temperature or heat stress (Saini and Aspinall, 1982), and low temperature (chill/frost) (Huang *et al.*, 1996; Subedi *et al.*, 1998). In many cases, B deficiency is at least partially responsible for the induction of floret sterility and low grain set and its impact may be exacerbated by environmental factors (Rawson, 1996a; Rerkasem, 1996a). In wheat, B deficiency causes poor anther and pollen development and low grain set (Cheng and Rerkasem, 1993). In vitro germination tests also showed that B was required for pollen germination and tube growth in wheat (Cheng and Rerkasem, 1993).

The wheat (Subedi *et al.*, 1997; Pant *et al.*, 1998) and barley (Ambak and Tadano, 1991; Jamjod and Rerkasem, 1999) plants with male sterility and grain set failure due to B deficiency may at the same time have more tillers and greater weight of the straw. This seems to agree with a lower requirement for B for vegetative than reproductive growth. In addition to the greater functional requirement of B in the anthers and carpel, sensitivity to B deficiency of reproductive development in wheat and other *Triticeae* cereals may also be related to B supply to these organs during critical times.

Nadim *et al.* (2012) reported that growth and yield response of wheat variety Gomal-8 was evaluated using micronutrients and their application methods. The trial was laid out in a randomized complete block design with split plot arrangements. Five different micronutrients were placed in the main plot while their three application methods were assigned to sub-plots. Results revealed that application of boron @ 2 kg ha^{-1} produced higher crop growth rate (23.58 g m^{-2} day^{-1}), net assimilation rate (2.82 mg m^{-2} day^{-1}), number of tillers (234.5 m^{-2}), number of grains (52.92 $spike^{-1}$) and grain yield (3.14 t ha^{-1}). The use of iron @ 12 kg ha^{-1} also showed encouraging results similar to boron. Among various application methods, side dressing at 4 weeks after sowing (WAS) showed the best results as compared to soil application and foliar spray. Higher leaf area index and crop growth rate were obtained with the application of zinc @ 10 kg ha^{-1}. Also, different micronutrients had significant interaction with application methods for physiological and agronomic traits including several tillers, leaf area index (LAI), crop growth rate (CGR), net assimilation rate (NAR) and grain yield. Side dressing best interacted with boron for producing a higher number of tillers, grains per spike, net assimilation rate and grain yield. This method showed a better combination with iron for a higher number of tillers, LAI and grain yield.

Debnath *et al.* (2011). Conducted atrial to evaluate the effect of different rates of boron application on wheat cv. Bijoy was studied through a field experiment at Bangladesh Agricultural University (BAU) farm, Mymensingh during 2009-10 rabi season. The BAU farm belongs to Old Brahmaputra Floodplain agro-ecological zone (AEZ 9). Texturally the soil was silt loam, with 7.2 pH, 0.81% organic matter and 0.15 mg kg^{-1} available boron content. The experiment was laid out in a randomized complete block design with five boron rates and four replications. Boron rates were 0, 0.75, 1.5, 2.25 and 3.0 kg ha^{-1}, with boric acid as a source. Every plot received blanket doses of 115 kg N, 25 kg P, 75 kg K and 15 kg S ha^{-1} from urea, TSP, MoP and gypsum, respectively. Treatment receiving B @ 2.25 kg ha^{-1} produced the highest grain yield (4.22 t ha^{-1}), which was statistically identical with that obtained with 1.75 kg B ha^{-1}. However, the crop response curve showed 1.90 kg ha^{-1} to be the optimum boron rate for the maximization of wheat yield. The lowest grain yield (2.84 t/ha) was recorded with control treatment. There was a positive relationship between grain yield and number of grains $spike^{-1}$. Boron had a significant influence on N, P, K, S and B uptake by the crop which, indeed was more influenced by crop yield and less by nutrient concentration, except N and B uptake where concentration had more influence than yield.

The yield of wheat in Bangladesh is low which can be attributed to boron deficiency (Jahiruddin *et al.*, 1992). Rerkasem *et al.* (1993) found that in B deficient wheat, the pollen did not accumulate starch and the nuclei, when present, was abnormal. Boron requirement may vary among plant species and within the genotype of a species (Rerkasem, 2002). Boron deficiency causes grain sterility in wheat as reported by Li *et al.* (1978) in China, Mandal and Das (1988) in India, Rerkasem *et al.* (1989) in Thailand, Sthapit (1988) in Nepal and Jahiruddin *et al.* (1992a) in Bangladesh. Reproductive growth, especially flowering, fruit and seed set are more sensitive to B deficiency than vegetative growth (Noppakoonwong *et al.*, 1997).

The range between deficiency and toxicity of B is quite narrow and an application of B can be extremely toxic to plant at concentrations only slightly above the optimum rate (Gupta *et al.*, 1985). As general guidelines for B fertilizer recommendations, Cooke (1982) suggests that when the hot water-soluble B in the soil is less than 0.5 mg kg^{-1}, the deficiency is likely to occur and all crops are to be treated with B; when it is 0.5-1.0 mg kg^{-1} B, deficiency may appear and insurance dressings are to be considered; when it is more than 1.0 mg kg^{-1} B, a deficiency is unlikely and B treatment is not necessary; and when it is 3-5 mg kg-1 B, crops may be poisoned from excess B. Thus, a careful and judicious application of boron is necessary for achieving higher and sustainable crop yield. Riaz and Muhammad (2011), to evaluate the response of wheat, rice and cotton to B application throughout Pakistan during 2005-08, conducted Thirty-one field experiments. Boron was applied at 1 kg ha-1 as Borax decahydrate (11.3% B) at different times along with recommended doses of N, P and K. The results revealed that B application at sowing time to wheat increased significantly the number of tillers plant-1 (15%), number of grains spike-1 (11%), 1000-grain weight (7%) and grain yield (10%) over control. Among the treatments, B application at sowing time showed best results followed by B application at 1st irrigation and booting stage. In rice (coarse), B application before transplanting substantially increased number of tillers hill-1 (21%), plant height (3%), panicle length (10%), and several paddy grains panicle-1 (17%), 1000-grain weight (11%) and paddy yield (31%) over control. The response of fine rice to B application was similar for all yield parameters as in coarse rice. In cotton, B application considerably increased plant height (3%), a number of mature bolls plant-1 (12%), seed weight boll-1 (8%) and seed cotton yield (9%) over control. Although, B application at all stages significantly increased yield parameters tested B application at sowing time was best among all treatments.

Debnath (2012) conducted a pot culture experiment on rice to study the critical limit of boron in soils of the old alluvial zone of West Bengal. The hot water-soluble (HWS) boron in these soils was found to be positively and significantly correlated with organic carbon, clay content and per cent dry matter yield of rice, boron concentration in plant tissues and B uptake by shoots. A negative relationship was also observed between WHS boron and silt and sand content. The critical concentration of soil available boron and plant tissues boron was worked out to 0.38 and 15.0 mg kg^{-1}, respectively. Soil containing available B below the critical limit responded appreciably to B fertilization. A negative response to boron application was also observed at its higher level. The average dry matter yield increases with increasing level of boron application up to 1.5 mg kg^{-1} in boron deficient soils. The response to boron application in rice on boron deficient soils was found to be 52.5%.

The Response of Boron Application to Vegetables

Broccoli *(Brassica oleracia* var. Brotrytis L. Cymosa Lam.) is a vitamin-rich winter vegetable. It is rich in carotene and ascorbic acid and contains appreciate quantities of thiamin, riboflavin, niacin and iron (Thomson and Kelly, 1985; Rahman, 1988). Broccoli is environmentally better adapted and can tolerate comparatively high temperature than cauliflower (Rashid, 1993). Unlike cauliflower, broccoli produces smaller flowering shoot from the leaf axile if the main apical flower bud is removed. Consequently, a field of broccoli may be harvested over a considerable period. Boron deficiency is now widespread in Bangladesh. This element deficiency causes many anatomical, physiological and biological changes. Hollow stem disorder is a major problem for broccoli production and is commonly associated with B deficiency (Shelp, 1992; Rahman *et al.*, 1995).

The present (Firoz *et al.*, 2008) investigation was undertaken to find out the response of three broccoli varieties to different levels of boron in the Chittagong hilly region. There was a significant and positive effect of boron application on the yield of the crop and 1.0 kg B/ha was found to be an optimum rate. The 1.0 kg B/ha rate produced the highest yield (512.3g/plant) followed by 2.0 kg B/ha showing 508.5 g/plant and the B control did the lowest (445.4g/plant). All other characters remained unaffected by B application. A field experiment was conducted by Naga *et al.* (2013) during rabi-2010 to find out the response of foliar application of micronutrients on vegetative and reproductive growth attributes, in two varieties of tomato viz- Utkal Kumari and Utkal Raja. The treatments consisted of boron, zinc, molybdenum, copper, iron, manganese, the mixture of all and control and the experiment was laid us in RBD with three replications. All the Micronutrients except manganese (@ 50ppm) were applied @ 100ppm in three sprays at an interval of ten days starting from 30 days after transplanting.

All the treatments resulted in improvement of plant growth characteristics viz. plant height, several primary branches, compound leaves, tender and mature fruits per plant and seed yield characteristics like recovery percentage, 100 seed weight, seed yield per plant and seed yield per hectare in both the varieties out of which application of micronutrients mixture showed the maximum effect. In tomato cv. Utkal Kumari, maximum growth rate (85.7 %) was observed with the application of zinc, followed by application of micronutrients mixture (78.2 %) and boron (77.5 %). Tomato cv. Utkal Raja, the maximum increase in branches per plant was observed with the application of manganese (148.7 %) followed by micronutrient combination (144.1 %). Highest seed recovery rates of 0.53 and 0.55 per cent recorded in the varieties Utkal Kumari and Utkal Raja, respectively by application of micronutrients mixture. The highest seed weight was observed with the application of micronutrients mixture (Utkal Raja) and boron (Utkal Kumari). In both varieties, application of micronutrients mixture gave maximum seed yield followed by boron treatment, in respect both the parameters, while the lowest yield was obtained in the control.

Boron deficiency in fresh-market tomatoes is a widespread problem that reduces yield and fruit quality (Davis *et al.*2003). As reported by Smit and Combrink (2004), at too low B levels in the root zone, the leaves of tomato are brittle and appear pale-green, a considerable fraction of flowers abscise and the fruits lack firmness, a problem that is worsened during storage. According to Smit and Combrink (2004), the above symptoms appeared at a B concentration level of 0.02 mg L^{-1} in the nutrient solution supplied to a crop grown on quartz-sand, but a B concentration of 0.16 mg L^{-1} seemed to be optimal for tomato and levels of up to 64 mg L^{-1} did not cause any toxicity symptoms. Furthermore, a suboptimal boron supply may considerably reduce fruit set, especially if no other means for pollination (*e.g.* vibration) are applied (Smit and Combrink 2005). In another study, it was shown that an enhanced boron supply (B foliar spray at 300 mg L^{-1}) was associated with a less frequent incidence of the physiological disorder shoulder check crack (Huang and Snapp 2004a), the visible symptoms of which are described under 'nutrition and fruit quality'. Davis *et al.* (2003) reported that the delivery of B either through the nutrient solution (1 mg L^{-1}) or by foliar spraying (1.87 mg L^{-1}) of boron chelated with mannitol, to tomato grown in river sand, was associated with increased plant growth and tissue K, Ca and B concentrations. In the above study, foliar spraying with boron significantly enhanced fruit B and K concentrations in comparison with no boron supply, which indicates firstly that B is translocated from the leaves to the fruit and secondly that B is also involved in K translocation within the plant. Smit and Combrink (2004) have reported enhanced uptake of Ca, Mg, Na, Zn and B with higher B levels in the root zone. Another recently investigated aspect related to boron nutrition in tomato is the interaction between boron and salinity or water stress. According to Ben-Gal and Shani (2002, 2003), under conditions of simultaneous boron deficiency and salt or water stress, the extent of growth suppression is determined by the factor imposing the most severe stress and not by the addition of the effects of both restrictive factors. Hence, a dominant-stress-factor model following the Liebig-Sprengel law of the minimum may be used to describe the responses of tomato to simultaneous exposure to boron and salinity or boron and water shortage. Furthermore, Ben-Gal and Shani (2002) found that the yield response of tomato to boron nutrition correlates better with B concentration in the irrigation water and soil solution than with the levels of boron in the plant tissue. According to Alpaslan and Gunes (2001), soil boron concentrations of 5 mg kg^{-1} or higher are expected to impose boron toxicity symptoms.

The present investigation was carried out by Sathya *et al.* (2013) at Agricultural College and Research Institute, Madurai to evaluate the effect of various levels of application of boron on availability of B and primary nutrients in the soil. There were twelve treatments (soil application of B as borax @ 0, 5, 10, 15, 20 & 25 kg ha^{-1} and foliar application of B as borax @ 0.25 and 0.50 per cent spray twice at 50th and 80th, 60th and 90th and 50th and 90th days after transplanting) which were replicated thrice in a randomized block design (RBD). The results revealed that soil application of B as borax @ 20 kg ha^{-1} recorded the highest available B irrespective of growth stages of tomato. Availability of B decreased with the advancement of crop growth stages of tomato. Similarly, available nitrogen, phosphorus and potassium status was also found to be high in the above treatment.

The Response of Boron Application to Oil Seeds

Mojtaba (2010) reported that the poor drainage of saline soils may be responsible for the excessive accumulation of B in the soil solution. Toxicity symptoms often occur because of additions of B via irrigation water and lack of drainage. The present study, as a pot experiment with *pistachio* plants, was conducted to study the effect of increasing boron application on nutritional status and shoot growth under greenhouse conditions. Seven treatments consisted of B levels (0, 2.5, 5, 10, 20, 40 and 60 mg/kg soil as H_3BO_3) Treatments were arranged in a factorial manner in a completely randomized design with three replications. Results of this study clearly show that relatively high B content might reduce the adverse effects of B and consequently improve plant growth. It is suggested that this finding be verified under field conditions. Moreover, in future studies of boron stress should have attentions to soil pH, separation of young and old leaves, modelling of B adsorption by sorts of local soils and role of B in operative enzymes of toxin metabolism.

According to Slavko *et al.* (2001) Boron (B) remobilization, mannitol and glucose concentrations, and the effect of B application on changes in soluble carbohydrates were investigated in various organs of bearing 'Manzanillo' olive trees (*Olea europaea* L. 'Manzanillo'). Following foliar 10B application to leaves of various ages, there was significant 10B export out of the treated leaves, and significant 10B enrichment in nontreated adjacent organs, including inflorescences and fruit. Results demonstrated that B can be remobilized from leaves of various ages and that foliar-applied B is phloem mobile in olive. Soluble carbohydrate analysis determined that mannitol and glucose are the predominant sugars in all organs analyzed and that the mannitol concentration in the leaves is adequate to account for all B transport. This is consistent with observations in other species, where the presence of mannitol is known to facilitate phloem B transport through the formation of a mannitol-B complex. Previous reports have indicated that B application can alter carbohydrate metabolism. In the present study, the foliar B application significantly suppressed glucose concentration in the leaf petioles of all ages and increased mannitol in petioles of the current-year-developed leaves.

Rashid *et al.* (2012) A field experiment was carried out in non-Calcareous Floodplain Soil of Spices Research Sub-Station, Lalmonirhat under AEZ 2 during the rabi season of 2007-08 and 2008-09. The objectives were to evaluate the effect of boron on the yield of mustard and to screen out the suitable variety tested against different boron levels for maximizing yield. Three varieties of mustard viz., BARI Sharisha-11, 13, and 14 and 5 levels of boron (0, 0.5, 1.0, 1.5 and 2.0 kg/ha) along with a blanket dose of N120 P35 K65 S20 Zn3.0 kg/ha were used in the study. Results revealed that BARI Sharisha-11 performed better with 1.5 kg B/ha which produced 1.82 t/ha seed. However, from regression analysis, a positive but quadratic relationship was observed between seed yield and boron levels. The optimum dose of boron was appeared to be 1.7 and 1.6 kg B/ha for Lalmonirhat during 2007-08 and 2008-09, respectively. The mustard crop is very much responsive to boron application (Mengel and Kirkby, 1987). There are numerous reports on the positive response of mustard to B fertilization (Islam, 2005; Hossain *et al.*, 1995 and Saha *et al.* 2003). There is a good scope to introduce high yielding varieties of mustard with proper management of fertilizer.

Amery et al. (2011) to determine the response of open-pollinated sunflower (Helianthus annuus L. cv. Shumos) to boron fertilizer carried out a study. Spring and autumn grown plots were sprayed with boron (0, 50, 100, 150, 200, and 250 mg L^{-1}) in a randomized complete block design. The results showed that boron and green Leaf area/LAI measured at the end of seed filling were linearly related in the spring crop. In the autumn crop, the relationship between boron and dry matter yield was linear, and applications above 100 mg L^{-1} significantly increased dry matter compared to the control. For the autumn crop, 100, 200, and 250 mg L^{-1} significantly increased seed number compared to the control, but only the 150 mg L^{-1} treatment increased seed weight significantly. Boron tended to decrease the empty seed percentage, and for the spring crop, this was incremental and linear with applications above 150 mg L^{-1} leading to significant reductions in empty seed percentage. Seed yield increased linearly in the autumn crop in response to boron, and 200 and 250 mg L^{-1} applications gave significant increases in yield compared to the control.

Lei *et al.* (2010) conducted field experiments were conducted to study the inheritance of boron efficiency in oilseed rape (*Brassica napus* L.) by evaluating the boron (B) efficiency coefficient (BEC, the ratio of the seed yield at below the critical boron level to that at the boron-sufficient level) with 657 F2:3 lines of a population derived from a cross between a B-efficient cultivar, Qingyou 10, and a B-inefficient cultivar, Bakow. Qingyou 10 had high BEC as well as high seed yield at low available soil B. On the contrary, Bakow produced low seed yield at low B status. Boron deficiency decreased the seed yield of the F2:3 lines to different extents and the distribution of BEC of the population showed a bimodal pattern. When the 657 F2:3 lines were grouped into B-efficient lines and B-inefficient lines according to their BEC, the ratio of B-efficient lines to B-inefficient lines fitted the expected ratio (3:1), indicating that one major gene controlled the B-efficiency trait.

127 F2:3 lines selected from the population at random, with the distribution of BEC similar to that of the overall population, were used to identify the target region for fine mapping of the boron efficiency gene.

Eight B-efficient cultivars and two B-inefficient cultivars were identified from 211 cultivars of rapeseed screened using a two-step method as reported by Wang and Lan (1995). The B efficiency coefficient (BEC), i.e., the ratio of the seed yield at below the critical soil B level (HWSB = 0.25 mg kg^{-1}) to that at the sufficient B level, was used to determine B efficiency. B efficiency may be attributed to increased B uptake efficiency, B utilization efficiency or re-translocation efficiency, which are all genetically controlled. Xu *et al.* (1998) reported that B efficiency was a dominant trait in B. napus. An F_2 population obtained from a cross between a B-efficient cultivar, Qingyou 10, and a B-inefficient cultivar, Bakow, was used to study the inheritance of B efficiency in *B. napus* and the ratio of B-efficient to inefficient individuals in the population fitted the expected ratio of 3:1 (Xu *et al.*, 2001). Quantitative trait locus (QTL) analysis detected one major locus and three minor loci associated with B efficiency (Xu *et al.*, 2002). To estimate the inheritance of B efficiency more accurately and facilitate fine mapping of the B-efficient gene, an F2:3 population, developed from a cross of the same parents with the F2 population, was planted in a field with low B and adequate B levels to further examine the genetic basis of B efficiency of *B. napus*.

The response of boron application and *Rhizobium* inoculation to legume/pulses

Crop growth and yields

In a Greenhouse study, a positive response of soybean (*Glycine max* L.) to applied B in acid soils was recorded (Singh and Sinha, 1987; Data *et al.*, 1994). Similarly, in field trial also, application of sodium tetraborate at the rate of 20 kg ha^{-1}, increased the grain and straw yield of soybean from 1438 to 23.2 and 22.8 to 35.8 q ha^{-1}, respectively in some acid soils (Diwvedi *et al.*, 1990). Yields of black gram (*Phaseolus mungo* L.) and soybean were also increased in response to B application in some calcareous soils. The yields of groundnut (*Arachis hypogaea* L.) and sunflower (*Helianthus annus* L.) were found to be increased to the tune of 4.2 and 3.2 q ha^{-1}, respectively at 2.5 and 1.5 ka ha^{-1} applied level of B (Sinha *et al.*, 1991). B, which also limits chickpea productivity, is a less important factor than Zn (Ahlawat *et al.*, 2007). B, in acidic soils, is a major reducer of chickpea yields in some regions (Srivastava et al., 1997). In comparison with others crops, the response of the crop to the application of B is higher in chickpea than in some cereals (Wankhade *et al.*, 1996); although differences between chickpea cultivars concerning B deficiency have also been observed (Ahlawat *et al.*, 2007). The application of B is important when the concentration of B in the soil is less than 0.3 mg kg^{-1} (Ahlawat *et al.*, 2007). According to Ankerman and Larger (1974) soils have low B availability when the concentration of B in the soil is less than 0.6 mg kg^{-1} (hot water extraction) and according to Sillanpää (1972), the soil may have a deficiency of B when their concentration in the soil is less than 0.5 mg kg^{-1} depending on the conditions, the extraction time and the soil. B deficiency also causes flower drop and, subsequently, poor podding of chickpeas (Srivastava *et al.*, 1997) and poor yields. A B deficiency can be caused by high pH in the soil, the availability of B decreases when the pH is larger than 6.5-7.0 (Sims, 2000), which occur in highly leached sandy soils or low organic matter soils.

According to Mahler and Shafii (2009) to Critical soil test boron (B) levels and consequent fertilizer recommendations for spring peas have been traditionally based on values developed for alfalfa, sweet clover, and other legumes commonly grown in other regions of the USA. The purpose of this study was to define the relationship between soil test B values and spring pea yields. Seventy-eight field studies using B were conducted between 1981 and 2001 on Mollisols and Alfisols in northern Idaho and eastern Washington. Soil sampling to the effective rooting depth of 0.6m for spring peas results in the best relationship between soil test B and spring pea yield; however, samples taken to the depth of 0.3m result in the same B critical value (0.5 mg/g). Based on these field studies, the data demonstrate that the 0.5 mg/g critical soil test value of B is important for spring pea yield optimization. Mondal *et al.* (2012) conducted an experiment for the evaluating the effect of irrigation and boron on growth and yield of mungbean variety 'Samrat' during summer seasons of 2009 and 2010 at Jaguli Instructional Farm, Bidhan Chandra Krishi Viswavidyalaya, West Bengal on sandy loam alluvial soil. The experiment was laid out in a split-plot design with four levels of irrigation in the main plot and four levels of boron in sub-plot reported that both irrigation and boron treatments influenced crop growth characters, yield attributes and seed yield of mungbean. Significant interaction effects were observed among the treatments. Maximum seed yield of 879 kg/ha was obtained with three irrigations at branching, pre-flowering and pod filling stages and was almost 38% higher over the rainfed treatment. Among the various levels of boron, foliar spray of 0.2% borax at flowering stage consistently influenced all the growth and yield parameters. A treatment combination involving three irrigations at branching, pre-flowering and pod filling stages along with a foliar spray of 0.2% borax at flowering stage recorded maximum mungbean seed yield (898 kg/ha) which was comparatively higher over other treatment combinations

Valenciano *et al.* (2010) reported that response of chickpea to the applications of Zn, B and Mo was studied in pot experiments with natural conditions and acidic soils in northwest Spain from 2006 to 2008 following a factorial statistical pattern (5 × 2 × 2) with three replicates. Five concentrations of Zn (0, 1, 2, 4 and 8 mg Zn pot^{-1}), two concentrations of B (0 and 2 mg B pot^{-1}), and two concentrations of Mo (0 and 2 mg Mo pot^{-1}) were added to the pots. Chickpea responded to the Zn, B and Mo applications. There were differences between soils. The mature plants fertilized with Zn, with B and with Mo had a greater total dry matter production. Harvest Index (HI) improved with the Zn application and with the Mo application. The highest HI was obtained with the $Zn_4 \times B_2 \times Mo_2$ treatment (60.30%) while the smallest HI was obtained with the $Zn_0 \times B_0 \times Mo_0$ treatment (47.65%). The Zn, B and Mo applications improved seed yield, mainly due to the number of pods per plant.

During 1993-94 Saha *et al.* (1996) conducted a pre-kharif [pre-monsoon] seasons trail on yellow sarson [*Brassica campestris* var. sarson] was given 0, 2.5 or 5.0 kg borax and 0, 1 or 2 kg sodium molybdate/ha applied as soil, 66% soil + 33% foliar or foliar applications and the residual effects were studied on summer green gram [*Vigna radiata*]. In both years green gram seed yield was highest with a combination of 5 kg borax + 2 kg sodium molybdate. Soil application gave higher yields than foliar or soil + foliar application. In a field with boron-deficient soil in 1994/95 in Nepal, chickpeas cv. Kalika was given no fertilizers, complete fertilizer (P, K, S, B, Zn, Mo, Cu, Mn and Fe) or the complete fertilizer minus each of the trace elements. Flower abortion was the highest and no seed was produced in the treatment given no B (Srivastava *et al.*1996). Islam *et al.* (1995) conducted a field experiment in rabi [winter] 1992/93 on grey floodplain soils at Bogra, Bangladesh, *Cicer arietinum* cv. Nabin was given 20 kg N + 60 kg P + 40 kg K + 5 kg Zn + 1 kg B + 1 kg Mo/ha (complete fertilizer package, CFP) with or without an additional 1 kg B/ha, NPKS alone or with 5 t cattle manure/ha or 7 other treatments which included the CFP with a different fertilizer missing in each treatment. The omission of N, P, K, S, Zn, Mo and B gave seed yields of 73, 67, 50, 85, 96, 80 and 7% of the yield with CFP, respectively. Applying NPKS only or NPKS + cattle manure gave 5 and 8% of the yield with CFP. CFP + 1 kg B/ha gave 37% higher yields than with CFP alone. In the following year (1993/94) application of 2 kg, B + 1 kg Mo/ha gave the optimum *C. arietinum* yield which was 514% higher than that obtained from 2 kg B alone.

(Shamima and Farid, 2003) conducted Field experiments during the winter (rabi) seasons of 2000-01 and 2001-02 on a Grey Terrace Soil in Bangladesh to study the effect of different nutrients on yield, yield components and nutrient uptake by garden pea cv. BARI Motorsuti-1 grown in Grey Terrace soil. Application of different nutrients caused significant increases in yield and nutrient uptake. The highest fresh pod yield and protein content were achieved by the treatment of N_{30} P_{50} K_{40} S_{20} Mo_1 B_1 and it was not significantly different from the N_{30} P_{50} K_{40} S_{20}, N_{30} P_{50} K_{40} S_{20} Mo_1 and N_{30} P_{50} K_{40} S_{20} Mo_1 B_1 Zn_1 treatments. Uptake of N, P, K, S, and Zn by shoot and seed was also highest under the treatment N30 P50 K40 S20 Mo1 B1 and not significantly different from N_{30} P_{50} K_{40} S_{20} Mo_1 and N_{30} P_{50} K_{40} S_{20} Mo_1 B_1 Zn_1. Addition of Mo, B and Zn with NPKS did not show any significant change in pod yield. Application of N_{30} P_{50} K_{40} S_{20} Mo_1, N_{30} P_{50} K_{40} S_{20} Mo_1 B_1 and N_{30} P_{50} K_{40} S_{20} Mo_1 B_1 Zn_1 kg/ha were not economical, but treatment N_{30} P_{50} K_{40} S_{20} kg/ha proved to be the most economically profitable for garden pea production.

Nutrient acquisition in chickpea needs to be efficient, because it is mainly grown as a post-rainy season, rainfed crop, and generally on soils inferior in physical characteristics and poor infertility. Nutrient deficiencies have been reported to cause yield losses of varying magnitude in chickpea, e.g., 22-50% due to iron (Fe), around 10% due to sub-optimal nodulation and hence nitrogen (N) deficiency, 29-45% due to phosphorus (P), up to 100% due to boron (B), and 16-30% due to sulphur (S). Yield losses due to salinity are equally large but are difficult to estimate because of its heterogeneous occurrence. In chickpea, genotypic differences in morpho-physiological (including root size) and functional (exudates) root traits, and in nodulation capacity for increased nitrogen fixation have been identified. Genotypic differences in response to an application of Fe, B and zinc (Zn) have also been found among chickpea genotypes. A drought-tolerant chickpea genotype ICC 4958, which has a relatively large root system, acquired more P than other genotypes during the vegetative period in a pot experiment at ICRISAT. The recent thrust on identifying QTLs for root size should facilitate progress in incorporating useful root traits through marker-assisted selection in desirable agronomic backgrounds. Selection for nodulation capacity in released cultivars has resulted in high nodulating chickpea genotypes that produced 10% higher yield than the control varieties. Information on targeted crop improvement for higher nutrient-use efficiency for P, S, Zn, B and Fe is not readily available. Methods to screen for tolerance to salinity are available, but sufficiently high levels of tolerance have not yet been found in germplasm or wild relatives of chickpea to warrant breeding for salinity tolerance. Use of alternative approaches, such as mutation to generate genetic diversity or introgression of alien genes from other crops (transgenic) are thus required, and these remain long-term objectives (Ali *et al.*, 2002). Field experiments were conducted by Ali and Mishra (2001) to evaluating four genotypes of kabuli chickpea namely, BG 267, BG 1003, HK 89-96 and L 550 subjected to a foliar spray of boron and molybdenum at 50 and 60 days after sowing (DAS) along with the control (water spray) were conducted during 1996-98 in Kanpur, Uttar Pradesh, India on sandy loam soil with 265 kg N, 18 kg P, and 285 kg K/ha with pH 7.1. Of the four genotypes of chickpea, L 550 gave the highest seed yield (1608 and 1538 kg/ha) followed by BG 267. The seed yield of L 550 was significantly superior to BG 1003 and HK 89-96 during 1996-97 only. L 550 had better nodulation (18.4 and 17.7 nodules/plant) than others. Foliar spray of 0.2% borax at 50 and 60 DAS increased the grain yield significantly in both the years resulting in the highest seed yield of 1796 and 1590 kg/ha. A significant response to the foliar spray of 0.1% ammonium molybdate was also observed during 1997-98. Among the reasons, the original near-neutral pH of many of the soils was no doubt a consideration and the use of acid-sensitive legumes to supply N in rotations in the early stages of agricultural development also played a part. The introduction of the pH meter at about the same time as N fertilizers found widespread popularity replacing legumes in rotations facilitated the measurement of soil acidity and removed the focus from the real problems of soil acidity, namely, toxic levels of Al^{3+} and Mn^{2+} and deficiencies of nutrients such as Ca, Mg, N, S, P, B and Mo (Sumner *et al.*, 2002).

Haby and Kalra (2002) Reported that soil acidity and fertility management are critical for crop production on Coastal Plain soils of the southern and southeastern USA. Grasses and legumes are the primary crops grown on these soils. Hybrid bermudagrass (*Cynodondactylon*) and bahiagrass (*Paspalum notatum*) are the major warm-season grasses while annual ryegrass (*Lolium multiflorum*) is the dominant cool-season grass. Bahiagrasstolerates acidity and low soil fertility. Bermuda grasses tolerate moderate soil acidity levels and readily respond to N, P, and K fertilizer application. Annual ryegrasses are highly responsive to limestone, N, and P. High rates of ammonium N applied for grass production rapidly increase acidity due to nitrification in these low-buffer-capacity soils. Liming these soils to increase pH improves P use efficiency and forage production. Fertilizer P increased soil test P 38%, as soil pH was raised from 4.5 to 6.2 in a Lilbert loamy fine sand (thermic Arenic Plinthic Paleudult) in Texas, USA. Forage legume crops respond favourably to applied P, K, and B on limed Ultisols. Limestone applied to strongly acid soils increased lucerne yields greater than 6.2 t ha-1. The critical 0- to 15-cm-depth, soil test P level was 20 mg kg-1 for lucerne using the pH 4.2 NH_4OAc-EDTA extractant on limed coastal plain soils. Lucerne response to applied B was 3.9 t ha-1 and was economical to 4.2 kg ha-1. Regression analysis predicted that raising the level of hot water-soluble B from 0.3 to 0.7 mg kg-1 in the 5- to 15-cm depth increased lucerne dry matter yield by 9 t ha^{-1}. The level of hot-water soluble B in the 0- to 5-cm and in the 0- to 15-cm depths did not correlate with yield.

The effects of *Rhizobium* seed inoculation (100 g/kg of seeds), N (urea at 50 kg/ ha) and B (1, 2 and 4 kg/ha) fertilizer application, on the nodulation, dry matter production and yield of chickpea (cultivars Hyprosola and Nabin) were studied in Bangladesh [date not given]. The number of total and effective nodules per plant, dry weight of nodules per plant and grain yield were significantly improved by *Rhizobium* inoculation. In Hyprosola, all treatments improved the nodulation, dry matter production and yield. *Rhizobium* inoculation alone produced the highest total nodules per plant (26.0), several effective nodules per plant (22.8), dry weight of nodules per plant (153 mg) and grain yield (632 kg/ha). The sole application of 50 kg N/ha resulted in the highest dry weight of tops per plant (3914 mg). In Nabin, *Rhizobium* inoculation+1 kg B/ha obtained the highest total nodules per plant (24.3), effective nodules per plant (18.4) and dry weight of tops per plant (4023 mg). The greatest dry weight of nodules per plant was recorded for *Rhizobium* inoculation alone. Grain yield was highest with *Rhizobium* inoculation+1 and 2 kg B/ha (900 kg/ha). N or B at higher rates was detrimental to nodule formation. The number of total and effective nodules was strongly and positively correlated with the dry weight of nodules and plant tops, as well as grain yield. A strong correlation existed among the dry weight of nodules, dry weight of plant tops and grain yield (Solaiman, 1999).

Azevedo *et al.* (2002) The objective of this work was to evaluate the effect of boron on nodule development in pea *Pisum sativum* L. plants cultivated in lowland soils under greenhouse conditions from May to July 1998. Samples of Alluvial Soil, Low Humic Gley, Humic Gley and artificially drained Bog Soil were collected from a layer 0-20 cm deep. The samples were previously cultivated with radish plants that received lime, P, K, S, Cu and Zn and doses of B (0, 0.25, 0.5, 1.5, 3.0 and 6.0 mg dm^{-3}). Before pea cultivation, only K was applied (100 mg dm^{-3}). The experimental design was randomized in a factorial scheme 4 x 6 with three replicates. Seeds were inoculated with *Rhizobium leguminosarum* strains (BR 618 and BR 619). To verify native strains, three pots of each soil were cultivated without inoculated seeds with same lime and fertilizers doses and 0.5 mg dm-3 of boron. After 45 days, the experiment was harvested. Native strains were not found in the evaluated soils. Boron affected nodule development and nitrogenase activity in the pea plants in Low Humic Gley and artificially drained Bog Soil.

Singh *et al.* (2002) The field experiments were conducted in sandy loam calcareous soil in Uttar Pradesh, India during 1995-96 to study the effect of boron application on yield of pea (cv. Rachna) and black gram (cv. PV-19). B was applied as borax at the rates of 1, 2, 3 and 4 kg borax/ha, with a control. Application of borax up to 4 kg borax/ha significantly increased the grain yield of black gram. The maximum yield was 15.42 q/ha and minimum grain yield of 1.65 q/ha was found in the control. The additional grain yield over the control was 280, 431, 899 and 1377 kg/ha at 1.0, 2.0, 3.0 and 4.0 kg borax/ha, respectively. Application of B progressively increased the grain yield of a pea from 510 to 1843 kg/ha up to 4 kg borax/ha. Bharti *et al.* (2003) Reported that the effects of B (0, 1.5 and 2.5 kg/ha) and *Rhizobium* (strains DG-48, JTG-1, RG-3, MTG-2 and BTG-1) application on the yield and nutrition of chickpea cv. BG 256 were determined in a field experiment conducted in Bihar, India during the winter of 1997-98. The mean grain yield, and grain and straw N and B content increased, whereas straw yield decreased with increasing B rate. *Rhizobium* strains DG-48 and JTG-1 were superior over all the other strains tested in improving the quantity and quality of chickpea. A field experiment was conducted in Muzaffarpur, Bihar, India, during the 1997/98 *Rabi* season on chickpea cultivar BG 256 to study the effect of B on *Rhizobium sp.* strain DG-48, JTG-1, RG-3, BTG-1 and MTG-2. The treatments comprised 0, 1.5, and 2.5 kg B/ha. The number, dry weight, nitrogenase activity, leghaemoglobin content, and active iron content of the nodules increased with B application. B at 1.5 kg/ha was optimum for most of the characteristics. DG-48 and JTG-1 had the best nodule parameters and the efficiency was enhanced with the application of 1.5 kg B/ha (Bharti *et al.*, 2003).

Ali *et al.* (2002) Nutrient acquisition in chickpea needs to be efficient, because it is mainly grown as a post-rainy season, rainfed crop, and generally on soils inferior in physical characteristics and poor infertility. Nutrient deficiencies have been reported to cause yield losses of varying magnitude in chickpea, e.g., 22-50% due to iron (Fe), around 10% due to sub-optimal nodulation and hence nitrogen (N) deficiency, 29-45% due to phosphorus (P), up to 100% due to boron (B), and 16-30% due to sulphur (S). Yield losses due to salinity are equally large but are difficult to estimate because of its heterogeneous occurrence. In chickpea, genotypic differences in morpho-physiological (including root size) and functional (exudates) root traits, and in nodulation capacity for increased nitrogen fixation have been identified. Genotypic differences in response to an application of Fe, B and zinc (Zn) have also been found among chickpea genotypes. Selection for nodulation capacity in released cultivars has resulted in high nodulating chickpea genotypes that produced 10% higher yield than the control varieties. Information on targeted crop improvement for higher nutrient-use efficiency for P, S, Zn, B and Fe is not readily available. Among micronutrients, many workers (Redondo et al., 2003, Rego et al., 2005) have discussed the role of boron on nodulation, growth and yield of different legumes as well as groundnut. Boron plays an indispensable role in nodule development. Bolaños *et. al.*, 1996 reported that pea (*Pisum sativum*) nodules that developed under low boron conditions were not functional and became prematurely senescent. Three and four week old boron deprived nodules showed impairments in nodule development due to degeneration of cell walls and membranes, which surround intracellular bacteroids. Because of which *rhizobia* inside nodules showed little or no ability to fix N_2. Boron deficiency exerted a stronger negative effect on the nodule setting resulting in decreased number of nodules and reduced specific nodulation rate in soybean. Such reports have pointed out that boron deficiency affects nodulation and nitrogen fixation in legumes. Survey studies conducted in different states of India have revealed that the deficiency of boron ranges from 1-84% with a mean of 33% (Singh, 2006). From a soil survey conducted in Karnataka, Sahrawat *et. al.*, 2007 reported that about 54-96% soils are deficient in boron content. Positive responses on nodulation in different crops due to boron application in B deficit soil have been reviewed in the recent past.

Boron (B) deficiency has a strong effect on molecular and cellular plant-bacteria interactions during the development of the legume–*rhizobia* symbiosis, leading to reduced infection and early necrosis of nodules, resembling a pathogenic-like rather than a symbiotic interaction. Therefore, induction of pathogenesis-related (PRs) proteins was investigated herein legume root nodules and results suggest that under B deficiency, the rhizobia–legume dialogue fails and the bacterium is recognized as a pathogen by the plant, which reacts to prevent infection by inducing at least these two identified PR10 proteins (Reguera, *et al*. 2010). The legume–rhizobia symbiotic interaction is largely dependent on boron (B), a micronutrient required for plants and other organisms in micromolar concentrations. Boron deficiency leads to aberrant nodule organogenesis(Reguera *et al.*, 2009), with an important reduction of infection(Bolan˜os *et al.*, 1996; Redondo- Nieto *et al.*, 2001), and a block of maturation of endophytic bacteria toN_2-fixing bacteroids (Bolanos *et al.*, 2001; Redondo- Nieto *et al.*, 2007), which could be at least partially due to a recognition of the bacterium as a potential pathogen. Inoculation of seeds through *Rhizobium* to utilize the resources efficiently has been a successful trend for quite some time. In Mungbean too several strains have been tested at several locations throughout the country, still, the scope is tremendous to find a more suitable strain depending upon the local conditions that can cause considerable crop yield.

The association of mung bean rhizobia with plants is a critical factor. Inoculation of rhizobia with a high capacity of nodulation and nitrogen fixation is an established practice and agronomic technique (Hadad and Loynachan, 1985; Rangarajan and Purushothaman, 1995) Usually grain legumes grown on marginal lands and poor yields in such soils are due to lack of effective and specific strains of *Rhizobium* in the rhizosphere (Subba Rao and Tilak, 1977). The requirement of nitrogen on marginal lands is high. Hence, there is a need to meet the nitrogen requirement through *Rhizobium*. Lalnde *et al.* (1986) by using total nitrogen balance method, in the wax bean, estimated that in the presence of an efficient strain, up to 74 per cent nitrogen present in the pods comes from nitrogen fixation, indicating that this process is an important one. Therefore, the presence of efficient and specific strains of *Rhizobium* in the rhizosphere is one of the most important requirements for proper crop establishment. Sahu and Behera (1972) from Bhubaneswar Agricultural Research Station reported that the number of nodules in green gram in a sandy loam soil increased largely by inoculation in presence of phosphate. Desmukh and Joshi (1973) found that the dry matter accumulation of cowpea increased with *Rhizobium* inoculation. Gupta et al. (1976) observed a significant increase in nodulation.

Idris and Sandhu (1979) conducted pot experiments and reported that inoculation of green gram with all the strains of *Rhizobium phaseoli* improved the number of nodules per plant and fresh weight of nodules as compared to uninoculated plants. Inoculation with each test strain increased the dry matter yield of shoots and roots per plant by 13 to 42 per cent. The dry matter yield of shoots, roots and pods together significantly correlated with the nitrogen-fixing efficiency. Basak and Goyal (1980) concluded that *Rhizobium* inoculation increased the nodulation and seed yield over un-inoculated control in *Vigna radiata* and Vigna *mungo*. Prasad and Ram (1992) studied the effect of inoculation on *Vigna radiata* seeds with a Rhizobial strain and found a significant increase in the number and dry weight of root nodules. Inoculation with mixed cultures was more effective than a single strain. Plant dry weight positively correlated with modulation and nodule leghaemoglobin contents.

Srivastava and Verma (1984) stated that seed inoculation with *Rhizobium* increased plant dry weight, several green leaves and branches per plant in green gram in both the years (1977 and 1978). Gill *et al.* (1985) noticed that the effect of inoculation with *Rhizobium phaseoli* and application of 56.09 kg P_2O_5 with 56.09 kg K_2O per hectare had a significant effect on plant height. Inoculation significantly increased the number of primary branches per plant also. A significant increase in dry matter accumulation of green gram (13%) with nitrogen application followed by the seed inoculation with *Rhizobium* culture was observed under both the fertility levels. Growth was better at a high level of phosphorus and potassium than their lower levels. Application of nitrogen or *Rhizobium* inoculation were equally effective in increasing the number of nodules in green gram (Maiti *et al.*1988). The difference in nodule count in Mungbean to different strains was significant but the same was insignificant in grain yield at Betinan Research station, Philippines. The effect of five strains of *Bradyrhizobium* japonica and P fertilizer in the un-inoculated control plots (Penaranda *et al.* 1988).

Basu *et al.* (1989) reported a significant difference in the yield of Mungbean cultivar due to inoculation of two *Rhizobium* strains viz. Jca-1 and M-10. They also reported the superiority of *Rhizobium* inoculation over control. Pal and Jana (1991) found that inoculation significantly improved the different growth parameters of green gram and also the number of pods per plant, number of seeds per pod and the 1000-seed weight which in turn resulted in increased seed yield over the un-inoculated treatment by 10-15 per cent. Rhizobium inoculation increased the dry matter accumulation by 10.5 per cent compared to uninoculated plants (Shukla and Dixit 1996).

Ardesna *et al.* (1993) pointed out that *Rhizobium* inoculated plants showed significantly greater plant height, a number of branches per plant, root nodules per plant and pods per plant, which ultimately improved the grain yield. Srivastava and Poi (2000) conducted field experiments to determine symbiotic efficiencies of green gram after inoculation with a native *Bradyrhizobium* strain and residual effects of 7 *Bradyrhizobium* strains(HG-13per1, M-10, kuthi AR-1, JCa-1, Caj-3, NK-4, Caj-6per1) in neutral pH soils in Mohanpur, West Bengal, India. Inoculation with M-10 strain resulted in highest dry matter production and nitrogen fixation. Sharma (2001) reported that out of three *Rhizobium* strains (Ludhiana, local isolated and IARI) of *Bradyrhizobium* tested at Palanpur, H.P., India, the local isolate outperformed the others about crop growth rate, relative growth rate days to maturity and grain yield. Similar results of the differential response of Mungbean to different *Rhizobium* strains were reported by Saxena *et al.* (1996). The effects of *Bradyrhizobium* sp. Inoculation on yield components of gram cv Pusa Baisakhi was investigated in silty clay loam soils of Palanpur, H.P., India. The *Bradyrhizobium* strains were Ludhiana, IARI and local isolate. The highest test weight, biological yield. Grain yield and harvest index were obtained with the local isolate followed by Ludhiana and IARI isolates. Inoculation with all the isolates produced higher yields compared to control (Sharma, 2003). Venkateswarlu (2004) reported of better nodule and nitrogen fixation by *Rhizobium* inoculation. Herridge *et al.* (2005) in Australia concluded that low nodulation and nitrogen fixation of commercial Mungbean is likely due to suppressive effects of nitrate andperor insufficient numbers of *Bradyrhizobium* in the soil. They reported 6% increase in yield due to inoculation.

Sahu and Behera (1972) at Agricultural Research Station Bhubaneswar found that the grain yield of Mungbean grown in a sandy loam soil was increased to a similar extent by inoculation or phosphate application and to a much greater extent by inoculation in combination with phosphate. Kushwaha and Srivastava (1978) reported that the 1000-seed weight of Mungbean significantly increased by *Rhizobium* inoculation compared with the un-inoculated control. *Rhizobium* inoculation increased seed yields of *Vigna radiata* and *Vigna radiata* Idris and Sandhu,1980). Prasasd and Ram (1982), Prasad and Ram (1984) reported a significant increase in grain yield of Mungbean by *Rhizobium* inoculation. Gill *et al.* (1985) studied that effect of phosphorus combined with *Rhizobium phaseoli* on the yield of *Vigna radiata.* Inoculation with *Rhizobium phaseoli* significantly increased the number of branches per plant, pods per plant, seeds per pod, straw yield, grain yield and harvest index. Maiti *et al.* (1988) reported that the grain yield of green gram increased by 5-10 per cent with *Rhizobium* inoculation. The increase in yield was mainly due to the increase in pods per plant. The seed yield and yield attributes except for the test weight of green gram were significantly influenced by *Rhizobium* inoculation. The increase in seed yield was 7.7% as compared with un-inoculated control (Jat and Rathore, 1994). Naidu and Ram (1996) concluded that *Rhizobium* inoculation had a positive and significant effect on nodule weight, dry matter production, number of pods per plant and seed yields over un-inoculated control.

Crop Quality

Grasses (monocotyledons) have a lower requirement of B compared to other crop species (Shive, 1941). In general, monocotyledons require only about one fourth as much B as require by dicotyledons for normal growth (Berger, 1945). All crops do not respond equally to B even in deficient soils (Mandal *et al.*, 1956). The *Leguminosae* and *Brassicae* have relatively higher B requirement. There is a small range between boron deficiency and boron toxicity in plants. Boron has been shown to play a role in carbohydrate metabolism, sugar translocation, pollen germination, hormone action, normal growth and functioning of the apical meristem, nucleic acid synthesis, and membrane structure and function (Howe, 1998). Symptoms of boron deficiency include discontinuance of root and leaf growth, bark splitting, retardation of enzyme reactions, reduced pollen germination, and even death (WHO, 1998). The initial stages of boron toxicity in plants include yellowing of leaf tips progressing into the leaf blade. Death of chlorotic tissue occurs followed by leaf loss. This ultimately results in a loss of photosynthetic capacity and a loss in plant productivity (Lovatt and Dugger, 1984; WHO, 1998). Sensitive plants can tolerate irrigation waters with only 0.3 mg/L boron, while very tolerant plants may be able to survive where 4 mg/L boron irrigation water is used (Keren and Bingham, 1985). Boron's involvement in hormone synthesis and translocation, carbohydrate metabolism, cell membrane functions, enzyme reactions and DNA synthesis probably contributed to additional growth and yield (Ratna Kalyani *et al.* 1993).

Mungbean (*Vigna radiata*) represents important luguminase in population nutrition in tropic and subtropic areas. Being low in fat (<1.5%), rich in proteins (22.1%) with a 39-66% of the biological value and total digestibility of 76% (Chaturvedi and Ali, 2002) it has got a very respectable position among pulses, particularly in India. Due to a higher percentage of nitrogen in seeds, it requires sufficient amount of available nitrogen in soil. Legumes, including peas, generally have a greater B requirement than cereals (Mahler and Shafii 2007; Shorrocks 1997). In legumes such as peas, B is required for normal growth processes, nodule development, and the subsequent nitrogen (N_2) fixation process (Bolanos *et al.* 1996; Dell and Huang 1997). The Pacific Northwest is considered a B-deficient region. However, little regional information on critical soil B levels for spring peas are available. Application of 0.25 to 0.50 kg B ha^{-1}significantly increased pod yield of groundnut (Kumar and Sen, 2004). Most of the B fertility recommendations used in the inland Pacific Northwest are based on research conducted on alfalfa (*Medicago sativa*) in the north-central states and Canada. In fact, there is much more published information on B toxicities than deficiencies in pea production.

Yang (1994) reported that the symptoms of boron deficiency in broad bean (faba bean) are described. The critical levels of boron in upland faba bean plants and soil were 15 and 0.47 mg/kg, respectively. Sodium borate was successfully used to prevent physiological symptoms. Bagheri *et al.* (1996) that the inheritance of tolerance to high concentrations of soil boron in pea (*Pisum sativum*) was studied in five cross combinations and their reciprocals. The genotypes used in the crosses were Alma (moderately tolerant to boron), Pennant (sensitive) and Pig16, SA310 and SA132 (tolerant). Segregation patterns for boron response in F2 populations and F3-derived families were established by visual assessment of leaf damage. The segregation ratios were explained in terms of two major gene loci interacting in an additive manner with incomplete dominance at each locus. Evaluation of selected tolerant and susceptible families indicated that tolerant families contained a significantly lower concentration of boron in shoots than susceptible families.

The fertility status of red (Alfisols) and black (Vertisols) soils at the title location was assessed. The fertility of the red soil was lower than that of the black soil, although the high clay content of the black soil may result in waterlogging. It is suggested that these soils are better utilized for deep-rooted crops such as cotton, legumes, and potatoes and shallow-rooted crops such as maize and sunflower. High divalent cation contents should sustain higher yields of oilseed crops (Upperi and Dasog, 1994). Mehta (1995) reported that boronated superphosphate increased yields and improved quality of groundnuts, cotton, Medicago, garlic, vegetables, legumes and fruit crops. In separate experiments, the beneficial effects of S fertilizers on 30 crops were also demonstrated on several S deficient soils. Winiarska (1993) found that chemical composition of leachates from fly ash applied with sludge at 0, 30, 60, 90, 120 and 240 t/ha and mineral fertilizers under cereals and legumes was studied by lysimeter. Excessive salinity, sulphates and boron in water were observed, being at a maximum in the second year of the experiment. Heavy metal content was low but increased over successive years. A total of 1-2% Mg, K and Na, 14% S and 23% B relative to initial contents were leached over five years.

Nuttall *et al.* (2003) reported that the soil salinity, sodicity, and high extractable boron (B) are thought to reduce wheat yields on alkaline soils of south-eastern Australia; however, little quantitative information on yield penalties to edaphic constraints is available. The relationships between the wheat yield of a B-tolerant cultivar and soil physicochemical conditions in the Victorian Mallee were explored using ridge regression analysis, using natural variation in the field. Wheat yields in the survey ranged from 1.3 to 6.1 t/ha, with low yields attributed to inadequate soil water supply during pre-anthesis growth. Crop sequences, fallow-wheat, and pulse-wheat left greatest soil water before sowing, of the wheat crop, and lucerne-wheat the least. A descriptive model explained 54% of the variation in wheat yield, with rainfall around anthesis, available soil water in the 0.10-0.40 m layer, nitrate in the 0-0.10 m layer at sowing, and salinity and sodicity in the 0.60-1.00 m layers being important factors. Subsoil salinity (ECe) and sodicity (ESP) appear to be effective surrogates for estimating the likelihood of water extraction in the deep subsoil. The analyses suggest that subsoils need to have an ECe < 8 dS/m and ESP < 19% for crops to make use of water deep in the profile. Although soluble B ranged from 2 to 52 mg/kg in the 0.60-1.00 m layer of the alkaline soils considered, B appeared to have little correlation with root growth, water extraction, or yield of wheat, which has been attributed to B-tolerance of the cultivar tested and/or the overbearing effect of high Na^+ in these soils.

Considerable quantities of flue gas desulfurization products (FGDs) are generated when coal is burned for the production of electricity, and these products have the potential to be reused rather than discarded. The use of FGDs as soil amendments could be important in the overall management of these products, especially on acidic soils. Glasshouse studies were conducted to determine shoot concentrations of Ca, S, K, Mg, P, B, Zn, Cu, Mn, Fe, Al, Na, Mo, Ni, Cd, Cr, and Pb in lucerne (Medicago sativa), white clover (*Trifolium repens*), orchard grass (*Dactylis glomerata*), tall fescue (*Festuca arundinacea*), switchgrass (*Panicum virgatum*), and eastern gamagrass (*Tripsacum dactyloides*) grown in acidic (pH 4) soil (Typic Hapludult) amended with various levels of three FGDs and the control compounds $CaCO_3$, $CaSO_3$, and $CaSO_4$. Shoot concentrations of Ca, S, Mg, and B generally increased as levels of soil-applied FGD increased. The concentrations of Mn, Fe, Zn, and Cu were lower in shoots, especially when soil pH was high (> 7). Shoot concentrations of the trace elements Mo, Ni, Cd, Cr, and Pb were not above those reported as normal for foliage. The overall concentrations of most minerals remained near normal for shoots when plants were grown in FGD amended acidic soil (Clark and Baligar, 2003).

Laboratory experiments were conducted by Dias *et al.* (2003) to investigate the effects of residues of black oats (*Avena nuda*), oilseed radish (*Raphanus sativus*), velvet beans (*Mucuna pruriens* var. utilis) and pigeon pea (*Cajanus cajan*) on the adsorption of sulfate, borate, zinc and copper by an acidic Oxisol. The plant materials used were collected at flowering stage. Zinc, copper, boron and sulfate were prepared at the following concentrations: Zn (4, 8, 16, 32 and 64 mg/kg); Cu (2, 4, 8, 16, 32 and 64 mg/kg); B (1, 2, 5, 10 and 20 mg/kg); and S (10, 20, 50, 100 and 200 mg/kg). At least for short term experiment, all plant residues increased Cu^{2+} and Zn^{2+} adsorptions by the soil. On the other hand, the presence of plant residue decreased the adsorption of S and B by soil. These results indicated that for short term effect plant residues decreased the availabilities of Cu and Zn through metal-organic complex reactions and increased availabilities of S and B through competition with organic anions by the adsorption sites on the soil.

The effects of modifying boron (B) and calcium (Ca^{2+}) concentrations on the establishment and development of *rhizobial* symbiosis in *Pisum sativum* plants grown under salt stress were investigated. Salinity almost completely inhibited the modulation of pea plants by *Rhizobium leguminosarum* bv. viciae 3841. This effect was prevented by addition of Ca^{2+} during plant growth. The capacity of root exudates derived from salt-treated plants to induce *Rhizobium* nod genes was not significantly decreased. However, bacterial adsorption to roots was highly inhibited in plants grown with 75 mM NaCl. Moreover, *R. leguminosarum* 3841 did not grow in minimal media containing such salt concentration. High Ca^{2+} levels enhanced both *rhizobial* growth and adsorption to roots and increased nodule number in the presence of high salt. Nevertheless, the nodules developed were not functional unless the B concentration was also increased. Because B has a strong effect on infection and cell invasion, these processes were investigated by fluorescence microscopy in pea nodules harbouring an *R. leguminosarum* strain that expresses green fluorescent protein. Salt-stressed plants had empty nodules and only those treated with high B and high Ca^{2+} developed infection threads and exhibited enhanced cell and tissue invasion by *Rhizobium*. Overall, the results indicate that Ca^{2+} promotes nodulation and B nodule development leading to an increase of salt tolerance of nodulated legumes (El-Hamdaoui *et al.*, 2003). Sahu and Behera (1972) reported that root and shoot nitrogen contents and grain protein contents in Mungbean grown in a sandy loam soil were increased to a similar extent by inoculation or phosphate application and to a much greater extent by inoculation with phosphate. Soil nitrogen was increased by 18 per cent with inoculation. Deshmukh and Joshi (1973) studied the data on the production of leaf protein which revealed that inoculation of cowpea with *Rhizobium* significantly increased the yields of crude protein, but did not affect protein extraction. Gupta *et al.* (1979) found that *Rhizobium* inoculation significantly increased seed protein contents in green gram. Idris and Sandhu (1979) conducted pot experiments and reported that inoculation of Mungbean with all the strains of *Rhizobium phaseoli* increased nitrogen-fixing efficiency of plants by 31 to 41 per cent.

Prasad and Ram (1982) studied the effect of inoculation on *Vigna radiata* seeds with a *Rhizobium* strain and found a significant increase in nitrogen, protein and carbohydrate contents of nodules. Srivastava and Varma (1982) pointed out that seed inoculation with *Rhizobium* increased seed protein content of green gram. A field trial with *Vigna radiata* cultivars showed that inoculation with two *Rhizobium phaseoli* strains increased crude protein contents but inoculation with mix culture was more effective than single strain (Prasad and Ram, 1984). Further, the *Rhizobium* inoculation increased the protein content as well (Naidu and Ram, 1996). Singh *et al.* (1994) reported that inoculation of seeds with *Rhizobium* culture significantly increased the content and total uptake of nitrogen, phosphorus and crude protein of seeds. Kuzmanovic *et al.* (2003) in their field experiment compared four *Bradyrhizobium* species (2001, 2801, 5338, and 12758) obtained Brazil and New Zealand and found that strain 2801 was superior to other strains which showed maximum increase in grain yield and total nitrogen uptake by grain. *Rhizobium* inoculation was found to be superior regarding the concentration of metheonine in the grains of Mungbean. Inoculation also increased the protein content of grains (Shahi *et al.* 2002). Out of three Rhizobial strains (Ludhiana, IARI and local isolate), the seed protein content was highest with inoculation of Ludhiana isolate whereas straw protein content was not affected by Bradyrhizobium inoculation. The result was reported by Sharma (2003) after his investigation in silty clay loam soil of Palanpur, H.P., India on green gram cultivar Pusa Baisakhi

Soil Fertilizer after Post-harvest Pulses

The activity of rhizobial bacteroids inside the root nodules often results in large amounts of N_2 fixed, ranging from 25 to 201 kg N ha-" in tropical grain legumes (Dakora & Keya, 1997). Sheikh *et al.*, 2009 found that when mungbean was cultivated in a Transplant Aman-Mustard-Mungbean cropping pattern soil pH, organic matter, total N, P, S, Zn and B content was increased. A field experiment was conducted by Shamsuddoha *et al.* (2011) at the research farm of Sher-e-Bangla Agricultural University, Dhaka, Bangladesh during the period from February to May 2008 to study the effects of sulfur and boron on nutrients of mungbean and the soil health. The variety of BARI mung-6 was used as the test crop. In most of the cases of sulfur and boron application, the individual treatments 8 kg S and 2 kg B ha^{-1} showed the highest nutrient concentration and maximum uptake of N, P, K, S and B by the seed and the stover of mungbean. Side by side sulfur and boron with the 8 kg S plus 2 kg B ha^{-1} treatment combination revealed the highest nutrient concentration and maximum uptake of N, P, K, S and B by the seed and the stover of mungbean. In case of S treatments the highest pH, organic matter, total N, available P, exchangeable K, available S and available B was recorded from S8 (8 kg S ha^{-1}) and the lowest pH, organic matter, total N, available P, exchangeable K, available S and available B was observed from S0 (no sulfur) treatment. In case of B application the highest pH, organic matter, total N, available P, exchangeable K, available S and available B was recorded from B_2 (2 kg B ha^{-1}) and the lowest pH, organic matter, total N, available P, exchangeable K, available S and available B was found from B_0 (no boron) treatment. For the combined application of S and B the highest pH, organic matter, total N, available P, exchangeable K, available S and available B was recorded from S_8B_2 (8 kg S ha^{-1} and 2 kg B ha^{-1}) treatment combination causing better soil health and the lowest pH, organic matter, total N, available P, exchangeable K, available S and available B was observed from S_0B_0 (no sulphur and boron). Individually 8 kg S and 2 kg B ha^{-1}, but as combined 8 kg S plus 2 kg B ha^{-1} treatments revealed the highest nutrient concentration and maximum uptake of N, P, K, S and B by the seed and the stover of mung bean and also showed the improvement in the nutrient status of post-harvest soil. Bilen *et al.* (2011). Found that a randomized block design with three replications was used in this experiment. Field experiments were conducted to evaluate the effects of B levels (0, 1, 3, 6 and 9 kg ha^{-1} B) on soil microbial population in cultivated wheat (*Triticum vulgare* cultivar Dogu-88) soils. Statistical results showed a significant ($p < 0.01$) differences between B applications and microbial population and between B applications and microbial respiration in 0 to 30 and 30 to 60 cm soil depths. The highest population of bacteria, fungi, actinomycetes and CO_2-C production were observed at 3 kg ha^{-1} B level in different growing periods of the plant and different soil depths. Urease, phosphatase and dehydrogenase enzyme activities showed a significant ($p < 0.01$) positive correlation with B applications. The highest urease activity was observed in 6 kg ha^{-1} B level and the highest phosphatase and dehydrogenase enzyme activities were observed in 3 kg ha^{-1} B level in harvest period in both soil depths. According to Singh *et al.* (2011) found that the application of 50% of recommended dose NPK fertilizers along with the application of 5 t FYM + 250 kg lime + 20 kg S + 1 kg B ha^{-1}, enhanced the organic carbon and available N, P, K and S contents in the soil.

References

Adams, J.A., Hamjah, Z. and Swift, R.S. Availability and uptake of boron in a group of pedogenetically-related Centerbury, *New Zealand soils. Aust. J. Soils*, **29**: 415-423, 1991.

Ahlawat, I.P.S., Gangaiah, B. and Ashraf, Z.M. Nutrient management in chickpea. In: Chickpea breeding and management Yadav, S.S., Redden, R., Chen, W., Sharma, B., *CAB International, Wallingford, Oxon, United Kingdom*. 213-232, 2007.

Ahmad, R. and Irshad, M. Effect of boron application time on yield of wheat, rice and cotton crop in Pakistan, *Soil & Environment*, 30 (1): 50, 2011.

Ali, M., Krishnamurthy, Y.L., Saxena, N.P., Rupela, O.P., Kumar, J. and Johansen, C. Scope for genetic manipulation of mineral acquisitions in chickpea, *Plant Soil*, **245**: 123-134. 2002.

Ali, M. and Mishra, J.P. Effect of foliar nutrition of boron and molybdenum on chickpea, *Indian J Pulses Res*, **14**: 41-43, 2001.

Ali, A., Zia, M.S., Hussain, F. and Khan, M.B. Boron requirement of rice and its management for rice production, *Pakistan J. Soil Sci.*, **11**: 68-71, 1996.

Aitken, R.L., Jeftercy, A.J. and Coptern, B.L. Evaluation of selected extractants for boron in some queensland soils, *Aust. J. Soil Research*. **25**: 263-273, 1987.

Alpaslan, A. and Gunes, A. Interactive effects of boron and salinity stress on the growth, membrane permeability and mineral composition of tomato and cucumber plants, *Plant Soil*, 236: 123–128, 2001.

Alwitt, R.S. The point of zero charge of pseudoboehmitte, *J. Colloid Interface Sci,* **40:** 195-198, 1972.

Ambak, K. and Tadano, T. Effect of micronutrient application on the growth and occurrence of sterility in barley and rice in a Malaysian deep peat soil. Micronutrients in Agriculture: Pakistani Perspective. NationalFertilizer Development Centre, Islamabad, Pakistan. 51p. 1991.

Amery, M.M., Hamza, J.H. and Fuller, M.P. Effect of Boron Foliar Application on Reproductive Growth of Sunflower *Helianthus annuus* L, *International Journal of Agronomy,* **5**: 2011.

Angers, D.A. and Eriksen-Hamel, N.S. Full-inversion tillage and organic carbon distributionin soil profiles: a meta-analysis. *Soil Sci. Soc. Of Am. Jour.* **72**(5): 1370-1374, 2008.

Anitha M.S., Kumar K.S., Nair, A., Shivaprasad, K.M., Naidu, C.R. and Sarkar Dipak, L.G.K. Soil Boron and Its Fractions in Agro-Climatic Zones of Karnataka, *Clay Research,* **2**(1). 25-33, 2013.

Ankerman, D. and Langer, S. Soil and plant analysis. *A & L Agricultural Laboratories,* Inc., New York, United States, 1974.

Anonymous. *Agricoop.nic.in/imagedefault/trade/pulses%20profile.doc,* 2011.

Anonymous. Prepration of GPS and GIS based soil fertility map of the selected districts of country. Project Report, 2012.

A.O.A.C., 1950. Official and tentative methods of analysis of the association of official agricultural chemists, 7th Edition, Washington, 910.

Arai, Y., Sparks, D.L. Phosphate reaction dynamics in soils and soil minerals: a multiscale approach, *Adv Agron,* **94**: 135–179, 2007.

Ardesna, R.B., Modhwadia, M.M., Khanpara, V.D. and Patel, J.C. Response of greengram to nitrogen, phosphorus and *Rhizobium* inoculation, *Indian Journal of Agronomy,* **38**(3): 490-492, 1993.

Aref, Farshid. The Effect of Boron and Zinc Application on Concentration and Uptake of Nitrogen, Phosphorous and Potassium in Corn Grain, *Indian Journal of Science and Technology,* 4(1):785-791, 2011.

Arora, S. and Chahal, D. S. Available boron content in some benchmark soils of Punjab under different moisture regimes in relation to soil characteristics, *Agropedology,* **15** (2): 90-94, 2005.

Awomi, T.A., Singh, A.K., Kumar, M., Bordoloi, L.J. Effect of phosphorus, molybdenum and cobalt nutrition onyield and quality of mungbean (*Vigna radiata* L.) in acidic soil of northeast India, *Indian J Hill Farm,* **25**(2): 22–26. 2012.

Azevedo, W.R., Faquin, V., Moreira, F.M. de S., Oliviera, J.A.C., Lisboa, C.C., Moreira, F.M., Azevedo, W.R. and Oliviera, J.A.C. Boron effects on nodule development in Pea Plants cultivated in lowland soils, *Pesquisa- Agropecuiria-Brasileira,* **37** (8): 1137-1143, 32.

Bagheri, A., Paull, J.G. and Rathjen, A.J. Genetics of tolerance to high concentrations of soil boron in peas (*Pisum sativum* L.), *Euphytica* ,**87**(1): 69-75, 1996.

Balasundaram, C.S., Lakshminarasimhan, C.R., Rajakannu, K. and Kanakadoss, A. Water soluble boron status of Tamil Nadu soils. *Madras Agric. J.* **59**: 426-430, 1972.

Basak, M.K. and Goyal, S.K. Studies on the biology of the legumes *Rhizobium* symbiosis, modulation and cross inoculation trails with the tree legumes and cultivated legumes, *Annals of Arid Zone,* **19**(4): 427-431, 1980.

Basu, T.K., Barui, A.K. And Bandopadhayay, S. Nodulation, growth and seed yield of mung in response to *Rhizobium* inoculation and nitrogen application, *Indian Journal of Mycological Research,* **27**: 153-157, 1989.

Bendale, J.R., Narayan, N. and Kibe, M.M. Trace elements contents of black cotton soils of a few citrus growing tracts of Bombay state, *Poona Agric. Coil. Mag.,* **42**: 1-4, 1951.

Ben-Gal, A. and Shani, U. Water Use and Yield of Tomatoes under combined excess boron and water stress, *Plant and Soil,* 256:179-186, 2003.

Ben-Gal, A. and Shani, U. Yield, transpiration and growth of tomatoes under combined excess boron and salinity stress, *Plant and Soil,* 247:211-221, 2002.

Berger, K.C. and Troug, E. Boron availability in relation to soil reaction and organic matter content, *Soil Sci. Soc. Am. Proc.,* **10** : 113-116, 1945.

Berger, K.C. and Troug, E. Boron determination in soils and plants, *Ind. Eng. Chem. Anal Ed.,* **11**: 540-544, 1939.

Beyrouty, C A., Vanscoyoc, G.E. and Feldkamp, J.R. Evidence supporting specific adsorption of boron on synthetic aluminum hydroxides, *Soil Sci. Soc. Am. J.,* **48**: 284–287, 1998.

Bhandari, A.R., and Randhawa, N.S. Distribution of available micronutrients in soils of apple orchards in Himanchal Pradesh, J. *Inidan Soc. Soil Sci.,* **33**: 171-174, 1985.

Bharti, N., Murtuza,M., Singh, A.P. Effect of boron on symbiotic nitrogen fixation in chickpea (*Cicer arietinum*). *Journal of Research, Birsa Agricultural University,* **15**(1): 19-23, 2003.

Bhatnagar, R.S., Attri,S. C., Mathur,G.S. and Chaudhary,R.S. Boron adsorption equilibrium in soils, *Annals Arid Zone,* **18**: 86–95, 1979.

Biggar, J. W. and Fireman, M. Boron adsorption and release by soils, *Soil Sci. Soc. Am. Proc.,* **24:** 115–120. 1960.

Bilen, S., Murat, B. And Sougata, B. The effects of boron management on soil microbial population and enzyme activities, *African Journal of Biotechnology* ,**10**(27): 5311-5319, 2011.

Bingham, F.T. Am. Chem. Soc.,Washington, DC, USA. Boron in cultivated soils and irrigation waters. In: Trace Elements in the Environment. Advances in Chemistry, **123**: 130–138. 1973,

Bingham, F.T., Page, A.L., Coleman, N.T. and Flach, K. Boron adsorption characteristics of selected soils from Mexico and Hawaii, *Soil Sci. Soc. Am. J.,* **35**: 546-550, 1971.

Blesa, M.A., Maroto,A.J., Gand Regazzoni, A.E. Boric acid adsorption on magnetite and zirconium dioxide, *J. Colloid Interface Sci.,* **99**: 32–40, 1984.

Blevins, D.G. and Lukaszewski, K.M. Boron in plant structure and function, *Annu Rev Plant Physiol Plant Mol Biol,* **49**:481–500, 1998.

Blevins, D.G. and Lukaszewski, K.M. Proposed physiologic functions of boron in plants pertinent to animal and human metabolism, *Environ Health Perspect,* **102**(7):31–33,1994.

Bloesch, P.M., Bell, L.C. and Hughes, J. D. Adsorption and desorption of boron by goethite Aust. *J. Soil Res.* **25:** 377–390, 1987.

Bolaños, L., Cebrián, A., Redondo-Nieto., M- Rivilla, R. and Bonilla, I. Lectin-like glycoprotein PsNLEC-1 is not correctly glycosylated and targeted in boron deficient pea nodules, *Mol Plant Microbe Interact,* **14**: 663–670, 2001.

Bolaños, L., Brewin, N.J. and Bonilla, I. Effects of boron on *Rhizobium*legume cell-surface interactions and nodule development, *Plant Physiol* , **110**: 1249–1256, 1996.

Bonilla, I., Mergold-Villase˜nor, C., Campos, M.E., S´anches, N. and P´erez, H. The aberrant cell walls of boron deficient bean root no dules have no covalently boundhydroxyproline/proline-rich proteins, *Plant Physiol,* 1997.

Bouycucos, G.J. Hydrometer method improved for making particle size analysis of soils. *Agron. J.,* **54**: 464-465, 1962.

Brady, N. and Weil, R. R. The nature and properties of Soils, *Pearson Education Upper Saddle River, New Jersey,* 321 – 766, 1999.

Brown, P.H., Bellaloui, N., Wimmer, M.A., Bassil, E.S., Ruiz, J., Hu, H., Pfeffer, H. and Dannel, F. V. Boron in plant biology. *Plant Biol* **4**: 205-223, 2002.

Brown, P.H. and Shelp, B.J. Boron mobility in plants, *Plant Soil,* **193**: 85 -101, 1997.

Bray R.H. and Kurtz, L.T. (1945) Determination of total organic and available forms of Phosphorus in soils. *Soil Sci.,* **59**, 39-45.

Cakmak, I. and Römheld, V. Boron deficiency-induced impairments of cellular functions in plants, *Plant Soil,* **193**: 71–83, 1997.

Cakmak, I., Kurz, H., Marschner, H. Short-term effects of boron, germanium and high light intensity on membrane permeabilityin boron deficient leaves of sunflower, *Physiol. Plant.,***95**: 11–18, 1995.

Cakmak, I. and Römheld, V. Boron deficiency-induced impairments of cellular functions in plants. *Plant Soil* **193**, 71–83, 1997.

Camacho-Cristóbal, J.J., Herrera-Rodríguez, M.B., Beato, V.M., Rexach, J., Navarro-Gochicoa, M.T., Maldonado, J.M. and González-Fontes, A. The expression of several cell wall-related genes in Arabidopsis roots is down-regulated under boron deficiency, *Env. Exp. Bot.,* **63**: 351-358, 2008.

Carpena,Ramo´n. O., Elvira Esteban, M., Jose´ Sarro., Jesu s Pen alosa., Agustı´n Ga´rate., Juan J. Lucena. and Pilar Zornoza. Boron and calcium distribution in nitrogen-fixing pea plants, *Plant Science.,* **151** (2000): 163–170.

Carpita, N.C. The biochemistry of the 'growing' plant cell wall, In *Physiologyof Cell Expansion During Plant Growth.,* 1987.

Car twright, B., Tiller, B.A., Zarcinos, B.A. and Spour L.R. The chemical assessment of boron status of soils, *Aust. J. Soil Research.* **21**: 321-323, 1985.

Chakraborty, S.K., Sinha, H. and Prasad, R. Distribution of boron and molybendum and their relationship with certain properties of soils from Assam, *J. Indian soc. Soil Sci.*, **30**: 92-93, 1982.

Chaturvedi, S.K. and Ali, M. Poor man's meat need fresh fillip, The Hindu, *Survey of Indian Agriculture.*, 2002.

Chaudhry, F. M., Latif, A., Rashid, A. & Alam, S. M. Response of the rice varieties to field application of micronutrient fertilizers, *Pakistan Journal of Industrial and Scientific Research.*, **19**. 134-139, 1976.

Chaudhry, F.M., Latif, A., Rashid, A., and Alam, S.M. Response of rice varieties to field application of micronutrient fertilizers, *Pak. J. of Sci. and Ind.*, **19**: 134-139, 1977.

Cheng, C. and Rerkasem, B. Effects of boron on pollen viability in wheat, *Plant Soil.*, **155/156**: 313–315, 1993.

Choi, W. W. and Chen, K.Y. Evaluation of boron removal by adsorption on solids, *Environ. Sci. Technol.*, **13**: 189–196, 1979.

Chopra, S.L. and Kanwar, J.S. (1982) *Analytical Agricultural chemistry* .Kalyani publisher, New Delhi.

Chude, V. The profile distribution of total and extractable boron in cacao-growing soils in southwestern Nigeria. *Plant and Soil* **107**: 293-295, 1988.

Chesin, L. and Yien, C.H. (1950) .Turbid metric determination of available sulphates, *Proc.soil.sci.soc.am.*, **14**,149-151.

Clark, R.B. and Baligar, V.C., Mineral concentrations of forage legumes and grasses grown in acidic soil amended with flue gas desulfurization products, *Communications-in-Soil-Science-and-Plant-Analysis.*, **34**: 11-12, **48**: 1681-1707, 2003.

Cooke, G.W. Fertilizing for Maximum Yield, *Granada, UK.*, **465**, 1982.

Couch, E.L. and Grim, R.E. Boron fixation by illites. *Clays Clay Miner.*, **16**: 249–256, 1968.

Cruz, M.C.P., Nakamura, A.M. and Ferreira, M.E. Adsorcao de boro pelo solo: efeito da concentracao e do pH, *Pesquisa Agropecuaria Brasileira, Brasilia.*, 22(3): 621-626, 1987.

D′enarie J., Cullimore,J. Lipo-oligosaccharide nodulation factors: a minireview new class of signaling molecules mediating recognition and morphogenesis, *Cell.*, **74**: 951–954, 1993.

Da Silva, A.R., de Andrade, J.M.V. Efeito micronutrientes no chochamento do trigo de seqqueiro e nas culturas de soja e arroz, em latossolo vermelhoamarelo [Influence of micronutrients on the male sterility, on upland wheat and on rice and soybean yields, in red-yellow latosol], *Pesq. Agropec. Bras. Brasilia.*, **18**: 593–601, 1983.

Das, D.K. Micronutrients their behavior in soils and plants, First Edition, *Kalyani publishers, Ludhiana*, **75**, 2000.

Das, D.K. In: Micronutrients: Their behaviour in soils and plants. 2nd ed. Kalyani publishers. **167**, 2007.

Das S.K., Sahoo R.K., Muralidhar J. and Nayak B.K. Mineralogy and geochemistry of profiles through lateritic nickel deposits at Kansa, Sudinda, Orissa. Journal Geological Society of India, **53**: 649-668, 1992.

Dakora F.D. and S.O. Keya Contribution of legume nitrogen fixation to sustainable agriculture in Sub-Saharan Africa, *Soil Biology and Biochemistry*, **29:** 809-817, 1997.

Datta, S.P. and Rathore, G.S. Status of available boron in some soils of Northern hill region of Chattisgarh, Madhya Pradesh, *J. Res. (BAU)* **4**: 37-42, 1992.

Datta, S.P., Rattan, R,K" Suribabu, K. and Datta,S.C. Fractionation and colorimetric determination of boron in soils, *Journal of Plant Nutrition and Soil Science* **165**, 179-184. 2002.

Davis, T.M., Sanders, D.C., Nelson, P.V., Lengnick, L. and Sperry, W.J. Boron improves growth, yield, quality and nutrient content of tomato. *J Am Soc. Hort. Sci.*, 128(3):441-446, 2003.

Debnath, M. R., Jahiruddin,M., Rahman, M. M. and Haque,M. A. Determining optimum rate of boron application for higher yield of wheat in Old Brahmaputra Floodplain soil. J. Bangladesh, *Agril. Univ.*, **9**(2): 205–210, 2011.

Debnath, P. Assessment of critical limit of available boron for rice in old alluvial zone of West Bengal. *ORYZA- An International Journal on Rice.*, **49**(4): 267-272, 2012.

Dell, B., Huang, L.B. Physiological response of plants to low boron. *Plant and Soil*, **193**: 103-120, 1997.

Dell, B., Malajczuk, N. Boron deficiency in eucalypt plantations in China. *Can. J. For. Res.*, **24**, 2409–2416, 1994.

Deshmukh, M.G. and Joshi, R.N. Effect of *Rhizobium* inoculation on the extraction of protein from the leaves of cowpea. *Indian Journal of Agricultural Sciences*, **43**(6): 539-542, 1973.

Dey, A., Dwivedi, B.S., Meena, M.C. and Datta, S.P. *Journal of the Indian Society of Soil Science.*, **61**(3) :179-187: 2013.

Dias, A.C.B., Pavan, M.A., Miyazawa, M. and Zocoler, D.C. Plant residues: short term effect on sulphate, borate, zinc and copper adsorption by an acid oxisol. *Brazilian-Archives-of-Biology-and-Technology.*, **46**(2): 199-202, 2003.

Dwivedi, G.K., Dwivedi, M. and Pal, S.S. Modes of application of micronutrients in acid soil in soyabean- wheat crop sequence. *J. Indian Soc. Soil Sci.*, **38**: 458-463, 1990.

El-Hamdaoui, A., Redondo-Nieto, M., Torralba, B., Rivilla, R., Bonilla, I. and Bolaños, L. Influence of boron and calcium on the tolerance to salinity of nitrogen-fixingpea plants, *Plant Soil*, **251**: 93-103, 2003.

Elrashidi, M.A. and O'Connor, G.A. Boron sorption and desorption in soils, *Soil Sci. Soc. Am. J.*, **46**: 27-31, 1982.

Elseewi, A.A. Some observations on boron in water, soils and plants at various locations in Egypt, *Alex. J. Agric. Res.*, **22**: 463–473, 1974.

Elseewi, A.A. and Elmalky, A.E. Boron distribution in soils and waters of Egypt, *Soil Sci. Soc. Am. J.*, **43**: 297–300, 1979.

Evans, C.M. and Sparks, D.L. On the Chemistry and mineralogy of boron in pure and mixed system: a review, *Communication on in Soil Science and Plant Analysis*, **4**:827-846, 1983.

Firoz, Z.A., Jaman, M.M., Alam, M.S. and Alam, M.K. Effect of boron application on the yield of different varieties of broccoli in hill valley, *Bangladesh J. Agril. Res.*, 33(3): 655-657, 2008.

Fleet, M.E.L. Preliminary investigations into the sorption of boron by clay minerals, *Clay Miner.*, **6**: 3–16, 1965

Fricke, F. and Leonhardt, I. Isoelektrischer Punkt und Pufferl¨osung, *Die Naturwissenschaften.*, **37**: 428, 1950.

Furlani, A.M.C., Carvalho, C.P., de Freitas, J.G. and Verdial, M.F. Wheat cultivar tolerance to boron deficiency and toxicity in nutrient solution, *Scientia-Agricola.*, **60**(2): 359-370, 2003.

Gajbhiye, K.S., Benerjee, N.K. and Goswami, N.N. Correlation study of water soluble B with Ece and silt plus clay in Non- Saline and low saline soils, *J. Indian Soc. Soil Sci.*, **28**: 251-253, 1980.

Gandhi, S.C. and Mehta, B.V. Water soluble boron contents of soils of Gujarat and Saurashtra, *J. Indian Soc. Soil Sci.*, **6**: 95-101, 1958.

Ganguly, B. Note on seedless ness in some wheat varieties caused by boron deficiency, *Indian J. Agric. Sci.*, **43**: 384–386. 1979.

Gestring, W.D. and Soltanpour, P.N. Evaluation of the ammonium bicarbonate-DTPA soil test for assessing boron availability to alfalfa, *Soil Sci. am. J.*, **48**: 96-100, 1984.

Ghani, M.O. and Haque, A.K.M.F. Studies of the boron status pf some Bengal soils, *Indian J. Agric. Sci.*, **15**: 257-262, 1945

Ghatak, R., Jana, P.K., Sounda, G., Ghosh, R.K. and Bandyopadhyay, P. Effect of boron on yield, concentration and uptake of N, P and K by wheat grown in farmer's field on red and laterite soils of Purulia, West Bengal, *Indian-Agriculturist.*, **50**(1/2): 15-77, 2006.

Gili, A.A., Trucco, R., Niveyro, S., Balzarini, M., Estelrich, D., Quiroga, A. and Noellmeyer, E. Soil texture and carbon dynamics in savannah vegetation patches of Central Argentina, *Soil Science Society of America Journal.*, **74**(2):647-657, 2010.

Gill, M.A., Naimat Ali. And Nayyar, M.M. Relative effect of phosphorus combined with potash and *Rhizobium phaseoli* on the yield of *Vigna radiata* (mung), *Journal of Agricultural Research, Pakistan*, **23**(4): 279-282, 1985.

Goldbach, H.E., Yu, Q., Wingender, R., Schulz, M., Wimmer, M., Findeklee, P., Baluska, F. Rapid response reactions of roots to boron deprivation, *J. Plant Nutr. Soil Sci.*, **164**: 173-181, 2001.

Goldbach, H. E. Influence of boron nutrition on net uptake and efflux of 32P and 14C-glucose in *Helianthus annuus* roots and cell cultures of *Daucus carota, J. Plant Physiol.*, **118**: 431–438, 1985.

Goldber, S. CRC Press, Boca Raton, USA. Chemistry and mineralogy of boron in soils. In Boron and its Role in Crop Production. *Ed. U C Gupta*. 3–44, 1993.

Goldberg, S. and Glaubig,R.A. Boron adsorption on aluminum and iron oxide minerals, *Soil Sci. Soc. Am. J.*, **49**: 1374–1379, 1985.

Goldberg, S., Corwin, D.L., Shouse, PJ. and Suarez, D.L. Prediction of boron adsorption by field samples of diverse textures, *Soil Science Society of America Journal*, **69**:1379-1388, 2005.

Goldberg, S., Forster, H.S. and Heick, E.L. Boron adsorption mechanisms on oxides, clay minerals and soils inferred from ionic strength effects, *Soil Sci. Soc. Am. J.*, **57**: 704–708, 1993.

Goldberg, S. Reaction of boron with soils, *Plant Soil*, **193**:35-48, 1997.

Grewal, J.S., Bhumbla, D.R. and Randhawa, N.S. Available micronutrient status of Punjab, Haryana and Himanchal Pradesh, *J. Indian Soc. Sci.*, **17** : 27-31, 1969.

Gu, B. and Lowe, L.E. Studies on the adsorption of boron on humic acids, *Can. J. Soil Sci.*, **70**: 305–311, 1990.

Gu, B. and Lowe, L.E. Observations on the effect of a soil polysaccharide fraction on boron adsorption by clay minerals, *Can. J. Soil Sci.*, **72**: 623– 626, 1992.

Gupta, N., Trivadi, S.K., Bansal, K.N. and Kaul, R.K. Vertical distribution of micronutrient cations in some soil series of Northern Madhya Pradesh, *J. Ind. Soc. Soil Sci.*, **51**(4): 517-522, 2003.

Gupta, R.P., Chahal, V.P.S., Chahal, D.S. and Kalra, M.S. Effect of inoculation and pelleting on mung, *Journal of Research*, P.A.U., Ludhiana., **13**(4): 395-397, 1976.

Gupta, U.C. Boron nutrition of crops, *Adv. Agron.*, **31**: 273– 307, 1979.

Gupta, U.C. A simplified method for determining hot water soluble boron on podzol soils, *Soil Sci.*, **103**: 424-428, 1967.

Gupta, U.C. Relationship of total and hot water soluble boron. And fixation of added boron to properties of podzal soils, *Soil Sci. Am. Proc.*, **32:** 45-48, 1968.

Gupta, U.C., Jame, Y.W., Campbell, C.A., Leyshon, A.J., Nicholaichuk, W. Boron toxicity and deficiency : a review, *Can. J. of Soil Sci.*, **65**(3)381-409, 1985.

Haby, V.A., Kalra, Y.P. Soil fertility and management of acid Coastal Plain soils for crop production. Fine tuning soil and plant analysis for economics and environmental betterment. 2001. International Symposium on soil and plant analysis, Edmonton, Alberta, Canada, 21-27 July 2001, *Communications in Soil Science and Plant-Analysis*, **33**(15-18): 2497-2520, 2002.

Hadad, M. A., and Loynachan,T. E. Abundance and characterization of cowpea miscellany Rhizobium from Sudanese soils, *Soil Biol. and Biochem.*, **17**:717-721, 1985.

Hansch, R. and Mendel, R.R. Physiological functions of mineral micronutrients (Cu, Zn, Mn, Fe,Ni, Mo, B, Cl), *Curr. Opin. Plant. Biol.*, **12**: 259–266, 2009.

Harada, T and Tamai, M . Some factors affecting behavior of boron in soil.I.Some soil properties affecting boron adsorption of soil, *Soil Sci. Plant Nutr.*, **14**: 215-224, 1968.

Harder, H. Einbau von Bor in detritische Tonminerale, *Experimente zur Erklarung des Borgehaltes toniger Sedmente. Geochim. Cosmochim.*, **21**: 284-294, 1961.

Hatcher, J.T., Bower, C.A., Clarck, M. Adsorption of boron by soils as influenced by hydroxialuminum and surface area, *Soil Science, Baltimore,* **104**(6): 422-426, 1967.

Herridge, D.G., Robertson, M.J., Cocks, B., Peoples, M.B., Holland, J.F. and Heuke, L. Low nodulation and nitrogen fixation of mungbean reduce biomass and grain yields, *Australian Journal of Experimental Agriculture.,* **45**(2/3): 269-277, 2005.

Hingston, F. J. Reactions between boron and clays:Aust, *J. Soil Res.,* **2**: 83- 95, 1964.

Hopmans, P., Flinn, D. W. Boron deficiency in *Pinus radiate* D. Don and the effect of applied boron on height growth and nutrient uptake, *Plant Soil.* **79**, 295–298, 1984.

Hossain, M. A., Jahiruddin, M. and Khatun, F. Response of wheat and mustard to manganese, zinc and boron in Calcareous Soil, *Bangladesh J. Crop Sci.,* 6(1&2): 51-56, 1995.

Hou, J., Evans, L.J. and Spires, G.A. Boron fractionation in soils, *Communication in Soil Science and Plant Analysis.* **25**, 1841-1853, 1994.

Hou, J., Evans, L.J. and Spires, G.A. (1994). Boron fractionation in soils, *Communication in Soil Science and Plant Analysis.* 25, 1841-1853.

Howe, P.D. A review of boron effects in the environment, *Biological Trace Element Research.* **66**: 153-166, 1998.

Howlader M. I. A., Islam, M. S., Rafiquzzaman, S., Chowdhury, M.M. M. A. A. and Rashid, M. M.. Yield response of mungbean, t. Aus and t. Aman rice to npk fertilizers in ganges tidal floodplain, *J. Soil. Nature,* **1**(3): 11-19, 2007.

Hu H, Penn SG, Lebrilla CB, Brown PH (1996) Isolation and characterisation of soluble B-complexes in higher plants. The mechanism of phloem mobility of boron, *Plant Physiol.,* 113, 649-655.

Huang, J. S. and Snapp, S. S. The effect of boron, calcium and surface moisture on shoulder check, a quality defect in fresh market tomato, *J. Amer. Soc. Hort. Sci.,* 129(4): 599-607, 2004.

Huang, L.B., Bell, R.W., Dell, B. Boron supply into wheat (Triticum aestivum L. cv. Wilgoyne) ears whilst still enclosed within leaf sheath, *J. Exp. Bot.,* **52**: 1731–1738, 2001.

Huang, L.B., Pant, J., Bell, R.W., Dell, B., Deane, K. Effects of boron deficiency and low temperature on wheat sterility. *Sterility Wheat Subtropical Asia ACIAR Proc.* **72**: 90-101, 1996.

Hunt, C. D. Dietary boron: An overview of the evidence for its roles in immune function. *Trace Elem. Exp. Med.* **16**, 291–306, 2003.

Hunt, C.D. Dietary boron modified the effects of magnesium and molybdenum on minera metabolism in the cholecalciferol-defi cient chick, *Biol Trace Elem Res.,* **22**: 201–220, 1989.

Hunt, C.D. Dietary boron: an overview of the evidence for its role in immune function, *J. Trace Elem Exp Med.,* **16**: 291–306, 2003.

Hussain, F. and Yasin, M. Soil fertility monitoring and management in Ricewheat system, *Annual Report, LRRP, NARC, Islamabad.* 1-33, 2004.

Idris, Mohammad and Sandhu, G.R. *Rhizobium* inoculation as an urd in the mungbean cultivation, *J. Sci. Res.*, **31**(3-4): 165-173, 1979.

In, J., Martens, D.C., Zelasny, L.W. Plant availability of applied and native boron in soils with diverse properties. Plant and Soil, **105**: 127-132, 1988.

Islam, M. B. Requirement of boron for mustard, wheat, and chickpea based rice cropping patterns. Ph. D. Dissertation, Department of Soil Sci, *Bangladesh Agricultural University, Mymensingh*, 2005.

Islam, M.A., Saha, U.K., Islam, M.S. and Saha, R.R. Fertilizing chickpea for high yield with special emphasis on boron and molybdenum in grey floodplain soils of Bangladesh. *Annals-of-Bangladesh-Agriculture*. **5**(1): 43-49, 1995.

Jackson, J.F. Borate control of energy-driven protein secretion from pollen and interaction of borate with auxin or herbicide—a possible role for boron in membrane events. **99**: 221– 29, 1991.

Jackson, M.L. (1973) Soil chemical Analysis. Prentice Hall Inc. Enlewood cliffs, USA.

Jahiruddin, M., Hoque, M.S., Haque, A.K.M. and Roy, P.K. Influence of boron, copper and molybdenum on grain formation in wheat. *Crop Res*. **5**(1): 35-42, 1992.

Jamjod, S., Rerkasem, B., Genotypic variation in responses of barley to boron deficiency. Plant Soil 215, 65–72. 184 B. Rerkasem, S. *Jamjod / Field Crops Research*. **89** (2004): 173–186, 1999.

Jasmund, K. and Linder, B. Experiments on the fixation of boron by clay minerals. *Proc. Int. Clay Conf. 1972*, 399-412, 1973.

Jat, R.L. and Rathore, P.S. Effect of sulphur, molybdenum and *Rhizobium* inoculation on greengram (*Phaseolus radiatus*), *Indian Journal of Agronomy*. **39**(4):651-654, 1994.

Javaid, A. and Bajwa, R. Field evaluation of effective microorganisms (EM) application for growth, nodulation, and nutrition of mung bean. *Turk J Agric* . 35: 443-452, 2011.

Jobbágy, EG., Jackson, RB. The vertical distribution of soil organic carbon and its relation to climate and vegetation. *Ecological Applications* **10**:423-436, 2000.

John, M.K. A batch handling technique for hot water extraction of boron from soils, *Soil Sci. Soc. Am. Proc.*, **37**: 332-333. 1973.

John, M.K. A batch handling technique for hot water extraction of boron from soils, Soil Sci. Soc. Am. Proc., **37**: 332-333, 1973.

Kannenberg, E.L., Brewin, N.J. Host-plant invasion by *Rhizobium*: the role of cell-surface components, *Trends Microbiol*. **2**: 277–283, 1994.

Kanwar, J.S. and Randhawa, N.S. Micronutrient research in soils and plants in *India. A review, I.C.A.R., New Delhi*, 1974.

Keren, R and Talpaz, H. Boron adsorption by montmorillonite as affected by particle size, *Soil Sci. Soc. Am. J.*, **48**: 555-559, 1984.

Keren, R. and Bingham, F.T. Boron in waters, soils and plants, *Advances in soil science*. **1**: 229-276, 1985.

Keren, R. and Sparks, D.L. Effect of pH and ionic strength on boron adsorption by pyrophyllite, *Soil Sci. Soc. Am. J.*, **58**: 1095-1100, 1994.

Keren, R. and Mezuman, U. Boron adsorption by clay minerals using a phenomenological equation. *Clays Clay Miner.* **29**: 198-203, 1981.

Keren, R., and Gast, R.G. pH-dependent boron adsorption by montmorillonite hydroxi-aluminum complexes, *Soil Sci. Soc. Am. J. Madison*, 47(6): 1116-1121, 1983.

Keren, R., F.T. Bingham, and J.D. Rhoades. Plant uptake of boron as affected by boron distribution between liquid and solid phases in soil, *Soil Sci. Soc. Am. J.*, **49**: 297-302, 1985.

Keren, R., Gast, R.G. and Bar, Y. B. pH dependent boron adsorption by Na-montmorillonite, *Soil Sci. Soc. Am. J.*, **45**: 45-48, 1981.

Keren, R., Gast, R.G. Effects of wetting and drying, and of exchangeable cations, on boron adsorption and release by montmorillonite, *Soil Sci. Soc. Am. J.*, **45**: 478-482, 1981.

Keren, R., Grossl, P.R and Sparks, D.L. Equilibrium and kinetics of borate adsorption-desorption on pyrophyllite in aqueous suspensions, *Soil Sci. Soc. Am. J.*, **58**: 1116-1122, 1994.

Kishor, P., Ghosh, A.K., Claramma, P.V., Influence of Tillage on Soil Physical Environment. International, *Journal of Agronomy and Plant Production*. **4** (10): 2592-2597, 2013.

Kluge, R. and Podlesak, W, Relationship between available boron leven on fine particles and humus content in soil, *Archiv fur Und Pflanzen bau Und Bodenkunde*. **25**: 595-600, 1981.

Koenig, R.A. and Johnson, S.R. (1942) colorimetric determination of P in biological material. Ind. Eng. Chem. Anal., **14**:155-156.

Kubota, j., Berger, K.C. and Troug, E. Boron movement in soil, *Soil Sci. Soc. Am. Proc.*, **13**: 130-134, 1948.

Kuchenbuch, R., Claassen, N., Jungk, A. Potassium availability in relation to soil moisture, *Plant and Soil*, **95**(2): 21-231, 1986.

Kumar, R., Rawat, K. S. and Yadav, B. Vertical distribution of Physico-Chemical properties under different topo-sequence in soils of Jharkhand. *Journal of Agricultural Physics*. **12**(1): 63-69, 2012.

Kumar, M., Sen, N.L. Effect of zinc, boron and gibberellic acid on yield of okra [Abelmoschus esculentus (L.) Moench], *Indian Journal of Horticulture*, **62**: 308-309, 2005.

Kumar, R., Singh, Y.V., Singh, Surendra., Latare, A. M.., Mishra, P.K. and Supriya. Effect of phosphorus and sulphur nutrition on yield attributes, yield of mungbean (*Vigna radiata* L. Wilczek), *Journal of Chemical and Pharmaceutical Research*, **4**(5):2571-2573, 2012.

Kumar, A., Prasad, S. and. Kumar, S.B. Effect of boron and sulphur on performance of gram (*Cicer arietinum*), *Indian Journal of Agronomy*, **51** (1): 57-59, 2006.

Kushwaha, J.P.S. and Singh, V. Boron content of soil and leaves of sweet orange (*Citrus Sinensis L.*) of osback orchards in Agra region of U.P., *Curr. Agric.*, **5**: 164-168,1981.

Kushwaha, P.D. and Srivastava, G.P. Effect of nitrogen, phosphorus and inoculation on the biochemical composition of mung grain, *Indian Journal of Agricultural Chemistry*. **11**(1): 43-48, 1978.

Kushwaha, Kumar, A ., Singh, S. and Singh, R.N. Available Nutrients and Response of Lentil *(Lense esculenta)* to Boron application in Rainfed Upland Soils of Ranchi, *Journal of the Indian Society of Soil Science*, 57(2): 219-222, 2009.

Kuzmanovic, Dj., Milicic, B. and Rasulic, N. Influence of inoculation on crop yield and content of nitrogen in grain of *Vigna radiata* (L. Wilczek) and *Vigna mungo* (L. Heer). *10th Congress of Yugoslav Society for Soil Science*. Belgrade (Yugoslavia). [22-26 Oct., 2001]. **52**(1-3): 7-12, 2003.

Lei, S., Wang, Y.H., Nian, F.Z., LU, J.W., Meng, J.L. and XU F.S. Inheritance of Boron Efficiency in Oilseed Rape, *Pedosphere,* **19**(3): 403–408, 2009.

Li H, C. X., Yang, Y., Xu, J., Gu, J. Development and genetic mapping of microsatellite markers from whole genome shotgun sequences in *Brassica oleracea*. Mol Breeding, 28: 585–596, 2010.

Li, B., Li, H., Kui, W.H., Chao, M.C., Jern, W.S., Li, H.P., Chu, W.J. and Wang, C.L. Studies on cause of sterility of wheat. *J. Northeastern Agric. College 3(in Chinese with English translation).*, 1–19, 1978.

Lindsay, W.L. Inorganic equilibria affecting micronutrients in soils. In: Micronutrients in Agriculture, J.J. Mortvedt *et al.* (ed.), *2nd ed. Soil Science Society of America, Madison, Wisconsin USA.*, 89- 144, 1991.

Lovatt, C.J. and Dugger, W.M. In: E. Frieden. Biochemistry of the Essential Ultratrace, Elements, *Plenum Pub. Corp.*, Boron. 389-421, 1984.

Luengo, C., Brigante, M., Antelo, J. and Avena, M. Kinetics of phosphate adsorption on goethite: comparing batch adsorption and ATR-IR measurements, *J Colloid Interface Sci.*, **300**: 511–518, 2006.

Mahler R.L. and Shafii, B. Relationship between soil-test boron and lentil yields in the inland pacific northwest, *Communications in Soil Science and Plant Analysis*. 38(9&10): 1193-1202, 2007.

Mahler, R. L. and Shafii, B. Relationship between Soil Test Boron and Pea Yields in the Inland Pacific Northwest, *Communications in Soil Science and Plant Analysis*, Volume, 40(13): 2603-2615, 2009.

Maiti, S., Das, C.C., Chatterjee, B.N. and Sengupta, K. Response of greengram and lentil to *Rhizobium* inoculation. *Indian Journal of Agronomy*. **33**(1): 92-96, 1988.

Mandal S.C., Ali, M.A. and Mukherjee, H.N. Boron requirements of some Bihar soils, *J. Indian Soc. Soil. Sci.*, **4**: 79-85, 1956.

Mandal, A.B. and Das, A.D., Response of wheat (*Triticum aestivum*) to boron application, *Indian J. Agric. Sci.*, **58**: 681– 683, 1988.

Mandal, B. and De, D.K. Depth wise distribution of extractable boron in some acidic incptisols of *India. Soil Sci.*, **155**: 256-262, 1993.

Mandal, S.C., Ali, M.A. and Mukherjee, H.N. Boron requirements of some Bihar soils. *J. Indian Soc. Soil. Sci.* **4**: 79-85, 1956.

Marzadori, C., Antisari, L.V., Ciavatta, C. and Sequi, P. Soil organic matter influence on adsorption and desorption of boron. *Soil Sci. Soc. Am. J.*, **55**: 1582-1585, 1991.

Matoh, T. Boron in plant cell walls, *Plant Soil.* **193**, 59–70, 1997.

Mattigod S.V., Frampton, J.A. and Lim, C.H. Effect of ion-pair formation on boron adsorption by kaolinite, *Clays Clay Miner.*, **33**: 433-437, 1985.

Maurice, J. Relationship between water extractable boron and pH, Organic matter influence on adsorption and desorption of boron, *Soil Sci. Am. J.*, **55**: 1582-1585, 1977.

Mc Phail, M., Page, A .L and Bingham, F.T. Adsorption interactions of monosilicic and boric acid on hydrous oxides of iron and aluminium, *Soil Sci. Soc. Am. Proc.*, **36**: 510-514, 1972.

Mengel, K. and Kirkby, E.A. Principles of plant nutrition. 4th ed. International Potash Institute, Worblaufen-Bern, Switzerland, 1987.

Metwally, A. I., El-Damaty, A.H. and Yousry, M. Anion adsorption as a possible mechanism of boron retention by soils, *Egypt J. Soil Sci.*, **14**: 23- 31, 1974.

Mezuman, U. and Keren, R. Boron adsorption by soils using a phenomenological adsorption equation, *Soil Sci. Soc. Am. J.*, **45**: 722-726, 1981.

Miljkovic, N. S., B. C. Mathews, and M. H. Miller. The available boron content of the genetic horizons of some Ontario soils. I. The relationship between water-soluble boron and other soil properties, *Can. J. Soil Sci.*, **46**:133-145, 1966.

Miljkovic, N. S., B. C. Mathews, and M. H. Miller. The available boron content of the genetic horizons of some Ontario soils. I. The relationship between water-soluble boron and other soil properties, *Can. J. Soil Sci.*, **46**:133-145, 1966.

Mishra, Peeyush., Singh, S. K., Srivatava, P.C. and Singh, Sobaran. Distribution of molybdenum and boron in Some Soils of Northern Alluvial Plain of U.P. and Uttaranchal in Relation to Soil Characteristics *Agropedology*:**16**(1):60-62, 2006.

Misra, R., Munankarmi, R.C., Pandey, S.P., Hobbs, P.R., Sterility work in wheat at Tarahara in the eastern Terai of Nepal. In: Mann, C.E., Rerkasem, B. (Eds.), Boron Deficiency in Wheat, *Wheat Special Report No. CIMMYT, Mexico,* **11**: 67–71, 1992.

Mojtaba, K., Leila, D., Hamidreza, M. and Ahmad,T. Effects of high boron concentration on boron uptake and growth of pistachio seedlings. 19th World Congress of Soil Science, Soil Solutions for a Changing World1, *Brisbane, Australia,* 2010.

Mondal, C., Bandopadhyay, P., Alipatra, A., Banerjee, H. Performance of summer mungbean Vigna radiate (L). Wilczek under different irrigation regimes and boron levels, *Journal of Food Legumes.* 25(1): 37-40, (2012).

Nable, R. O., Cartwright, B. and Lance, R. C. M. Genotypic differences in boron accumulation in barley: relative susceptibilities to boron deficiency and toxicity. In Genetic Aspects of Plant Mineral Nutrition. Ed. N El Bassam, *Kluwer AcademicPublishers, Dordrecht, The Netherlands.* 243–251, 1990.

Nadim, M. A., Awan, I. U., Baloch, M. S., Khan, E. A., Naveed, K. and Khan, M. A. Response of wheat (*Triticum aestivum* L.) to different micronutrients and their application methods, *The Journal of Animal & Plant Sciences,* **22**(1): 113-119, 2012.

NagaSivaiah, K., Swain, S.K., Raju, B. and SandeepVarma, V. Influence of micronutrients application on growth and seed yield in tomato (*Lycopersicon esculentum* Mill). *International Journal of Development Research.* 3(11): 191-195, 2013.

Naidu, M.V.S., Hanuman R. Effect of sulphur and *Rhizobium* inoculation on *Rhizobium* and non-rhizosphere microflora, *Agropedology.* **5**: 53-58, 1995.

Naidu, M.V.S., Hanuman R. Effect of sulphur and *Rhizobium* inoculation on dry matter, grain yield and protein content in greengram (*Vigna radiata* L.), *Legume Research.* **19**(1): 10-14, 1996.

Nand Kumar, T., Ganguly, K., Sharma, P. and Gulati, A. Food and Nutritional security status in Indian opportunity for investment partnership. ADB sustainable development working paper series 16. Manila Philippiner; *Asian development bank,* 2010.

Nathani, G.P., Shankarnarayan, H.S. and Mathurn, C.M. Status of water soluble boron in soils of Rajasthan, *J. Indian Soc. Soil sci.,* **18**: 341-344, 1970.

Nicholaichuk, W., Leyshon, A.J., Jame, Y.W. and Campbell, C.A. Boron and salinity survey of irrigation projects and the boron adsorption characteristics of some Saskatchewan soils, *Can. J. Soil Sci.,* **68**: 77-90, 1988.

Noppakoonwong, R.N., Rerkasem, B., Bell, R.W., Dell, B. and Loneragan, J.F. Prognosis and diagnosis of boron deficiency in black gram (*Vigna mungo* L. *Hepper*) in the field by plant analysis. *Kluwer Aca. Pub., Dordrecht, The Netherlands.* 89- 94, 1997.

Nuttall, J.G., Armstrong, R.D., Connor, D.J. Evaluating physicochemical constraints of Calcarosols on wheat yield in the Victorian southern Mallee, *Australian-Journal-of-Agricultural-Research.,* **54** (5): 487-497, 2003.

O'Neill, M. A., Ishii, T., Albersheim, P., Darvill, A. G. Rhamnogalacturonan II: Structure and function of a borate cross-linked cell wall pectic polysaccharide, *Ann. Rev. Plant Biol.,* **55**: 109–139, 2004.

Okazaki, E. and Chao, T.T. Boron adsorption and desorption by some Hawalian soils, *Soil sci.,* **105**: 255-259, 1968.

Osodeke V.E., Nwotiti I.L. and Nuga B.O. Sesquioxides distribution along a toposequence in umudike area of southeastern Nigeria, *Electronic journal of environment, Agriculture and food chemistry.* **4**(6): 1117-1124, 2005.

Page, N.R. and Paden, W.R. Boron supplying power of several South Carolina soils, *Soil. Sci.,* **77**: 427-434, 1954.

Pal, A.K. and Jana, P.K. Effect of phosphorus, *Rhizobium* inoculation and irrigation on growth, yield and water-use efficiency of summer greengram, *Indian Journal of Agronomy,* **36**(4): 536-540, 1991.

Pal, Y. R., Gilkes, J. and Wong, M. T. F. Soil factors affecting the availability of potassium to plants for Western Australian soils: a glasshouse study, *Australian Journal of Soil Research,* **39**(3): 611 – 625, 2001.

Pandey, N. and Gupta, B. The impact of foliar boron sprays on reproductive biology and seed quality of black gram, *Journal of Trace Elements in Medicine and Biology*. **27**: 58– 64, 2013.

Pant, J., Rerkasem, B., Noppakoonwong, R. Effect of water stress on the boron response of wheat genotypes under low boron field conditions, *Plant Soil.* **202**: 193–200, 1998.

PARC. Annual Report Effect of Boron ob rice productivity, *Pak. Agri. Res. Coun. Islamabad*: **97**: 2002.

Parfitt, R.L. Phosphate reactions with natural allophone, ferrihydrite and goethite, *J. Soil Sci.,* **40**: 359–369, 1989.

Parks, W.L and White, J.L. Boron retention by clay and humus systems saturated with various cations, *Soil Sci. Soc. Am. Proc.,* **16**: 298-300, 1952.

Parr, A. and Loughman, B.C. Boron membrane functions in plants. In: Metals and Micronutrients: Uptake and utilization by plants, edited by Robb, D.A. and Pierpoint, *W.S. London: Academic Press,* 87-107, 1983.

Penaranda, F.L., Mendoza, H.T., Sanlao, G.R., Bod, T., Candy, L.S. and Bongolan, F.F. Screening of different strains of *Rhizobium* on soybean, mungbean and peanut under Zamboanga del Sur condition (Philiines), *Philiines Journal of Crop Science (Philiines)*. 13: 11, 1988.

Ponnamperuma, F. N., Caytan, M.T. and R.S. Lantin. Dilute hydrochloric acid as an extractant for available zinc, copper, and boron in rice soils, *Plant Soil.* **61**: 297-310, 1981.

Prasad, J., and Ram, H. Effect of zinc, coer and Rhizobium inoculation on microbial population in soil and yield of greengram (Phaseolus radiatus), International Journal of Tropical Agriculture. 10(2): 157-160, 1992.

Puri, A.N. (1930) A new method of estimating total carbonates in soils, *Imprical Agricultural Research Pusa Bulletin.,* **206:**7-11.

Quddus, M. A., Rashid, M. H., Hossain, M. A. and Naser H. M. Effect of zinc and boron on yield and yield contributing characters of mungbean in low ganges river floodplain soil at madaripur, Bangladesh, *Bangladesh j. Agril. Res.,* **36**(1): 75-85, 2011.

Rahman, M.S., Uddin, M.Z., Zamman, M.M. and Chowdhury, M.M.U. Annual Report, *Agricultural Research Station, BARI, Raikhali, Rangamati.* 1:1994-95, 1995.

Rahman, H.M. Shabjee-o-phul (Vegetable and flower). Mrs Amina Khatun, 3/D Jigatala, Dhaka. 2:19, 1988.

Rakshit, A., Bhadoria, P.B.S. and Ghosh, D. Influence of Boron on NPK uptake of rice (*Oryza sativa* L) in acid alluvial soils of Coochbehar, *West Bengal. & Ecol.,* **20**(1): 188-190, 2002.

Ram, S., Bist, L. D., Sirohi, S. C. Internal fruit necrosis of mango and its control, *Acta Hort.,* **231**, 805–813, 1989.

Ranganathan, D.S. and Perumal, R Effect of micronutrients with /without organics and biofertilizers on growth and development of tomato in inceptisol and alfisol, *South Indian Hort.*, 43(3&4): 89-92, 1995.

Rashid, A. and Din, J. Differential susceptibility of chickpea cultivars to Iron chlorosis grown on calcareous soils of Pakistan, *Journal of Indian Society of Soil Science.* **40**: 488-492, 1992.

Rashid, A. Secondry and micronutrients in Soil Science. A. Rashid and K.S. Memon (Managing Authors*), National Book Foundation, Islamabad, Pakistan.* 341-385, 1996.

Rashid, A., Muhammad, S., Rafique, E. Gynotypic variation in boron uptake and utilization in rice and wheat. In: Goldbach, H.E., Rerkasem, B., Wimmer, M.A., Brown, P.H., Thellier, M., Bell, R.W., All Aspects of Plant and Animal Boron Nutrition, *Kluwer and Plenum Academic Publishers*, 305–310, 2002.

Rashid, A., Yaseen, M., Ashraf, M. and Mann, R.A. Boron deficiency in calcareous reduces rice yield and impairs grain quality, *Internationa Rice research Notes (IRRI)*, **29**: 58-60, 2004.

Rashid, M. H. L., Hasan, M. M., Ahmed, M., Rahman, M. T. And Rahman, K. A. M. M. Response of mustard to boron fertilization, *Bangladesh J. Agril. Res.*, 37(4): 677-682, 2012.

Rashid, M.M. Shabjeer chash (Vegetable cultivation). Begum Shehala Rashid. BARI Residential Area, *Joydebpur, Dhaka*. 97, 1993.

Ram, L.R., Pedogenesis characterization and classification of specially associated soil of Prakashan district A.P., *Ph. D. Thesis* Submitted in B.H.U., 2012.

Ratna Kalyani, R., Sree, D.V., Satyanarayana, N.V. and Madhava, R. K.V. Effect of foliar application of boron on crop growth and yield of pigeonpea (*Cajanus cajan* (L.) *Millsp.*), *Indian Journal of Plant Physiology*. **36**(4): 223–226, 1993.

Rawson, H.M. Hypothesis of why sterility occurs in wheat in Asia. In: Rawson, H.M., Subedi, K.D. (Eds.), Sterility in Wheat in Subtropical Asia: Extent, Causes and Solution, *ACIAR Proceedings*. **72**: 132–134. 1996

Redondo, N.M., Wilmot, A.R., El-Hamdaoui, A., Bonilla, I., Bolanos, L. Departamento de Biologia, Facultad de Ciencias, Universidad Autonoma de Madrid. 28049-Madrid, Spain. Relationship between boron and calcium in the N2-fixing legume-rhizobia symbiosis, *Plant,-Cell-and-Environment*. **26**(11): 1905-1915, 2003.

Redondo-Nieto, M., Pulido, L., Reguera, M., Bonilla, I. and Bolaños, L. Developmentally regulated membrane glycoproteins sharing antigenicity with rhamnogalacturonan II are not detected in nodulated boron deficient *Pisum sativum. Plant Cell Environm.*, **30**:1436–43. 2007.

Redondo-Nieto, M., Rivilla, R., El-Hamdaoui, A., Bonilla, I. and Bolaños, L. Boron deficiency affects early infection events in the pea-*Rhizobium* symbiotic interaction, *Aust. J. Plant Physiol*. **28**: 819-823, 2001.

Redondo-Nieto, M., Wilmot, A., El-Hamdaoui, A., Bonilla, I., Bola˜nos L. Relationship between boron and calcium in the N2-fixing legume-rhizobia symbiosis, *Plant Cell Environm.*, **26**: 1905–1915, 2003.

Rego, T.J., Wani, S.P., Sahrawat, K.L. and Pardhasaradhi, G. Global Theme on Agroecosystems, International Crops Research Institute for the Semi-Arid Tropics (ICRISAT), Patancheru 502 324, Andhra Pradesh, India. Macro-benefits from boron, zinc and sulfur application in Indian SAT: a step for Grey to Green Revolution in agriculture, *Global-Theme-on-Agroecosystems-Report*. **16**(20): 2005.

Reguera, M.M., Wimmer, P., Bbustos, H. E., Goldbach, L., Bolaños, L. and Bonilla, I. Ligands of boron in Pisum sativum nodules are involved in regulation of oxygen concentration and rhizobial infection, *Plant, Cell Environ.*, 33: 1039-1048. 2010.

Rerkasem, B. A general survey of the incidence of wheat sterility. In Rawson HM, Subedi K, Eds.Sterility in wheat in subtropical Asia: extent, causes and solutions. Proceedings of a workshop 18-21 September 1995, *Lumle Agricultural Research Centre, Pokhara, Nepal. ACIAR*, **72**: 8-12, 1996.

Rerkasem, B. and Jamjod, S. Boron deficiency in wheat: a review, *Field Crops Research*. **89**: 173-186, 2004.

Rerkasem, B. and Jamjod, S. Genotypic variation in crop response to low boron and implication for plant breeding, *Plant Soil*. **193**: 169-180, 1997.

Rerkasem, B. Boron and plant reproductive development. In Rawson, H.M., Subedi, K. Eds.Sterility in wheat in subtropical Asia: extent, causes and solutions. Proceedings of a workshop 18-21 September 1995, *Lumle Agricultural Research Centre, Pokhara, Nepal. ACIAR*, **72**:32-35, 1996

Rerkasem, B. Boron nutrition of crops and genotypic variation in boron efficiency. In: Boron in Plant and Animal Nutrition, Goldbatch et al. *Kluwer Aca. Pub., New York*. 269-280, 2002.

Rerkasem, B., Jamjod, S. and Niruntrayagul, S. Increasing boron efficiency in many international bread wheat, durum wheat, triticale and barley germplasm will boost production on soils low in boron, *Field Crop Res.*, 86: 175–184, 2004.

Rerkasem, B., Saunders, D.A. and Dell, B., Grain set failure and boron deficiency in wheat in Thailand, *J. Agric. (CMU)*, **5**: 1–10, 1989.

Ryan, J.G. (1989). The systematic of lithium, beryllium and boron in young volcanic rocks, Doctoral dissertation, Columbia University.

Saha, A., Mandal, B.K., Mukhopadhyay, P. Residual effect of boron and molybdenum on the yield of succeeding mungbean in summer, *Indian-Agriculturist*. **40**(1): 11-16, 1996.

Saha, P.K., Saleque, M. A., Zaman,S. K. And Bhuiyan,N. J. Response of mustard to S, Zn, and B in Calcareous Soil, *Bangladesh J. Agril. Res.*, 28(4): 633-636, 2003.

Sahrawat, K.L., Wani, S.P., Rego, T.J., Pardhasaradhi, G. and Murthy, K.V.S. Widespread deficiencies of sulphur, boron and zinc in dryland soils of the Indian semi-arid tropics, *Current Science*. 93: 1428-1432, 2007.

Sahu, S.K., and Behera, B. Note on effect of *Rhizobium* inoculation on cowpea, groundnut and greengram, *Indian Journal of Agronomy*. **17**(4): 359-360, 1972.

Saini, H.S. and Aspinal, D. Effect of water deficit on sporogenesis in wheat. Annals of Botany, *Triticum aestivum L*, **48**: 623-633, 1981.

Saini, H.S., Aspinal, D. Sterility in Wheat (*Triticum aestivum* L.) induced by water deficit or high temperature: possible mediation by abscisic acid, *Aus. J. of Plant Physiology*, **9**: 529-537, 1982.

Sakal, R., Singh, A.P. and Sinha, R.B. Differential susceptibility of maize varities to boron deficiency in a calcareous soils, *J. Indian Soc. Soil Sci.*, **37**: 582-584, 1989.

Sakal, R., Singh, B.P., Singh, A.P. and Sinha, R.B. Critical limit of boron in soils and plants for the response of black gram to applied boron in calcareous soils, *J. Indian Soc. Soil Sci.*, **37**: 725-727, 1985.

Saltali, K., Bilgili, A.V., Tarakcioglu, C. and A. Durak. Boron adsorption in soils with different characteristics, *Asian Journal of Chemistry*. **17**: 2487-2494, 2005.

Sarkar, D., Kumar De, D., Das, Ruma and Mandal, Biswapati. Removal of organic matter and oxides of iron and manganese from soil influences boron adsorption in soil, *Geoderma*, 2013.

Sarkar, A.K., Chakraborty, S. Effect of micronutrient on yield of major field crops grown in the Gangetic delta region, *Indian Agric.*, **24**: 259–266, 1980.

Sathya, S., Mahendran, P.P. and Arulmozhiselvan, K. Influence of soil and foliar application of borax on fractions of boron under tomato cultivation in boron deficient soil of Typic Haplustalf, *African Journal of Agricultural* Research. 8(21): 2567-2571, 2013.

Saxena, A.K., Mittal, S., Dwivedi, A., Shashikala and Tilak, K.V.B.R. Competitiveness and symbiotic efficiency of potential inoculant strains of Brady, *Rhizobium* sp. (*Vigna*). *Microbiological Research*. **151**(2): 219-224, 1996.

Schalscha, E.B., Bingham, F.T., Galindo, G.G., Galvan, H.P. Boron adsorption by volcanic ash soils in Southern Chile, *Soil Sci. Baltimore*, **116**(2): 70-76, 1973.

Schmucker, T.H., Uber den Einfluss von Borsäure auf Pflanzen, insbesondere keimender Pollenkorner. Planta, 23: 264-289. 1935

Shafiq, M., Ranjha, A.M., Yaseen, M., Mehdi, S.M. and Hannan, A. Comparison of Freundlich and Langmuir adsorption equations for boron adsorption on calcareous soils, *Journal of Agricultural Research*. **46**:141-148, 2008.

Shahi, D.K., Sharma, A. and Singh, L. Improving protein quality of greengram (*Vigna radiata* L. Wilczek) through use of *Rhizobium* inoculation, phosphorus and sulphur, *Journal of Research Birsa Agricultural University*. **14**(2): 171-174, 2002.

Shamima, N., Farid, A.T.M. Nutrient uptake and yield of garden pea as influenced by various fertilizer treatments, *Thai-Journal-of-Agricultural-Science*. **36**(2): 185-192, 2003.

Shamsuddoha, A. T. M., Anisuzzaman1, Sutradhar, G. N. C., Hakim, M. A. and Bhuiyan, M. S. I. Effect of Sulfur and Boron on Nutrients in Mungbean (*Vigana radiata* L.) and Soil Health. *Bangladesh J. of Bot.*, 224-229, 2011.

Sharma, S. Growth physiological and yield aspects of mung bean (*Vigna radiata*) as affected by inoculation treatment by different strains of Brady *Rhizobium* culture, *Research on crops*. **2**(2): 112-115, 2001.

Sharma, S. Response of various isolates of Brady *Rhizobium* inoculation on protein content and the yield attributes of greengram [*Vigna radiata* (L.) Wilczek]. *Legume Research*. **26**(1): 28-31, 2003.

Sheikh, M.H.R., Hannan, M.S.A., Huda, A. and Rahman, M. T. Sustainable crop production retaining soil fertility and environment through mustard-mungbean-T. Amancropping pattern, *J. Soil. Nature*. **3**(2):10-14, 2009

Shelp, B. J. Physiology and biochemistry of boron in plants. Chapt. 4 in U. C. Gupta, ed. Boron and its role in crop production. CRC Press, Boca Raton, FL. (In press). 1992.

Shive, J.W. Significant role of trace elements in the nutrition of plants, *Plant physiol.*, **16**: 435-445, 1941.

Shorrocks, V. M. The occurrence and correction of boron deficiency, *Plant Soil*. **193**: 121–148, 1997.

Shukla, S.K. and Dixit, R.S. Effect of *Rhizobium* inoculation, plant population and phosphorus on growth and yield of summer greengram, *Indian Journal of Agronomy*. **41**(4): 611-615, 1996.

Shuman, L.M., Bandel, V.A., Donohue, S.J., Isaac, R.A., Lippertr, R.M., Sims, J.T. and Tuckor, M.R. Comparison of Mehlich-1 And Mehlich -3 Extractable soil boron with hot water extractable boron, *Commun. Soil Sci. Plant Anal.*, **23**: 1-14, 1992.

Sillanpaa, M. Trace elements in soils and agriculture, *Food and Agriculture Organization of the United Nations*, Rome, Italy, 1972.

Sims, J.R., and Bingham, F.T. Retention of boron by layer silicates, sesquioxides and soil materials: III. Iron and alumunium coated layer silicates and soil materials, *Soil sci. Am. Proc.*, **32**: 369-373, 1968.

Sims, J.R., and Bingham, F.T. Retention of boron by layer silicates, sesquioxides, and soil materials: I Sesquioxides, *Soil Sci. Soc. Am. Proc.*, **31**: 728-732, 1967.

Sims, T.T. Soil fertility evaluation. In: Handbook of soil science (Summer M.E., ed), *CRC Press LLC, Boca Raton, Florida, USA.*, 113-154, 2000.

Singh, T.B., Patra, S.K. and Devi, K. N. Study on Sesquioxide Content and Lime Requirement of Lateritic Soils, *International Journal of Bio-resource and Stress Management*. **4**(4):518-522, 2013.

Singh, R.N., Singh, S. and Kumar, B. Interaction Effect of Sulphur and Boron on Yield, Nutrient Uptake and Quality Characters of Soybean *(Glycine max* L. Merill) Grown in Acidic Upland Soil, *Journal of the Indian Society of Soil Science*, 54(4): 516-518, 2006.

Singh, M. Equilibrium adsorption of boron in soils and clays, *Geoderma*, **5**: 209-217, 1971.

Singh, M.V. Emerging boron deficiency in soils and crops in India and its management. 18th World Congress of Soil Science, *Philadelphia*, 2006.

Singh, R., Yadav, D.S., Maurya, M.L. and Singh, R. Effect of boron application on yield of pea (*Pisum sativum* L.) and black gram (*Vigna mungo Hepper*) in calcareous soil, *Fertiliser-News*. **47**(2): 67-68, 2002.

Singh, R.N. and Sinha, H. Evaluation of critical limits of available boron for soyabean and maize in acid red loam soil, *J. Indian Soc. Soil Sci.*, **35**: 456-459, 1987.

Singh, R.P., Sinha, S.D. and Prasad, R.B. Effect of boron on seed setting in wheat under North Bihar conditions, *Indian J. Agron.*, **21**: 100–101, 1976.

Singh, S.K., Singh, Room and Singh, H.P. Effect of sulphur on growth and yield of summer mung, *Legume Research.* **17**(1): 53-56, 1994.

Singh, S.S. Boron adsorption equilibrium in soils, *Soil Sci.*, **98**: 383-387, 1964.

Sinha R.B., Sakal, R., Singh, A.P. and Bhogal, N.S. Response of some field crop to boron application in calcareous soils, *J. Indian Soc. Soil. Sci.*, **39**: 118-122, 1991.

Slavko, P., Nacer, B., Carl, G., Hening, H., and Patrick, H.B. Boron Transport and Soluble CarbohydrateConcentrations in Olive, *J. Amer. Soc. Hort. Sci.*, 126(3): 291–296, 2001.

Smit, J. N. and Combrink, N. J. J. The effect of boron levels in nutrient solutions on fruit production and quality of greenhouse tomatoes, *S. Afr. J. Plant Soil.* 21(3): 188-191. 2004.

Smith, J.N. and Combrink Pollination and yoeld of winter grown green house tomatoes as affected by boron nutrition, cluster vibration and relative humadity, *South African journal of plant and soil.* (2005).

Soares, M.R., Casagrande, J.C. and Alleoni, L.R.F. Adsorção de boro em solos ácricos em funçao da variação do pH, *Revista Brasileira de Ciencias do Solo.* **32**:111-120, 2008.

Solaiman, A.R.M. Influence of *Rhizobium* inoculant, nitrogen and boron on nodulation, dry weight and grain yield of chickpea, *Annals-of-Bangladesh-Agriculture.* **9**(1), 75-84, 1999.

Spaink, H.P. Root nodulation and infection factors produced by rhizobial bacteria, *Annu. Rev. Microbiol.* **54**: 257–288, 2000.

Spiegel, H., Dersch, G., Hosch,J. and Baumgarten, A. Tillage effects on soil organic carbon and nutrient availability in a long-term field experiment in Austria, *Die Bodenkultur.* **58** (1–4) :2007 .

Srinivasa, R.Ch. and Wani, S.P. and Sahrawat, K.L. and Rego, T.J. and Pardhasaradhi, G *Zinc, boron and sulphur deficiencies are holding back the potential of rainfed crops in semi-arid India: Experiences from participatory watershed management. International Journal of Plant Producti.* **2** (1). 89-99, 2008.

Srivastava, A.K. and Poi, S.C. Response of greengram (*Vigna radiata*) and black gram (*Vigna mungo*) to Brady *Rhizobium* as well as the residual effects of Brady *Rhizobium* strains on the same crops, *Advances in Plant Sciences.* **13**(2): 431-435, 2000.

Srivastava, S.P., Yadav C.R., Rego, T.J., Johansen, C., Saxena, N.P. and Ramakrishna, A. Diagnosis of boron deficiency as a cause of flower abortion and failure of pod set in chickpea in Nepal, *International-Chickpea-and-Pigeonpea-Newsletter.* **3**: 29-30, 1996.

Srivastava, S.P., Yadav, C.R., Rego,T.J., Johansen, C., Saxena, N.P. Diagnosis and alleviation of boron deficiency causing flower and pod abortion in chickpea

(*Cicer arietinum* L.) in Nepal. In: Boron in soils and plants. Developments in Plant and Soil Sciences 76 (BellR.W., Rerkasem B., eds). *Kluwer Academic Publishers, Dordrecht, The Netherlands*. 95-99, 1997.

Sthapit, B.R. Studies on wheat sterility problem in the Hills, Tar and Tarai of Nepal. Technical Report No. 16/88. *Lumle Agricultural Research Centre, Pokhara, Kaski, Nepal*. 1988.

Su, C. and Suarez, D.L. Coordination of adsorbed boron: *A* FTIR spectroscopic study, *Sci. Technol*. **29**: 302-311. 1995.

Subba, R. N.S. and Tilak, K.U.B.R. *Rhizobium culture*-their role in pulse production. Souvenier Bell Directorate of pulses, *Department of Agriculture, Government of India, Lucknow*, 31-34, 1977.

Subedi, K.D. Wheat sterility in Nepal: a review. In: Mann, C.E., Rerkasem, B. (Eds.), Boron Deficiency in Wheat, *Wheat Special Report No.* **11**. *CIMMYT, Mexico*, 57–64, 1992.

Subedi, K.D., Budhathoki, C.B., Subedi, M. Variation in sterility among wheat (*Triticum aestivum*) cultivars in response to boron deficiency in Nepal, *Euphytica*, **95**: 21–26. 1997.

Subedi, K.d., Gregory, P.J., Summerfield, R.J. and Gooding, M.J. Cold temperatures and boron deficiency caused grain set failure in spring wheat (*Triticum aestivum L.*), *Field Crop Research*. **57**: 277-288, 1998.

Subbiah, B.V. and Asija, G.L. A rapid procedure for the determination of available nitrogen in aoils. *Curr.Sci.*, **25**: 259-260, 1956.

Sumner, M.E., Yamada, T., Kalra, Y.P. Farming with acidity. Fine tuning soil and plant analysis for economics and environmental betterment. 2001 International Symposium on soil and plant analysis, Edmonton, Alberta, Canada, 21-27 July 2001, *Communications-in-Soil-Science-and-Plant-Analysis*. **33**(15-18): 2467-2496, 2002.

Takano, J., Miwa, K., von W, N., Fujiwara, T. Boron dependent endocytosis and degradation of boron transporter at BOR1, *Plant Cell Phys.*, **46**, S151–S151, 2005.

Takano, J., Miwa, K., Yuan, X., von Wiren, N., Fujiwara, T. Endocytosis and degradation of BOR1, a boron transporter of *Arabidopsis thaliana*, regulated by boron availability, *Proc. Natl. Acad. Sci. Unit. States Am.*, **102**, 12276–12281, 2005.

Takano, J., Noguchi, K., Yasumori, M., Kobayashi, M., Gajdos, Z., Miwa, K., Hayashi, H., Yoneyama, T. and Fujiwara, T. *Arabidopsis* boron transporter for xylem loading, *Nature*, **420**: 337–340, 2002.

Takano, J., Wada, M., Ludewig, U., Schaaf, G., von Wiren, N., Fujiwara, *T.* The *Arabidopsis* major intrinsic protein NIP5;1 is essential for efficient boron uptake and plant development underboron limitation, *Plant Cell*, **18**: 1498–1509, 2006.

Tandon, H. L. S. Micronutrients in soils, crops and fertilizer, *Fer. Dev. & Con. Org.* New Delhi, India, 41-50, 1999.

Thomson, H.C. and Kelly, W.C. Vegetable crops. 5th edition, *Mengrow Hill Book Co.,* 18:280-281, 1985.

Toner, C.V. and Sparks, D.L. Chemical relaxation and double layer model analysis of boron adsorption on alumina, *Soil Sci. Soc. Am. J.,* **59**: 395-404, 1995.

Tripathi, P. K., Singh, M. K., Singh, J. P. and Singh, O. N. Effect of *rhizobial* strains and sulphur nutrition on mung bean (*Vigna radiata* (l.) wilczek) cultivars under dryland agro-ecosystem of Indo-Gangetic plain, *African Journal of Agricultural Research,* 7(1): 34-42, 2012.

Tripathi, D., Singh, K. and Upadhyay, G.P. Distribution of micronutrients in some representative soil profiles of Himachal Pradesh, *J. Ind. Soc. Soil Sci.,* **42**(1): 143-145, 1994.

Troug, E. Lime in relation to availability of plant nutrient, *Soil Sci.,* 65: 1-7, 1948.

Udo, E. J. Profile distribution of Iron sesquoxides contents in selected Nigeriasoils, *Journal Agricultural Science of Cambridge.* **95**: 191–198. 1980.

Upperi, S.N. and Dasog, G.S. Fertility status and management of red and black soils of the Regional Research Station, Raichur (India), *Karnataka-Journal-of-Agricultural-Sciences.* **7**(3): 346-348, 1994.

Valenciano, J. B., Boto, J. A. and Marcelo, V. Response of chickpea (*Cicer arietinum* L.) yield to zinc, boron and molybdenum application under pot conditions, *Spanish Journal of Agricultural Research.* 8(3): 797-807, 2010.

Van Brussel, A., Bakhuizen, R., Van Spronsen, P. C., Spaink, H. P., Tak, T., Lugtenberg, B. J. J. and Kijne, J. W. Induction of pre-infection thread structures in the leguminous host plant by mitogenic Lipooligosaccharides of *Rhizobium. Science,* **257**: 70-72. 1992.

Venkateswarlu, B. Response of greengram and pigeonpea to Brady *Rhizobium* inoculation and soil moisture conservation practice, *Indian Journal of Microbiology.* **44**(3): 215-217, 2004.

Walkley, A. and Black, LA. An examination of the degtjareff method for determining soil organic matter, and a proposed modification of the chromic acid titration method. *Soil Science* 37, 29-38, 1934.

W.H.O. (World Health Organization). Environmental health criteria boron. *International programme on chemical safety, Geneva, Switzerland.* **204**: 105–106, 1998.

W.H.O. Environmental Health Criteria: Boron (7740-42-8). *World Health Organization.* Document No. **204:** 1998.

Wang, Y.H. and Lan, L.F. A study on the boron efficiency of rape Brassica napus, *Journal of Huazhong Agricultural University. Sup. Sum.,* **21**:71-84, 1995.

Wankhade, S.G., Dakhore, R.C., Wanjari, S.S., Patil, D.B., Potdukhe, N.R. and Ingle, R.W. Response of crops to micronutrients, *Indian J. Agric Res.,* **30**: 164-168, 1996.

Wei, Y. Z., Bell, R. W., Yang, Y., Ye, Z. Q., Wang, K. and Huang, L. B. Prognosis of boron deficiency in oilseed rape (*Brassica napus*) by plant analysis, *Aust. J. Agric. Res.,* **49**: 867–874, 1998.

Wild, A. and Mazaheri, A. Prediction of the leaching rate of boric acid under field conditions, Geoderma, **22**: 127-136, 1979.

William L. Peacock and Peter Christensen, L. Drip irrigation can effectively apply boron to San Joaquin Valley vineyards, *California Agriculture volume 59, number.* **3**: 188-191, 2005.

Williams, C.H. and Steinbergs, A. Soil sulphur fractions and chemical !ndices of available sulphur in some Australian soils. *Australian Journal of Agricultural Research* **10,** 340-352, 1959.

Winiarska, Z. Effect of bituminous coal furnace wastes treated with sludge on the composition of infiltrating waters in a lysimeter experiment, *Pamietnik-Pulawski,* **103:** 99-122, 1993.

Xu, F. S., Wang, Y. H. and Meng, J. L. Mapping boron efficiency gene(s) in Brassica napus using RFLP and AFLP markers, *Plant Breeding*. **20**(4): 319–324, 2001.

Xu, F. S., Wang, Y. H., Ying, W. H. and Meng, J. L. Inheritance of boron nutrition efficiency in Brassica napus, *J. Plant Nutr.*, **25**(4): 901–912, 2002.

Xu, F.S., Wang, Y.H. and Li, J.C. Inheritance of rapeseed (Brassica napus L.) boron efficiency in F1 generation, *Plant Nutr. Fert. Sci. In Chinese.* **4**(3): 305–310, 1998.

Yang, S.C. The diagnosis and control of the absence of boron in broad bean. *Bulletin-of-Agricultural-Science-and-Technology*. **4**(10), 1994.

Yermiyahu, U., Keren, R. and Chen, Y. Boron sorption on composted organic matter, *Soil Sci. Soc. Am. J.*, **52**: 1309-1313, 1988.

Yermiyahu, U., Keren, R. and Chen, Y. Boron sorption by soil in the presence of composted organic matter, *Soil Sci. Soc. Am. J.*, **59**: 405-409, 1995.

Zhao,D., Shaohong W., Erfu D. and Yunhe Yin. Effect of climate change on soil organic carbon in Inner Mongolia, *International Journal of Climatology*, 2014.

Heavy Metal Detoxification and Signal Transduction Pathway

Yaman

Department of Agronomy, School of Agriculture
Lovely Professional University, Jalandhar, Punjab, India, 144411
Email: yamannaik21@gmail.com

Abstract

The problem of heavy metals is increasing day by day due to the increase in industrial activities and other anthropogenic activities which increase their concentration in the soil to a higher degree and from which the plants, which influence its growth and development, are taken up. The toxicity of heavy metals in plants disrupts the various physiological and metabolic activities. Various cellular mechanisms are used for the detoxification and tolerance of heavy metal plants. There are two different approaches, such as physiological and biological approaches to the detoxification of heavy metals in soil. Timely activation of the tolerance mechanism in plants is very important for maintaining its growth under stressed conditions. Root exudates play a very important role in the immobilization of heavy metals, and extracellularly mycorrhiza comes into action for their detoxification. Heat shock proteins (HSPs) have a very efficient role to play in the repair of damaged proteins. Phytochelatins, metallothioneins, organic acids and amino acids help to chelate heavy metals, which restrict metal accumulation in a full reactive form in cytosols. Plant cells can develop proteins that sense the presence of metals and transmit a signal to detoxify metals. In signal transduction, metals that cause stress to plants because signaling perception by some external signals, such as miRNA or Ca, through which the signals are quickly transmitted by MAP kinases to the responsible transcription factors.

Introduction

The sustainability of agriculture depends on land and water resources. Agriculture is very important for human survival, but there is maximum exploitation of these natural resources and the soil is severely degraded or polluted due to anthropogenic activities. Pollution is caused by emissions, effluents and solid discharges from industry, vehicle exhaustion and metals from smelting and mining, and by non-point sources such as dissolved salts (normal and artificial), pesticide utilization, industrial and municipal waste disposal in agriculture, and by the high utilization of fertilizers (McGrath et al., 2001; Nriagu and Pacyna, 1998). Every source of pollution has harmful effects on plants, creatures and, at last, on human well-being, yet those who add heavy metals to soil and water are truly concerned about their perseverance in the environment and the cancer-causing nature to humans. They can't be obliterated naturally yet they just change from one oxidation state or natural complex to another. In this way, the contamination of heavy metals poses an incredible potential danger to nature and human well-being. Metals are normally available at different levels in the crust of the Earth. The consumption of non-renewable energy sources, the mining of metallic minerals, the waste of metropolitan areas and the exorbitant use of pesticides and fertilizers lead to the rapid release of metals and metalloids into different parts of the environment. Some metals such as copper (Cu), iron (Fe), manganese (Mn), nickel (Ni) and zinc (Zn) are very important for plant cells, but can also be toxic to plants when used at higher concentrations. Higher metals, whether of common natural origin or anthropogenic action, may become harmful to soil microflora. The level of toxic heavy metals depends on the metallic component and its presence in the soil. The presence of heavy metals in the soil depends on abiotic factors such as metal concentration, humidity and soil pH and biotic factors such as microflora that releases metals. Different metals such as Zn, Cu and Mn are important for plants at low levels but become harmful at higher doses, yet different metals have never been shown to be essential for the advancement of living beings and are poisonous even at low levels, i.e. Hg, Pb, Cd. Heavy metals are deposited in the environment as a result of the use of sludge, municipal compost, use of pesticides, fertilizers, emissions from industrial waste incinerators, vehicle exhausts, metallic rock or mining residues and other smelting in various industries (Garbisu and Alkorta, 2003, Halim et al., 2003).

The availability of metals and metalloids in areas where they are not required and present at a much higher level than the required quantity may pose a problem to the ecosystem and human beings that we can call contaminants. Metal concentrations in soil range from <1 mg / kg of soil to >100,000 mg /kg of soil. Higher concentrations of certain metals debase soil fertility, reduce crop production along with quality, and cause environmental, human and animal health risks. Heavy metals produce various stimuli for plants that respond in a variety of ways, such as chelation and compartmentalization of metal ions, immobilization, exclusion, and stress response mechanisms such as stress protein and ethylene (Andrades-Moreno et al., 2014). The Heavy Metal Resistance System in plants has been extensively monitored during extensive examinations over several years, during which various points of view were considered as defilement destinations. Basel Heavy metal resistance is likely to be found in all plant species. In this way, they run a perplexing framework consisting of take-up/efflux, transport/sequestration and chelation. These key components are firmly embedded in the homeostasis of basic metal micronutrients. Much of the attribute of these elements divide the plant kingdom into two groups: (hyper) accumulating and non-accumulating plants. Most plants can be referred to as non-accumulator plants. Overall, they all need to be adapted to heavy metals for food and development purposes in metallurgical soils, respectively. Subsequently, they need to have a finely tuned mechanism for living with even toxic heavy metals. The least complex methodology is to stay away from soil metals or to prevent metals from moving into shoots. Besides, elements for the collection and sequestration of fundamental metals are frequently used. This may, however, cause impedance with plant metal homeostasis and may lead to toxic symptoms. Symptoms such as weak chlorophyll amalgamation resulting in chlorotic leaves, altered proportions of chlorophyll a and b and photosynthetic action (Küpper et al., 2007), plant dwarfism or root ultrastructure effects (Barcelo et al., 2004). In any case, it is not clear whether these effects are either the cause or the result of metabolic disturbances in heavy metal exposed plants. It is therefore very important to examine the tolerance and toxicity of metals at cellular and molecular levels. There is hyper-tolerance, which is a special case of metal tolerance. Metal hyper-tolerant plants, except for hyper-accumulators, may reject metals from their tissues to reduce metal accumulation, particularly in their over-the-ground tissues. This is the key distinction for the hyperaccumulation of plants. All things considered, the hyperaccumulation of metal is related to hyper resistance, which reveals another detoxification system. Metal hyperaccumulation plants are represented by a shoot/root ratio of metal accumulation > 1. Plants have advanced a wide range of versatile reactions to heavy metal toxicity. These include immobilization, exclusion, chelation and compartmentalization of metal particles, often involving metal-binding ligands. Several such ligands have been distinguished in plants and contain organic acids, amino acids, peptides and polypeptides. Polypeptides incorporate metallothioneins (MTs), small, encoded genes, cysteine-rich polypeptides, and phytochelatins (PCs), which are interestingly enzymatically synthesized, cysteine-rich peptides. To maintain soil and water quality, we need to keep them away from contamination, continuous efforts have been made to create innovations that are easy to use, sustainable and financially feasible. Physiochemical methodologies have been widely used for the treatment of polluted soil and water, particularly on a small scale. There are many problems related to the curing of contaminated soils on a larger scale due to high costs and side effects. Some microorganisms are used for the remediation of contaminated soils. Microorganisms are not harmful to environmental stability and are also low-cost. In modern-day use of plants for the healing of heavy metals contaminated soil, attention is paid to everyone called phytoremediation. There are many species of plants that can take heavy metals out of the soil and accumulate in their tissues. Progress has been made in the mechanical and practical application of bioremediation and phytoremediation components. This review will provide an overview of the tolerance mechanism, with more attention being paid to the level of cells and signalling pathways induced by plants.

Chelation of Heavy Metals

The chelation of these metals with ligands is the general mechanism for tolerance against the toxicity of heavy metals in plants. The symptoms of toxicity due to heavy metals can be seen when these metals interact with molecules at the cellular level. The interaction of heavy metals with cells causes death due to the generation of free radicles by binding of heavy metals to the protein sulfhydryl group. There are several other mechanisms for the detoxification of heavy metals in plants that provide tolerance to heavy metal stress. There is a presence of the few species of plants that represent increased oxidative stress than the homeostatic and evasion mechanisms that show that they can escape the impact of stress. To withstand heavy metal stress, plants adopt several strategies to avoid harmful effects on their growth.

Heavy Metals Origin in Soil

Geochemical Origin of Heavy Metals

Approximately 99 per cent of the total element in the earth crust consists of the ten significant elements named O, Si, Al, Fe, Ca, Na, K, Mg, Ti and P. The rest of the elements are known as trace elements. The concentration of the individual trace element is < 1000 mg / kg or 0.1%. The trace elements are mainly present in the primary minerals of igneous rocks. At the time of crystallization, these trace elements become an important part of the primary minerals by isomorphic substitution. A greater number of elements are available in clays and shales because they absorb metal ions very quickly. Ag, As, Cd, Cu, Pb, Mo, U, V and Zn, all these metals and metalloids high in black shales. The sediments from which they are framed act as both an adsorbent for heavy metals and a substrate for various microbes. The last catalyzes the formation of a reduced condition as a result of the increased deposition of heavy metals due to precipitation of metal sulfides.

Different Sources of Heavy Metals in Soil

Although there is the availability of heavy metals naturally in the soil there are also some human activities which lead to the accumulation of heavy metals in the soil cause degradation of the soil which as follows:

Mining Activities

In the modern world, the demand for metals is increasing as a result of their utilization in manufacturing units. These metals are available as ore bodies in the earth crust. These minerals are very important because they act as a vast reservoir of different minerals and metals. As a result, the mining activities and the overuse of these ore bodies lead to the extraction of metals using very well-developed mineral extraction technology. During mining activities, an enormous quantity of waste materials is produced, in particular, by tailings which are small milled pieces of rock and left-over parts of the ores which are not expelled by the separation process. As a result, these smaller ore particles are transported from the mining areas to the soil ecosystem with the help of wind and water and are deposited in the soil. Where oxidation of these particles occurs and due to various weathering activities, they are very well distributed through the soil and contaminate it. This causes very harmful effects on micro-fauna crops and soils.

Agricultural Practices

The results of various agricultural practices have also increased the accumulation and contamination of heavy metals. This is more of a problem in areas where intensive farming is prevalent. Day after day, there is an increase in the world's population, which creates pressure on the government. To feed the nations. As a result, various improved technologies and several agrochemicals have been adopted to increase crop production. For more yield farmers to apply excess doses of fertilizers and various pesticides in the field, taking into account soil health and environmental stability, which affects the sustainability of agriculture. The use of excess fertilizer causes the deposition of certain heavy metals in the soil. For example, the continuous use of phosphate fertilizer, which leads to the deposition of cadmium in the soil, which has a very serious effect on crop growth. Cd, even at a smaller concentration, poses a serious threat to crops and microbes present in the soil. According to the Lowestoft Laboratory of the Ministry of Agriculture, Fisheries and Food (MAFF 1986), the major sources are as follows:

- **Impurities in fertilizers:** Cd, Cr, Mo, Pb, U, v and Zn
- **Pesticides contain:** Hg, Pb, Mn, Cu, As and Zn
- **Sewage sludge:** Cd, Ni, Cu, Pb and Zn
- **Manures from intensive animal production (pigs & poultry):** As, Cu and Zn
- **Refused derived composts (generally not use in agriculture):** Cd, Cu, Ni, Pb and Zn
- **Wood preservatives:** As, Cu, Cr
- **Corrosion of metal objects (galvanized metal roofs and wire fences):** Cd and Zn

Sewage and Sludge

The treatment of wastewater in domestic and industrial areas produces the final product of sewage and sludge. Sewage and sludge are also used in agricultural fields because they contain the essential nutrients needed by plants for their normal growth and development. To maintain sustainability in agriculture, sewage and sludge have a very important role to play. Apart from these qualities, they also contain certain harmful substances such as toxic heavy metals and organic micro-pollutants (PAHs, PCBs and pesticides). The concentration of heavy metals in sewage and sludge depends on the areas from which it is produced. Sewage and sludge are produced in industrial areas with a higher concentration of heavy metals than in domestic areas. Various heavy metals used in sewage and sludge viz, Cd, Cu, Ni and Zn. Among these, Cu and Zn are also essential for crop growth, but only when applied at a lower concentration. Cd is very harmful to crop growth even at very low concentrations.

Fossil Fuel Combustion

By burning fossil fuels, the suspension of different types of heavy metals in the environment or soil depends on the type of fossil fuel. Heavy metals released from fossil fuels are Pb, Cr, Cd, As, Zn, Sb, Cu, Ba, Mn, U and V. These non-renewable materials are accumulated over a longer period during the process of formation of raw coal and petroleum. When coal is burned and this is hidden in the ash that is deposited in the soil after it has been transported from the area of combustion to a place where agriculture takes place, which then poses a problem to crop production. For example, large contamination of the environment and soil is caused by the deposition of lead (Pb) after the combustion of petrol (gasoline) which acts as the main source of Pb. Vehicles that use lead-containing petrol contaminate the environment by exhaustion as aerosol particles with a diameter of 0.01-0.1 mm. These aerosol particles form clusters after reactions with different other pollutants present in the air which increase the diameter, i.e. 0.3-1 mm. Complex compounds like 2PbBrCl-NH4Cl.

Metallurgical Industries

Metallurgical industries appear, for several reasons, to be the main source of heavy metal contamination in the environment and soil. For example, releasing aerosols and dust particles into the air deposit in the soil after mixing in the rainwater. Release of liquid waste from these metal industries directly to rivers that ruin the quality of the water and are very hazardous to aquatic bodies as well. They are also deposited in the soil after flooding or by using that water to irrigate crops. There is also dumping of waste material into the soil, which then leaches into the deeper layers of the soil and ruins the health of the soil. It also pollutes groundwater and water aquifers. Various heavy metals such as V, Mn, Pb, Cr, Cu, Sn, Si, Te, Ge, Sb, Cd, As, Ag and many others are used for the production of special alloys and steel. As a result, the utilization of these in the industry has led to a direct or indirect problem of heavy metal toxicity.

Electronics

For the manufacturing of the various electronic appliances such as semiconductors, cables and many more there is the utilization of many heavy metals like Cu, Zn, Au, Ag, Pb, Sn, Y, W, Cr, Se, Sm, Ir, In, Ga, Ge, Re, Sn, Tb, Co, Mo, Hg, Sb, As and Gd according to the MAFE, 1986. During the production of these appliances, there is the emission of some pollutants which contaminate the environment and their waste material release into the soil cause heavy metal deposition which affects the stability od soil ecosystem. There is also the addition of some other appliances such as capacitors and transformers contained polychlorinated biphenyls (PCBs), which is major organic pollutants in soils for a longer period.

Disposal of Waste Materials

Household, municipal and industrial waste and its disposal to the soil and water bodies cause heavy metal toxicity. The dumping of solid municipal waste into the soil is the serious issue related to the increase in heavy metal concentration in soil. All this municipal waste acts as a storehouse for many heavy metals like Cd, Cu, Pb, Sn and Zn. There dumping cause leaching of these into the lower layer of the soil which is then polluting the groundwater. Temporary waste stockpiles can cause more soil pollution which can not be found until analysis at a later time when the land is no longer useful for this purpose.

Other Heavy Metal Sources

Some other sources, which have the potential to cause heavy metal pollution of the soil and environment, which as following (MAFE 1986):

- Printing and Graphics: Se, Pb, Cd, Zn, Cr and Ba
- Polymer stabilizers: Cd, Zn, Sn and Pb
- Catalysts: Pt, Sm, Sb, Ru, Co, Rh, Re, Pd, Ni and Mo
- Pigments and paints: Pb, Cr, As, Sb, Se, Cd, Ba, Zn and Co
- Batteries: Pb, Sb, Zn, Cd, Ni and Hg
- Additives in fuels and lubricants: Se, Te, Pb, Mo and Li

Soil-Plant Relationship of Heavy Metal

Soil-Plant System

There is an important interrelationship which influences the movement of heavy metals from the soil to the plant system as shown in the figure below. Various contaminants are present, such as soil fertilizer application, pesticides, sewage sludge, which increase the concentration of heavy metal in the soil and plant. Heavy metals present in the soil move through the plant. Roots and accumulate in the parts of the plant and the products harvested. Some metals are leached into a deeper layer of soil and deposited in groundwater, which ultimately re-uses plants when irrigation is applied after pumping water from the tube wells (fig. 1).

Fig. 1. The soil-plant system showing the key components concerned with the dynamics of heavy metals

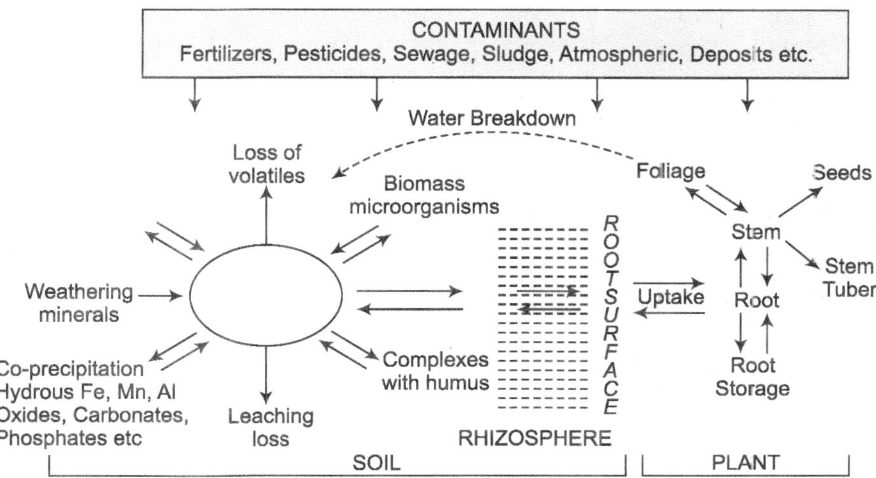

(Source: Modified from Peterson and Alloway 1979)

Metals Uptake by the Plants

There is a varying amount of heavy metals taken up by plants, depending on the different factors, which are as follows:

- The concentration of the heavy metals present in soil solution
- Mobility of the heavy metals from the soil to the rhizosphere
- At which amount the entry of heavy metal from the root surface into the root
- Movement of metals from the root system to shoot part

Larger uptake of metals by plants can be seen in mobile ions, while uptake of immobile ions depends on root interception. Mycorrhiza is a fungus colonized on the root surface, which increases the availability of plant nutrients, especially in the case of phosphorus. Active and passive processes for metal absorption by plants are involved. The active absorption of ions from the soil into the plant system occurred against the concentration gradient. Active absorption requires metabolic energy for the transport of ions from cell metabolism either directly or indirectly. There are two concepts such as carrier concept to explain the mechanism of active transport, and the other is cytochrome pump. There is no requirement for metabolic energy in the passive absorption of metals, which depends on the diffusion of ions in soil solution and enters into roots. Absorption in a passive process does not depend on temperature and metabolic inhibitors. It depends on the concentration of ions present in the soil solution. The concept of Active Ion Absorption includes mass flow theory, contact exchange theory, carbonic acid exchange theory, and Donnan equilibrium. There are different mechanisms for plant uptake of particular heavy metals. For example, plants through passive absorption absorb lead (Pb). On the other hand, active and passive processes for the transport of copper, molybdenum and zinc into the plant system are involved. For all ions there is the same mechanism that is absorbed in the roots, the maximum number of ions compete with each other for transport to the root system, as Cu and H+ restrict zinc absorption. Another example is the inhibition of the copper ion by zinc, NH_4^+, Ca and K. Some organic compounds are released to the rhizosphere by plant roots, such as root exudates and mucilage, which enhances microbial activity near the rhizosphere and mobilizes immobile elements that are strongly adsorbed by soil colloids. There is a difference between the metal uptake of plants under field conditions and the plants, which are grown in pots under greenhouse conditions (same soil uptake in pots under field conditions). This is due to the availability of metals under field conditions because the soil is the largest buffer, microclimate and due to the difference in soil moisture.

Essentiality of the Trace Elements

There are three criteria of essentiality given by the Arnon and Stout (1939) which was refined by Arnon (1954):

- The plant cannot complete its life cycle under the deficiency of the essential element
- The deficiency of the particular element is only be corrected by supplying of that element
- The elements are directly involving in the metabolism of the plant

Apart from C, H, O, N, P, K, and S, the elements which are essential for plants are Al, B, Br, Ca, Cl, Co, Cu, F, Fe, I, K, Mg, Mn, Mo, Na, Ni, Rb, Si, Ti, V, and Zn

(Kabata-Pendias and Pendias, 1992). Trace elements are also called micronutrients which are very important for the enzymatic activity of the plant. Under their deficiency plant growth and development is greatly affected. They are required in very low quantity. Careful handling of the micronutrients is very important because they produce toxicity symptoms if they are used in more amount as they are required. When the application of micronutrients is increased over the lower critical limit, then there is luxury consumption not affect to crop yield. If micronutrients supply gets over the upper critical limit which then causes toxicity to the plants and lowering the crop yield.

Toxicity of Heavy Metals in the Plants

Plants required an optimum quantity of all essential nutrients for better growth and development. Higher concentrations, even in the case of essential elements, cause signs of toxicity in plants. Some causal mechanisms are caused by the high concentration of essential and non-essential metals as follows.

(i) Alteration in the permeability of the cell membrane: Ag, Au, Br, Cd, Cu, Hg, F, I, Pb and UO2

(ii) Sulphydryl (-SH) groups reaction with cations: Ag, Hg and Pb.

(iii) Heavy metals compete with essential metabolites for sites: Sb, As, Se, Te, W and F

(iv) Affinity for reacting with phosphate groups and active group of ADP or ATP: Al, Be, Zr, lanthanides and generally all heavy metals.

(v) Replace the important cations: Li, Rb, Se and Sr

(vi) Occurring of the phosphate and nitrate sites: Arsenate, borate, selenate, tungstate.

The toxicity symptoms of heavy metals vary depending on the plant species, as some plant species are tolerant to heavy metal stress, while some are highly affected. However, some metals are very harmful to plants and all living organisms, such as cadmium, mercury, nickel, lead, cobalt and silver to a higher degree. Food products containing one of these metals are very hazardous to human health.

Heavy Metals Induce Damage to the Soil Microorganisms

Various reports show that the presence of heavy metals in the soil has a very harmful effect on soil microorganism activity. There is a higher concentration of copper and zinc near the forest area affecting the decomposition of the conifer litter which is severely affected by the release of plant nutrients. In heavy metal contaminated soil, there is an inhibitory effect on microbes by limiting respiration and dehydrogenase activity. Due to the long-term persistence of heavy metals in the soil, some microorganisms develop tolerance to heavy metals. The sewage and sludge used in leguminous crops contain those metals which affect the nitrogen fixation in the roots of the bacteria, which reduces the nitrogen content of the crop and also reduces the protein content. Because nitrogen is a very important part of amino acids. To maintain microbial activity in the soil, it is very important to find ways of helping to detoxify heavy metals and to maintain sustainability in agriculture.

Detoxification of Heavy Metals in soil

There are two approaches for detoxification of heavy metals in the soil which as follows:

Physiological Approach

Excavation Method

This involves the disposal and burial of polluted soils in special landfill sites (Conder et al., 2001; Jing et al., 2007; Lombi et al., 2002; Neilson et al., 2003; Bennett et al., 2003). Although the excavation method is the most widely used approach to reclaim contaminated soil (Lombi et al., 2001), it does not remediate the soil (Neilson et al., 2003).

Capping of the Contaminated Soil

Covering contaminated soil means covering the polluted soil with better nutritious soil. In this, we've got to take care of the proper depth. The depth at which the roots of the plant can not absorb heavy metals from the covered contaminated soil. However, this method can not solve the root-level problem because heavy metals are then leached to the lower depth of the soil and mixed with groundwater, which is then raised by tube wells and used for irrigation in the soil.

Fixation and Inactivation of the Contaminated Soil

Fixation and inactivation of contaminated soil mean that the toxic form of heavy metals is converted into a non-toxic form by making it fixed with the soil colloids through which it becomes immobile and less available for plant uptake. There are some chemical immobilizing agents available that help to make heavy metals less mobile. For example, zeolite, gravel sludge, alkaline materials, sewage sludge, compost, phosphate and lime stabilized municipal biosolids.

Washing of the Soil

This means washing of the contaminated soil with water at higher pressure. In this first of all heavy metals are solubilize in the contaminated soil with the help of some solubilizing agents like utilization of acids (HCL and HNO_3), chelators (EDTA, Nitriloacetic acid, DTPA etc.) and biosurfactant. After this soil is flushed out with the help of water pumps. But this method has not cost-effective and farmers are not able to adopt this method to reclaim the polluted soil. There are also other physiological methods which involve: thermal treatment (Jing et al., 2007), precipitation or flocculation followed by sedimentation, exchange of ions, reverse osmosis and microfiltration (Lone et al., 2008). All these methods are not more practicable due to their high cost. This method also disturbs the structure of the soil and reduce the soil health because flushing out the soil with pumps at high pressure affects the microbial activity in the soil.

Biological Approach

Use of the Microorganisms to Remediate the Polluted Soil

Due to increased pressure of the heavy metals in the soil or its persistence, there are several soil microbes which develop resistance against various heavy metals. There have been numerous reports which revealed that this resistance mechanism is the plasmid-mediated and maximum present in eubacterial groups of microbes. Some aerobic microbes which also develop resistance like *Staphylococcus* sp., *Escherichia coli, Pseudomonas aeruginosa* and *Bacillus* sp (Nakahara et al., 1977; Marques et al., 1979; Harnett and Gyles, 1984; Schwarz and Hobel, 1989; Belliveau, et al.,1991; wang and Shen, 1995). There is also study on some anaerobic microbes which develop resistance against mercury stress which are Bacteroides and *Clostridium* sp.

Resistance mechanism developed in microorganisms also chromosomal based which is a greater complex system than plasmid-mediated resistance system. Plasmid encoded resistance mechanism generally the toxic-ion efflux mechanism. In the microbial cells, there is a development of resistance to prevent the damage to the sensitive cellular components from heavy metal stress. As a result, some microbes are somehow able to survive under a stressed condition.

Phytoremediation of Heavy Metal Contaminated Soil

Phytoremediation means utilization of the plants for removing the heavy metal contamination from the soil, water and air. This method uses the knowledge and integration of the various disciplines like plant physiology, soil chemistry and soil microbiology for decontaminating the soil. There are several contaminants which impose a threat to the environment and agricultural sustainability like heavy metals, radionuclides, petroleum hydrocarbons, chlorinated solvents, PCBs, PAHs, organophosphate insecticides, explosives and surfactants (Khan et al., 2004). Use of the green plants is very useful for decontaminating both organic and inorganic contaminants. Such plant species withstand under high concentration of heavy metals because they can uptake large quantity of heavy metals via roots and store in their specified tissues without expressing any kind of toxicity symptoms. Plants which are utilized for the phytoremediation having some unique characteristics which as follows:

- Having the ability to uptake greater quantity of heavy metals from the deeper layer of the soil through roots.
- Then translocate the extracted heavy metals to the specified organs where they will be stored.
- Having the ability to produce a higher quantity of plant biomass even under toxicity without showing the symptoms.

Some plant species also having other features to show the tolerance mechanisms which detoxify the heavy metals in their shoots. Example of plant species i.e. *Brassica juncea* use for the phytoremediation due to its ability to accumulate a high quantity of Ni and Cd in its above-ground parts.

Tolerance Mechanism for Heavy Metals in Plants

There is a presence of essential and nonessential heavy metals in the soil ecosystem. They are very harmful to the plant growth if their concentration is increased up to the toxic level. Even some heavy metals also show toxicity if they present in very low concentration. When these heavy metals bind with the proteins through sulphydryl groups then plants show the toxicity symptoms which results in a reduction of the plant growth. Enhanced concentration of heavy metals in plant parts also increase the production of reactive oxygen species (ROS) which leads to oxidative stress. Due to the increasing problem of heavy metals and persistence in the soil for a longer time i.e. metalliferous soils plants adapt themselves by expressing some tolerance mechanism against some common metals present in all plants. Plants have some mechanisms at the cellular level for detoxifying the heavy metals in their system and tolerate the stress to a higher extent. The way through which plants show tolerance in which cells initially protect the sensitive part of the cell and avoid any damage to its homeostasis. Extracellularly in some cases, mycorrhizas come in action for protection. The plasma membrane also plays a better role for tolerance by lowering the extraction of heavy metal from soil or by enhancing the efflux pumping of heavy metals which are entered to the cytosol. For repairing the damaged proteins by stress, heat shock proteins or metallothioneins plays a very important role. Organic acids, amino acids or peptides cause chelation of heavy metals. Vacuolar compartmentalization also takes place for tolerance in plants (fig. 2).

Fig. 2. Biochemical mechanisms involved in heavy metal tolerance

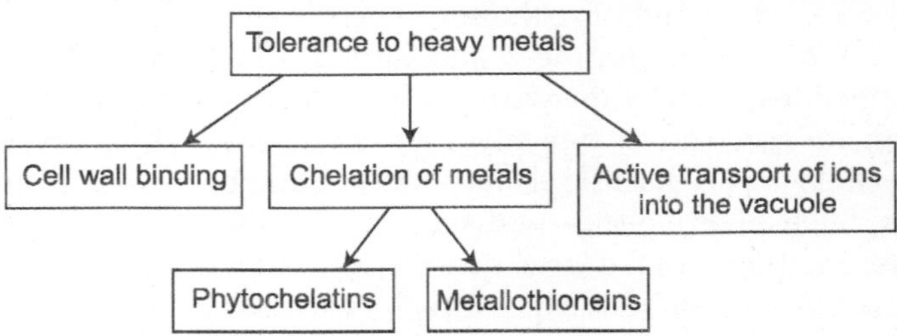

Table 1: Mechanisms involved in heavy metal tolerance

Metals	Mechanism Involved
Copper, Zinc	Mycorrhizae
Lead	Vacuolar Compartmentalization
Cadmium	Phytochelatins and Heat shock proteins
Arsenic	Plasma Membrane (Reduced Uptake)
Chromium	Metallothioneins
Others	Organic acids and amino acids

Mycorrhizae

There is a presence of a large number of microorganisms in the rhizosphere such as different bacteria and various fungi like mycorrhizal fungi. There is the colonization of the bacteria with the roots and at the same time presence of mycorrhizal fungi enhance the availability of several metals and also the uptake of heavy metals. For instance, a strain of *Xanthomonas maltophyla* bacterium can be exposed for catalyzing the reduction and precipitation of highly mobile Cr ions from significantly less movable environmentally less hazardous compounds (Blake et al., 1993). Heavy metals mobility and solubility increase the dissimilatory nonreducing activity of bacteria. Presence of some microbes in the rhizosphere increases the solubility of zinc in the soil which leads to the higher deposition of the Zn in the shooting part of *T. caerulescens*. There have been several pieces of research which show that there is a symbiotic association of mycorrhiza with plant roots acts as a tolerance mechanism against heavy metals. The tolerance mechanism which is express by the fungi at the cellular level same as that shown by the higher plants in which there is a binding of the metals to extracellular components. There are some examples like *Pisolithus tinctorius* is a fungus which is tolerant against Cu and Zn due to the binding of these metals with the extrahyphal slime whereas other example is detoxification of the cadmium through its binding with cell walls and transfer it to the vacuole in case of *P. involutus*. There are some other tolerance mechanisms like in ectomycorrhiza which helps the host plants by inhibiting the movement and uptake of heavy metals by the roots. A hyphal sheath which absorbs the heavy metals leads to decrease their reach to the apoplast because of hydrophobicity of fungal sheath, chelation by fungal exudates and adsorption on the outer mycelium. In the maximum number of the plant's species, there is the colonization of the Arbuscular mycorrhizal fungi with the plant roots which then having the special features to help in enhancing the useful soil microbes in the soil ecosystem. This fungus also helps to promote the growth and development of the plants due to its ability to increase the availability of essential plant nutrients especially in the case of immobile nutrient like phosphorus. Several studies have been reported that vesicular-arbuscular mycorrhiza (VAM) can buffer the stress due to heavy metals. Hence, its remedial characteristics show that both VAM-mediated nutritional effects and the effect of VAM fungus o metal issuance at the link of the soil and plant are includes in the discarding of the heavy metals.

Root Exudates and Cell Wall

Here is the tolerance mechanism in which there is the binding of the metals with the cell wall but this mechanism is quite controversial. Adsorption of the heavy metals on the roots has a limited capacity because there is direct contact of the metals with the roots. It is difficult to say about the specificity of this mechanism against which metals. One example of *Silene vulgaris* spp. *humilis* shows the tolerance due to its potential to accumulate different metals in the epidermal cell walls or by binding the metals with the proteins. Root exudates also play a better role in providing tolerance against heavy metal toxicity. According to the review from various works of literature it has been noticed that there is a decrease in the uptake of Ni from the soil with the help of Ni-chelating histidine and citrate which have the potential to accumulate in the exudates of the roots of non-hyper accumulating plants and detoxify the Ni toxicity. In the soils where the availability of the Fe and Cu is low, there can be an increase in Fe and Cu uptake by the dicots if the reduction of Fe^{3+} and Cu^{2+} takes place with the help of ferric reductase genes which are extracted from the *Zea mays*.

Tolerance Provided by Plasma Membrane

Under the heavy metal contaminated soils, the first living structure which is affected is the cell membrane. Under the toxicity of Cu, the plasma membrane is highly affected by increased leakage from the plant cell. Other reported that the root part of *Silene vulgaris* and wheat is highly affected under the high concentration of the Cu du to the ion leakage results damage to the plasma membrane. Due to the high concentration of cadmium in the soil, the activity of ATPase in the cell membrane is reduced especially in case of wheat and sunflower roots. In Nitella, Cu-induced changes in cell permeability were attributed to nonselective conductance increases and inhibition of the light-stimulated H^{+}-ATPase pump (Demidchik et al., 1997). Thus, to tolerate the heavy metal stress by the plants includes the safety of plasma membrane and reduce the leakage of the cell. There is also an improved tolerance mechanism in which the plasma membrane has to play a very vital role by inhibiting the entry of heavy metals into the cell via efflux mechanism. In bacteria, the maximum of the tolerance mechanism depends upon the efflux of toxic ions dependent on the energy. Another system for reducing the metal levels within the cell at the cell membrane includes the active efflux of ions of metal. Latest research has shown that plants have various classes of metal transporters that must be associated with a metal take up and homeostasis generally and hence play a very important role in tolerance against heavy metals. These involve the heavy metal CPx-ATPases, Nramps and cation diffusion facilitator family (Williams et al., 2000), the ZIP family (Guerinot, 2000).

Organic Acids and Amino Acids

Release of the organic acids increases the absorption by roots through the heavy metals mobilization. There is a prevalence of various organic acids having the low molecular weights which can affect the release of cadmium from the soil because there is the formation of complex compounds by combining the organic acids with cadmium. Various carboxylic and amino acids like citric, malic and histidine act as ligands for heavy metals which detoxify the heavy metals and also provide tolerance.

Heat Shock Proteins

The temperature which is greater than the required optimum temperature for any living organisms then there is an increase in the expression of heat shock proteins (HSPs). These proteins are classified according to their molecular weight and present in all the groups of living organisms. These heat shock proteins are coming into action when the plants faced any environmental stress including heavy metal stress. These proteins are also called as 'molecular chaperones' which acts as a sealing agent for repairing the damaged proteins under stress condition. For example, in the rice crop, there is enhanced in the levels of mRNAs for the HSPs having the lower molecular weight under the heat and heavy metal stress. Under the higher concentration of copper in the soil there is an increase in the expression of HSP17 in the root part of *Armeria maritima*. In the heavy metals contaminated soil with different metals again HSP17 comes into action by increasing its expression through enhancing the cell clusters in *Silene vulgaris* and *Lycopersicon peruvianum*. As per the ultrastructural studies it is found that by providing heat stress before heavy metal stress gives a tolerance effect by protecting the damage to the cell membrane.

Phytochelatins

Phytochelatins plays a very vital role in the detoxification of the heavy metals in the plants. Heavy metal detoxification and tolerance mechanism provided by the chelation of the heavy metals in the cytosol with the help of high-affinity ligands. These ligands constitute the organic acids, amino acids, peptides, phytochelatins and metallothioneins. For providing tolerance against the cadmium stress, the phytochelatins play a very important role in the plants. PC synthase, the genes which are responsible for its production are found in Arabidopsis and yeast. It is shown that in Brassica juncea Cd accumulation is accompanied by a rapid induction of PC biosynthesis and that the PC content is theoretically sufficient to chelate all Cd taken up; this protects photosynthesis but did not prevent a decline in transpiration rate (Haag-Kerwer et al., 1999). Different studies have been shown that accumulation and detoxification of the heavy metals happen along with peptide structure and biosynthetic regulation of PC synthase (Mishra and Dubey, 2006; Pal and Rai, 2009). Several reports which revealed that glutathione inhibits the biosynthesis of Pc synthase which show their better role in heavy metals detoxification in some fungi, algae and plants. In case of the certain bacteria and fungi, there is enhanced in the tolerance to heavy metal due to overexpression of PCs genes.

Metallothioneins

These are the minute proteins which helps in binding the proteins omnipresent in the cyanobacteria, eukaryotes and certain multicellular organisms. Metallothioneins are found in the BGA, algae and higher plants. Metallothioneins which are present in plants are different from those found in mammals because of their difference in amino acid sequence, charge, the total content of cysteine, peptide spacing patterns and length of linker domains. In plants, there is a prevalence of highly complex metallothioneins and in mammals, less complex metallothioneins are found. Under the toxicity of cadmium and mercury in plants. MTs are very important and plays a very useful role. Homeostasis of MTs also influenced by the copper and zinc. In the *Focus vesiculosus*, sixty-seven-amino acids Mt with sixteen cysteine residues are copper inducible and helps in binding the both Cd(II) and Cu(II) in vitro. In the mustard crop which is under the cadmium stress found that expression of MT2 was found to be late relative to the PC synthesis. For the metabolism of the metals, MTs have a very crucial role but there is no evidence of its appropriate function, there might be a different function for various metals.

Vacuolar Compartmentalization

It is one of the tolerance mechanism in which there is the transportation of the heavy metal into the vacuole for reduction of the toxicity in the cytosol. The ABC transporter which helps in the transporting of cadmium and PCs in the vacuole but there is also report that the vacuole is also essential in the accumulation of some other heavy metals having other tonoplast transport system. It is also an important mechanism for reducing the toxicity of lead (Pb) and provide tolerance to the plants. Other than the Cd-PC accumulation process, the better example of metal tolerance through vacuolar compartmentalization can be seen in the case of zinc (Zn).

Signal Transduction of Heavy Metal Stress

For detoxification of the heavy metals in the plants first of all plant has to sense the availability of the metal in its system then any tolerance mechanism gets started to prevent it from any type of damage. Plants sense the changing in the cytoplasmic and disturbing in the cellular components at which plants prepared to itself for the detoxification process. In this part, we will go through the heavy metal-induced signal transduction as a result of over metal exposure. In the signal transduction, there are metals which cause stress to plants called plant stressor like cadmium cause signalling perception with the help of some outer signals i.e. Ca or miRNA through which the signals are transmitted quickly by MAP kinases to the responsible transcription factors.

Dual Role of Reactive Oxygen Species (ROS)

Reactive oxygen species (ROS) has a dual role which reacts with the macromolecules and also acts as a signal transduction molecule. At the time of stress, there is an excess release of H_2O_2 which cause greater damage to the cell. However, H_2O_2 is also an important signalling molecule, which regulates plant development, hormone signalling, programmed cell death, and biotic and abiotic stress response and tolerance (Matilla, 2012). For maintaining the growth and development of the plants under heavy metal stress it is very important to control the production of reactive oxygen species (ROS). ROS accumulation due to the Cd toxicity is interceded and controlled by NADPH oxidase and calcium but there is also use of miR398. When there is excess exposure of the metals at that time there is a production of superoxide (O2 •-) from oxygen with the help of NADPH oxidase present on the cell membrane. Under the high concentration of cadmium in the plants which restrict the miR398 expressions which generally affects the Cu/Zn superoxide dismutase genes i.e. CSD 1 and CSD2. The decrease in the expression of miR398 leads to enhancing the expression of CSD 1 and CSD 2 which results in the superoxide detoxification into low toxic H2O2. This also causes reactive oxygen species (ROS) detoxification and accumulation of antioxidant which then lowering down the accumulation of ROS in the plant cells. Cd-induced production of ROS can also be mediated by a different process, involving protein phosphatases, Ca2? channels, and cGMP, as found in pea (Romero-Pertas et al., 2004).

Reactive Oxygen Species (ROS) and Stress Response

For detecting the ROS generation and its accumulation in the plant cell under the heavy metal stress by the oxidative signal inducible 1 protein for H2O2 and EX1 (Executer 1) and EX2 proteins for superoxide (Rentel MC, Lecourieux D, Ouaked F, Usher SL, Petersen L, Okamoto H, Knight H, Peck SC, Grierson CS, Hirt H, 2004 & Lee KP, Kim C, Landgraf F, Apel K, 2007). These proteins can increase the quick signal transduction of a downstream mitogen-activated protein kinase (MAPK) cascade. A mitogen-activated protein kinase cascade is a progression of phosphorylation ventures from MKKKs (MAPK kinase) through MKKs (MAPK-kinase) to MPKs. The signals which are originated from the MAPK cascade will be additionally transduced to transcription factors which can tie to specific cis-regulatory components and incite metal responsive gene expression. In the plants which are tolerant to metal toxicity, there is an activation of the effective metal tolerance mechanism by MAPK cascades while on the other hand in metal sensitive plants MAPK cascades regulate trans and cis-regulatory mechanisms which are not effective against metal tolerance.

Signal Transduction In Plant With Response To Heavy Metals Toxicity

Heavy metal toxicity is one of the abiotic stresses that cause deleterious health effects both in animals and plants. Owing to their higher reactivity they directly affect growth, development, senescence and energy synthesis process. The signalling network eventually leading to transcriptional activation of the metal-responsive gene to combat the stress. *Sorghum* is one of the most important staple foods for the world's poorest and most food-unsecured people across the semi-arid tropics. The root system of *Sorghum vulgare* is fibrous in nature which leads to the accumulation of metals more in the root. But interestingly, grains of the plant remain free from contamination because transfer and transportation ratios remain low for the metals within the plants under field condition.

Due to rapid industrial development and urbanization, there is high emission of toxic ions into the environment (Nriagu and Pacyna, 1988). The soil is the primary recipient of a myriad of waste products and chemicals. Heavy metals contamination of soils results from the application of phosphate fertilizers, sewage, sludge and wastewater effluents. The signaling responses of heavy metals stress comprise of a signal transduction network which is activated by sensing of heavy metal and synthesis of stress-related proteins and signaling molecules. This signalling network eventually leading to transcriptional activation of the metal-responsive gene to combat the stress (Maksymiec, 2007). Many signal transduction pathways converge on a synthesis of stress-related genes and their cross-talk, ROS signaling and mitogen-activated protein kinase (MAPK) phosphorylation cascade.

The metal-dependent gene expression, particularly to Cd, Pb, Al and Zn may be regulated on multiple levels which involve direct action on DNA, chromatin modifications, transcription, RNA processing, translation and post-translational events. The metal-dependent modulation of gene expression has not received much attention of investigation, though most of such study been focused on the gene or the protein products that confer plant to tolerance to metals. The recent advancements in the field of transcriptomics, proteomics and metabolism have widened over knowledge on the metal response in plants. Genes induced by metal stress are classified into two classes; [a] gene encoding for protein gives direct protection against metal stress such as detoxifying enzymes and other functional proteins and; [b] regulating gene expression and signal transduction i.e., transcription factors and kinases.

Metal Receptors

Receptor-like protein kinase mediates the perception of extracellular signals. For instance, in barley, the gene coding for lysine motif receptor-like kinase (HvLysMR1) was shown to be induced by Cr, Cd and Cu during leaf senescence (Ouelhadj *et al.*, 2007). Similarly, a putative receptor protein kinase was involved in signaling of Cd by rice roots.

Ca-CALMODULIN SYSTEM

The expression of HvLysMR1 is also mediated by Ca level. Ca^{+2} signaling features in responses to several abiotic stress signaling. Heavy metals modify the stability of Ca channels thereby increasing calcium flux into the cell. Studies have demonstrated that certain metals like Ni, Cd causes disbalance in intracellular calcium level and thus interfere with calcium signaling by substituting Ca in calmodulin regulation (Ghelis *et al.*, 2000). Cd and Cu induce calcium accumulation in rice roots (Yeh *et al.*, 2007). Ca^{+2} acts as a secondary messenger interacting with calmodulin to initiate the signal leading to downregulation of genes associated with heavy metal transport, metabolism and tolerance. Similarly, Ca calmodulin is also involved in the response to other heavy metal toxicity, like Ni and Pb; transgenes tobacco expressing NtCBP4 (Nicotiana tabacum calmodulin-binding protein) tolerate higher levels of Ni^{+2} but are hypersensitive to Pb^{+2}, indicating foe exclusion of Ni^{+2} but the accumulation of more Pb^{+2} than wild –type plants (Arazi *et al.*, 1999).

ROLE OF ROS AND MAPK CASCADE

Cd produces reactive oxygen species (ROS) directly via the Fenton and Haber-weiss reaction, while indirectly by inhibiting anti-oxidant enzymes (Romero Puertas *et al.*, 2007). H_2O_2 levels increased in response to Cd and Cu treatment in *Arabidopsis* (Maksymiec and Krupa 2006), besides Hg treatment in tomato and response to Mn toxicity in barley (Cho and Park, 2000). Similarly, the treatment of tobacco cells and Scots Pine Mots with Cd and lupine roots with Pb caused H_2O_2 production (Olmos *et al.*, 2003; Przymusinski *et al.*, 2007). The oxidative burst in tobacco due to Cd was shown to be mediated by calmodulin and /or calmodulin-dependent protein. ROS can induce anti-oxidant defence mechanism directing via antioxidant responsive element (ARE) found in the promoter region of such gene: one such gene is from CAT1 (CATALASE1) whose gene expression was upregulated during Cd toxicity (Smeets, 2008). Cytosolic ascorbate peroxidase such as APX1 & APX2 as well as plastidic iron superoxide dismutase (FSD1) is activated by H_2O_2 and affected by metal stress (Smeets, 2008). ROS are known to activate scavenging mechanism via redox-sensitive transcription factor or via activation of kinase cascades which subsequently activate transcription factors that trigger gene transcription (Scandalios, 2005). MAPKs are one of the largest families of serine-threonine kinases in higher plants that transduce extracellular signals regulating cellular processes. The MAPK pathways are activated in response to Cd (Jonak *et al.*, 2004), Cu (Yeh *et al.*, 2003) and iron stress. MAPK cascade is a well-known response in plant subjected to both biotic and abiotic stress. At the end of the cascade of phosphorylation, MAPKS phosphorylated different substrate in different cellular compartments, thus it allows transduction of information to downstream targets. Cd and Cu activated four different MAP kinase (S1MK, MMK2, MMK3 & SAMK) in alfalfa (Jonak *et al.*, 2004). Cd also induced ATMEKK1 in kinase in Arab (Suzuki *et al.*, 2001) and OsMAPK2 in rice (Yeh *et al.*, 2004). Yeh *et al.* (2007) reported that Cd and Cu-induced MAPK activation required the involvement of Calmodulin dependent protein kinase (CDPK) and phosphatidyl-inositol 3-kinase (PI3 kinase); this reflects that Cd and Cu signaling pathway induces ROS production as well as calcium accumulation. These signallong pathways eventually converge in the regulation of transcription factors that activate genes needed for stress adoption particularly the genes that activate metal transporters and biosynthesis of chelating compounds.

Phytohormone Signaling

Plant hormones assume a critical role in the adaptation to abiotic stress since there is a regulation of hormone (plant growth regulator) synthesis in the presence of heavy metals. The signaling pathways involving ABA, SA and auxin participate in response to metals; cis-DNA regulatory elements were detected in metal-induced genes (Suzuki *et al.*, 2001). Similarly, auxin-responsive mRNA was detected in Cd-treated *Brassica* plants (Fusco *et al.*, 2005). Cd and Cu activate the synthesis of ethylene by up-regulating ACC synthase expression and activity. These heavy metals also are shown to induce accumulation of Jasmonic acid in *Phaseolus* plant (Makysmiec *et al.*, 2005). SA protected the barley roots from lipid peroxidation caused by Cd toxicity (Metwally *et al.*, 2003). The transcription activation of SAMT gene associated with SA synthesis was detected in pea plants subjected to Hg.

Nitric Oxide Signalling

Nitric oxide (NO) is an important signalling molecule influencing plant development process and defence response (Crawford and Guo, 2005). Cd-treated soybean cells have been shown to release NO, thus demonstrating the role of NO in alleviating heavy metal-induced stress (Kopyra *et al.*, 2006). Three was NO accumulation observed in plastids of *Arabidopsis* cells exposed to Fe; NO acts denovostream of Fe and upstream of P2A-type phosphatase to promote the increase of mRNA coding for Ferritin (*AtFer1*) (Arnaud *et al.*, 2006) (Table 2).

Table 2: Examples of genes and proteins induced in plants by heavy metals

Genes/proteins induced	Metal involved	References
Ish, gsh2, gr1-glutathione metabolic genes	Cu, Cd	Xiang & Oliver (1998)
PR-1 protein caused by SA, TMV	Cu	Yan *et al.*, 1999
OsMSRMK2-multiple-responsive MAPK gene induced by elicitor, drought, H_2O_2, ABA	Cu, Hg, Cd	Agrawal *et al.*, 2003
CSAP-Cd-stress associated 51kDa protein	Cd	Mittra *et al.*, 2004
HSP	Hg	Didierjean *et al.*, 1996

Genes Controlling Metal-Dependent Transcription

The transcription profiling of plats treated with Cd, Pb and Zn indicate for the existence of metal-induced transcription factors (Kovalchuk *et al.*, 2005). Cd-induced transcripts for bZIP (basic region leucne zipper), Myb and zinc finger transcription factors were detected in *Arabidopsis thaliana,* and *Brassica juncea* (Suzuki *et al.*, 2001; Fusco *et al.*, 2005). The multiple MYB genes of *Arabidopsis* not only expressed in response to Cd but also salinity stress and phytohormones like ABA, GA, IAA, JA, SA and ethylene. The Cd-induced bZIP transcription factor-OBF5) in *Arabidopsis* binds to the promoter region of glutathione transferase gene-GST6, which is induced by auxin, SA and oxidative stress (Suzuki *et al.*, 2001). In yet another study it was demonstrated that Zn treatment of *Arabidopsis* induced transcription factors-bHLH, while the expression of another transcription factor –WRKY and GATA-type decreased in the presence of an excess of Zn (van de Mortel *et al.*, 2006). ROS likely plays a significant role in activating metal-induced transcription factor in plants. Organism like yeast and animals possess specific metal-induced transcription factor (MTF) which bind to metal responsive element (MRE) present in promoters of a metal responsive element (MRE) present in promoters of metal-responsive genes (Giedroc *et al.*, 2001;). The cis-acting elements associated with MRE are found within promoters of a few plants genes inducing metallothionein-like genes (Perry *et al.*, 2005; Qi *et al.*, 2007). There are two cis-DNA elements reported in plants to be functional in metal response: the iron-dependent regulatory sequences (IDRS) mediates iron-regulated transcription of genes related to the iron acquisition, and the one identified within the promoter region of PvSR2 gene from *Phaseolus vulgaries* (Qi *et al.*, 2007), PvSR2 gene encodes a heavy metal stress-related protein whose expression is strongly stimulated by Cd, As, Hg and Cu (Zhang *et al.*, 2001b). Apart from the effect of heavy metal on a transcription of specific genes, there also affect the transcription process indirectly by substitution of Zn in RNA polymerase or Zn-requiring transcription factors (Robinson *et al.*, 1997). Several metals such as Mn, Ni, Mg, Cu, Co, Cd, Pb bind to DNA directly causing conformational changes (Duguid *et al.*, 1995); these changes impair the genes present within the DNA region affecting the transcription related events. For instance, Cr stress altered gene expression in animal cells by forming chromium-DNA adducts and Cr-DNA cross-links, thereby disturbing the transcription activator complexes (Wei *et al.*, 2004). Such metal-induced to damage to DNA has been reported in Cd-treated soybean, tobacco, broad bean (Fojtova and Kovarik 2000, Gichner *et al.*, 2004, Sobkowiak and Deckert 2004), Pb-treated lupine and Al-treated barley (Pan *et al.*, 2001). Heavy metal is also known to alter DNA repair processes; process; Cd interaction with DNA inhibiting mismatch repair in yeast and human cells. This inactivation of DNA repair machinery restricts DNA to exhibit its basic coding function in plants.

Chromatin Modification

The chromatin structure can be influenced by the position of nucleosomes on DNA through an ATP dependent remodelling complex. Such repositioning of nucleosomes to another part of DNA exposes the binding site resulting in gene expression. Induction of transcripts homologous to coding for ATP-dependent chromatin remodelling (SWI/SNF) and putative histone deacetylase (PDAC) was detected in Cd-treated Brassica plant (Fusco *et al.*, 2005) as well as Al-treated Saccharum species, indicating for the metal-induced chromatin modifications in plants. The chromatin modification can also be manifested by post-translational modification of histone protein such as acetylation, methylation, phosphorylation and ubiquitination (Figure a). Instances are known whereby chromatin modification was effected by replacing major histone proteins with their variants (Hsieh & Fischer, 2005). Ni and Cu inhibit histone acetylation via action of HAT (histone acetyltransferases) enzyme, which is linked to silencing of gene expression (Kang *et al.*, 2004, Yan *et al.*, 2003) Besides, the post-translational process involving microRNA plays a key role in the regulation of at least one plant metal-dependent gene i.e.., Cu, Zn-SOD (Yamasaki *et al.*, 2007)(Figure b).

Fig. b. Schematic representation of modulation of gene expression in plants subjected to heavy metal

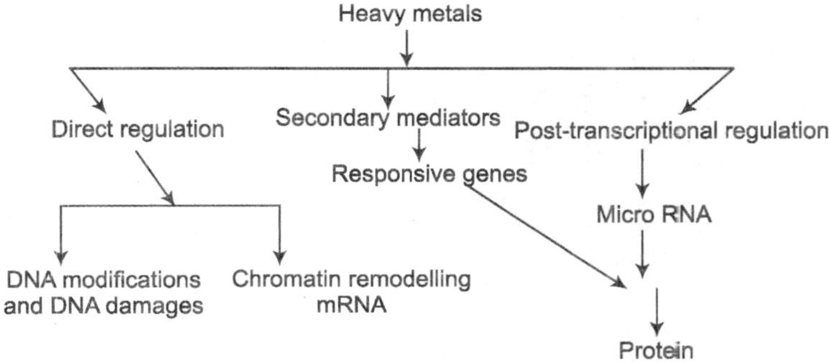

(**Source:** drawn by the author, unpublished, 2020)

Heavy Metal Chelation in the Cytosol

G. Morton and H. Drew to give the word "chelate", subsequent from the Greek term chela, meaning "great claw" of the lobster or other crustaceans. The word chelate suggests the approach in which an organic compound "clamps" onto the cationic elements, which it chelates. For a compound to be called a true chelating agent, it must have certain chemical characteristics. This chelating compound must consist of at least two sites capable of donating electrons (coordinate covalent bond) to the metal it chelates. For true chelation to occur the donating atoms(s) must also be in a position within the chelating molecule so that a formation of a ring with the metal ion can occur. The term sequestered deals additional with the accomplishment of chelation or complexing, not the actual chemical understanding or meaning. The term "complexed" originates from combinations of minerals and organic compounds that do not gather the guidelines of a true chelate.

There are five categories of compounds that are usually diverse with minerals and used in agricultural foliar and soil-applied applications:

[A] Synthetic Chelates
[B] Ligno Sulfonates
[C] Humic or Fulvic acids
[D] Organic Acids
[E] Protein (Amino acids)

[A] ***Synthetic Chelating Agents:*** The most ordinary form is EDTA that is typically worn in agricultural mineral formulations as the disodium salt of EDTA. Synthetically chelated minerals among the strongest forms of chelation used in commercial agricultural applications. Usually used forms of minerals are; calcium EDTA, magnesium EDTA, zinc EDTA, manganese EDTA and copper EDTA. Forms of synthetic iron chelates diverge rather, according to the circumstance in which they are utilized. One accepted form is iron EDDHA (ethylenediamine [di (o-hydroxyphenyacetic) acid]). The efficiency of artificial chelating agents for foliar application, it is recognized that in a nutrient solution it is necessary to have a surplus of overall mineral about chelating agent so that the chelating agent does not participate for the mineral with the plant (Table 3).

Table 3: EDTA Advantage and Disadvantage

EDTA-Advantage	EDTA-Disadvantage
EDTA chelates are extremely stable and can be speckled with just about any phosphate-containing fertilizer at basically any pH	EDTA is a synthetic compound, not formed by plants.
EDTA is incredibly resistant to microbial degradation; consequently, it remains quite constant in soils.	Synthetic chelating agents can compete with the plant for a mineral. EDTA is known for its strong affinity toward calcium and although debatable, some think that because of its insistence in a plant, it is probable that a zinc EDTA foliar applied, could be winning for zinc, but the EDTA could fight for calcium once in the plant.

[Source: Kumar *et al.*, 2012]

[B] ***Ligno Sulfonates:*** Measured to be a water-soluble, non-toxic polymer. Polymer typically means that the molecule is reasonably large. Lignin, a most important polymeric constituent of woody tissues in higher plants, is composed of repeating phenyl propane units and usually amounts to 20-30 % of the dry weight of wood. It is a consequence of the pulp and paper industry. The lignin resulting from softwoods is dissimilar that from hardwoods, consequently, their reactivity differs fairly. Ligno sulfonates and mineral combinations are fairly frequently referred to as sequestered or complexed. With all the obtainable sites on the lingo sulfonate molecule, there would probably be 2 or 3 obtainable to make a true chelate; though, the size of the molecule to several potential chelate sites makes these molecule incompetent chelators. Even though it is likely to make extremely effectual mineral formulations from lingo sulfonate, the majority of commercially utilized sources of this material are not formulated for effectual foliar uptake (Table 4).

Table 4: Ligno Sulfonate- Advantage and Disadvantage

Ligno Sulfonates Advantages	Ligno Sulfonates Disadvantage
The best advantage of lingo based minerals is its lower cost	The main disadvantage of lingo sulfonate for foliar application is the size of the molecule. There are many different grades of lignin and thus different molecular weights. Some areas are high as 21,000.
Another advantage is the polymer is biodegradable and supportive of soil microorganism	With proper sulfonation, lower MW lignin sulfonates do exist
Ligno sulfonate also has surfactant properties	

[Source: Kumar *et al.*, 2011]

[C] ***Humic or Fulvic acids***: In foliar applications, the fulvic acids in NUE amino minerals may not play a straight role in accessibility and movement of minerals in the plant. Though, they accomplish act as plant stimulants and perhaps precursors to plant hormones (Table 5).

Table 5: Humic acid and Fulvic acid

Humic and Fulvic Advantages	Humic and Fulvic Disadvantages
Humic and fulvic acids are price effectual compounds to adjoin to foliar and soil-applied minerals	In foliar applications, humic has a definite disadvantage of being too large in size
When they are "small" adequate to efficiently move into the plant, they can provide a valuable resource of precursors to significant plant chemicals	In trace mineral mixtures humic molecules are inclined to resolve out in the container, mostly due to their molecular size and of their reduced suspension of these complex colloids in the water at low pH

[Kumar *et al.*, 2012]

[D] *Organic Acids*: These compounds are one of the two groups of compounds that are essential for the transportation and solubility of divalent elements in plants. The organic acids can be called "anionic organic acids" because of their negative charge. One such organic acid in agriculture is citric acid. Although organic acids do not directly chelate monovalent elements, they are associated with plant systems by loosely held ionic attractions.

In foliar applications, these anionic organic acids play an important role in converting cationic minerals into nonionic compounds for increased leaf penetration and movement into the plant. Plants can manufacture many types of organic acids such as citric, malic, fumaric, succinic and others. Some of these acids are utilized to transport minerals in the plant. Other are excreted the roots to solubilize and take needed minerals into the plant system. NUE Amine Minerals contain a group of these natural organic acids, derived from a fermentation process (Table 6).

Table 6: Organic acid-Advantage and Disadvantage

Organic Acids-Advantage	Organic acids-Disadvantages
Organic acids are one of the usual systems a plant utilizes to solubilizing and translocate minerals.	Commercially produced individuals organic acids are quite expensive compared to some other chelating and complexing agents
The groups of organic acids in NUE Amino minerals have been proven in field demonstrations to be more effective in correcting mineral deficiencies on some crops than mineral applications without these compounds	Compared to synthetic chelating agents and some amino acids chelating compounds organic acids chelates form weaker bonds with minerals, particularly at raising pH levels.
By neutralizing the charge of minerals, organic acids create mineral uptake additional effective.	

[Source: Kumar *et al.*, 2012, 2014]

[E] ***Protein (Amino Acids):*** Amino acids are the second type of natural compounds that plants produce to solubilize and translocate minerals. Plants manufacture these compounds to make mineral biologically available in the cell. Even when unchelated or uncomplexed minerals are sprayed on plants, the mineral must combine with organic compounds such as an amino acid before the plant can effectively utilize it. As with organic acids, amino acid compounds also play a role in mineral uptake into plant tissue, because of the increase permeability effect of the amino acid on the cuticle (Table 7).

Table 7: Amino acid –Advantage and Disadvantage

Amino Acids-Advantages	Amino Acids-Disadvantages
Amino acids are one of the natural systems a plant utilizes to translocate and utilize minerals	Enzymatically hydrolyzed protein is expensive to produce
The groups of amino acids in NUE Amino minerals have been proven in field demonstrations to be more effective in correcting mineral deficiencies on some crops than mineral applications without these compounds	The actual combination of properly prepared amino acid/peptide compounds and minerals is not 100% complete with simple tank mix procedures.
By neutralizing the charge of minerals, amino acids make mineral uptake more effective	

[Source: Kumar *et al.*, 2012]

Inside the cell, heavy metal ions that are not instantly necessary metabolically may attain toxic concentrations and plant cells have evolved various mechanisms to store excess metals to stop their participation in unwanted toxic reactions. If the toxic metal concentration exceeds a certain threshold inside the cells, an active metabolic process contributes to the production of chelating compounds. Specific peptides such as PCs and MTs are used to chelate metals in the cytosol and to sequester them in specific subcellular compartments. A large number of small molecules are also involved in metal chelation inside the cells, including organic acids, amino acids and phosphate derivatives.

[1] *Phytochelatins:* PCs are the best-characterized heavy metal chelators in plants, particularly in the context of Cd tolerance. PCs are a family of metal-binding peptides with the general structure $(\gamma\text{-Glu-Cys})_n\text{Gly}$ (n=2-11). They are present in plant and fungi. The cysteine thiol groups allow PCs to chelate metals and form complexes with a molecular weight of 2.5-3.6 kDa. PCs are synthesized in the cytosol and then transported as complexes to the vacuole. Their synthesis is fast activated in the presence of heavy metals such as Cd, Cu, Zn, Ag, Hg and Pb. Synthesis involves the chain extension of GS by PCS a constitutively expressed cytosolic enzyme whose activity is controlled post-translationally because the metal ions chelated by PCs are required for enzyme activity. Due to their metal likeness, PCs are thinking to be involved in the homoeostasis and trafficking of essential metals such as Cu and Zn and the detoxification of heavy metals, but they do not seem to be involved in Hyperaccumulation. A confirmation has been reported for the role of PCs in heavy metal tolerance. They have a clear function in the reply of plants and yeast to Cd, e.g., they are induced rapidly in *Brassica juncea* following the intracellular accumulation of Cd, thus protecting the photosynthetic apparatus despite a decline in transpiration and leaf expansion. Besides, the Cd compassion of various *A. thaliana* mutants correlated with their ability to accumulate PCs. Cd and Cu treatment also induce the transcription of genes involved in the synthesis of GS, the precursor of PCs.

[2] *Metallothioneins* and *Ferritins:* MTs have been established in many organisms, even though the MTs in plants differ considerably from those found in mammals and fungi. As they contain mercaptide groups they can bind metal ions. In plants, MTs are induced by a variety of abiotic stresses but are also expressed during development. In wheat and rice, MTs are induced by metal ions, such as Cu and Cd and by abiotic stresses such as temperature extremes and nutrient deficiency. Plant MTs sequester surplus of metals by coordinating metal ions with the multiple cysteine thiol groups and have a particular affinity for Zn^{+2} and Cu^{+2}. MTs possibly have diverse functions in response to dissimilar heavy metals and could also contribute to additional antioxidant defence mechanism and plasma membrane repair. Ferritins are ever-present multimeric proteins that can store up to 4500 iron atoms in a central cavity. Animal Ferritins can also store other metals, including Cu, Zn, Cd and Al. Whereas plant Ferritins have only been exposed to store Fe. Plants Ferritins are synthesized in responses to various environmental stresses, including photoinhibition and iron overloading. Ferritins are therefore a front-line defence mechanism against free iron-induced oxidative stress. The chief function of plant Ferritins is not to store and release iron, as previously reported, but to scavenge free reactive iron and prevent oxidative damage.

[3] *Organic Acids, Amino Acids and Phosphate Derivatives:* Organic acids and amino acids can attach heavy metals and may consequently be deployed in reaction to metal toxicity. Organic acids such as malate, citrate and oxalate confer metal tolerance by transporting metals through the conducting tissue xylem and sequestrating ions in the vacuole, but they have multiple additional roles in the cell. Citrate, which is synthesized in plans by enzymes citrate synthase, has an elevated ability for metal ions than malate and oxalate and even though its principle role in chelate Fe^{+2}. It also has a strong affinity for Ni^{+2} and Cd^{+2}. Malate is a cytosolic Zn-chelator in zinc-tolerant plants. Amino acids and derivatives are capable to chelate metals conferring to plants resistance to toxic levels of metal ions. Histidine is measured as the most important free amino acids in heavy metal metabolism. Due to existence of carboxyl, amino and imidazole groups, it is a flexible metal chelator, which can confer Ni tolerance and enhance Ni transport in plants when supplied in the growth medium, possibly reflecting its normal role as a chelator in root exudates. Phytate (Myo-inositol hexakisphosphate) is the standard form of preserve phosphorous in plants and is frequently localized in the roots and seeds. The molecule comprises six phosphate groups which permit the chelation of multiple cations, including Ca^{+2}, Mg^{+2} and K^{+}, but also Fe^{+2}, Zn^{+2} and Mn^{+2}. The allocation of phytates and its capacity to chelate multiple metal species propose it could be mobilized as a detoxification strategy. Extraction methods are mainly based on the ratio of a given reagent (extractants) to a given weight of soil sample, and on the chemical feature of the extractants, i.e. with or without a chelating agent. Some extractants commonly extract a wide range of soil elements while others are applied for single elements. Amount of Pb extracted was poorly correlated with soil physicochemical properties, irrespective of the type of extractants. The multiple regression analysis established for EDTA gave the best fit. The accessibility to plants of soil elements is ruled by more than a few natural factors such as soil reaction, soil physicochemical properties, temperature and climate. Redundant side reactions can be made more selective for specific elements using complexions.

Conclusion

Plants have evolved miscellaneous resources to support physiological concentrations of essential metal ions and to decrease contact to non-essential heavy metals. Several methods are ubiquitous since they are also essential for universal metal homeostasis and they reduce injure caused by the eminent concentration of heavy metals in plants by detoxifications, consequently conferring tolerance to heavy metals stress. Many plants naked to lethal concentrations of metal ions effort to avoid or decrease uptake into root cells by restricting metal ions apoplast, binding them to the cell wall or cellular exudates, or by inhibiting long-distance convey. If this fails, metals before in the cell are addressed using a choice of storage space and detoxification strategies, counting metal transport, chelation, trafficking and sequestration into the vacuole. When this tackle is worn-out, plants turn on oxidative stress protection mechanisms and the amalgamation of stress-related proteins and signaling molecules, such as heat shock proteins, hormones and reactive oxygen species. When metal ions are accumulated in surplus within the cytosol, plants have to take away them to diminish their toxic belongings. Plants react to high intracellular concentrations of metal ions using efflux pumps either to export the ions to the apoplast or to compartmentalize them within the cells. The major storage compartment for metal ions in the vacuole, which in plants accounts for up to 90% of the volume. Several families of intracellular transporters concerned in this procedure have been recognized in plants and yeast and they come into view to be extremely discriminating. The reply to heavy metal stress involves a complex signal transduction system that is activated by sensing the heavy metal and is characterized by the production of stress-related proteins and signaling molecules and lastly the transcriptional commencement of precise metal-responsive genes to offset the stress. The relevant signal transduction passageway includes the Ca-calmodulin system, hormone, ROS signaling and the mitogen-activated protein kinase (MAPK) phosphorylation cascade. Dissimilar signalling pathways may be used to respond to dissimilar heavy metals. The increasing concentration of the heavy metals like Cd, Pb, As, Hg and Ni etc. in the soil and environment impose greater hazard to the human, animals and plant growth and development. Toxicity of heavy metals in the plants disturbs the physiology and metabolism of the plants which highly reduce the overall yield of the crop and cause economic damage to the farmers. There are several sources through which the concentration of the heavy metals increases in the soil and environment. Sources like mining activities, different agricultural practices like excess use of the fertilizers, combustion of fossil fuels, disposal of industrial waste like from metallurgical and electronics industries to the water and soil and also sewage and sludge had potential contaminants which toxify the soil. Once these metals are accumulated in the plants or in cells which adversely affects all the internal functioning of the plant. Hence this paper highly concentrates on the mechanisms which are activated under stress condition to provide resilience and tolerance to withstand under those conditions. All the mechanisms which are adopted by the plants aim to control the accumulation and detoxification of heavy metals. For every species of the plants, there are different tolerance mechanism which restricts the uptake of heavy metals in the cytosol by directing it to an apoplastic space. Some plant species even adopt two or more tolerance mechanisms to combat heavy metal caused damage. To tolerate the various heavy metal stress plants developed several cellular mechanisms. Extracellularly mycorrhiza plays a very important role in heavy metal tolerance. This fungus also helps to promote the growth and development of the plants because of its ability to mobilize the immobile nutrients like phosphorus and make it available for plant growth. There is also the presence of improved tolerance mechanism in which plasma membrane restrict the movement of heavy metals into the cell via efflux tolerance mechanism. Vacuolar compartmentalization involves in the transportation of the heavy metals to the vacuole under the higher concentration of metals where these are detoxified. For repairing the damaged proteins heat shock proteins (HSPs) have a very efficient role. Phytochelatins, metallothioneins, organic acids and amino acids help in the chelation of heavy metals which restrict the metal accumulation in the full reactive form in cytosol. The better understanding of the signal transduction pathway which is activated under the heavy metal stress condition is very essential for tolerance mechanism.

References

Afzal, M., Shabir, G., Tahseen, R., Islam, E. U., Iqbal, S., Khan, Q. M., & Khalid, Z. M. (2014). Endophytic Burkholderia sp. strain Ps JN improves plant growth and phytoremediation of soil irrigated with textile effluent. *CLEAN–Soil, Air, Water, 42*(9), 1304-1310.

Arnon, D. I., & Stout, P. R. (1939). The essentiality of certain elements in minute quantity for plants with special reference to copper. *Plant physiology, 14*(2), 371.

Arnon, D. I., Whatley, F. R., & Allen, M. B. (1954). Photosynthesis by isolated chloroplasts. II. Photosynthetic phosphorylation, the conversion of light into phosphate bond energy. *Journal of the American Chemical Society, 76*(24), 6324-6329.

Andrades-Moreno, L., Del Castillo, I., Parra, R., Doukkali, B., Redondo-Gómez, S., Pérez-Palacios, P., ... & Rodríguez-Llorente, I. D. (2014). Prospecting metal-resistant plant-growth promoting rhizobacteria for rhizoremediation of metal contaminated estuaries using Spartina densiflora. *Environmental Science and Pollution Research, 21*(5), 3713-3721.

Ahmad, I., Akhtar, M. J., Zahir, Z. A., & Jamil, A. (2012). Effect of cadmium on seed germination and seedling growth of four wheat (Triticum aestivum L.) cultivars. *Pakistan Journal of Botany, 44*(5), 1569-1574.

Ahn, S. J., Rengel, Z., & Matsumoto, H. (2004). Aluminum-induced plasma membrane surface potential and H+-ATPase activity in near-isogenic wheat lines differing in tolerance to aluminum. *New Phytologist, 162*(1), 71-79.

Akhtar, M. S., Chali, B., & Azam, T. (2013). Bioremediation of arsenic and lead by plants and microbes from contaminated soil. *Research in Plant Sciences, 1*(3), 68-73.

Andrades-Moreno, L., Del Castillo, I., Parra, R., Doukkali, B., Redondo-Gómez, S., Pérez-Palacios, P., ... & Rodríguez-Llorente, I. D. (2014). Prospecting metal-resistant plant-growth promoting rhizobacteria for rhizoremediation of metal contaminated estuaries using Spartina densiflora. *Environmental Science and Pollution Research, 21*(5), 3713-3721.

Antosiewicz, D. M. (2005). Study of calcium-dependent lead-tolerance on plants differing in their level of Ca-deficiency tolerance. *Environmental Pollution, 134*(1), 23-34.

Burges, A., Epelde, L., & Garbisu, C. (2015). Impact of repeated single-metal and multi-metal pollution events on soil quality. *Chemosphere, 120*, 8-15.

Assunção, A. G., Schat, H., & Aarts, M. G. (2003). Thlaspi caerulescens, an attractive model species to study heavy metal hyperaccumulation in plants. *New Phytologist, 159*(2), 351-360.

Arazi, T., Sunkar, R., Kaplan, B., & Fromm, H. (1999). A tobacco plasma membrane calmodulin-binding transporter confers Ni2+ tolerance and Pb2+ hypersensitivity in transgenic plants. *The Plant Journal, 20*(2), 171-182.

Agrawal, J., Sherameti, I., & Varma, A. (2011). Detoxification of heavy metals: state of art. *Detoxification of heavy metals*, 1-34.

Achard, P., Herr, A., Baulcombe, D. C., & Harberd, N. P. (2004). Modulation of floral development by a gibberellin-regulated microRNA. *Development, 131*(14), 3357-3365.

Agrawal, G. K., Iwahashi, H., & Rakwal, R. (2003). Rice MAPKs. *Biochemical and biophysical research communications, 302*(2), 171-180.

Ahmad, A., Hadi, F., & Ali, N. (2015). Effective phytoextraction of cadmium (Cd) with increasing concentration of total phenolics and free proline in Cannabis sativa (L) plant under various treatments of fertilizers, plant growth regulators and sodium salt. *International journal of phytoremediation, 17*(1), 56-65.

Al-Hakimi, A. M. A. (2007). Modification of cadmium toxicity in pea seedlings by kinetin. *Plant Soil and Environment, 53*(3), 129.

Alloway, B. J. (1995). Heavy metals in soils. Blackie Academic and Professional. *An Imprint of Chapman & Hall. Glasgow.*

Alloway, B. J., & Jackson, A. P. (1991). The behaviour of heavy metals in sewage sludge-amended soils. *Science of the Total Environment, 100*, 151-176.

Arazi, T., Sunkar, R., Kaplan, B., & Fromm, H. (1999). A tobacco plasma membrane calmodulin-binding transporter confers Ni2+ tolerance and Pb2+ hypersensitivity in transgenic plants. *The Plant Journal, 20*(2), 171-182.

Appenroth, K. J. (2010). Definition of "heavy metals" and their role in biological systems. In *Soil heavy metals* (pp. 19-29). Springer, Berlin, Heidelberg.

Beveridge, T. J., & Murray, R. G. (1976). Uptake and retention of metals by cell walls of Bacillus subtilis. *Journal of Bacteriology, 127*(3), 1502-1518.

Bennett, L. E., Burkhead, J. L., Hale, K. L., Terry, N., Pilon, M., & Pilon-Smits, E. A. (2003). Analysis of transgenic Indian mustard plants for phytoremediation of metal-contaminated mine tailings. *Journal of environmental quality, 32*(2), 432-440.

Belliveau, B. H., Starodub, M. E., & Trevors, J. T. (1991). Occurrence of antibiotic and metal resistance and plasmids in Bacillus strains isolated from marine sediment. *Canadian journal of microbiology, 37*(7), 513-520.

Barber, S. A. (1995). *Soil nutrient bioavailability: a mechanistic approach.* John Wiley & Sons.

Banuelos, G. S., Ajwa, H. A., Mackey, B., Wu, L., Cook, C., Akohoue, S., & Zambruzuski, S. (1997). *Evaluation of different plant species used for phytoremediation of high soil selenium* (Vol. 26, No. 3, pp. 639-646). American Society of Agronomy, Crop Science Society of America, and Soil Science Society of America.

Bothe, H. (2011). Plants in heavy metal soils. In *Detoxification of heavy metals* (pp. 35-57). Springer, Berlin, Heidelberg.

Brune, A., Urbach, W., & Dietz, K. J. (1994). Compartmentation and transport of zinc in barley primary leaves as basic mechanisms involved in zinc tolerance. *Plant, Cell & Environment, 17*(2), 153-162.

Bringezu, K., Lichtenberger, O., Leopold, I., & Neumann, D. (1999). Heavy metal tolerance of Silene vulgaris. *Journal of Plant Physiology, 154*(4), 536-546.

Blechert, S., Brodschelm, W., Hölder, S., Kammerer, L., Kutchan, T. M., Mueller, M. J., ... & Zenk, M. H. (1995). The octadecanoic pathway: signal molecules for the regulation of secondary pathways. *Proceedings of the National Academy of Sciences, 92*(10), 4099-4105.

Blaudez, D., Botton, B., & Chalot, M. (2000). Cadmium uptake and subcellular compartmentation in the ectomycorrhizal fungus Paxillus involutus. *Microbiology, 146*(5), 1109-1117.

Bernal, M. P., McGrath, S. P., Miller, A. J., & Baker, A. J. (1994). Comparison of the chemical changes in the rhizosphere of the nickel hyperaccumulator Alyssum murale with the non-accumulator Raphanus sativus. *Plant and Soil, 164*(2), 251-259.

Begonia, M. T., Begonia, G. B., Miller, G. S., & Gilliard, D. (2004). Effects of chelate application time on the phytoextraction of lead-contaminated soils. *Bulletin of environmental contamination and toxicology, 73*(6), 1033-1040.

Baker, A. J., & Walker, P. L. (1990). Ecophysiology of metal uptake by tolerant plants. *Heavy metal tolerance in plants: evolutionary aspects, 2*, 155-165.

Baker, A. J., & Brooks, R. (1989). Terrestrial higher plants which hyperaccumulate metallic elements. A review of their distribution, ecology and phytochemistry. *Biorecovery., 1*(2), 81-126.

Clemens, S., Palmgren, M. G., & Krämer, U. (2002). A long way ahead: understanding and engineering plant metal accumulation. *Trends in plant science, 7*(7), 309-315.

Cobbett, C. S. (2000). Phytochelatins and their roles in heavy metal detoxification. *Plant physiology, 123*(3), 825-832.

Chand, J., & Kumar, P. (2020). Yield attribute shift of mustard grown under cadmium contaminated soil. Plant Archives, 20(2), 3518-3523.

Chand, J., & Kumar, P. (2020). Biochemical shift of mustard grown under cadmium contaminated soil. Journal of Pharmacognosy and Phytochemistry, 9(3), 178-183.

Chand, J., & Kumar, P. (February 2021). Retrieved from http://www.ijrar.org/IJRARJFM1204.pdf. 'A Review on cost of production and cost Benefit Analysis in Rice Crop', IJRAR. International Journal of Research and Analytical Reviews(IJRAR), E-ISSN 2348-1269, P-ISSN 2349-5138, 8, Issue 1, Page No pp.672-679.

Chand, J., Kumar, P., Upadhyay, H. (March 2021). Retrieved from http://www.jetir.org/papers/JETIR2103212.pdf.Geographical Distribution of Various Crops Like Cereals, Legumes, Oilseed, Vegetables, Fodder and Forages, Commercial Crop, Condiments and Species, Medical and Aromatic Plant: An Overview. Int. J. Emerg. Technol. Innov. Res, ISSN: 2349-5162,8(3), Page No.1679-1725.

Conder, J. M., Lanno, R. P., & Basta, N. T. (2001). Assessment of metal availability in smelter soil using earthworms and chemical extractions. *Journal of Environmental Quality, 30*(4), 1231-1237.

ChitraMani, P. K. (2020). Evaluation of antimony induced biochemical shift in mustard. Plant Archives, 20(2), 3493-3498.

Cummings, D. E., Caccavo, F., Fendorf, S., & Rosenzweig, R. F. (1999). Arsenic mobilization by the dissimilatory Fe (III)-reducing bacterium Shewanella alga BrY. *Environmental Science & Technology, 33*(5), 723-729.

Cobbett, C. S. (2001). Heavy metal detoxification in plants: Phytochelatin biosynthesis and function. *Iubmb Life, 51*(3), 183-188.

Di Toppi, L. S., & Gabbrielli, R. (1999). Response to cadmium in higher plants. *Environmental and experimental botany, 41*(2), 105-130.

Diatloff, E., Forde, B. G., & Roberts, S. K. (2006). Expression and transport characterisation of the wheat low-affinity cation transporter (LCT1) in the methylotrophic yeast Pichia pastoris. *Biochemical and biophysical research communications, 344*(3), 807-813.

Dong, R., Formentin, E., Losseso, C., Carimi, F., Benedetti, P., Terzi, M., & Schiavo, F. L. (2005). Molecular cloning and characterization of a phytochelatin synthase gene, PvPCS1, from Pteris vittata L. *Journal of Industrial Microbiology and Biotechnology, 32*(11-12), 527-533.

Doty, S. L. (2008). Enhancing phytoremediation through the use of transgenics and endophytes. *New Phytologist, 179*(2), 318-333.

Das, P., Samantaray, S., & Rout, G. R. (1997). Studies on cadmium toxicity in plants: a review. *Environmental pollution, 98*(1), 29-36.

Grill, E., Winnacker, E. L., & Zenk, M. H. (1987). Phytochelatins, a class of heavy-metal-binding peptides from plants, are functionally analogous to metallothioneins. *Proceedings of the National Academy of Sciences, 84*(2), 439-443.

Garbisu, C., & Alkorta, I. (2003). Basic concepts on heavy metal soil bioremediation. *ejmp & ep (European Journal of Mineral Processing and Environmental Protection), 3*(1), 58-66.

Ghosh, M., & Singh, S. P. (2005). A review on phytoremediation of heavy metals and utilization of it's by products. *Asian J Energy Environ, 6*(4), 18.

Hall, J. Á. (2002). Cellular mechanisms for heavy metal detoxification and tolerance. *Journal of experimental botany, 53*(366), 1-11.

Hänsch, R., & Mendel, R. R. (2009). Physiological functions of mineral micronutrients (cu, Zn, Mn, Fe, Ni, Mo, B, cl). *Current opinion in plant biology, 12*(3), 259-266.

Halim, M., Conte, P., & Piccolo, A. (2003). Potential availability of heavy metals to phytoextraction from contaminated soils induced by exogenous humic substances. *Chemosphere, 52*(1), 265-275.

Haag-Kerwer, A., Schäfer, H. J., Heiss, S., Walter, C., & Rausch, T. (1999). Cadmium exposure in Brassica juncea causes a decline in transpiration rate and leaf expansion without effect on photosynthesis. *Journal of experimental botany, 50*(341), 1827-1835.

Ikenaka, Y., Nakayama, S. M., Muzandu, K., Choongo, K., Teraoka, H., Mizuno, N., & Ishizuka, M. (2010). Heavy metal contamination of soil and sediment in Zambia. *African Journal of Environmental Science and Technology, 4*(11), 729-739.

Jalmi, S. K., Bhagat, P. K., Verma, D., Noryang, S., Tayyeba, S., Singh, K., ... & Sinha, A. K. (2018). Traversing the links between heavy metal stress and plant signaling. *Frontiers in plant science, 9*, 12.

Järup, L. (2003). Hazards of heavy metal contamination. *British medical bulletin, 68*(1), 167-182.

Kumar, R., Mishra, R. K., Mishra, V., Qidwai, A., Pandey, A., Shukla, S. K., ... & Dikshit, A. (2016). Detoxification and tolerance of heavy metals in plants. In *Plant Metal Interaction* (pp. 335-359). Elsevier.

Jing, L., Ng, M. K., & Huang, J. Z. (2007). An entropy weighting k-means algorithm for subspace clustering of high-dimensional sparse data. *IEEE Transactions on knowledge and data engineering, 19*(8), 1026-1041.

Küpper, H., Parameswaran, A., Leitenmaier, B., Trtilek, M., & Šetlík, I. (2007). Cadmium-induced inhibition of photosynthesis and long-term acclimation to cadmium stress in the hyperaccumulator Thlaspi caerulescens. *New Phytologist, 175*(4), 655-674.

Kondo, N., Isobe, M., Imai, K., Goto, T., Murasugi, A., & Hayashi, Y. (1983). Structure of cadystin, the unit-peptide of cadmium-binding peptides induced in a fission yeast, Schizosaccharomyces pombe. *Tetrahedron Letters, 24*(9), 925-928.

Lin, Y. F., & Aarts, M. G. (2012). The molecular mechanism of zinc and cadmium stress response in plants. *Cellular and molecular life sciences, 69*(19), 3187-3206.

Lone, M. I., He, Z. L., Stoffella, P. J., & Yang, X. E. (2008). Phytoremediation of heavy metal polluted soils and water: progresses and perspectives. *Journal of Zhejiang University Science B, 9*(3), 210-220.

Millaleo, R., Reyes-Díaz, M., Ivanov, A. G., Mora, M. L., & Alberdi, M. (2010). Manganese as essential and toxic element for plants: transport, accumulation and resistance mechanisms. *Journal of soil science and plant nutrition, 10*(4), 470-481.

Mishra, S., & Dubey, R. S. (2006). Inhibition of ribonuclease and protease activities in arsenic exposed rice seedlings: role of proline as enzyme protectant. *Journal of plant physiology, 163*(9), 927-936.

Martınez, C. E., & Motto, H. L. (2000). Solubility of lead, zinc and copper added to mineral soils. *Environmental pollution, 107*(1), 153-158.

Marschner, H. (1991). Mechanisms of adaptation of plants to acid soils. *Plant and soil, 134*(1), 1-20.

McGrath, D. A., Smith, C. K., Gholz, H. L., & de Assis Oliveira, F. (2001). Effects of land-use change on soil nutrient dynamics in Amazonia. *Ecosystems, 4*(7), 625-645.

Mani, C. (2019). Impacts of Foliar Feeding Of Zinc And Polyamine On The Productivity Of Wheat. International Journal of Emerging Technologies and Innovative Research, 6(2), 354-372.

McLaughlin, M. J., Hamon, R. E., McLaren, R. G., Speir, T. W., & Rogers, S. L. (2000). A bioavailability-based rationale for controlling metal and metalloid contamination of agricultural land in Australia and New Zealand. *Soil Research, 38*(6), 1037-1086.

Nriagu, J. O., & Pacyna, J. M. (1988). Quantitative assessment of worldwide contamination of air, water and soils by trace metals. *Nature, 333*(6169), 134-139.

Neilson, J. P., Lavender, T., Quenby, S., & Wray, S. (2003). Obstructed labour: reducing maternal death and disability during pregnancy. *British medical bulletin, 67*(1), 191-204.

Ostrowski, S. R., Wilbur, S., Chou, C. H. S. J., Pohl, H. R., Stevens, Y. W., Allred, P. M., ... & Tylenda, C. A. (1999). Agency for Toxic Substances and Disease Registry's 1997 priority list of hazardous substances. Latent effects—carcinogenesis, neurotoxicology, and developmental deficits in humans and animals. *Toxicology and Industrial Health, 15*(7), 602-644.

Parádi, I. (2003). Influence of arbuscular mycorrhiza and cadmium on the polyamine contents of Ri T-DNA transformed Daucus carota L. root cultures. *Acta Biologica Szegediensis, 47*(1-4), 31-36.

Rivera-Becerril, F., Calantzis, C., Turnau, K., Caussanel, J. P., Belimov, A. A., Gianinazzi, S., ... & Gianinazzi-Pearson, V. (2002). Cadmium accumulation and buffering of cadmium-induced stress by arbuscular mycorrhiza in three Pisum sativum L. genotypes. *Journal of Experimental Botany, 53*(371), 1177-1185.

Rascio, N., & Navari-Izzo, F. (2011). Heavy metal hyperaccumulating plants: how and why do they do it? And what makes them so interesting?. *Plant science, 180*(2), 169-181.

Rauser, W. E. (1999). Structure and function of metal chelators produced by plants. *Cell biochemistry and biophysics, 31*(1), 19-48.

Siedlecka, A., & Krupa, Z. (1996). Interaction between cadmium and iron and its effects on photosynthetic capacity of primary leaves of Phaseolus vulgaris. *Plant physiology and biochemistry (Paris), 34*(6), 833-841.

Schalscha, B. E., & Ahumada, T. I. (1998). Heavy metals in rivers and soils of central Chile. *Water Science and Technology, 37*(8), 251-255.

Shruti, M., & Dubey, R. S. (2010). Heavy metal uptake and detoxification mechanisms in plants. *International Journal of Agricultural Research, 5*(7), 482-501.

Verbruggen, N., Hermans, C., & Schat, H. (2009). Mechanisms to cope with arsenic or cadmium excess in plants. *Current opinion in plant biology, 12*(3), 364-372.

Viehweger, K. (2014). How plants cope with heavy metals. *Botanical Studies, 55*(1), 1-12.

Williams, E. J., Williams, G., Gour, B., Blaschuk, O., & Doherty, P. (2000). INP, a novel N-cadherin antagonist targeted to the amino acids that flank the HAV motif. *Molecular and Cellular Neuroscience, 15*(5), 456-464.

Yadav, S. K. (2010). Heavy metals toxicity in plants: an overview on the role of glutathione and phytochelatins in heavy metal stress tolerance of plants. *South African journal of botany, 76*(2), 167-179.

Zhang, L., & Wong, M. H. (2007). Environmental mercury contamination in China: sources and impacts. *Environment international, 33*(1), 108-121.

Transporters for Metal Storage in Organelles

Priyanka Aley

Department of Agronomy, School of Agriculture,
Lovely Professional University, Jalandhar, Punjab, India, 144411
Email: priyankaaley12@gmail.com

Abstract

Metal is fundamental and very important for plants, we can say that because without these several physiological and biological processes, it can not occur in the plant. Plants need metal transporters to meet several basic capacities, ranging from metal assimilation to metal sequestration and capacity. Plants also need to manage harmful metals, such as chromium, mercury, lead and arsenic, or poisonous overabundance of fundamental metals, but some metals are also useful when present in small quantities. As we all know some metals are useful for plants to complete their life cycle, but the maximum amount of metal is not in the available form, they are present in organic form and the plant only absorbs metal in inorganic form, so that these metals are first converted and there are different transporters present in the plant cell that regulates the transport of metal al. This review focuses on the different genes and transporters, such as NRAMP, ZIP, VIT1, PSC and others, how they work, how different processes take place, the path through which the metals travel and where they stop, what happens there, how the metal converted into a complex as before iron absorption is converted into Fe2 + and Fe3 +, only then available for absorption. There is also a different path from where the metal travels with gene regulation. Specific metals have their way of travelling and also different types of movement type, for example, there are apoplastic and symplastic pathways through which metals travel and in which there are many special genes that help them to travel. Some genes regulate the mobility of certain genes which help the metals to be stored at a specific site, the genes are metal-specific and also plant-specific, as in graminae family plants, the way of transport, absorption, sequestration and storage is very different from that of grasses. After absorption of these metals, they help the plant to grow and develop as some metals help in the regulation of photosynthesis, the production of chloroplasts, photosynthesis, tissue formation and also enhance the activity of other metals and even compensate for the deficiency of other metals. However, if these metals are present in large quantities, they can cause toxicity to the plant and, in this situation, the transporters act as a barrier and start detoxifying. So these transporters not only transfer metals but also protect plants from an overdose of metals, they can work in both ways there are many more things discussed below in an elaborated form.

Introduction

As we all know the plant is a living thing, we've learned different things about the plant since childhood. The plant can grow, make its food, die and reproduce. Plants have their human-like metabolism, which enables them to perform different internal activities and plays an important role in balancing the earth system. Without a plant, there will be no flora and fauna on which we depend directly or indirectly. But how can the plant perform functions of this kind, what is the true process behind all of these. Similarly, as the human plant breathes, it excretes the waste and reproduces the young. Metals are transported in plants as trace metals or heavy metals through the soil pores in the form of ions. Plants must maintain the basic components of the rhizosphere, for the most part, by their underlying foundations. 17 components are especially applicable as we can say they are the soul of plants without which the life cycle of plants is incomplete. Both the lack and abundance of these components lead to weakened development, often reflected as high production misfortune in agribusiness. Plants have developed refined and tightly controlled components for the take-up and translocation of basic components. In the same way, plants take some heavy metal along with them, for example, cadmium, chromium and arsenic. Ongoing work has shown that sub-atomical administrative pathways to maintain the homeostasis of the basic components are finely managed in unique crosstalk. Also, destructive components are taken along pathways similar to basic components. As a reaction to this metal, chelators like muginiacs come from the phytosiderophores and nicotianamine family. Phytosideraphores considered being the first distinguished iron chelators to be attached to the rhizosphere after discharge to assembling low-solvent ferric iron. Iron translocation is one of the main processes in plants performed by nicotianamine because it acts as an iron chelator. In any case, future research has shown that the two MAs and NAs can chelate iron as well as other metal supplements, such as copper, manganese and zinc, are engaged in the homeostasis of these metal supplements. Furthermore, it has been shown that the control of the creation of nicotianamine influences the uptake and translocation of the destructive cadmium component. The uptake and translocation of essential supplements are managed by explicit carrier proteins. Subsequently, the link between the basic supplements suggests that the transport framework must be covered. The extent to which components cooperate during securing and translocation are still insufficiently understood, even though it is relevant to the decision on the quality of yield and quantity in agribusiness. The transition of metal particles is fundamental to the biochemical capacity of clusters by being consolidated into or partnering with proteins to inspire works in living cells. Key elements of metal cofactors in plant mitochondria include digestion, electron transport, detoxification, and ATP combination of responsive oxygen species (ROS). For example, the trading of electrons in the electron transport chain (ETC) supported by copper and iron, tricarboxylic corrosive (TCA) protein cycles, which occur in that process, uses metal particle cofactors to catalyze essential metabolic substances. The metal filling of the other metal at the initiation of NAD-malic protein and succinyl-CoA is cobalt (Co). Cobalt (Co) separates the catalytic theory from the mitochondrial plant, in any case, it is not known whether there is an in vivo need to follow the proportions of Co for plant respiratory processing.

Metal particles, however, they may be, can also be deeply harmful to cells and the capacity of the organ cells. Redox-inert harmful metal cadmium has a strong affinity to oxygen, nitrogen, and sulfur particles and may inhibit compound action by directly obstructing protein capacity or by dislodging characteristic metal habitats. There are various reports of substantial metals. exhausting cell glutathione pools, causing reduced cancer prevention in the cell, and causing ROS amassing. Cadmium has been accounted for both the mitochondrial work of the Mitochondrial Plant as well as the Co Buildings in a roundabout way. For example, redox dynamic metal impulses, Cu and Fe cations can be cytotoxic in the same way as the hydroxyl anion they react with ROS or Fenton-type. Inhibiting effects of exogenous inclusion of Cu and Fe plant respiratory limits, As a result, the proximity of free metal cations, redox dynamic or inactive, in mitochondria may significantly add to the origin and spread of oxidative weight. Follow metals can enter the plant cell by spreading through the plasma layer, by endocytosis, and by the movement of unique metal transporters working in the plasma layer. A lot of consideration has been given to the last angle and many audit papers are available, plants need different nutrients such as iron, zinc, manganese, nickel, etc. After absorption, these metals are transported to different parts of the plant and this metal allocation plays an important role. So, as we all know, these metals are transported through the root to shoot xylem, phloem, and parenchyma cells inside the plants. Sometimes chelators such as nicotianamine, glutathione and phytochelatins also act as transporters of long-distance metal transport. Plant vacuoles that are present in the plant cells or can be said to act as temporary storage for fundamental and toxic metabolites, mineral supplements and harmful toxins. Healthy information based on total metabolism indicates that the transport of these particles and chelators, the Vacuum Sequestration Capacity / Limit (VSC) also makes a significant contribution to the long-distance transport and storage of metals in plants. Vacuoles are buffer pools where the sequestration limit is strongly acclimated to a changing ecological sign, through collaborations between ion chelators and tonoplast-confined transporters.

Metal Transporters

Intracellular organelles have a specific ability to interfere with the availability of metals and are subject to transporters in their layers. Metal conjugates, for example, phytochelatin buildings, are additionally moved to the vacuole. At first, the majority of transporter families were perceived for their association of metal particles in the vehicle over the plasma film (or digestion). Individuals in the family were therefore confined to the intracellular compartment (for capacity and appropriation). In the case of a carrier type, the route of transport inside or outside the cytoplasm is dared to be preserved, paying little attention to the film in which the carrier resides.

(a) **NRAMP:** These premises are commonly divalent cation/proton symporters (as depicted for mammalian NRAMP2 and the yeast homolog Smf1p, whereas one NRAMP was appeared to have a physiological job in Al^{3+} transport. Even though NRAMPs can ship a wide scope of divalent cations, as decided for the most part with heterologous articulation in *Xenopus laevis oocytes,* these transporters are frequently embroiled in the transport of Fe^{2+} and, to a lesser extent, Mn2+. In seeds, these transporters restrict to the globoids, and nramp3, nramp4 twofold mutants show abandons in the assembly of iron put away in globoids, which is basic during seed germination in iron-drained soil. AtNRAMP3 and AtNRAMP4 are likewise situated in mesophyll cells, the significant site for photosynthesis, where they are engaged with the assembly of Mn^{2+} from vacuoles this wellspring of Mn^{2+} is required for photosystem action when plants face imperfect manganese sustenance in the dirt. Shockingly, manganese sustenance influences neither the articulation nor the bounties of these transporters. The plants which are in Gramineae family have this gene NRAMP in the further two forms NRAMP3 and NRAMP4.

(b) ZIP: Divalent cation transporters are also called the members of this ZIP family. By the addition of bicarbonate, the transportation of Zn^{2+} is enhanced. Out of the four ZIPs in *S.cerevisiae,* Zrt3p restricts to the vacuole. The comparing quality is an immediate objective of the Zn responsive controller Zap1p, and articulation is likewise expanded by zinc consumption. ZRT3 isn't related to a development deformity in Zn2-drained medium except if the qualities encoding the two significant assimilatory transporters are additionally erased. Indeed, even in the triple take out, a conspicuous development phenotype is just seen inside a restricted window of supplemental Zn^{2+}. This phenotype is run of the mill for vacuole-restricted transporters because the transporter is just ready to give a fitness advantage if there is a satisfactory gracefully of metal in the vacuole to be activated. At low centralizations of Zn^{2+}, there isn't sufficient Zn^{2+} accessible to help develop and be put away, whereas, at high groupings of Zn^{2+}, an inner flexibly isn't important to help development. On the off chance that the vacuole is first stacked with Zn^{2+} by presenting cells to elevated levels of Zn^{2+} in the medium, this store of Zn^{2+} can support development without an outer flexibly.

(c) CDF: CDF has special characteristics it transports the catalyze metal/proton antiport, and the acidic lumen present in the vacuole in this manner influence metal transporters out of the cytosol. Given succession comparability, the CDF family can be separated into useful subgroups that relate to described physiological substrates. The transporters like Zrclp and Cotlp which is the subgroup of Zn^{2+} are responsible for resistance to excess Zn^{2+} gathering are also regulated by Zn^{2+}.

(d) VITI: Some of the traces are there in protein like VIT1 from *A.thaliana* are found in every kingdom, but absent in animal. These VIT1 traces found in protein within endoplasmic reticulum in the root nodules. All together for activation of Fe^{2} inside the seed AtNRAMP3 and AtNRAMP4 is very much needed, Fe^{2} must be preloaded into the compartment by AtVIT1. AtVIT1 is profoundly communicated in creating seeds, and AtVIT1 restricts to the vacuole layer. The nonattendance of this metal transporter does not influence the total centralization of iron in seeds yet leads to a distorted circulation of iron inside the seed, which is impeding when seeds from the freak are sprouted on basic soil (since iron is less dissolvable in basic soil, plant iron absorption is influenced). Due to its constitutive job in seed development, AtVIT1 articulation doesn't change in light of iron sustenance.

Regulations By Transporters

Cutting-edge, most tonoplast-limited metal transporters in plants have been identified to manage VSC and thus metal area between various organs or tissues (Table1).

Table 1: Mineral/metal transporters

Transporters	Minerals/metals	Reference
MTP1	Zn	Ricachenevsky et al. (2013)
MTP3	Zn	Arrivault et al. (2006)
Z1F1	Zn	Haydon and Cobbett(2007)
NRAMP4/NRAMP4	Fe	Thomine et al. (2003)
VIT1/VIT2	Fe, Zn	Kim et al. (2006)
FPN2(IREG2)	Fe, Co, Ni	Morrissey et al. (2009)
COPT5	Cu	Klaumann et al. (2011)
CAX2/CAX4	Cu	Korenkov et al. (2007)
HMA3	Cu	Ueno et al. (2010,2011)

The movement or remobilizing genes from vacuoles to cytosol like NRAMP3 and NRAM4 during mineral nutrient deficiency are managed by diminishing or reducing the VSC to release out the stored nutrient. In contrast, other genes including MTP3, VIT2, and CAX4 which manage metal sequestration within the vacuoles, are automatically directed upwards when excessive metals are accessible in the environment, and the improved articulation consequently leads to enlarging VSC and metal accumulation in vacuoles. Another accumulation of qualities including VIT1, MTP1, COPT5 remains unaffected by the metal status in the earth, anyway viable premature birth of these qualities altered metal task at either subcellular or tissue levels, suggesting that a consistent verbalization level of those characteristics engages fitting metal circulation. Metal hyper gatherers change their VSC by methods for a more conventionalist course than we have imagined, as they will when all is said in done utilize a couple of transporters which are similarly present in their relating non-hyper collectors yet have advanced with progressively complex guideline instruments. MTP1 is a key transporter tweaking the VSC of Zn. Heaped proof recommends that the articulation level of MTP1 is enormously improved in various metal hyper combinators which later on was shown to be inferable from genome amplification of value copies, multifaceted nature to a lone copy in their relative non-hyper gatherers A similar rule of TcHMA3 (Heavy Metal ATPases 3) was in the like manner seen among which quality amplification prompts much higher enunciation of TcHMA3 in the totalling ecotype Ganges of *Thlaspi caerulescens* than in the non-hoarding Prayon, in this way VSC of Cd in roots is altogether diminished and progressively metal is gone to translocate to shoots.

Regulation by Chelators

In plants, there is certain particles are presently known as Phytochelatins which remove out heavy metals through chelation and consequently framing stable PCs metal which gets hindered into the vacuoles. Nicotianamine is another kind of metal chelator which is universally present in higher plants. Modification and distribution of zinc and iron are also done by nicotinamide. Huge ratio of NA is present in the root tissue while hyperaccumulation of *Arabidopsis halleri* enhances the transportation of Zn to the shoot. On the other hand, several research is performed in which they purposed that NA sequestration in the vacuole is enhanced by the overexpression of AtZIFI in *Arabidopsis* as a result in the tissue of shoots there is over-accumulation of iron. As it also considers that presence of NA in vacuole exhaust or consume the cytosolic NA and hence intercellular mobility of iron takes place in the form of iron-nicotinamide. In detoxification and translocation of Ni sometimes amino acid histidine is responsible. Ni-histidine (Ni-his) is the form in which Ni is supplied, there is some increasing rate can be seen in the vesicles of tonoplast. The concentration of histidine is very high in the root of *T.caerulescens,* they suggested that if there is accumulation if histidine in the roots of the plant then automatically VSC contained by Ni gets reduced, hence regulate the hyperaccumulation and transportation of Ni in the shoots of the plants. The regulation of VSC by chelators are further divided into two categories-

1. Translocation into the vacuole is much more susceptible by the chelated metals.
2. By chelation of metals the sequestration of metal get reduced as a result VSC is also get reduced as a result of long-distance of metals like cadmium get easier.

These several studies concluded that the long distance transportation of metals enhanced when chelation of metals is done by NA and PCs. And VSC is also enlarged up to some extent due to this sequestration.

Regulation By VSC (Vacuole Sequestration Capacity)

Most of the iron found in the chloroplast which regulates the function in photosynthesis, OsVIT1 and OsVIT2 these are the two metal transporters localized in the tonoplast. Sequestration of Fe and Zn in the plant vacuole over the tonoplast is the role of these two (OsVIT1 and OsVIT2). If anything happens and there will be slight disruption in the function of these two genes then the accumulation of iron and zinc get decreased. These two genes are very important for translocation of iron and zinc. The study of this also depicts that zinc transportation is mediated into the vacuole by AtMTP3. If the zinc concentration increases, AtMTP3 get activated and sequestration of VSC and zinc takes place into the vacuole of the root. There is also another zinc transporter AtZIF1 this is tonoplast- localized gene. The excessive expression of this AtZIF1 showed accumulation of zinc is get decreased in shoots of plant and increase zinc concentration in root to shoot ratio. In terms of transportation of copper ios enhanced by the gene OsHMA3 which is present in the tonoplast of the plant, it translocates the cadmium into vacuoles. If the OsHMA3 shows high expression the amount of cadmium gets decreased, so OsMHA3 regulates the VSC which is very helpful for the ling-distance transportation of cadmium from roots to shoot. Now, if we talk about metal hyperaccumulation and non-hyper accumulation. In comparison to non-hyper accumulation, hyperaccumulators the accumulation of the metals in the shooting part can be seen. It also is shown that by the regulation of VSC metal transportation is very common and not only transportation but VSC induce detoxification of metals in hyperaccumulation.

Regulation By NA (nicotianamine)

Copper accumulation in the root of a plant at the time of growth leads to the severe deficiency of copper, which increase as the new leaves are produced. So, the translocation of copper from root to shoot is done by nicotinamide (NA) through the sap of xylem. At the pH of 7 (neutral) and mild basic, the NA chelation properties are its peak. In the meristematic tissue there is a high concentration of NA, and help in the metal uptake, but it does not work properly at the time of Fe absorption. But it does not mean it is not involved in the regulation of Fe. NA keeps the Fe in that way so that it can easily solubilize and get distributed evenly in all parts of the cell but if Fe increases and leads to the toxicity in a cell then NA starts detoxification by chelating the iron and maintain it into the vacuole. NA has some special characteristics that it can easily bond with different metals like and form stable complexes e.g. it bond with manganese, iron, cobalt, zinc and nickel. And various further studies indicated that-

1. Rather than simple metals, the complex ones are more stable at the pH above 6.5 and NA act as a chelator of metals.
2. And among all the complex metal, copper is more stable in its complex form when it bonds with NA and forms Cu-NA.
3. NA can chelate the iron

At the time of nickel transportation NA does the same first it make a bond with Ni i.e. Ni-NA bond after that it is transported to various parts of plants it makes the process easy in hyper accumulated plants. In the phloem sap of *Ricinus hypocotyls,* the NA is detected. Actually, in that particular sap, Fe is bounded there with a protein specially ITP (iron transport protein) so NA removes or shuttles the Fe from ITP by entering in the symplastic pathway. By using HPLC we can find out the total content of NA present in the plant. We can say that NA has the characteristics it can regulate, distribute and even detoxify the Fe if needed. But in the case of Gramineae family plants rather than NA, they transform this NA into PS. PS has also the power of chelation they are synthesized in the soil, by chelation it solubilizes the ferric hydroxides. This is the process by which metals are transported.

Regualtion By NRAMP/VIT1

The several studies concluded that Fe^{2+} transportation in *Arabidopsis* is also done by VIT1 at the stage of development of seed and embryo. So, for proper distribution of Fe in an embryo is regulated by VIT1 which enhances the viability of seedling even if the amount of Fe is low. But at the storage time Fe in a vacuole, first it as to be converted in the form of Fe^{3+}. Now talking about NRAMP it is further divided into two groups NRAMP3 and NRAMP4 already present in the vacuole of the cell. The work of these genes started after the Fe is stored in vacuole it is again taken out from vacuole. In the case of wild type seed Fe is still present in the vacuole in the form of Fe-phytate globoids, but disappear at germination time but on the other hand NRAMP3 and NRAMP4 these globoids remain in the seed and can be seen it indicates that Fe was not mobilized from the vacuole. So, it is confirmed that Fe is first transported into the vacuole by VIT1 and by NRAMP3 and NRAMP4 the Fe is released out from vacuole at the time of germination (Fig. 1).

Process-

1. Transportation and sequestration of Fe.
2. Then this Fe is transported into the vacuole by the regulation of VIT1.
3. Inside the vacuole Fe bonds with phytate and nicotinamide
4. Then by NRAMP3 and NRAMP4 the Fe is again remove out from vacuole at the time of germination
5. But in the mitochondria, Fe is complex with ferritin and decrease the oxidative stress.

Process how Essential Metals are Transported

First of all diagrammatic view of copper, iron and zinc transporters as shown in (fig1) and then the brief discussion-

Fig. 1. Zinc, iron and copper transportation

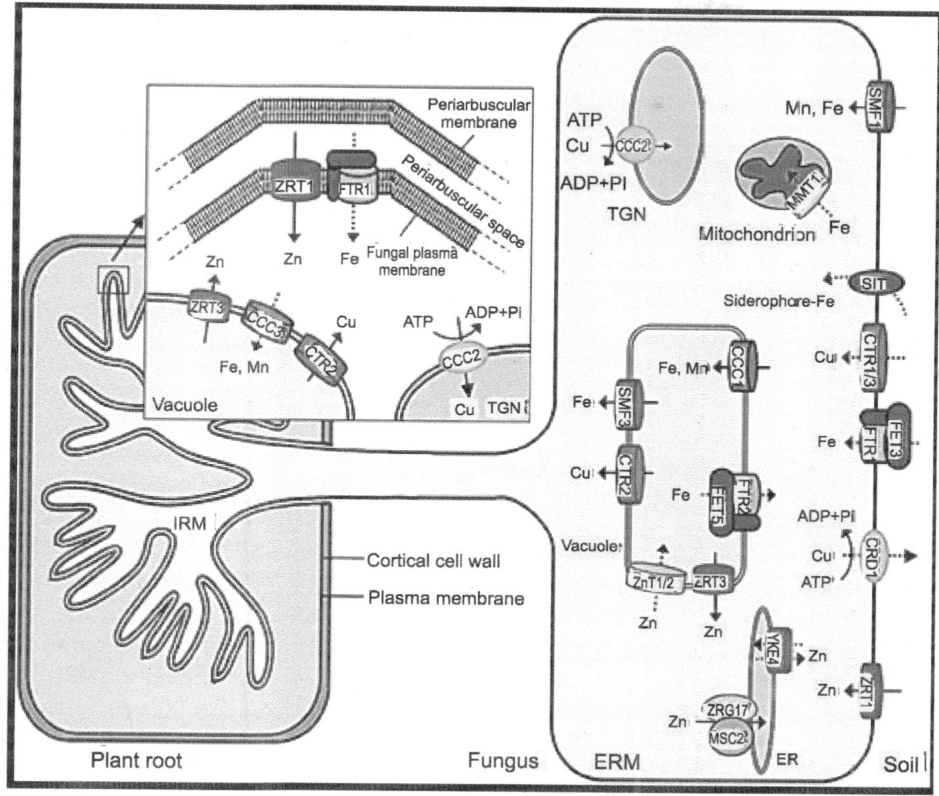

How Iron Uptake takes Place and Transported to Plants?

The Fe is fundamental and has a specific role in plant development. Simultaneously, Fe is exceptionally receptive and poisonous using the Fenton response. Thus, plants firmly control Fe homeostasis and take action if the plant is Fe deficient just as Fe over-burden. Fe is found abundantly in the earth crust. The availability of Fe is very less in the roots of the plants. Fe is oxidized easily if the soil pH is very high or the soil condition is aerobic. But most of Fe is in the form of ferric oxide which is insoluble in nature. But plants have to uptake the Fe to complete its life cycle, so when the pH of the soil is low the ferric form of Fe is released out from oxides and readily convert in the form of available iron, Fe which can easily uptake by the roots of the plant. As we said earlier several genes are involved for the uptake of Fe. In the epidermis, the rate of chelation and overexpression of genes get increased to make it into the available form of Fe, several plants belonging to the different family has their way to convert it. Like as we stated above Fe^{3+} is reduced in the form of Fe^{2+} by the regulation of membrane-bound in plants which do not belong to Gramineae family. While in grasses Fe^{3+} bond with PSs (phytosiderophore) secreted by the plants and again transported into the root. There is another strategy used by species which do not belong to Gramineae family i.e. reduction strategy. If iron is deficient than in the rhizosphere protons are released by ATPase, it makes the Fe into a more soluble form by lowering the pH of the soil then by the help of AtIRT1 metal transporter into the epidermal cells of roots. But sometimes due to over presence of cadmium and cobalt the deficiency of zinc is triggered. Co and Cd disturb the Fe movement and comes between the pathway. The genes which translocate the Fe are plant-specific we can say because several plants have various genes for the transportation of Fe. In rice, there is different gene can be seen in the roots and shoots phloem i.e. OsIRT1-GUS. It activates suddenly if the plant is going through Fe deficiency. This gene has a specific role in the transportation of Fe(II) and in case of castor bean at the time when it is only 7 days old there is some trace of protein which binds Fe(II) and in the plant barley, there is special system PETRIS (positron-emitting tracer imagining system which controls Fe uptake and transportation.

Storage

For flowering and development of seeds, Fe has to bond with NA and form Fe-NA which is must for plants without this the flower become deformed and the sterility will be affected. But in the development of seed Fe-NA is little not possible, then there is another gene that comes forward 'YSLs'. YSL1 is the form of YSLs that present in the veins of the leaves, its role is loading of seeds and it's done only by this gene no other genes can compensate. As we know VIT1 is responsible for Fe^{2+} transportation in vacuole at the time of seed embryo development. The homeostasis, storage, movement of iron is a very lengthy and complex process too.

Zinc Transportation in Plants and its Movement within the Plants

Transportation and dissemination of zinc in plants are affected by the amount of zinc content in soil effortlessly. Zinc is considered as a micronutrient and its importance is very much needed by the plant as iron. Zinc is very essential for the activity of other 300 enzymes like some proteins, DNA and RNA polymerase without Zn it cannot be possible. Zinc enhances the various activities in the plant. The number of processes like regulation of hormone, synthesis of tryptophan, repair of PSII, CO_2 concentration is maintained in mesophyll etc. are not complete if Zn is not involved. Various research and reviews concluded that regulation of Rubisco is also enhanced by zinc and prevent the wheat plant from any adverse heat stress. Different authors have also discussed the zinc importance in their papers. But, nowadays zinc deficiency is increasing day by day and creating a problem in the field of agriculture. Many cultivated lands which are preferred for the production of cereal crops are facing problems because half of the land percentage is zinc deficient by which the growth and development of plants get affected. It not only affects then flora but also affects the fauna residing in soil by feeding this kind of soil their metabolism gets affected. Now again talking about plants zinc deficiency cause some serious issue like growth will be stunted, as we know the mode of absorption in plants are via root but how these minerals come from their primary source are the parent material when these parent material weathers they break into particles and as the day passed they get solubilize in the soil. But the zinc in the soil is of organic form and plants take these compounds in inorganic form means in available form and its availability also rely on the physicochemical properties of the soil, biological activity(microorganisms) and plant root. Zinc distribution in the soil affected by soil pH which directly affects the solubility of the zinc in soil. When the pH increases Zn adsorption is triggered and slight it into the site of cation exchange as a result the availability of zinc gets reduced. When the pH of the soil is "between" 5.5 to 7.0 the concentration of zinc gets decreased in soil solution. Another factor which affects the zinc uptake is soil moisture also. After uptake, there is the transportation of zinc takes place via various enzymes but mobility is affected by redox potential of soil. The requirements of zinc are different in different plants so as there concentration and this is because differences in uptake of zinc, the concentration of zinc absorbed by root to shoots and sequestration of zinc. But these differences are must for the better function and living of the cells, zinc is taken in the form of divalent cation via roots. Sometimes the organic form of zing also absorbed by roots of the plant but we know plants only take the inorganic form of compounds but it only depends upon the ligand secreted by plant roots. But there are two types of approach is there to uptake the divalent zinc

1. Ist strategy- the reductants, organic acid and H are effluxes which triggered and increase the solubility of zinc phosphate and hydroxides which are in complex forms and release out Zn^2 for the further absorption by an epidermal cell of roots. By the epidermal cell or by mucilage organic acid are released which also include citric and malic acid.
2. 2nd strategy- formation of zinc complex when phytosiderophores are effluxed into the epidermal cells of the root.

But this mechanism of absorption does not occur in cereal crops roots. The molecular weight of phytosiderophores is very low by which the binding is much easier. S-adenosylmethionine is condensed and NA is formed by which phytosiderophores are biosynthesized. The release of organic and amino acid by plant root and we know zinc is only absorbed when it is in there divalent form through a mechanism like diffusion and mass flow via roots. Water as a solvent works here then only zinc is uptake and also the concentration of zinc around RCPM (root cell-plasma membrane). After entering into the root zinc has to across various tissue like

a. Epidermis of root
b. Cortex
c. Pericycle
d. Endodermis

But the thickness of the layer of this tissue decrease along with the depth or the centre part. The process is known as radial transport because the minerals and water move towards the region of stele. In the cell membranes, special transporter protein is present on specific tissue, the two routes are followed at the time of transportation of zinc to xylem i.e. apoplastic and symplastic movement. In the cells of roots, there is a connection between cytoplasm is connected by plasmodesmata by this ion of minerals and water move through these cell wall and spaces between them which is one of the reason of zinc transportation in roots but if we look at the structure of cell of roots in the case of apoplast movement there is some obstruction created by suberin lamellae present in the Casparian strips. And these layer does not allow any mineral and water pass through it as we can say they are highly impermeable in nature so, when this layer about to come to the zinc change its route from apoplastic to symplastic. But both processes enhance and contribute to the mobility of zinc. The zinc enters into the root by the apoplast movement by moving through cell wall plasma membrane and after this further movement is occur through symplastic movement. Now after this process zinc enter to the pericycle and parenchyma of xylem after it crosses the Casparian border but this is not the only obstruction there are furthermore at the time of nutrient transfer as we call it xylem loading. It is the main step in the export of metal in the root. In the xylem, there is parenchyma which causes hyperpolarization of the membrane when there is the frequent activity of H^+-ATPase which create some disturbance in the mobility of positive ions in the cytosol. Then an active process occurs where the zinc is loaded from pericycle cells to the xylem (apoplastic. Then here one transporter is already available for efflux of zinc this transporter is HMA group, it is located on a cell of pericycle. Xylem sap is a little more acidic because in plasma membrane there is rapid activity and overexpression of plasma membrane H^+-ATPase. The metal ions are transported after the conversion metal to complex metals along with histidine (His), organic acids, nicotianamine (NM). In the above study we already discussed the nicotinamide (NA), it is a well-known chelator. At the surface of some useful cells are there which regulate the mobility of solutes in a symplastic way. After zinc enters into the phloem, next step of mobility in different parts is processed by both short-distance and long-distance pathway, but zinc is more mobile in phloem rather than xylem because it consists chelating solutes in very high amount i.e. organic acids etc. first zinc has to be converted into an ionic form in the tissue of phloem. The phloem has a high concentration of solutes we already discussed but on the other hand, xylem consists of the low amount of solutes in it. Zinc transportation is also engaged with a group of protein transporters they are, heavy metal ATPase family, ZIP family, metal tolerance protein family etc. but out of all these HMA family involved in the zinc efflux inside the cytosol. The concentration of zinc varies according to the overexpression of metal proteins

Copper Translocation within the Plants

In the earth, copper is abundant up to 60mg per kg and its deficiency is also very common in the plants. There is process biofortification it triggers the phytoavailability of copper by making some changes in the physiological process like the uptake of copper by plants, its accumulation and sequestration. Copper is mainly present in two forms CuI AND CuII it can bind with the various substrate. Copper transport the electrons into mitochondria and chloroplast.

Fig. 2. It represents the absorption and movement of the copper into the root of the plant

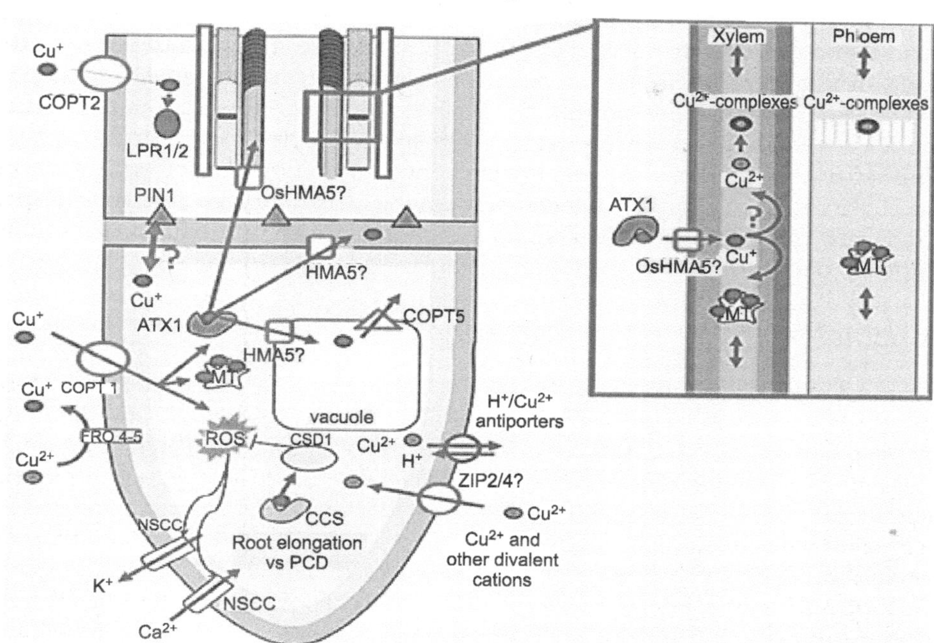

According to this Diagram

a. First of all, copper has to be converted by COPT proteins like other metals into Cu^+

b. Then the efflux of Cu^+ is regulated by antiporters i.e. H^+ and Cu^{2+}

c. Then it travels into the cytosol, their generation of ROS takes place opening NSCCs when Cu^{2+} induced and allows Ca^{2+} which enhances the root growth.

This is just a short process in points about the absorption and mobility of copper in the root. Now after absorption by roots it is transported by the transporter (COPT). This group of transporters contain three transmembrane domain i.e. histidine, TMD2 and TMD3 and cysteine Cu^+ binding. The Cu(I) transportation is regulated by the gene COPT4, and COPT1 is a plasma membrane specific. After this, the copper enters into the cytosol of the cell and the generation of OH. In the pericycle of roots and vascular tissue, the copper is transported in the form of Cu(I) at the stage of vegetative and reproductive. Reoxidation of Cu(I) and Cu(II) has been occurring in the dicot plants at the time of translocation from root to shoot because it is long-distance transportation and needs more energy. But in the sap of xylem, this copper is transported in the form of Cu(II) by regulation of chelators. In there, some amino acids are also present which bind with copper and make them more stable complex metal which is very useful for copper mobility. At the xylem sap the pH increases, NA and histidines rapidly create a bond with copper. Copper transportation in rice is regulated by the DMA (deoxymugineic acid) which also acts as a chelator. After this the zinc is translocating in mitochondria, it is not much discussed this but in *Arabidopsis,* cysteine family is responsible for the allocation of copper from cytosol to the mitochondria of the plant cell. The accumulation of copper in plants is situated inside the chloroplast, vacuole and the cell wall but the cell is just a minor site as we can say primary site it is made up of cellulose, hemicellulose, pectins in which proteins are embedded in the centre. Copper is first absorbed, accumulated and then transported to the various sites of plant for further process. Copper regulates the reproductive growth of the plants and aids in root metabolism and also helps in the utilization of proteins. If copper is deficient then there will be various symptoms are shown by the plants like distortion occur at the younger leaves, necrosis and plant will be stunted.

Conclusion

Plants need all kind of minerals and metals for their proceeds with digestion and plant to comprise of different transporters for translocation of the metal at the spot of capacity. The above-written things conclude that how much these plant transporters are important for transportation of metals in a plant. They not only regulate the metal but in case if metal concentration increases and toxic metal occur then these transporters also start detoxification and prevent the plant from any further toxicity. We know how a plant relies on these metals to complete its life cycle. The metals sequestration, transportation and storage are done by various genes and before this metals are converted into readily available form because the plant only takes the inorganic form of metals. These metals travel various parts of the plant and their cell by both apoplastic and symplastic pathway and in the cells, there are different genes already present which helps the metal and give instruction to them where to stop, and when to move. Not only this various metal have their way along with specific gene and this gene is plant-specific. The genes are plant are different and have a different function.

Reference

Alscher, R. G., Erturk, N., & Heath, L. S. (2002). Role of superoxide dismutases (SODs) in controlling oxidative stress in plants. *Journal of experimental botany, 53*(372), 1331-1341.

Arduini, I., Godbold, D. L., & Onnis, A. (1996). Cadmium and copper uptake and distribution in Mediterranean tree seedlings. *Physiologia Plantarum, 97*(1), 111-117.

Assunção, A. G. L., Martins, P. D. C., De Folter, S., Vooijs, R., Schat, H., & Aarts, M. G. M. (2001). Elevated expression of metal transporter genes in three accessions of the metal hyperaccumulator Thlaspi caerulescens. *Plant, Cell & Environment, 24*(2), 217-226.

Bargagli, R., Monaci, F., & Agnorelli, C. (2003). Oak leaves as accumulators of airborne elements in an area with geochemical and geothermal anomalies. *Environmental Pollution, 124*(2), 321-329.

Becher, M., Talke, I. N., Krall, L., & Krämer, U. (2004). Cross-species microarray transcript profiling reveals high constitutive expression of metal homeostasis genes in shoots of the zinc hyperaccumulator Arabidopsis halleri. *The Plant Journal, 37*(2), 251-268.

Binder, B. M., Rodríguez, F. I., & Bleecker, A. B. (2010). The copper transporter RAN1 is essential for biogenesis of ethylene receptors in Arabidopsis. *Journal of Biological Chemistry, 285*(48), 37263-37270.

Birkenbihl, R. P., Jach, G., Saedler, H., & Huijser, P. (2005). Functional dissection of the plant-specific SBP-domain: overlap of the DNA-binding and nuclear localization domains. *Journal of molecular biology, 352*(3), 585-596.

Blaby-Haas, C. E., Padilla-Benavides, T., Stübe, R., Argüello, J. M., & Merchant, S. S. (2014). Evolution of a plant-specific copper chaperone family for chloroplast copper homeostasis. *Proceedings of the National Academy of Sciences, 111*(50), E5480-E5487.

Bleecker, A. B., & Kende, H. (2000). Ethylene: a gaseous signal molecule in plants. *Annual review of cell and developmental biology, 16*(1), 1-18.

Bossolini, E., Wicker, T., Knobel, P. A., & Keller, B. (2007). Comparison of orthologous loci from small grass genomes Brachypodium and rice: implications for wheat genomics and grass genome annotation. *The Plant Journal, 49*(4), 704-717.

Broadley, M. R., White, P. J., Hammond, J. P., Zelko, I., & Lux, A. (2007). Zinc in plants. *New phytologist, 173*(4), 677-702.

Broomfield, R. (1996). A quasi-experimental research to investigate the retention of basic cardiopulmonary resuscitation skills and knowledge by qualified nurses following a course in professional development. *Journal of Advanced Nursing, 23*(5), 1016-1023.

Brun, L. A., Maillet, J., Hinsinger, P., & Pépin, M. (2001). Evaluation of copper availability to plants in copper-contaminated vineyard soils. *Environmental Pollution, 111*(2), 293-302.

Burzyński, M., Migocka, M., & Kłobus, G. (2005). Cu and Cd transport in cucumber (Cucumis sativus L.) root plasma membranes. *Plant science, 168*(6), 1609-1614.

Carrillo-González, R., Šimůnek, J., Sauve, S., & Adriano, D. (2006). Mechanisms and pathways of trace element mobility in soils. *Advances in agronomy, 91*, 111-178.

Catty, P., Boutigny, S., Miras, R., Joyard, J., Rolland, N., & Seigneurin-Berny, D. (2011). Biochemical characterization of AtHMA6/PAA1, a chloroplast envelope Cu (I)-ATPase. *Journal of Biological Chemistry, 286*(42), 36188-36197.

Chaignon, V., Di Malta, D., & Hinsinger, P. (2002). Fe-deficiency increases Cu acquisition by wheat cropped in a Cu-contaminated vineyard soil. *New Phytologist, 154*(1), 121-130.

Chen, C. C., Chen, Y. Y., Tang, I. C., Liang, H. M., Lai, C. C., Chiou, J. M., & Yeh, K. C. (2011). Arabidopsis SUMO E3 ligase SIZ1 is involved in excess copper tolerance. *Plant physiology, 156*(4), 2225-2234.

Chen, E. L., Chen, Y. A., Chen, L. M., & Liu, Z. H. (2002). Effect of copper on peroxidase activity and lignin content in Raphanus sativus. *Plant Physiology and Biochemistry, 40*(5), 439-444.

Chmielowska, J. A. G. N. A., Deckert, J. O. A. N. N. A., & Diaz, J. O. S. É. (2008). Activity of peroxidases and phenylalanine ammonia-lyase in lupine and soybean seedlings treated with copper and an ethylene inhibitor. *Biol Lett, 45,* 59-67.

Chu, H. H., Chiecko, J., Punshon, T., Lanzirotti, A., Lahner, B., Salt, D. E., & Walker, E. L. (2010). Successful reproduction requires the function of Arabidopsis Yellow Stripe-Like1 and Yellow Stripe-Like3 metal-nicotianamine transporters in both vegetative and reproductive structures. *Plant physiology, 154*(1), 197-210.

Claus, J., Bohmann, A., & Chavarría-Krauser, A. (2013). Zinc uptake and radial transport in roots of Arabidopsis thaliana: a modelling approach to understand accumulation. *Annals of botany, 112*(2), 369-380.

Cobbett, C., & Goldsbrough, P. (2002). Phytochelatins and metallothioneins: roles in heavy metal detoxification and homeostasis. *Annual review of plant biology, 53*(1), 159-182.

Colangelo, E. P., & Guerinot, M. L. (2004). The essential basic helix-loop-helix protein FIT1 is required for the iron deficiency response. *The Plant Cell, 16*(12), 3400-3412.

Colzi, I., Arnetoli, M., Gallo, A., Doumett, S., Del Bubba, M., Pignattelli, S., ... & Gonnelli, C. (2012). Copper tolerance strategies involving the root cell wall pectins in Silene paradoxa L. *Environmental and Experimental Botany, 78,* 91-98.

Colzi, I., Doumett, S., Del Bubba, M., Fornaini, J., Arnetoli, M., Gabbrielli, R., & Gonnelli, C. (2011). On the role of the cell wall in the phenomenon of copper tolerance in Silene paradoxa L. *Environmental and Experimental Botany, 72*(1), 77-83.

Colzi, I., Pignattelli, S., Giorni, E., Papini, A., & Gonnelli, C. (2015). Linking root traits to copper exclusion mechanisms in Silene paradoxa L.(Caryophyllaceae). *Plant and Soil, 390*(1), 1-15.

Company, P., & González-Bosch, C. (2003). Identification of a copper chaperone from tomato fruits infected with Botrytis cinerea by differential display. *Biochemical and biophysical research communications, 304*(4), 825-830.

Cornu, J. Y., Staunton, S., & Hinsinger, P. (2007). Copper concentration in plants and in the rhizosphere as influenced by the iron status of tomato (Lycopersicon esculentum L.). *Plant and Soil, 292*(1), 63-77.

Cosgrove, D. J. (2001). Wall structure and wall loosening. A look backwards and forwards. *Plant physiology, 125*(1), 131-134.

Curie, C., Cassin, G., Couch, D., Divol, F., Higuchi, K., Le Jean, M., ... & Mari, S. (2009). Metal movement within the plant: contribution of nicotianamine and yellow stripe 1-like transporters. *Annals of botany, 103*(1), 1-11.

Curie, C., Cassin, G., Couch, D., Divol, F., Higuchi, K., Le Jean, M., ... & Mari, S. (2009). Metal movement within the plant: contribution of nicotianamine and yellow stripe 1-like transporters. *Annals of botany, 103*(1), 1-11.

Deinlein, U., Weber, M., Schmidt, H., Rensch, S., Trampczynska, A., Hansen, T. H., ... & Clemens, S. (2012). Elevated nicotianamine levels in Arabidopsis halleri roots play a key role in zinc hyperaccumulation. *The Plant Cell, 24*(2), 708-723.

Del Pozo, T., Cambiazo, V., & González, M. (2010). Gene expression profiling analysis of copper homeostasis in Arabidopsis thaliana. *Biochemical and biophysical research communications, 393*(2), 248-252.

Deng, F., Yamaji, N., Xia, J., & Ma, J. F. (2013). A member of the heavy metal P-type ATPase OsHMA5 is involved in xylem loading of copper in rice. *Plant physiology, 163*(3), 1353-1362.

Dima, O., Morreel, K., Vanholme, B., Kim, H., Ralph, J., & Boerjan, W. (2015). Small glycosylated lignin oligomers are stored in Arabidopsis leaf vacuoles. *The Plant Cell, 27*(3), 695-710.

Dinh, N. T., Vu, D. T., Mulligan, D., & Nguyen, A. V. (2015). Accumulation and distribution of zinc in the leaves and roots of the hyperaccumulator Noccaea caerulescens. *Environmental and Experimental Botany, 110*, 85-95.

Dong, J., Kim, S. T., & Lord, E. M. (2005). Plantacyanin plays a role in reproduction in Arabidopsis. *Plant Physiology, 138*(2), 778-789.

Dumville, J. C., & Fry, S. C. (2003). Solubilisation of tomato fruit pectins by ascorbate: a possible non-enzymic mechanism of fruit softening. *Planta, 217*(6), 951-961.

Eticha, D., Stass, A., & Horst, W. J. (2005). Cell-wall pectin and its degree of methylation in the maize root-apex: significance for genotypic differences in aluminium resistance. *Plant, Cell & Environment, 28*(11), 1410-1420.

Fang, Y., Fu, D., Tang, W., Cai, Y., Ma, D., Wang, H., ... & Shen, X. (2013). Ubiquitin C-terminal Hydrolase 37, a novel predictor for hepatocellular carcinoma recurrence, promotes cell migration and invasion via interacting and deubiquitinating PRP19. *Biochimica et Biophysica Acta (BBA)-Molecular Cell Research, 1833*(3), 559-572.

Feng, H., Zhang, Q., Wang, Q., Wang, X., Liu, J., Li, M., ... & Kang, Z. (2013). Target of tae-miR408, a chemocyanin-like protein gene (TaCLP1), plays positive roles in wheat response to high-salinity, heavy cupric stress and stripe rust. *Plant molecular biology, 83*(4-5), 433-443.

Ferreres, F., Figueiredo, R., Bettencourt, S., Carqueijeiro, I., Oliveira, J., Gil-Izquierdo, A., ... & Sottomayor, M. (2011). Identification of phenolic compounds in isolated vacuoles of the medicinal plant Catharanthus roseus and their interaction with vacuolar class III peroxidase: an H2O2 affair?. *Journal of experimental botany, 62*(8), 2841-2854.

Festa, R. A., & Thiele, D. J. (2011). Copper: an essential metal in biology. *Current Biology, 21*(21), R877-R883.

Fry, S. C. (1998). Oxidative scission of plant cell wall polysaccharides by ascorbate-induced hydroxyl radicals. *Biochemical Journal, 332*(2), 507-515.

Fry, S. C., Miller, J. G., & Dumville, J. C. (2002). A proposed role for copper ions in cell wall loosening. In *Progress in Plant Nutrition: Plenary Lectures of the XIV International Plant Nutrition Colloquium* (pp. 57-67). Springer, Dordrecht.

Garcia-Molina, A., Andrés-Colás, N., Perea-García, A., del Valle-Tascón, S., Peñarrubia, L., & Puig, S. (2011). The intracellular Arabidopsis COPT5 transport protein is required for photosynthetic electron transport under severe copper deficiency. *The Plant Journal, 65*(6), 848-860.

Garcia-Molina, A., Andrés-Colás, N., Perea-García, A., Neumann, U., Dodani, S. C., Huijser, P., ... & Puig, S. (2013). The Arabidopsis COPT6 transport protein functions in copper distribution under copper-deficient conditions. *Plant and Cell Physiology, 54*(8), 1378-1390.

Grotz, N., Fox, T., Connolly, E., Park, W., Guerinot, M. L., & Eide, D. (1998). Identification of a family of zinc transporter genes from Arabidopsis that respond to zinc deficiency. *Proceedings of the National Academy of Sciences, 95*(12), 7220-7224.

Gupta, N., Ram, H., & Kumar, B. (2016). Mechanism of Zinc absorption in plants: uptake, transport, translocation and accumulation. *Reviews in Environmental Science and Bio/Technology, 15*(1), 89-109.

Harada, E., Sugase, K., Namba, K., Iwashita, T., & Murata, Y. (2007). Structural element responsible for the Fe (III)–phytosiderophore specific transport by HvYS1 transporter in barley. *FEBS letters, 581*(22), 4298-4302.

Hell, R., & Mendel, R. R. (Eds.). (2010). *Cell biology of metals and nutrients*. Berlin, Heidelberg: Springer.

Herbik, A., Giritch, A., Horstmann, C., Becker, R., Balzer, H. J., Baumlein, H., & Stephan, U. W. (1996). Iron and copper nutrition-dependent changes in protein expression in a tomato wild type and the nicotianamine-free mutant chloronerva. *Plant Physiology, 111*(2), 533-540.

Himelblau, E., & Amasino, R. M. (2000). Delivering copper within plant cells. *Current opinion in plant biology, 3*(3), 205-210.

Hinsinger, P., & Courchesne, F. (2008). Biogeochemistry of metals and metalloids at the soil-root interface. *Biophysic–chemical processes of heavy metals and metalloids in soil environments. Wiley, Hoboken*, 267-311.

Huang, J., Zhang, Y., Peng, J. S., Zhong, C., Yi, H. Y., Ow, D. W., & Gong, J. M. (2012). Fission yeast HMT1 lowers seed cadmium through phytochelatin-dependent vacuolar sequestration in Arabidopsis. *Plant physiology, 158*(4), 1779-1788.

Hübner, R., Depta, H., & Robinson, D. G. (1985). Endocytosis in maize root cap cells. *Protoplasma, 129*(2-3), 214-222.

Jordanov, J., Courtois-Verniquet, F., Neuburger, M., & Douce, R. (1992). Structural investigations by extended X-ray absorption fine structure spectroscopy of the iron center of mitochondrial aconitase in higher plant cells. *Journal of Biological Chemistry, 267*(24), 16775-16778.

Kampfenkel, K., Van Montagu, M., & Inzé, D. (1995). Effects of iron excess on Nicotiana plumbaginifolia plants (implications to oxidative stress). *Plant Physiology, 107*(3), 725-735.

Kerkeb, L., & Krämer, U. (2003). The Role of Free Histidine in Xylem Loading of Nickel inAlyssum lesbiacum and Brassica juncea. *Plant physiology, 131*(2), 716-724.

Kim, S. A., Punshon, T., Lanzirotti, A., Li, L., Alonso, J. M., Ecker, J. R., ... & Guerinot, M. L. (2006). Localization of iron in Arabidopsis seed requires the vacuolar membrane transporter VIT1. *Science, 314*(5803), 1295-1298.

Kim, S. A., Punshon, T., Lanzirotti, A., Li, L., Alonso, J. M., Ecker, J. R., ... & Guerinot, M. L. (2006). Localization of iron in Arabidopsis seed requires the vacuolar membrane transporter VIT1. *Science, 314*(5803), 1295-1298.

Klaumann, S., Nickolaus, S. D., Fürst, S. H., Starck, S., Schneider, S., Ekkehard Neuhaus, H., & Trentmann, O. (2011). The tonoplast copper transporter COPT5 acts as an exporter and is required for interorgan allocation of copper in Arabidopsis thaliana. *New Phytologist, 192*(2), 393-404.

Kochian, L. V. (1991). *Mechanisms of micronutrient uptake and translocation in plants. eds. Micronutrients in Agriculture* (No. REP-9212. CIMMYT.).

Korenkov, V., Hirschi, K., Crutchfield, J. D., & Wagner, G. J. (2007). Enhancing tonoplast Cd/H antiport activity increases Cd, Zn, and Mn tolerance, and impacts root/shoot Cd partitioning in Nicotiana tabacum L. *Planta, 226*(6), 1379-1387.

Krämer, U. (2010). Metal hyperaccumulation in plants. *Annual review of plant biology, 61*, 517-534.

Kramer, U., Cotter-Howells, J. D., Charnock, J. M., & Baker, A. J. (1996). M.; Andrew, C.; Smith, J. *Nature, 379*, 635-638.

Krämer, U., Talke, I. N., & Hanikenne, M. (2007). Transition metal transport. *FEBS letters, 581*(12), 2263-2272.

Krzesłowska, M. (2011). The cell wall in plant cell response to trace metals: polysaccharide remodeling and its role in defense strategy. *Acta Physiologiae Plantarum, 33*(1), 35-51.

Krzesłowska, M. (2011). The cell wall in plant cell response to trace metals: polysaccharide remodeling and its role in defense strategy. *Acta Physiologiae Plantarum, 33*(1), 35-51.

Krzesłowska, M. (2011). The cell wall in plant cell response to trace metals: polysaccharide remodeling and its role in defense strategy. *Acta Physiologiae Plantarum, 33*(1), 35-51.

Krzesłowska, M., Lenartowska, M., Mellerowicz, E. J., Samardakiewicz, S., & Woźny, A. (2009). Pectinous cell wall thickenings formation—a response of moss protonemata cells to lead. *Environmental and Experimental Botany, 65*(1), 119-131.

Li, X. K., Ji, W. J., Zhao, J., Wang, S. J., & Au, C. T. (2005). Ammonia decomposition over Ru and Ni catalysts supported on fumed SiO2, MCM-41, and SBA-15. *Journal of Catalysis, 236*(2), 181-189.

Liao, M. T., Hedley, M. J., Woolley, D. J., Brooks, R. R., & Nichols, M. A. (2000). Copper uptake and translocation in chicory (Cichorium intybus L. cv. Grasslands Puna) and tomato (Lycopersicon esculentum Mill. cv. Rondy) plants grown in NFT system. I. Copper uptake and distribution in plants. *Plant and soil, 221*(2), 135-142.

Longnecker, N. E., & Robson, A. D. (1993). Distribution and transport of zinc in plants. In *Zinc in soils and plants* (pp. 79-91). Springer, Dordrecht.

Lu, L., Liao, X., Labavitch, J., Yang, X., Nelson, E., Du, Y., ... & Tian, S. (2014). Speciation and localization of Zn in the hyperaccumulator Sedum alfredii by extended X-ray absorption fine structure and micro-X-ray fluorescence. *Plant physiology and biochemistry, 84,* 224-232.

Lyons, G. H., Stangoulis, J. C., & Graham, R. D. (2004). Exploiting micronutrient interaction to optimize biofortification programs: the case for inclusion of selenium and iodine in the HarvestPlus program. *Nutrition Reviews, 62*(6), 247-252.

Malone, C., Koeppe, D. E., & Miller, R. J. (1974). Localization of lead accumulated by corn plants. *Plant Physiology, 53*(3), 388-394.

Marschner, H. (2011). *Marschner's mineral nutrition of higher plants.* Academic press.

Mäser, P., Thomine, S., Schroeder, J. I., Ward, J. M., Hirschi, K., Sze, H., ... & Guerinot, M. L. (2001). Phylogenetic relationships within cation transporter families of Arabidopsis. *Plant Physiology, 126*(4), 1646-1667.

Meyers DE, Kopittke PM, Auchterlonie GJ, Webb RI (2009) Characterization of lead precipitate following uptake by roots of Brassica juncea. Environ Toxicol Chem 28:250–255

Michaud, A. M., Bravin, M. N., Galleguillos, M., & Hinsinger, P. (2007). Copper uptake and phytotoxicity as assessed in situ for durum wheat (Triticum turgidum durum L.) cultivated in Cu-contaminated, former vineyard soils. *Plant and Soil, 298*(1), 99-111.

Morrissey, J., & Guerinot, M. L. (2009). Iron uptake and transport in plants: the good, the bad, and the ionome. *Chemical reviews, 109*(10), 4553-4567.

Morrissey, J., Baxter, I. R., Lee, J., Li, L., Lahner, B., Grotz, N., ... & Guerinot, M. L. (2009). The ferroportin metal efflux proteins function in iron and cobalt homeostasis in Arabidopsis. *The Plant Cell, 21*(10), 3326-3338.

Murata, Y., Harada, E., Sugase, K., Namba, K., Horikawa, M., Ma, J. F., ... & Kusumoto, S. (2008). Specific transporter for iron (III): Phytosiderophore complex involved in iron uptake by barley roots. *Pure and Applied Chemistry, 80*(12), 2689-2697.

Pádua, M., Aubert, S., Casimiro, A., Bligny, R., Millar, A. H., & Day, D. A. (1999). Induction of alternative oxidase by excess copper in sycamore cell suspensions. *Plant Physiology and Biochemistry, 37*(2), 131-137.

Pádula, M., Aubert, S., Casimiro, A., & Bligny, R. (1996). Arrest of mitochondrial biogenesis in copper-treated sycamore cells. *FEBS letters, 398*(2-3), 248-252.

Peng, J., & Gong, J. (2014). Vacuolar sequestration capacity and long-distance metal transport in plants. *Frontiers in plant science, 5,* 19.

Pich, A., Scholz, G., & Stephan, U. W. (1994). Iron-dependent changes of heavy metals, nicotianamine, and citrate in different plant organs and in the xylem exudate of two tomato genotypes. Nicotianamine as possible copper translocator. *Plant and Soil, 165*(2), 189-196.

Pilon, M. (2009). Copper homeostasis. *New Phytol, 182*, 799-816.

Polle, A., & Schützendübel, A. (2003). Heavy metal signalling in plants: linking cellular and organismic responses. In *Plant responses to abiotic stress* (pp. 187-215). Springer, Berlin, Heidelberg.

Printz, B., Lutts, S., Hausman, J. F., & Sergeant, K. (2016). Copper trafficking in plants and its implication on cell wall dynamics. *Frontiers in plant science, 7*, 601.

Puntel, L. A., Sawyer, J. E., Barker, D. W., Dietzel, R., Poffenbarger, H., Castellano, M. J., ... & Archontoulis, S. V. (2016). Modeling long-term corn yield response to nitrogen rate and crop rotation. *Frontiers in plant science, 7*, 1630.

Ranathunge, K., Steudle, E., & Lafitte, R. (2005). A new precipitation technique provides evidence for the permeability of Casparian bands to ions in young roots of corn (Zea mays L.) and rice (Oryza sativa L.). *Plant, Cell & Environment, 28*(11), 1450-1462.

Rascio, N., & Navari-Izzo, F. (2011). Heavy metal hyperaccumulating plants: how and why do they do it? And what makes them so interesting?. *Plant science, 180*(2), 169-181.

Richau, K. H., Kozhevnikova, A. D., Seregin, I. V., Vooijs, R., Koevoets, P. L., Smith, J. A. C., ... & Schat, H. (2009). Chelation by histidine inhibits the vacuolar sequestration of nickel in roots of the hyperaccumulator Thlaspi caerulescens. *New Phytologist, 183*(1), 106-116.

Saha, S. K., Pietras, E. M., He, J. Q., Kang, J. R., Liu, S. Y., Oganesyan, G., ... & Cheng, G. (2006). Regulation of antiviral responses by a direct and specific interaction between TRAF3 and Cardif. *The EMBO journal, 25*(14), 3257-3263.

Samardjieva, K. A., Gonçalves, R. F., Valentão, P., Andrade, P. B., Pissarra, J., Pereira, S., & Tavares, F. (2015). Zinc accumulation and tolerance in Solanum nigrum are plant growth dependent. *International journal of phytoremediation, 17*(3), 272-279.

Samardjieva, K. A., Tavares, F., & Pissarra, J. (2015). Histological and ultrastructural evidence for zinc sequestration in Solanum nigrum L. *Protoplasma, 252*(1), 345-357.

Sarret, G., Saumitou-Laprade, P., Bert, V., Proux, O., Hazemann, J. L., Traverse, A., ... & Manceau, A. (2002). Forms of zinc accumulated in the hyperaccumulatorArabidopsis halleri. *Plant physiology, 130*(4), 1815-1826.

Sarret, G., Saumitou-Laprade, P., Bert, V., Proux, O., Hazemann, J. L., Traverse, A., ... & Manceau, A. (2002). Forms of zinc accumulated in the hyperaccumulatorArabidopsis halleri. *Plant physiology, 130*(4), 1815-1826.

Sauvé, S., McBride, M. B., Norvell, W. A., & Hendershot, W. H. (1997). Copper solubility and speciation of in situ contaminated soils: effects of copper level, pH and organic matter. *Water, Air, and Soil Pollution, 100*(1), 133-149.

Stephan, U. W., & Grün, M. (1989). Physiological disorders of the nicotianamine-auxothroph tomato mutant chloronerva at different levels of iron nutrition II. iron deficiency response and heavy metal metabolism. *Biochemie und Physiologie der Pflanzen, 185*(3-4), 189-200.

Steudle, E. (1994). Water transport across roots. *Plant and Soil, 167*(1), 79-90.

Tan, Y. F., O'Toole, N., Taylor, N. L., & Millar, A. H. (2010). Divalent metal ions in plant mitochondria and their role in interactions with proteins and oxidative stress-induced damage to respiratory function. *Plant physiology, 152*(2), 747-761.

Thomine, S., Lelièvre, F., Debarbieux, E., Schroeder, J. I., & Barbier-Brygoo, H. (2003). AtNRAMP3, a multispecific vacuolar metal transporter involved in plant responses to iron deficiency. *The Plant Journal, 34*(5), 685-695.

Ueno, D., Milner, M. J., Yamaji, N., Yokosho, K., Koyama, E., Clemencia Zambrano, M., ... & Ma, J. F. (2011). Elevated expression of TcHMA3 plays a key role in the extreme Cd tolerance in a Cd-hyperaccumulating ecotype of Thlaspi caerulescens. *The Plant Journal, 66*(5), 852-862.

Ueno, D., Yamaji, N., Kono, I., Huang, C. F., Ando, T., Yano, M., & Ma, J. F. (2010). Gene limiting cadmium accumulation in rice. *Proceedings of the National Academy of Sciences, 107*(38), 16500-16505.

Vert, G., Grotz, N., Dédaldéchamp, F., Gaymard, F., Guerinot, M. L., Briat, J. F., & Curie, C. (2002). IRT1, an Arabidopsis transporter essential for iron uptake from the soil and for plant growth. *The Plant Cell, 14*(6), 1223-1233.

Wintz, H., Fox, T., Wu, Y. Y., Feng, V., Chen, W., Chang, H. S., ... & Vulpe, C. (2003). Expression profiles of Arabidopsis thaliana in mineral deficiencies reveal novel transporters involved in metal homeostasis. *Journal of Biological Chemistry, 278*(48), 47644-47653.

Woźny, A., Zatorska, B., & Młodzianowski, F. (1982). Influence of lead on the development of lupin seedlings and ultrastructural localization of this metal in the roots. *Acta Societatis Botanicorum Poloniae, 51*(3-4), 345-351.

Wu, T., Kamiya, T., Yumoto, H., Sotta, N., Katsushi, Y., Shigenobu, S., ... & Fujiwara, T. (2015). An Arabidopsis thaliana copper-sensitive mutant suggests a role of phytosulfokine in ethylene production. *Journal of experimental botany, 66*(13), 3657-3667.

Yamasaki, H., Hayashi, M., Fukazawa, M., Kobayashi, Y., & Shikanai, T. (2009). SQUAMOSA promoter binding protein–like7 is a central regulator for copper homeostasis in Arabidopsis. *The Plant Cell, 21*(1), 347-361.

Yang, H., & Jie, Y. (2005). Uptake and transport of calcium in plants. *Journal of plant physiology and molecular biology, 31*(3), 227.

Yruela, I. (2009). Copper in plants: acquisition, transport and interactions. *Functional Plant Biology, 36*(5), 409-430.

Yuan, H. M., Xu, H. H., Liu, W. C., & Lu, Y. T. (2013). Copper regulates primary root elongation through PIN1-mediated auxin redistribution. *Plant and Cell Physiology, 54*(5), 766-778.

Yuan, M., Li, X., Xiao, J., & Wang, S. (2011). Molecular and functional analyses of COPT/Ctr-type copper transporter-like gene family in rice. *BMC plant biology, 11*(1), 1-12.

Zhang, W., Curtin, C., Kikuchi, M., & Franco, C. (2002). Integration of jasmonic acid and light irradiation for enhancement of anthocyanin biosynthesis in Vitis vinifera suspension cultures. *Plant Science, 162*(3), 459-468.0

Metal Tolerance and Protein Involved in the Compartmentalization of the Heavy Metal: An Overview

Jagdish Chand
Department of Agronomy, School of Agriculture,
Lovely Professional University, Jalandhar, Punjab, India, 144411
Email: jagdishnaik464@gmail.com

Abstract

Heavy metal stress to plants manly crops is increasing day by day, which creates many problems in the process of consumption and transportation of farmers produce. Not all the heavy metal present in the soil is taken up by the plants very little quantities of the heavy metal is taken up by the plant's rest of the heavy metal uptake is chocked by the plants in their roots by producing some amino acids. Whenever plant takes some heavy metal then there is various type of the defence response expressed by the plant. The various types of genes control these defence mechanisms. There are the many reasons behind the daily increase in the concentration of heavy metal in the soil such as excessive use of the fertilizers, pesticides, and other agrochemicals etc. Due to these activities, the surface water and groundwater is also polluted by these contaminants and the traces of these contaminants are present in the harvested crops this may lead to the cause of the several types of the diseases and physiological disorders in the human being. One of the examples of heavy metal poisoning is the Minamata disease which is caused by methyl mercury poisoning in Japan (1965). This concentration of the methyl mercury is increased due to the disposal from the industries into the eater sources. There is also some essential heavy metal for plants, which are involved to complete their inner functions like (iron, zinc, and copper). Some heavy metal like Cadmium, mercury, and arsenic these are also known non- essential heavy metals so we have to keep them below their threshold toxicity level to avoid the toxicity. There may be the chances that the essential heavy metal can also cause damage to the plants if their concentration exceeds from their optimal level by offending the uptake of other nutrients.

Introduction

The living cell consists of several components and proteins are one of the important components which plays an important role. According to the Amm et. al., various functions played by the proteins include membrane fusion, structural support and protection, cellular regulations and cellular signalling. The structural of the protein is the important factor by which in determining the function of the protein in a specific way. Stability of protein largely depends on the surrounding physical and chemical environment of protein. The surrounding environment of the protein is affected by the various types of stress which are induced by the extreme temperature, reactive molecules, heavy metal (HM) ions and other stresses. According to some scientist like Goldberg, 2003; Amm et al., 2014; Zhou et al., 2016; stated that these stresses induce both the folding and misfolding process proteins. In the last few decades rapid industrialization, urbanization and excessive usage of the agriculture amendments in the fields due to these activities the concentration of the pollutants have been increased tremendously in our environment. By standing at one place plants are continuously confronted by the wide range of biotic and abiotic stresses which also includes the stress of heavy metals stated by the scientist named as (Al- Whaibi, 2011). The natural binding of the proteins is greatly affected by the stresses caused by the increased concentration of the heavy metal according to the scientist (Hossain and Komatsu, 2013), this may results in the alteration in the biological functioning of the proteins. Now a days new emerging stress is cadmium stress according to the (Chrestensen et al., 2000) Cd stress disrupt the stability interaction which is associated with the changes in the tertiary structure and cause loss of promising functions of that protein. In stress condition, the cellular protein homeostasis is affected by the aggregation of the nascent or non- native protein, which leads to the stress on the endoplasmic reticulum and decreased the cell viability. According to the Amm et al., (2014) there is a typical set of protein in the centre of the system known as a heat shock protein (HSPs). These proteins worked as a surveillance agent for maintaining the stability of the cell under the stress condition. According to the Scientists, Liu and Howell in (2016) stated that there is one process named as ER-associated degradation (ERAD) this process helps in the maintaining the misfolding and the accumulation of that protein which fails to maintain the stability of the protein. According to the (Jarup, 2003), there are 53 elements categorized as a heavy metal because of their density i.e. greater than 5 g/cm3 and all are placed in "d block". During the growth of angiosperm, there are 19 macro and micronutrients elements involved in basic plant metabolism stated by the scientist named as (Ernst, 2006). In addition to that some nutrients beneficial nutrients like Si and help in the maintenance of plant structures according to the (Epstein, 1999). Various types of harmful effects that are caused by the toxicity of the heavy metal like an accumulation of low biomass, symptoms of chlorosis, photosynthetic and growth inhibition, it also affects the water balance inside the plant body as well as nutrient assimilation, and senescence. Thus all of these harmful effects may lead to the death of the plant in their extreme condition. Some of the metal posses the great potential of half-life in like Pb which having the half-life of 150- 5000 years due to this high retention power these heavy metal remain in the soil as such stated by the scientist like Nanda Kumar et al., 1995; Yang et al., 2005). If the crop plants are grown in the toxic soil of heavy metal, therefore, the products we obtained from is also contaminated and those contaminated food products are used by the human population as a food on daily basis. Some scientist like (McLaughlin et al.,) stated that if we continue to consume this type of contaminated food products than it may cause various types of diseases like Dermatitis and cancer. Thereby taking this all scenario in view it can be concluded that there are huge demand and need of research work in this field to stop and decrease the heavy metal contamination in cropland as well as from other area includes water bodies, groundwater also. Among the biotic and abiotic stresses, about 70 per cent of damage to crop production is caused by the abiotic stresses thus it may consider as a great constraint for crop production stated by the (Acquaah, 2007; Jewell et al., 2010). If we are willing to find the new quantitatively and qualitatively methods for improving the crop production in the contaminated soil then, first of all, we need to understand the responses of the plant in stresses condition. On the other hand, plants have two vascular pumps in the tonoplast. These pumps depend on the two vascular pumps like (V- ATPase and V-P Pase) and the driven force of these pumps are proton motive force and primary ATP- dependent pumps. Some antioxidant enzymes like glutathione and APX (ascorbate peroxidase) are also helpful for the plant in stress condition. Some of the heat-sensitive proteins called molecular chaperons help maintain the plant cell stability under stress condition. Thus to summarize the role of plant protein in neutralizing the effects of heavy metal stress by compartmentalization and other means.

What are the Stress and their Types

Stress can be considered as a negative pressure for any kind that leads to disturbance of their normal growth and development. Some scientist stated that the stress is the negative effects of all the biotic and abiotic agents that alter the normal growth and development.

Stresses are divided into two types biotic and abiotic. Abiotic stresses include salinity, floods, extremes of temperature, heavy metal, floods etc. and biotic stresses include attack of the various kind of the microorganisms like fungi, bacteria, nematodes, oomycetes, sometimes herbivores (stray animals) are included in biotic stresses. All the kind of stresses directly or indirectly leads to the production of the ROS (Reactive oxygen species). These species include peroxides, superoxides, and nascent oxygen (Fig. 1).

Fig. 1. Agents of (ROS) production

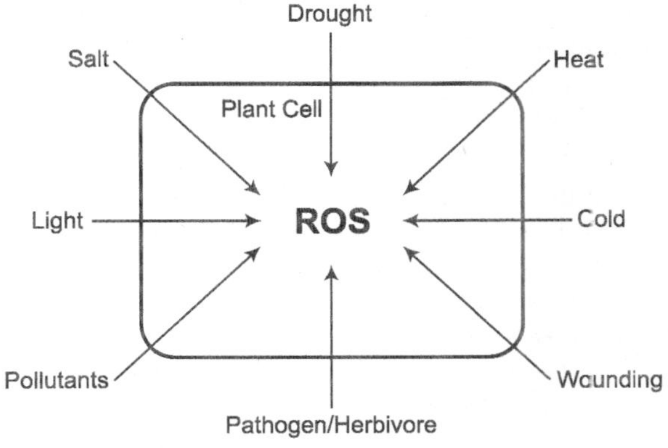

Source: https://www.google.com/search?

In nature plant is sessile and they have no option to change their position whenever they come under any kind of the stress but to overcome in that condition plant can build up many kinds of the proteins and the hormones to fight against these stress and to maintain the normal growth and development (plant cell homeostasis). Various kind of the proteins that are build-up by the plants to compart and detoxify these heavy metals from the cytosol to vacuoles for maintaining the level of the toxicity.

Source of the Heavy Metal in the Soil

For the identification of the heavy metal, the density of the metal ions is considered as the main factor, mainly the density of the heavy metal is at least 5 times higher than the density of the water. In the soil system, the presence of heavy metal is mainly due to the two reasons includes anthropogenic and natural sources. The rate natural sources of the heavy metal are not to fast but on the other hand, the anthropogenic or man-made sources increases the concentration of the heavy metal at very high speed. The heavy metal term for the group of the metal as well as the semimetals (metalloids). There are the various sources of the heavy metal in the soil includes;

1. **Natural sources:** The different concentration of the heavy metal present in the soil naturally. Due to the weathering of the parent material, these elements reach the topsoil by degrading from their parent material. There are the major ten elements that constitute the 99% content of the earth crust. These ten elements are (O, Si, Al, Fe, Ca, Na, K, Mg, Ti, and P). The rest of the elements present are in very small concentration (not more than 1.000 mg/kg) also known as a trace element.
2. **Anthropogenic source:** These types of sources mainly include the human activities that lead to an increase in the concentration of the heavy metal in the soil and make the soil more contaminated. There are the various component comes under the anthropogenic sources includes; Agricultural activities, industrial and the domestic effluents, fertilizers, pesticides, Municipal waste (Sewage and Sludge), spillage from the petrochemical, as well as the atmospheric deposition. These types of sources also increase the rate of the deposition of the heavy metal from the natural sources and make the soil more contaminated. Therefore these activities are the major issue of the concern to stop the contamination of the soil due to the metal and the semi metals. Sources come under this:
   (a) **Fertilizer:** Crop plant needed various types of the macro as well as the micronutrients to complete the plant the lifecycle. Sometimes the soil does not contain the adequate amount of the nutrients either macro as well as the micronutrients. So to fulfil the plant requirement these nutrients are supplied from outside as an external application in the form of the fertilizers. Theses fertilizers not only contains the specific nutrients sometimes some the fertilizers also supplied the trace quantities of the heavy metal includes (Pb, Hg, Cu, Mn, V, Cd). We continuously tend to increase the application of the fertilizers to the

soil, then the concentration of the heavy metal in the soil increases more than the permissible limits.

(b) **Pesticides:** In the agricultural system, various types of pesticides are used to tackle the pest problem. These pesticides contain some of the metal compounds approved by the United Kingdom in a report these compounds are mainly (Cu, Hg, Mn, Pb and Zn). There is some example of the pesticides includes Copper sulphate (Bordeaux mixture), and copper oxychloride.

(c) **Wastewater:** In the many parts of the world the application of the water from the municipal and the industrial wastewater. There is an estimate of about at least 20 million hectares of the cultivated land is irrigated by the wastewater. The concentration of heavy metal in the wastewater is relatively low as compared to the other sources but in case of the repetitive application of the same wastewater increases the concentration of the heavy metal from the permissible limits and makes the soil more contaminated.

(d) **Metal Mining:** The numbers of metal mining and the smelting industries are increases day by day. As their numbers are increases then it makes the soil more contaminated and increases the risk towards human life as well as the health of the ecological health.

Heavy Metal Uptake

According to the scientist like (Ma et al., 2001; Yang et al., 2002 for the uptake of the heavy metal there is some kind of the hyperaccumulators and that posses the ability to absorb the heavy metal from the different concentration of these metals. Various soil characters such as the Water content, Soil organic matter and pH of the soil also may affect the uptake of these heavy metals. A few scientists have detailed in contrast with pH affects i.e. 1) Discharge of the protons from the roots to the rhizosphere, this also leads to the further acidify the rhizosphere, this leads to further enhancing the metal dissolution. ii) there is also the growth of the plant species having the metal- accumulating capacity stated by the (Bernal et al., 1994; Peng et al., 2005; Kuriakose and Prasad, 2008). Hence organic substances and pH that are released from the rhizosphere enhance the uptake of the heavy metals by mobilizing it. According to the (Whiting et al., 2000) plant roots has the enhanced root proliferation and attributes associated with the overexpression of the genes under the heavy metal stress condition (Fig. 2).

Fig. 2. Uptake of heavy metals

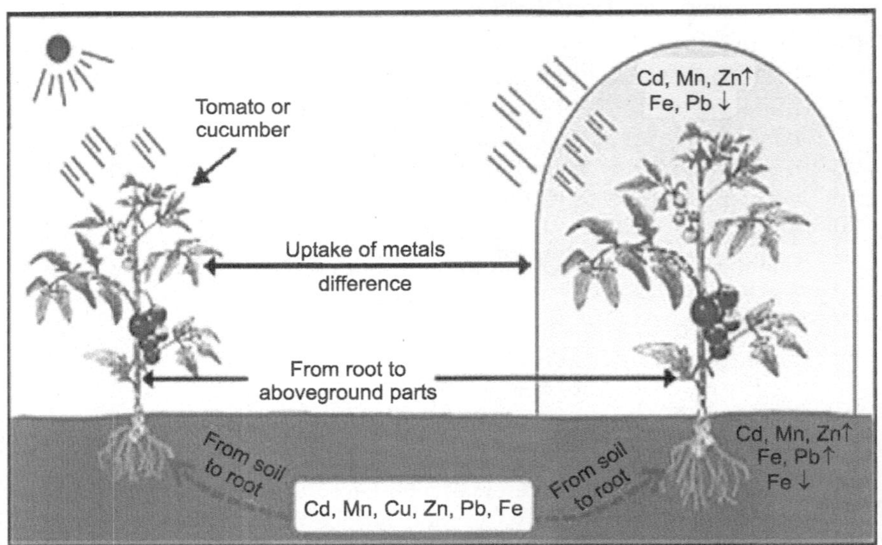

According to the scientist such as (Caille et al., 2005) an examination in the hyper gatherer *Pteris vittat* and non- hyper aggregator *Pteris tremula* has demonstrated that plasma film of roots cells of *P. that* had a high thickness of phosphate/arsenate transport than *Pteris tremula.*

Plant Protein Regulating the Plant Response of Metal Stress

Plants are considered ass as living things but they cannot able to move from one place to another they have to stand at the same place even in the unfavourable environmental condition. Due to the exposure of these unfavourable environment leads to the improvement in the ability to survive under adverse condition. There is a different mechanism that helps in the plant response to the abiotic heavy metal toxicity. Via;

1. Signal transduction in response to metals
2. The entry of metal ions from the soil into the root
3. Movement of metal from the root to the shoot
4. Metals chelated in the cytoplasm
5. Distribution of metal ions to other cell organelles

1. **Signal Transduction in response to metal:** When some plants or the organs of the plants gets exposed to the heavy metal stress or other stress, In that situation it is very difficult the examine the changes due to the stress, Even some times some of the pernicious damages couldn't be noticed in the signal transduction. For understanding the changes due to the stress condition there is a change in the protein abundance. Sometimes some gene expression, as well as the production of metabolites, can give an estimation or thorough understanding of the signaling pathway. Nature gives the

innate capacity to plants to respond to the metal stress condition through complex networks. Plants have their strategies to grow in the metal toxicity soil by synthesizing some proteins, molecules like metabolites and all of these are results in inducing the expression of the gene. Regulation and production of the proteins on metal toxicity soils is a different stimulation in the plant because these produced proteins which play an important role in improving the defence mechanism of plants against the toxicity. This was noticed in sunflower when grown in the cadmium stressed soils then there is the production of some protein is related to the defence-related mechanism. Thus, with the presence of the metal pressure, there is always a production of some kinds of proteins which directly results in regulation of the certain mechanism which helps to plant in that condition. There are the various kinds of the protein and other compounds that include Ca- calmodulin framework, Hormonal blend, ROS flagging and the MAPK (mitogen actuated proteins kinase) etc. These systems are initiated by a particular sort of metal toxicity. Hereafter is a complete synopsis of the pathways invigorated under abiotic stress.

2. **The entry of metal ions from the soil into the root:** Roots of the certain plants and their exudates permit the development of the metal particles in the rhizosphere. The entry of these heavy metals from the soil to roots is stopped by the plant up to a certain extent by releasing the certain chemicals which play an important role in inhibiting the entry of the metal into the root because these chemicals act as chelating agents so that they trap these metal in the rhizosphere. Sometimes the flow or the concentration of the heavy metal outside the plant roots is more than with this flow these metals ions as entered inside the plant roots. For example plants, roots release some nickel- chelating histidine and citrate which don't permit its entry inside the root hairs. In certain cases, some plants produce some amino acids in their roots to stop their further movement within the plant. Resistance toward to stop the entry the toxic metal ions inside the plant vary according to the plant species to plant species. Like in the root cell of tobacco genotype having the different chemical composition that helps to restrict the entry of the toxic manganese ions inside the plants. Thus the property of the cell wall somehow control the entry of toxic ions inside the plants but these site of stopping the entry of toxic ions are limited.
3. **Movement of metal from the root to the shoot:** Movement of the metal ions from roots to the shoot is accomplished by the conductive cells in the form of the metal conjugates. For example for the transportation of the copper ions from root to shot some of the derived substances of the amino acid-like methionine play an important role for the transportation and a chelating agent like histidine helps in the transportation of nickel ions. There are some of the transporter protein which helps in the transportation of the metal ions from root to shoot. To transfer the zinc in then xylem and also to translocate the cadmium from roots to shoot part Co groups an important role.
4. **Metals chelated in the cytoplasm:** There are some kinds of heavy metal which are not readily dissolved into the simpler compounds they required

some amount of time for this. During that time plant produces some of the metallothioneins and the phytochelatin chemical which acts as a binding agent for these metal ions by sequestering them to stable forms.

5. **Distribution of metal ions to other cell organelles:** This phenomenon is going to happen at that times when the concentration of these ions is going to increase from their normal concentration than plant uses their own constructed proteins like a transporter agent to transfer these high concentrated metal ions from the cytoplasm of the to other plants parts of the plants. Under this phenomenon sometimes plants produce some modified parts like salt glands at their leaves to overcomes against these types of conditions for their normal growth and development.

What are the Phytochelatins: Their Regulation, and Role against the Heavy Metal Stress

Whenever plant goes under the heavy metal stress plant produces some kind of protein named as phytochelatins. Those proteins help the plant to prevent the damage that happens due to the stress of metal toxicity, an example of the phytochelatins are cysteine- rich oligomers. Some scientist such as (Ashraf et al., 2010; Pochodylo and Aristilde, 2017) stated that whenever the plant comes under the heavy metal stress then the production of the heavy metal is the plant primary defence mechanism. Some other scientist like (Clemens, 2006; Emamyerdiam et al., 2015) also said that the synthesis of the phytochelatins plays an important and crucial role tolerance of plant against the heavy metal stress. It was reported that in some plants the biosynthesis of phytochelatin is started at a post-translational level under metalloid stress. According to some scientist like Pomponi et al., 2006; Furini, 2012; In some of the mutant plants if there is the high production of the phytochelatins increases the accumulation of the heavy metal without making plant tolerant. Thus (Furini, 2012) also stated that another additional role of the phytochelatins like they have involved antioxidant mechanism, sulfur metabolism and essential metal ion homeostasis. According to the (Hasan et al., 2016) Potential mechanism for dealing with the toxicity of heavy metal is induced by the circulation of toxic metal inside the cytosol. Beside the chelation, there is another method of detoxification of the heavy metal that includes the accumulation and also the stabilization of the heavy metal as high molecular weight (HMW) complexes with the PCs in the vacuole stated by the Jabeen et al., 2009; Furini, 2012). There is another method for the ameliorating the ions toxicity is the heavy metal sequestration. For the sequestration of the heavy metal ions first, these ions are arrested in the cytosol and then they are further transported to the vacuoles with the help of the transporters.

Regulations of the Phytochelatins

For the regulations of the biosynthesis pathway of the phytochelatins, there are a number of the pathways. From these numbers of the pathways, the GHS biosynthesis regulation studies in the plant of the Indian mustard (*Brassica juncea*) with an increase in the expression in the GHS biosynthesis enzymes that enhances the phytochelatin biosynthesis. According to some scientist like (Yong et al., 1999; Zhu et al., 1999), this GHS biosynthesis in Indian mustard leads to the tolerance of the plant against the CD stress or toxicity. According to the (Schafer et al., 1998) plants of the Indian mustard (wild type) respond against the exposure of the Cd toxicity with an increased level of GCS transcript. Thus all the studies show that for the regulation of the phytochelatins synthesis there is need of the activation of the enzymes and this enzyme is only activated whenever plants come under the stress condition.

Role of the Phytochelatins

According to the Rauser, 1995; 1999; Zenk, 1996; observation of the various types of studies are interpreted for the indication of the role of the phytochelatins in the homeostasis of essential metabolism of metal ions. In one lab experiment have shown that the complexes of the PC-Zn and PC-Cu reactivate with the Copper-dependent (Apo form) enzyme diamino oxidase and also reactivate with the Zn- dependent enzyme carbonic anhydrase respectively. In case of Zn and Cu complexes were no more effective as compared to the metal- requiring ions at that condition these phytochelatins are capable of donating the metal ions to those enzymes which need those enzymes. Phytochelatins play also an important role in the detoxification of the high level of the Cd toxicity usually more than that of 1 µm. According to some scientist like (Wagner, 1993) soil having no pollution from heavy metal also contains some trace amount of the Cd (0.3 µm). Thus, we can say that without phytochelatins the detoxification of the toxicity of heavy metal is not an easy task. They play an important role in their detoxification from that time when we never address their contribution in maintaining the concentration of the metal ions toxicity inside the plant's roots or sometimes outside the plant's roots i.e. in the rhizosphere.

Plant Response to Heavy Metal Stresses

In nature, plants are subjected to various types of natural changes. With the change in the environment, condition plants can't move accordingly. This changes in the climatic conditions lead to a trigger of a wide scope of physiological and biochemical adjustment and improvement. This adjustment and the improvement creates a well as develops a methodology that permits them to adapt against the uneven environmental condition and substantial metal harmfulness. Plants have their instruments that help plant to react with upgrading outer environmental condition and metal poisoning. This includes; 1) detecting of outside pressure boosts, 2) signal transduction and transmission of a sign into the plant and 3) activating suitable measures to balance the negative effect of stress upgrades by adjusting the physiological, biochemical, and atomic status.

Some Defence Mechanism Employed by Plants against HM Stress

The primary defence and the front line of the defence mechanism employed by the plants against the heavy metal stress are the physical barriers. Physical barriers include the development of some morphological structure includes the development of the thick cuticle, trichomes (biological active tissues), plant cell wall sources of barrier whenever the plant is going under the metal stress condition. Trichomes act as a storage purpose for metals and secret some of the secondary metabolites those help to overcome the metal stress. In some conditions when the heavy metals ions break the plant primary defence mechanism and reach to the plant tissue and cell than in the response of this situation plant initiate several types of cellular defence mechanism against the heavy metal ion toxicity. In the cellular defence, mechanism plant produces various types of the compound which includes low molecular weight protein, glutathione, phytochelatins, heat shock proteins, and also produces some kind of the amino acids such as histidine and pralines. The plant also produces some kind of such as jasmonic acid, salicylic acid, and ethylene. In certain condition, the concentration of the heavy metal in outside the environment is more than there is the possibility that these metal ions can also overcome the plant secondary defence mechanism. In that situation, the production of the ROS (Relative Oxygen Species) is increasing. This increase in the ROS leads to the early senescence of the plant. To overcome with situation plant has the other defence mechanism such as the development of the antioxidant defence. In the antioxidant defence mechanism, there is the production of the several antioxidant enzymes like guaiacol peroxidase (GPX), Superoxide dismutase (SOD), Catalase (CAT), ascorbate peroxidase (APX)

Fixing of Damaged Proteins

The Tamas et al., 2014, reported that proteins are mainly concentrated on heavy metals and he also reported that proteins are either complex structure with attached a side chain collecting of proteins or uprooted the fundamental particles from the metalloproteins and also encourage the disability of physiological capacities. (Bierkens, 2000; Tamas et al., 2014) demonstrated that heavy metals gathering with the local affirmations of proteins but again straining in collapsing process of beginning or non-local proteins that exhibit both qualitative inadequacies of the influenced proteins and they also develop proteotoxic totals. (Hossains et al., 2013) demonstrated that plants inquisitively react to annoy by setting up of the qualities related to cell strength or the ability to passes through the contaminated conditions. Gupta et al., 2010, reported that in the piece of plants many reaction occurs which includes a lot of quality proteins, those proteins named as pressure qualities, and these are instigated to blend a group of a protein called HSP's. However, these proteins in stress condition are instigated gathering of HSP's or presume a wide job in keeping up the cell homeostasis by accurate collapsing of early and stress amassed misfiled proteins, the total proteins or by advancing particular debasement of reducing proteins (Huttner et al., 2015). As we know that HSP's works as an atomic chaperons proteins which are engaged with "Housekeeping" in the inner side of the cell (Sorensen et al., 2003, Schoffl et al., 1999) demonstrated that a few classes of HSP's have been observed in the plant's cell or the HSP's proteins having 10 to 200 KD sub atomic loads are running and these are described as chaperons which took part in the recognize in the pressure conditions. In case the endoplasmic reticulum(ER) all the early polypeptides, right off the bat, are stable out by chaperones such as HSP40 and HSP70 like proteins, these proteins act as binding agents, for example, ERdj3 and restricting proteins (Bip) before they are properly altered and collapsed, which forestalls accumulation and enables their legitimate collapsing (Howell, 2013) (Figure 3).

Fig. 3: Fixing of the damaged proteins

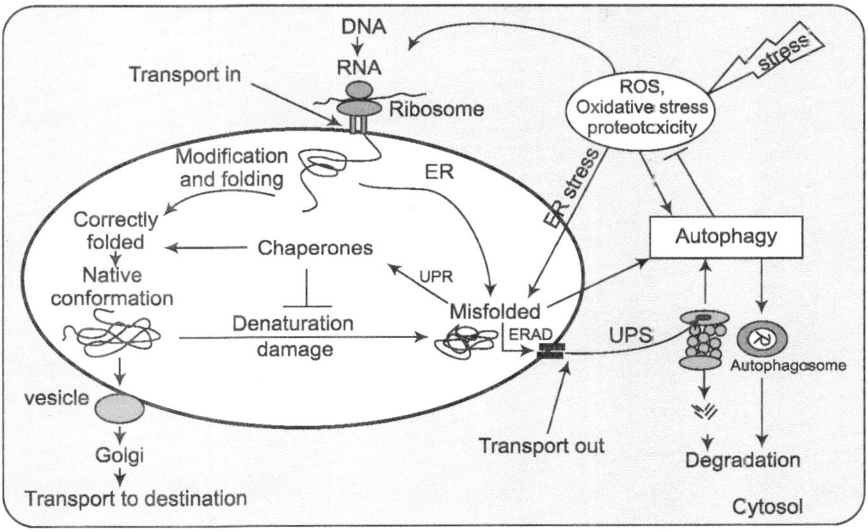

Heavy Metal Detoxification

Li et al., 2016a., reported that the cellular level of these metal stress causes oxidative stress by generating many types of ROS species. Therefore, these ROS species also encourage the damaged proteins to the DNA and also affect its reshaped mechanisms. On the other hand, there is a better in the protective mechanism of the plant cell to reduce the effects of the stress of the heavy metal and also manage the metal detoxification mechanisms to reduce their irregular toxic effects. In the plant mechanism, there is a system of protection mechanism in the plant body which is managed by the plant as a plan to block the uptake of harmful ion the cell by binding them to the cell wall or the cellular exudates ((Manara, 2012; Hansan et al., 2015). However, they also inhibited the huge distance transport (Fig. 4).

Fig. 4. Detoxification Chain

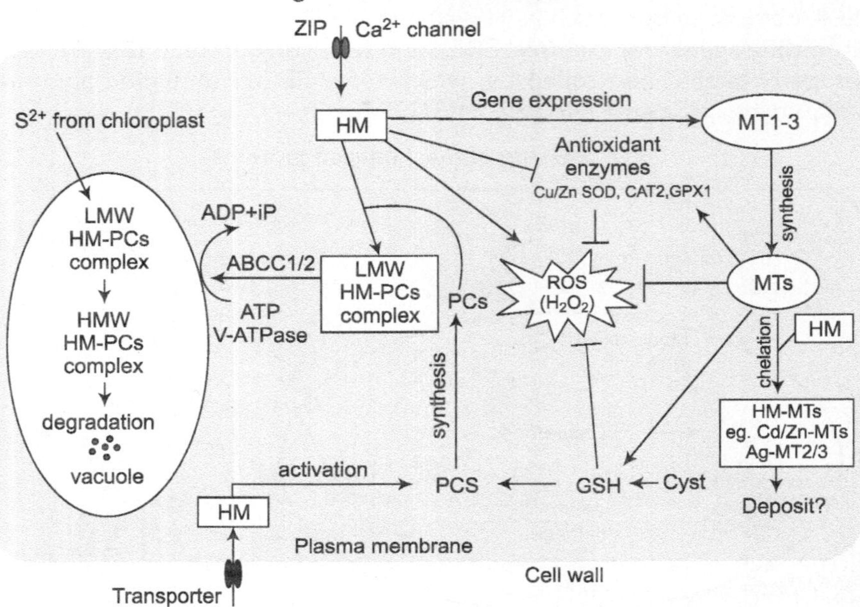

(Source: Zhao and Chengcai, 2011)

According to (Zhao and Chengcai., 2011) demonstrated that there are many processes which help in the detoxification strategies such as chelation, phytochelatins and metallothioneins all these processes occur in the cytosol and sequestration into the vacuole by the vacuolar transporter system.

Involvement of Vesicle Trafficking in Heavy Metal Transport

According to (Litenmaier & Kupper (2011) demonstrated that the vesicles shipping is likewise o be consequences worry in vacuolar compartmentalization of Heavy metals, however, there are only a few studies those recognize this issue. By using leaf mobile protoplasts of Cd/Zn hyperaccumulator Thalaspi caerulescens, the located Cd healthy vesicles structure in between the cytoplasm. Therefore, specific beginning or fate of these vesicles converted in uneven particles, they are probably playing an important role in vacuolar sequestration of Cd (Fan et al., 2011) suggested that Cd brought about prohibit of advance endocytosis and vesicles trafficking in the inner side of the cytoplasm of A. Thaliana root hairs. Moreover, on the other hand, big vesicles have been found, which prepared inside the cytoplasm. The SNARE proteins Sec22p/Sec22 has been proven to influence made up with caesium (Cs+), non-crucial cations, in yeast and plants. Reduction in the Saccharomyces cerevisiae Sec22p decreased the Cs awareness by using 1/2. However, this effect linked with the compromised vacuolar deposition. Thus, the same effect has been found in the loss of function mutant of A. Thaliana ortologue SEC22 reported by (Draxl et al., 2013). The unusual Zn- enriched vesicular compartment immersed in the cytoplasm of yeast cells were situated suggested by (Devirgillis et al., 2004). According to (Hasse& Beryersmann 1999) reported that the resemble zincosomes of a mammalian cell. The yeast zincosomes are stunning from vacuoles, formed independently of endocytosis and likely serve as transient Zn shops from these proteins reported by (Devirgillis et al., 2004). According to (Kawachi et al., 2009) demonstrated that the presence of Zn (zero. 5mM). Simultaneously, Zn structure has additionally have been proved in the root hairs extremely Zn treated Paulowina tomentosa (Azzarello et al., 2012). In the future work present whether plants cells possess zincosomes or Zn specific vacuoles with the transporter-like MTP1 (Kawachi et al., 2009). On the other hand example such as of vesicles, the exhibit is taken by their on transporter IRT2 whose proteins are fused with florescent A. Thailana (Vert et al., 2009). However, the IRTI proteins are converted into determined to localized to early endosomes/ trans-Golgi community in A. Thailana root hairs cells because of the endocytosis. IRTI cycles to the plasma membrane execute to perform uptake of Fe and cells because of endocytosis. According to (Barberon et al., 2011) reported that the IRT1 trafficking the plasma membrane and next turnover is moreover based that tightly handles the IRT1 tiers on the plasma membrane for the perfect iron and metallic homeostasis.

Responses of Vacuolar H+ Pumps to Toxic HM Ions

Particularly there are two number one proton pumps, V-ATPase and V-PPase develop the proton and generated a force on the tonoplast and permit some proton antiport, and the membrane acidic trapping and having the potential to transport. These heavy metals must be adjusted structurally and functionally to the cellular need for the accumulation to the various conditions. The controlled proteins accumulation must be transcriptional, post-transcriptional, post-translational and also put up translational mechanisms and interest, therefore in case of vacuolar pumps the best transcriptional and put up translational steps have been improved.

V-APTase

V-ATPase builds up an important mechanism in the reclamation the transcription subunit of the V-ATPase that is visible when Mesembryanthemum crystallinum is tolerant to sodium (salt) and also in the various abiotic conditions of stress which includes heavy metals stress demonstrated by the different scientist such as (Dietz et al., 2001; Finkemeier et al., 2003; Sharma et al., 2004). From the above observation, we can say that these the transcriptional regulation support in adjusting the activity of the V-ATPase under the situation of the stress and for the compensating the needs of the translational and it requires the offsetting the enhanced stressed proteins.

V-P Pase

In comparison to V-ATPase, there are few studies in contrast with the stress response of the V-PPase. There are many fewer studies done in the V-PPase stress condition. But in the case of V-PPase there are more studies done the scientists.

According to the (Brini et al., 2007; Muto et al.,) reported that the activity of the V-PPase are moreover express in those conditions when the plants show few specific abiotic conditions such as salinity, drought and flooding. Similar to the activity of the V-ATPase the activity of the V-PPase is decreased under the stress conditions of different elements like CO, Cd and Cu reported by (Maeshima 1991). There are some new studies are done two types of the isogenic (like CSVHP1;1 & CsVHPI;2) of V-PPase in crops such as C. sativa. The activity of the transcript accumulations of isogenic first (CSVHP; 1) represents extended response towards to the low potential of the Cu and Zn as compared with the second isogenic (CsVHPI; 2).

Role of HSPs in Plant Tolerance to HM Stress

The production of the heat shock proteins in the plant is for to maintain the plant growth and development under the abiotic condition and mainly in response to the production of the various heat shock proteins. The production of ROS (Relative Oxygen Species) is the direct confirmation of that the plant is under the stress condition. Thus we can say that there is a very close relationship between the Heat shock proteins and the ROS (Relative Oxygen Species). Heat shock proteins are the class of molecular chaperones they not only produced in the heavy metal stress condition but also produces in other condition (like extreme temperature, dehydration, salinity, high-intensity irradiation etc.) that leads to the production of the ROS stated by the various scientist like Swindell et al., 2007; Al-Whaibi- 2011; Xu et al., 2011). According to the Gupta et al., 2010 the aggregation of the HSP 70 is in the light and natural stressor condition in the wide range of the crop species. The activity of this family is mainly confined to the cytosol, endoplasmic reticulum (ER) and the mitochondria to maintain the homeostasis of the cell. According to the (Pratt and toft, 2003; Al- Whaibi, 2001), a great numbers of the study indicated that the proteins of the HSP 90 family present in the various plant species which assume a significant job in protein collapsing and controlling the transduction system, control in the cell cycle, protein corruption and also protein dealing. Some scientist like Wang et al., 2004 Gupta et al., 2010; Park and web optimization 2015; stated that these proteins found additionally in the relationship with the other few intercellular proteins, which includes; Calmodulin, actin, tubulin and some other receptors of the flagging kinase. According to Wang et al., (2004) this group of the protein shows the hereditary which adds the developmental adjustment in the plant in typical uneven condition or the stress condition.

Types of the HSPs

There are the various types of the HSPs such as HSP100, HSP90, HSP70, and HSP60, according to their different types their function is also different.

HSP100

This size of the HSP is the member of the superfamily of AAA ATPase. This types of HSP are produced by the plant when there is heavy production of the ROS in the plant. This condition emerges out when the plant is suffering from various types of stressor at a particular time. According to the scientist such as Lee et al., 2007; stated that some members of this class of HSP act a housekeeping protein to maintain the function of the cell.

HSP90

HSP 90 is similar to the HSP 100, besides that it HSP 90 works in association with the HSP 70, which results in the major function played by these two in the association. According to some scientist like Pratt and Toft 2003; Pratt et al., 2004 the major function played by the HSP90 is the folding of the protein, transduction of the signal and also controls the cell cycle. Hubert et al., 2003; liu et al., 2004; Thao et al., 2007) stated that the HSP plays an important function in to imparts the resistance to plant against the attack of the Phyto pathogens.

HSP70

The function of the HSP 70 is in the association with CO- chaperones. These proteins help in the to avoid the accumulation of the proteins those are misfolded and enhance the refolding of the proteins under the stress and non-stress condition stated by the scientist such as (Sung et al., 2001; Su and Li, 2008).

HSP60

The other name of the members of the HSP 60 is Chaperonins. These types of proteins are found mainly in the prokaryotes, and in case of the eukaryotes, they are found mainly in the plastids and mitochondria. According to the (parsell and Lindquist, 1993), these proteins act as a post-transcriptional regulation of many proteins. According to the Jackson- Constan et al., (2001) the HSP 70and HSP 60 in association manages the function of those proteins which helps in the chloroplast-mediated import signal.

Degradation of Metal-induced Denatured Proteins

Selective degradation of the damaged proteins and unwanted misfolded proteins controlled the plant growth, development and physiological processes. This process of degradation and the protolysis in the plant is accomplished by the various processes such as UPS (ubiquitin-proteasome system) or by the induction of the autophagosomes sated by the scientist like Lie and Howell, (2016). The first method of the degradation of unwanted protein is a very rapid and effective method at particular time. On the other hand, the second process is the biological self- destruction process which helps in maintaining the cellular homeostasis of the eukaryotic cell by transferring the damaged proteins during the developmental transition in the top the vacuoles stated by the (Lie and Bassham, 2012; Wen- Xing, 2012). According to the various scientist like Liu and Bassham, (2012), there are various types of autophagy which include, 1) Microautophagy, macroautophagy, chaperones- mediated autophagy and the organelle-specific autophagy.

Role of the Ubiquitin-Proteasome System (UPS)- Dependent Proteasome Activity and Metal Stress

In the harsh environment condition like a condition of the heavy metal stress, there is going to increase the free radicals that affect the quality of the proteins by creating various types of negative effects on the proteins like misfolding, denaturing, and damage to protein. So that in the stress conditions these affected proteins are not going to protect the plant because of that plants need such type of defence mechanism to repair the proteins if possible and removal of those proteins which are not repairable. In that condition plant ubiquitin-proteasome system (UPS) system play a vital role to adopt the plant system under that environmental condition stated by the (Stone, 2014). Plant ubiquitin-proteasome system (UPS) works in the plant nucleus as well as in the plant cytoplasm which helps in removing the maximum of the peptide for balancing the level of the all regulatory proteins and temporary remove the cellular regulator which may aggregate successive exposure towards the abiotic stress according to the (Lyzenga and stone, 2012). According to the (Genschik et al., 1992; Jungmann et al., 1993), the role of the plant UPS system has been identified at several years ago also manifested because of increased polyubiquitin genes expression. In the study when rice plant goes under the genetic analysis indicates the induction of the polyubiquitin expression in root and shoot when the plant is exposed to even low concentration of Cd toxicity. According to the (Yu et al., 2011), there is the activation of the apoptosis due to the change in the cellular protein homoeostasis because of increase in the concentration of the metal ions toxicity that disturbs the activity of the proteasomal activity and deposits the abnormal proteins in the cytoplasm. For instance whenever various species of the plant are exposed to the different concentration of the different metal ions at the same concentration to check the response of the activity of the ubiquitin. In this two different results may occur that in the first case there is the reduction in the activity of the ubiquitin when the plant gets exposed the Hg, Cu, Cd, Ni, Co, Pb, and Cr, at the concentration of 100 μM but in case of the Zn and Al there is no change at that concentration of the ions. According to the (Aina et al., 2007; Pena et al.,2007; 2008), there is the increase in the activity of the 26 S proteasome when the plant gets exposed to metal stress even at low concentration. According to the scientist like Goldberg, (2003) digestion of the proteins takes place in two steps that include first of identification of the proteins done by the ubiquitin and next to the identification of marked proteins are digested into small peptide from the proteolytic complex by the ubiquitylated. The function of the ubiquitylate is executed by the different enzymes (E1-E2-E3) in combination stated by the Maupin-Furlow, 2013). These three enzymes combine with the ubiquitin in three steps, in the first step the ubiquitin is combined covalently in an ATP- dependent reaction with ubiquitin-activating enzymes (E1), in the second step this activated enzyme is transferred by the transesterification from enzyme E1 to E2, at last in the third process the enzyme E3 helps in the catalyzing the of the transformation of ubiquitin to the target protein to complete the process.

The Response of the Autophagy in the Stress of the Meta

According to the scientist like Han et al., 2011; Wang et al., 2015; Xu et al., 2016; Yang et al., 2016; autophagy not only helps plant in against with the metal stress but it also helps the plant to adopt in the wide range of different environmental condition includes the; Drought, the lake of nutrient, heat stress, and sometimes against with the pathogen attack also. Now times there has been more deeply working is done to explore the working of the autophagy towards the stress in the plant as well as its working of adaptations is the same as inhuman and yeast. Apart from the other environment toxicants, heavy metal is recognized as strong inducers for the production of the ROS, these results in the regulation in the homeostasis.

Conclusion

Plants are considered as a living entity having disadvantages that they cannot move from one place to another place according to their comfort. So during the plant growth cycle have to face various types' stresses those includes: Biotic and Abiotic stresses. The major concern of this book chapter is heavy metal stress that comes under the abiotic stresses. In the soil system, the presence of heavy metal is mainly due to the two reasons includes anthropogenic and natural sources. These stresses cause various types of negative effect on plant growth and development. To stand against these stresses plants are blessed with the various types of mechanisms that help the plant to overcome against these conditions. Plants synthesise various types of chelating proteins and regulate various types of the hormones and mechanism to compartmentalize the concentration of the heavy metal as well as to stop the entry of these toxic metal (in roots as well as sometimes outside the roots i.e. in rhizosphere) from soil to inside the plants. In response to the heavy metal plant synthesis, the various types of the proteins and regulates the kinds of the mechanism that helps in the detoxify the negative effects of the heavy metal stress. Plant posses the various kinds of the primary, Secondary and cellular defence mechanism to choke the effect from the heavy metal stress condition. Besides the plant defence mechanism, it is the responsibility to stop the application of the harmful elements to the soil because it not only harmful for the crop plant but also harmful for us as well as our ecological environment. Thus care must be taken to minimize the contamination of the soil.

References

Acquaah, M. (2007). Managerial social capital, strategic orientation, and organizational performance in an emerging economy. *Strategic management journal, 28*(12), 1235-1255.

Ahsan Ullah, A. K. M. (2010). *Population migration and Asia: theories and practice*. Nova Science Publishers. Sharma, S. S., & Dietz, K. J. (2009). The relationship between metal toxicity and cellular redox imbalance. *Trends in plant science, 14*(1), 43-50.

Aina, R., Labra, M., Fumagalli, P., Vannini, C., Marsoni, M., Cucchi, U., ... & Citterio, S. (2007). Thiol-peptide level and proteomic changes in response to cadmium toxicity in Oryza sativa L. roots. *Environmental and experimental botany, 59*(3), 381-392.

Al-Whaibi, M. H. (2011). Plant heat-shock proteins: a mini-review. *Journal of King Saud University-Science, 23*(2), 139-150.Amm, I., Sommer, T., and Wolf, D. H. (2014). Protein quality control and elimination of protein waste: the role of the ubiquitin-proteasome system. Biochim. Biophys. AcaBBA. Molecul. Cell Res. 1843, 182–196. DOI: 10.1016/j.bbamcr.2013.06.031

Amm, I., Sommer, T., & Wolf, D. H. (2014). Protein quality control and elimination of protein waste: the role of the ubiquitin-proteasome system. Biochimica et Biophysica Acta (BBA)-Molecular Cell Research, 1843(1), 182-196.

Ashraf, M. (2010). Inducing drought tolerance in plants: recent advances. *Biotechnology advances, 28*(1), 169-183.

Azzarello, E., Pandolfi, C., Giordano, C., Rossi, M., Mugnai, S., & Mancuso, S. (2012). Ultramorphological and physiological modifications induced by high zinc levels in Paulownia tomentosa. *Environmental and experimental Botany, 81*, 11-17.

Barberon, M., Zelazny, E., Robert, S., Conéjéro, G., Curie, C., Friml, J., & Vert, G. (2011). Monoubiquitin-dependent endocytosis of the iron-regulated transporter 1 (IRT1) transporter controls iron uptake in plants. *Proceedings of the National Academy of Sciences, 108*(32), E450-E458.

Bernal, R. (1994). Statistical power and aquacultural research. *Aquaculture, 127*(4), 371-388.

Bierkens, J. G. (2000). Applications and pitfalls of stress-proteins in biomonitoring. *Toxicology, 153*(1-3), 61-72.

Bierkens, J., Van de Perre, W., & Maes, J. (1998). Effect of different environmental variables on the synthesis of Hsp70 in Raphidocelis subcapitata. *Comparative Biochemistry and Physiology Part A: Molecular & Integrative Physiology, 120*(1), 29-34.

Brini, F., Hanin, M., Lumbreras, V., Amara, I., Khoudi, H., Hassairi, A., ... & Masmoudi, K. (2007). Overexpression of wheat dehydrin DHN-5 enhances tolerance to salt and osmotic stress in Arabidopsis thaliana. *Plant cell reports, 26*(11), 2017-2026.

Chrestensen, C. A., Starke, D. W., & Mieyal, J. J. (2000). Acute cadmium exposure inactivates thioltransferase (Glutaredoxin), inhibits intracellular reduction of protein-glutathionyl-mixed disulfides, and initiates apoptosis. *Journal of Biological Chemistry, 275*(34), 26556-26565.

Clemens, S. (2006). Toxic metal accumulation, responses to exposure and mechanisms of tolerance in plants. *Biochimie, 88*(11), 1707-1719.

Clemens, S., & Ma, J. F. (2016). Toxic heavy metal and metalloid accumulation in crop plants and foods. *Annual review of plant biology, 67*, 489-512.

Cui, F., Liu, L., Zhao, Q., Zhang, Z., Li, Q., Lin, B., ... & Xie, Q. (2012). Arabidopsis ubiquitin conjugase UBC32 is an ERAD component that functions in brassinosteroid-mediated salt stress tolerance. *The Plant Cell, 24*(1), 233-244.

D'Alessandro, A., Taamalli, M., Gevi, F., Timperio, A. M., Zolla, L., & Ghnaya, T. (2013). Cadmium stress responses in Brassica juncea: hints from proteomics and metabolomics. *Journal of proteome research, 12*(11), 4979-4997.

DalCorso, G., Fasani, E., & Furini, A. (2013). Recent advances in the analysis of metal hyperaccumulation and hypertolerance in plants using proteomics. *Frontiers in plant science, 4,* 280.

DalCorso, G., Fasani, E., & Furini, A. (2013). Recent advances in the analysis of metal hyperaccumulation and hypertolerance in plants using proteomics. *Frontiers in plant science, 4,* 280.

DalCorso, G., Manara, A., & Furini, A. (2013). An overview of heavy metal challenge in plants: from roots to shoots. *Metallomics, 5*(9), 1117-1132.

Deng, Y., Srivastava, R., & Howell, S. H. (2013). Endoplasmic reticulum (ER) stress response and its physiological roles in plants. *International journal of molecular sciences, 14*(4), 8188-8212.

Devirgiliis, C., Murgia, C., Danscher, G., & Perozzi, G. (2004). Exchangeable zinc ions transiently accumulate in a vesicular compartment in the yeast Saccharomyces cerevisiae. *Biochemical and biophysical research communications, 323*(1), 58-64.

Dietz, P., & Leigh, D. (2001, November). DiamondTouch: a multi-user touch technology. In *Proceedings of the 14th annual ACM symposium on User interface software and technology* (pp. 219-226).

Dräxl, Stephan, Johannes Müller, Wei B. Li, Bernhard Michalke, Hagen Scherb, Burkhard A. Hense, Jochen Tschiersch, Ulrike Kanter, and Anton R. Schäffner. «Caesium accumulation in yeast and plants is selectively repressed by loss of the SNARE Sec22p/SEC22.» *Nature communications* 4, no. 1 (2013): 1-10.

Eapen, S., & D›souza, S. F. (2005). Prospects of genetic engineering of plants for phytoremediation of toxic metals. *Biotechnology advances, 23*(2), 97-114.

Emamverdian, A., Ding, Y., Mokhberdoran, F., & Xie, Y. (2015). Heavy metal stress and some mechanisms of plant defense response. Sci World J 2015: 756120.

Emanuel, E. (1999). Silicon. *Annual Review of Plant Physiology and Plant Molecular Biology, 50,* 641-664.

Ernst, W. H. (2006). Evolution of metal tolerance in higher plants. *For Snow Landsc Res, 80*(3), 251-274.

Fahr, M., Laplaze, L., Bendaou, N., Hocher, V., El Mzibri, M., Bogusz, D., & Smouni, A. (2013). Effect of lead on root growth. *Frontiers in plant science, 4,* 175.

Fan, J., Fang, G., Wang, X., Zeng, F., Xiang, Y., & Wu, S. (2011). Targeted anticancer prodrug with mesoporous silica nanoparticles as vehicles. *Nanotechnology, 22*(45), 455102.

Farias, J. G., Antes, F. L., Nunes, P. A., Nunes, S. T., Schaich, G., Rossato, L. V., ... & Nicoloso, F. T. (2013). Effects of excess copper in vineyard soils on the mineral nutrition of potato genotypes. *Food and Energy Security,* 2(1), 49-69.

Fidalgo, F., Azenha, M., Silva, A. F., de Sousa, A., Santiago, A., Ferraz, P., & Teixeira, J. (2013). Copper-induced stress in S olanum nigrum L. and antioxidant defense system responses. *Food and Energy Security,* 2(1), 70-80.

Finkemeier, I., Kluge, C., Metwally, A., Georgi, M., Grotjohann, N., & Dietz, K. J. (2003). Alterations in Cd-induced gene expression under nitrogen deficiency in Hordeum vulgare. *Plant, cell & environment, 26*(6), 821-833.

Furini, F., Persiani, C. A., & Toth, P. (2012, April). Aircraft sequencing problems via a rolling horizon algorithm. In *International Symposium on Combinatorial Optimization* (pp. 273-284). Springer, Berlin, Heidelberg.

Goldberg, A. L. (2003). Protein degradation and protection against misfolded or damaged proteins. *Nature, 426*(6968), 895-899.

Haq, M. A., Hasnain, A., & Azam, M. (2014). Characterization of edible gum cordia film: Effects of plasticizers. *LWT-Food Science and Technology, 55*(1), 163-169.

Hasan, M. A., Mazumder, M. H. H., Chowdhury, A. S., Datta, A., & Khan, M. A. (2015). Molecular-docking study of malaria drug target enzyme transketolase in Plasmodium falciparum 3D7 portends the novel approach to its treatment. *Source code for biology and medicine, 10*(1), 1-14.

Hasan, M. M., Sharma, A., Johnson, F., Mariethoz, G., & Seed, A. (2016). Merging radar and in situ rainfall measurements: An assessment of different combination algorithms. *Water Resources Research, 52*(10), 8384-8398.

Hernández, R., Vallés, C., Benito, A. M., Maser, W. K., Rius, F. X., & Riu, J. (2014). Graphene-based potentiometric biosensor for the immediate detection of living bacteria. *Biosensors and bioelectronics, 54*, 553-557.

Hossain, M. A., AL-Raqmi, K. A. S., Al-Mijizy, Z. H., Weli, A. M., & Al-Riyami, Q. (2013). Study of total phenol, flavonoids contents and phytochemical screening of various leaves crude extracts of locally grown Thymus vulgaris. *Asian Pacific journal of tropical biomedicine, 3*(9), 705-710.

Hossain, Z., & Komatsu, S. (2013). Contribution of proteomic studies towards understanding plant heavy metal stress response. *Frontiers in Plant Science, 3*, 310.

Hossain, Z., Makino, T., & Komatsu, S. (2012). Proteomic study of β-aminobutyric acid-mediated cadmium stress alleviation in soybean. *Journal of proteomics, 75*(13), 4151-4164.

Howell, L. L. (2013). Compliant mechanisms. In *21st century kinematics* (pp. 189-216). Springer, London.

Huttner, S. (2012). *Further development and application of the 3D microclimate simulation ENVI-met* (Doctoral dissertation, Universitätsbibliothek Mainz).

Imran, M., Garbe-Schönberg, D., Neumann, G., Boelt, B., & Mühling, K. H. (2017). Zinc distribution and localization in primed maize seeds and its translocation during early seedling development. *Environmental and Experimental Botany, 143*, 91-98.

Jabeen, R., Ahmad, A., & Iqbal, M. (2009). Phytoremediation of heavy metals: physiological and molecular mechanisms. *The Botanical Review, 75*(4), 339-364.

Järup, L. (2003). Hazards of heavy metal contamination. *British medical bulletin, 68*(1), 167-182.

Jewell, N. A., Cline, T., Mertz, S. E., Smirnov, S. V., Flaño, E., Schindler, C., ... & Durbin, J. E. (2010). Lambda interferon is the predominant interferon induced by influenza A virus infection in vivo. *Journal of virology, 84*(21), 11515-11522.

Kuriakose, S. V., & Prasad, M. N. V. (2008). Cadmium stress affects seed germination and seedling growth in Sorghum bicolor (L.) Moench by changing the activities of hydrolyzing enzymes. *Plant growth regulation, 54*(2), 143-156.

Li, J., Monroe, W., Ritter, A., Galley, M., Gao, J., & Jurafsky, D. (2016). Deep reinforcement learning for dialogue generation. *arXiv preprint arXiv:1606.01541.*

Li, M., Ahammed, G. J., Li, C., Bao, X., Yu, J., Huang, C., ... & Zhou, J. (2016). Brassinosteroid ameliorates zinc oxide nanoparticles-induced oxidative stress by improving antioxidant potential and redox homeostasis in tomato seedling. *Frontiers in plant science, 7*, 615.

Li, M., Welti, R., & Wang, X. (2006). Quantitative profiling of Arabidopsis polar glycerolipids in response to phosphorus starvation. Roles of phospholipases Dζ1 and Dζ2 in phosphatidylcholine hydrolysis and digalactosyldiacylglycerol accumulation in phosphorus-starved plants. *Plant Physiology, 142*(2), 750-761.

Liu, J. X., & Howell, S. H. (2016). Managing the protein folding demands in the endoplasmic reticulum of plants. *New Phytologist, 211*(2), 418-428.

Ma, J. F., Goto, S., Tamai, K., & Ichii, M. (2001). Role of root hairs and lateral roots in silicon uptake by rice. *Plant Physiology, 127*(4), 1773-1780.

MAESHIMA, M. (1991). H+-translocating inorganic pyrophosphatase of plant vacuoles Inhibition by Ca2+, stabilization by Mg2+ and immunological comparison with other inorganic pyrophosphatases. *European journal of biochemistry, 196*(1), 11-17.

Manara, A. (2012). Plant responses to heavy metal toxicity. In *Plants and heavy metals* (pp. 27-53). Springer, Dordrecht.

Manara, A. (2012). Plant responses to heavy metal toxicity. In *Plants and heavy metals* (pp. 27-53). Springer, Dordrecht.

Mani, C. (2019). Impacts Of Foliar Feeding Of Zinc And Polyamine On The Productivity Of Wheat. *International Journal of Emerging Technologies and Innovative Research,* 6 **(2)**, 354-372.

Mani, C., P. K. (2020). Evaluation of antimony induced biochemical shift in mustard. *Plant Archives, 20*(2), 3493-3498.

McLaughlin, M. J., Parker, D. R., & Clarke, J. M. (1999). Metals and micronutrients–food safety issues. *Field crops research, 60*(1-2), 143-163.

Mejstrik, V., & Benecke, U. (1969). The ectotrophic mycorrhizas of Alnus viridis (Chaix) DC and their significance in respect to phosphorus uptake. *New Phytologist, 68*(1), 141-149.

Muto, A., Ohkura, M., Kotani, T., Higashijima, S. I., Nakai, J., & Kawakami, K. (2011). Genetic visualization with an improved GCaMP calcium indicator reveals spatiotemporal activation of the spinal motor neurons in zebrafish. *Proceedings of the National Academy of Sciences, 108*(13), 5425-5430.

Pena, L. B., Pasquini, L. A., Tomaro, M. L., & Gallego, S. M. (2007). 20S proteasome and accumulation of oxidized and ubiquitinated proteins in maize leaves subjected to cadmium stress. *Phytochemistry, 68*(8), 1139-1146.

Peng, B., & Kimmel, M. (2005). simuPOP: a forward-time population genetics simulation environment. *Bioinformatics, 21*(18), 3686-3687.

Pines, J., & Lindon, C. (2005). Proteolysis: anytime, any place, anywhere?. *Nature cell biology, 7*(8), 731-735.

Pochodylo, A. L., & Aristilde, L. (2017). Molecular dynamics of stability and structures in phytochelatin complexes with Zn, Cu, Fe, Mg, and Ca: Implications for metal detoxification. *Environmental Chemistry Letters, 15*(3), 495-500.

Pomponi, M., Censi, V., Di Girolamo, V., De Paolis, A., Di Toppi, L. S., Aromolo, R., ... & Cardarelli, M. (2006). Overexpression of Arabidopsis phytochelatin synthase in tobacco plants enhances Cd 2+ tolerance and accumulation but not translocation to the shoot. *Planta, 223*(2), 180-190.

Pratt, W. B., & Toft, D. O. (2003). Regulation of signaling protein function and trafficking by the hsp90/hsp70-based chaperone machinery. *Experimental biology and medicine, 228*(2), 111-133.

Rauser, W. E. (1995). Phytochelatins and related peptides. Structure, biosynthesis, and function. *Plant physiology, 109*(4), 1141.

Ritchie, E. G., & Johnson, C. N. (2009). Predator interactions, mesopredator release and biodiversity conservation. *Ecology letters, 12*(9), 982-998.

Rodríguez-Celma, J., Rellán-Álvarez, R., Abadía, A., Abadía, J., & López-Millán, A. F. (2010). Changes induced by two levels of cadmium toxicity in the 2-DE protein profile of tomato roots. *Journal of Proteomics, 73*(9), 1694-1706.

Ruschak, A. M., Slassi, M., Kay, L. E., & Schimmer, A. D. (2011). Novel proteasome inhibitors to overcome bortezomib resistance. *Journal of the National Cancer Institute, 103*(13), 1007-1017.

Sadanandom, A., Bailey, M., Ewan, R., Lee, J., & Nelis, S. (2012). The ubiquitin–proteasome system: central modifier of plant signalling. *New Phytologist, 196*(1), 13-28.

Sarry, J. E., Kuhn, L., Ducruix, C., Lafaye, A., Junot, C., Hugouvieux, V., ... & Bourguignon, J. (2006). The early responses of Arabidopsis thaliana cells to cadmium exposure explored by protein and metabolite profiling analyses. *Proteomics, 6*(7), 2180-2198.

Schäfer, J., & Puchelt, H. (1998). Platinum-Group-Metals (PGM) emitted from automobile catalytic converters and their distribution in roadside soils. *Journal of Geochemical Exploration, 64*(1-3), 307-314.

Schöffl, F., Prandl, R., & Reindl, A. (1999). Molecular responses to heat stress. *Molecular responses to cold, drought, heat and salt stress in higher plants, 83*, 93.

Schröder, M., & Kaufman, R. J. (2005). ER stress and the unfolded protein response. *Mutation Research/Fundamental and Molecular Mechanisms of Mutagenesis, 569*(1-2), 29-63.

Shahzad, Z., Gosti, F., Frérot, H., Lacombe, E., Roosens, N., Saumitou-Laprade, P., & Berthomieu, P. (2010). The five AhMTP1 zinc transporters undergo different evolutionary fates towards adaptive evolution to zinc tolerance in Arabidopsis halleri. *PLoS Genetics, 6*(4), e1000911.

Sharma, S., & Sultana, S. (2004). Effect of Hibiscus rosa sinensis extract on hyperproliferation and oxidative damage caused by benzoyl peroxide and ultraviolet radiations in mouse skin. *Basic & clinical pharmacology & toxicology, 95*(5), 220-225.

Singh, B., Sharma, D. K., Kumar, R., & Gupta, A. (2009). Controlled release of the fungicide thiram from starch–alginate–clay based formulation. *Applied Clay Science, 45*(1-2), 76-82.

Singh, S., Parihar, P., Singh, R., Singh, V. P., & Prasad, S. M. (2016). Heavy metal tolerance in plants: role of transcriptomics, proteomics, metabolomics, and ionomics. *Frontiers in plant science, 6*, 1143.

Singh, S., Parihar, P., Singh, R., Singh, V. P., & Prasad, S. M. (2016). Heavy metal tolerance in plants: role of transcriptomics, proteomics, metabolomics, and ionomics. *Frontiers in plant science, 6*, 1143.

Sørensen, J. G., Kristensen, T. N., & Loeschcke, V. (2003). The evolutionary and ecological role of heat shock proteins. *Ecology letters, 6*(11), 1025-1037.

Srivastava, S., Srivastava, A. K., Suprasanna, P., & D'souza, S. F. (2013). Identification and profiling of arsenic stress-induced microRNAs in Brassica juncea. *Journal of experimental botany, 64*(1), 303-315.

Stone, N. J., Robinson, J. G., Lichtenstein, A. H., Bairey Merz, C. N., Blum, C. B., Eckel, R. H., ... & Wilson, P. W. (2014). 2013 ACC/AHA guideline on the treatment of blood cholesterol to reduce atherosclerotic cardiovascular risk in adults: a report of the American College of Cardiology/American Heart Association Task Force on Practice Guidelines. *Journal of the American College of Cardiology, 63*(25 Part B), 2889-2934.

Sun, C. W., & Callis, J. (1997). Independent modulation of Arabidopsis thaliana polyubiquitin mRNAs in different organs and in response to environmental changes. *The Plant Journal, 11*(5), 1017-1027.

Talke, I. N., Hanikenne, M., & Krämer, U. (2006). Zinc-dependent global transcriptional control, transcriptional deregulation, and higher gene copy number for genes in metal homeostasis of the hyperaccumulator Arabidopsis halleri. *Plant physiology, 142*(1), 148-167.

Tamás, M. J., Fauvet, B., Christen, P., & Goloubinoff, P. (2018). Misfolding and aggregation of nascent proteins: a novel mode of toxic cadmium action in vivo. *Current genetics, 64*(1), 177-181.

Tamás, M. J., Sharma, S. K., Ibstedt, S., Jacobson, T., & Christen, P. (2014). Heavy metals and metalloids as a cause for protein misfolding and aggregation. *Biomolecules, 4*(1), 252-267.

Tomas, M., Tinti, A., Bofill, R., Capdevila, M., Atrian, S., & Torreggiani, A. (2016). Comparative Raman study of four plant metallothionein is forms: insights into their Zn (II) clusters and protein conformations. *Journal of inorganic biochemistry, 156*, 55-63.

Uibopuu, A., Moora, M., Öpik, M., & Zobel, M. (2012). Temperate forest understorey species performance is altered by local arbuscular mycorrhizal fungal communities from stands of different successional stages. *Plant and Soil, 356*(1), 331-339.

van der Zaal, B. J., Neuteboom, L. W., Pinas, J. E., Chardonnens, A. N., Schat, H., Verkleij, J. A., & Hooykaas, P. J. (1999). Over expression of a novel Arabidopsis gene related to putative zinc-transporter genes from animals can lead to enhanced zinc resistance and accumulation. *Plant physiology, 119*(3), 1047-1056.

Verbruggen, N., Hermans, C., & Schat, H. (2009). Molecular mechanisms of metal hyperaccumulation in plants. *New phytologist, 181*(4), 759-776.

Vert, M. (2009). Degradable and bioresorbable polymers in surgery and in pharmacology: beliefs and facts. *Journal of Materials Science: Materials in Medicine, 20*(2), 437-446.

Vierstra, R. D. (2009). The ubiquitin–26S proteasome system at the nexus of plant biology. *Nature Reviews Molecular Cell Biology, 10*(6), 385-397.

Wagner, G. J. (1993). Accumulation of cadmium in crop plants and its consequences to human health. *Advances in agronomy, 51*, 173-212.

Wang, Q., Rowan, M. J., & Anwyl, R. (2004). β-amyloid-mediated inhibition of NMDA receptor-dependent long-term potentiation induction involves activation of microglia and stimulation of inducible nitric oxide synthase and superoxide. *Journal of Neuroscience, 24*(27), 6049-6056.

Whiting, S. N., Leake, J. R., McGRATH, S. P., & Baker, A. J. (2000). Positive responses to Zn and Cd by roots of the Zn and Cd hyperaccumulator Thlaspi caerulescens. *New Phytologist, 145*(2), 199-210.

Yaman, & Kumar, P. (2019). Effect Of Azospirillum And Foliar Application Of Iron (Fe) On Growth And Yield Attributes Of Barley. *International Journal of Emerging Technologies and Innovative Research,* 6 **(2)**, 373-393.

Yaman, & Kumar, P. (2019). Trichoderma Mediated Mitigation of Lead Toxicity in Mustard with Special Reference to its Yield and Attributes. *International Journal of Emerging Technologies and Innovative Research,* 6 **(2)**, 311-324.

Yaman, & Kumar, P. (2019). Trichoderma Mediated Mitigation of Lead Toxicity In Mustard. *International Journal of Emerging Technologies and Innovative Research,* 6 **(2)**, 343-353.

Yang, Z. (2007). PAML 4: phylogenetic analysis by maximum likelihood. *Molecular biology and evolution, 24*(8), 1586-1591.

Yong, J., & Zhou, X. Y. (1999). *Stochastic controls: Hamiltonian systems and HJB equations* (Vol. 43). Springer Science & Business Media.

Yuan, Y. X., Zhang, J., Wang, D. W., & Ling, H. Q. (2005). AtbHLH29 of Arabidopsis thaliana is a functional ortholog of tomato FER involved in controlling iron acquisition in strategy I plants. *Cell research, 15*(8), 613-621.

Zenk, M. H. (1996). Heavy metal detoxification in higher plants-a review. *Gene, 179*(1), 21-30.

Zhang, C. L., McKinsey, T. A., Chang, S., Antos, C. L., Hill, J. A., & Olson, E. N. (2002). Class II histone deacetylases act as signal-responsive repressors of cardiac hypertrophy. *Cell, 110*(4), 479-488.

Zhou, S., Okekeogbu, I., Sangireddy, S., Ye, Z., Li, H., Bhatti, S., ... & Thannhauser, T. W. (2016). Proteome modification in tomato plants upon long-term aluminum treatment. *Journal of proteome research, 15*(5), 1670-1684.

Zhu, Y. L., Pilon-Smits, E. A., Tarun, A. S., Weber, S. U., Jouanin, L., & Terry, N. (1999). Cadmium tolerance and accumulation in Indian mustard is enhanced by overexpressing γ-glutamylcysteine synthetase. *Plant physiology, 121*(4), 1169-1177.

Physiological and Molecular Mechanism of Metals Accumulation in Hyperaccumulator Plants

Chiter Mani

Department of Agronomy, School of Agriculture

Lovely Professional University, Jalandhar, Punjab, India, 144411

Email: chitermani@gmail.com

Abstract

There are approximately more than five hundred species of plants and about 0.2 per cent of angiosperm plants which shows the properties of hyperaccumulation of metals inside their body. As hyperaccumulators have the potential for a tolerance of heavy metals, therefore they are considered as model plants for the process of phytoremediation. Both the tolerance and hyperaccumulation of metals are inherited traits which are passed to their offsprings genetically. Heavy metals can be detoxified through some special kinds of cellular mechanisms which are possessed by plants which in return help the plants to tolerate stress done through the presence of excess availability of metals. Nowadays, due to the huge advancements in the field of science, gives us a lot of ideas for both the physiological and molecular discoveries of hyperaccumulation of metals and metal tolerance in plants. Although, the biological and ecological importance of metal hyperaccumulation is yet not crystal clear. There is an evolution in the functioning of adaptation of plants for drastic metallic situations and phytoremediation innovations which is due to the advancement in the physiological and molecular procedures. Because of the progression of molecular science, the information on the physiology of hyperaccumulation of metals has as of late created. Both the physiological and molecular procedures helped us to comprehend the qualitative gene guidelines framework and homeostasis of metal for plants. This review surveys both the capabilities and restrictions of molecular and physiological viewpoints and gives an expansive outline of most significant qualities which are associated to hyperaccumulation of metals and resilience, proof for the impact of overwhelming metal on the production of biomass, biochemical for plants, oxidation prevention agent protection framework and explore about the possibilities of plants for transgenesis.

Introduction

Consistently, we enjoy the staggering assorted variety of plants in the area of our surrounding. It is equally important for us to have aesthetic nature as much as it is a statement of the requirements of the endurance of the species of human. Plants possess some less degree of noticeable, physiological acknowledgement but developed a multitude of living things. In the course of recent years, our major physiological and molecular comprehension for plants is reformed, exploiting an emphasis on a couple of idiotypes. Analysts are permitted for expanding their degree inside into phylogenetic neighbourhood of the old-style model living beings and for communicating the molecular premise versatile significance, and also the development of normal phenotypic assorted variety because of ongoing advances in bioinformatics, sequencing advance, and molecular science. These improvements probably result to achieve a change of understanding of human for nature and advancement in a period portrayed by quickening human-induced worldwide alterations. This study centres around normally chosen hyperaccumulation of metal and related huge tolerance like developing the model attributes for this growing exploration discipline. Within a couple of plant species, there is an accumulation of concentration of both the transition metalloids or metals like Pb, Zn, As, Co, Tl, Cu, Ni, Sb, Mn, Se, Cd, etc. in over-the-ground biomass which is supposed to be multiple in number and magnitude order is greater than the further adjoining plants. An aggregate of around 500m plant species (◉0.2% total angiosperm) until this point in time has been accounted for such a remarkable degree of collection of possibly poisonous trace components, about for fixations can advent and likewise even surpass the convergences of the macronutrient cations K^+ and Ca^{2+}. Hyperaccumulation of metal is related with an unequivocally upgraded capacity for detoxifying the metal amassed in an over-the-ground group of cells, and subsequently along with high tolerance for metals. In the mid-nineties, the main broad investigations of the vegetation grown on soils contaminated with metals raised some fervor among biologists on account of the extremely high pressure of selection in the few locales and thus more tolerant races for metals show some evident race of advancement like in *Agrostis* and *festuca* spp of grasses. Due to this, Antonovics et al. came up for the conclusion: "Overwhelming resistance for metal in plants is most likely a case of more impressive development in real life than the moths showing melanism in the industrial areas". Therefore in resulting years, For example - nontolerant populaces plants of *Silene vulgaris* were contrasted with plants gathered from neighbourhood populaces on contrastingly sullied soils showing Ni, Cu, Zn, Co, Cd or more tolerance, individually. More tolerance is found to emerge to one specific metal greater than once in S. vulgaris and is represented by just a couple of significant gene loci, just like more races for metal tolerance of a few different species of plants. Those plants which have more tolerance for metals seek across the board metal excluder system of resistance of metal that is physiological resilience components acts for limiting the metal amassing in an over-the-ground group of cells. More accumulation of zinc was first found in 1865 in *Noccaea caerulescens*, likewise, more accumulation of nickel was first found in 1948 in *Alyssum bertolonii*/Brassicaceae which started to draw in expanding consideration in the late nineties as rates of the elective metal gatherer system. The extremely significant level of hypertolerance metal saw in species having more accumulation was one of the purposes behind this. For a deep insight of molecular and physiological hyperaccumulation and more tolerance of metals, zinc and cadmium species for hyperaccumulation such as *N. caerulescens* and *Arabidopsis halleri* have been most effectively utilized as these have a nearby phylogenetic relationship with *Arabidopsis thaliana.*

Both the hyper resistance & hyperaccumulation are scored easily and they are considered to be normally chosen, outrageous & have complex physiology of model attributes. Likewise, metals in soils include essential applicable particular weight on metalliferous soil as an intermediary & after some time becomes constantly equivalent & opens qualitatively by the means of basic diagnostic strategies. The plants for the accumulation of metals i.e. hyperaccumulation becomes an important inspiration for – utilizing such plants for research purpose, also the instruments required for physiological & molecular studies for the advancement of phytoremediation, for improving the production of the harvest supplements, advances in the bio-fortress & lastly phytomining. A significant alteration in the metal aggregation would occur when proper adjustments are made. Analysts intrigued through homeostasis of the plant metal to uncover the key arena of homeostasis of metals. A coordinated comprehension of homeostasis of plant metal is the result of the management of proper instrumentation of metal hyperaccumulators which in return brings advancements in the framework of approaches of science. The hyper tolerance of metals is a case of an outrageous abiotic stress tolerance attribute. Although, a particular bit of leeway of hyperaccumulation of metal in the abundance of physiological necessity remains tricky. Doubtlessly, hyperaccumulation of metals goes about natural barrier against herbivorous spp. as well as assault of pathogens. The availability of connections b/w stress caused by living entities & homeostasis of metal could be exploited in hyperaccumulation of metals. In general, these connections are probably going to have expansive importance in plants. At long lasts, metals for example – Cadmium and arsenic seems to have potential toxins & cancer-causing ability both in people as well as in creatures & can become a part of the food chain through the accumulation of contaminants which are going to rapidly affects expanding quantities of individuals. The endurance of hyperaccumulator plants at a particular concentration of metals inside their bodies might encourage the revelation of the solutions for crippling impacts of metal contamination on the health of human beings which are permitted by the molecular & physiological comprehension of exceptionally successful machinery for the detoxification process.

The Phenomenon of Hyperaccumulating Plants

Various plant species have developed higher resistance to trace elements which involve both metals & metalloids. The processes like translocation of trace metals from the root zone to shoot zone, restricting the entry are done in the tolerant plants which are also known as excluders. In any case, there are some uncommon plants which are also known as hyperaccumulators consolidate a higher degree of resistance to trace elements along with their gatherings. Now days, the hyperaccumulator plants increased intrigue due to their latent capacity utilization for the process of phytoremediation (Chaney et al., 1997) along with the bio-enrichment of the food crops (Broadley et al., 2007). Some extraordinary concentrations of the trace substances can be accumulated in their above-ground parts of hyperaccumulators without any noticeable poisonous side effects. Till dates, one of the most extraordinary models answered is presumably the *Sebertia acuminata* Latex that is supposed to have Nickel 26 per cent (w/w) (Jaffre et al., 1976). As said earlier approximately greater than five hundred spp. which constitutes about 0.2 per cent gymnosperms are recognized as plants for the process of hyperaccumulation of the trace elements & non-metals from which most of them are foliar fixations for hyperaccumulators for nickel i.e. seventy-five per cent (Brooks et al., 1974, 1977). There is an over-reporting in the process of hyperaccumulation (Mecnair et al., 2003). For instance, foliar fixations for a hyperaccumulation of copper & cobalt in various spp. from the belt of copper in Congo gave off an impression of being the after effect on the surface of leaves for the contamination of airborne, as opposed to taking up to the zones of root & translocation (Faucon et al., 2007).

Significant Procedures Occurring within Plants for the Major Accumulation of Trace Metals

Higher plants show certain paradox phenomena for the hyperaccumulation in plants. Following are few phenomena, for example – 1) Entry of metals over plasmalemma of the rhizosphere, 2) Translocation & loading in xylem 3) metal sequestration & detoxification both at cellular & entire plant level. Fabaceae & Cruciferae family individuals are the first to show the process of hyperaccumulation. Hyperaccumulation of metals & a significant quantity of plant species demonstrates the ability for aggregating at least two components which is present more than four hundred species of plants have been accounted so far out of which majority of plants aggregating nickel of some of them collect zinc, cobalt, copper, etc. Manganese & cadmium are also collected by some of these species. As so far now, there is no complete comprehension of symptoms of hyperaccumulation of metals in these plants. As a rule, the gathering capacity of a given particular metal is controlled by the take-up limit of plants. The significant procedures that are thought to impact the rates of a gathering of metals within plants are delineated in the following figure 1.

Fig. 1. Significant procedures occurring within plants for the major accumulation of trace metals

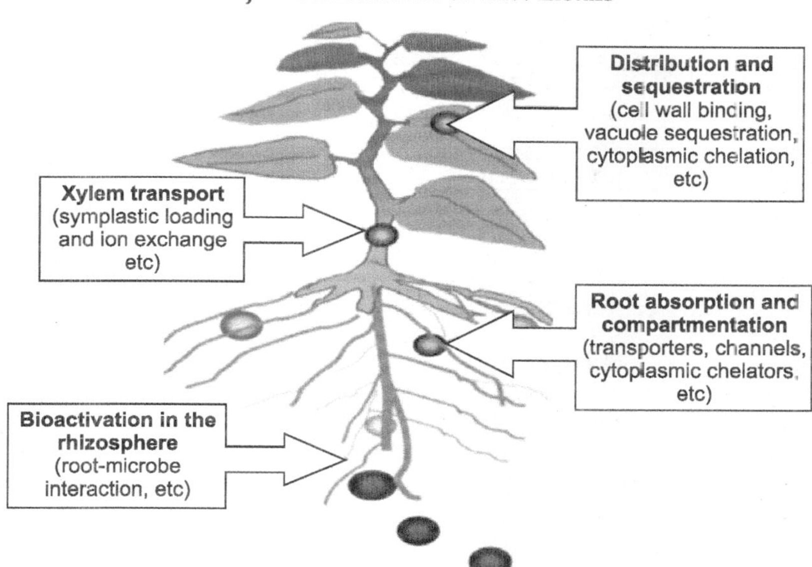

(Source:https://www.google.co.in/search?q)

Potential of Root Zone for the Trace Metal Bioactivation

There is an unprecedented capacity possessed by the hyper accumulated plants for the absorption of metals from the underground portion of soil & then gather them in shoot zone within higher & lower levels of metals. The majority of substantial metals show less portability inside the ground which results in poor consumption inside the rhizosphere, for example, the aggregation of zinc, as well as overall zinc in underground portion, shows no huge connection b/w them for *Thlaspi caerulescens.* However, there is an intact connection was noted for the aerial portion of zinc gathering & zinc levels for extraction of soil. There are certain factors like pH, a quantity of H_2O, natural & metal substances, Eh, different components in the root zone which influences both the take-up & bioavailability of overwhelming plant metals from the soil. The bioavailability of metals in the root zone is improved by roots of plants, microorganisms in soil & their collaboration with each other which involves the discharge of amino acids, phytochelatins, protons, acids which are organic in nature & other compounds.

Fig. 2. Phenomena of mobilization of heavy metal in the root zone by a microbe–root interaction

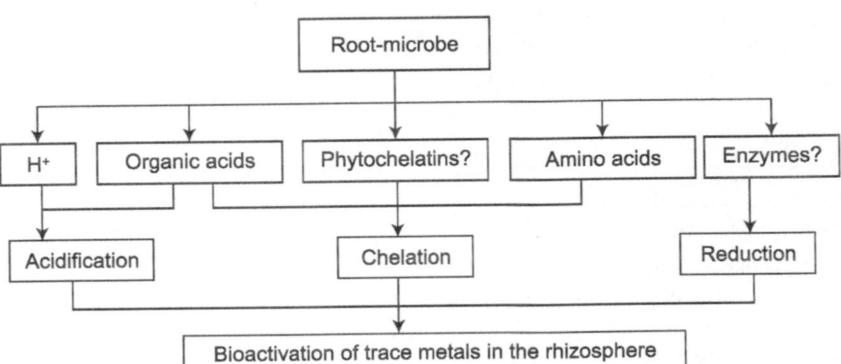

The increment of the disintegration of metal & acidification of the root zone is due to the proton emission. Under the conditions of solution culture, the discharge and development of plants of the hyperaccumulators of nickel like *Alyssum mutate* are due to the influence of pH (Bernel et al.). Although, there is no clear cut distinction for absolutely clarifying the improved disintegration of metal & the acidification of the root zone. When the plants are grown under copper & other metal contaminated soil in soil field test conditions, the pH is higher than the soil of root zone of copper collecting plant species like *Eisholtzia splendens*. The plasma layer of hydrogen pumps & hydrogen ATPase results in the working of roots for the expulsion of roots. Various factors that affect the molecular base, as well as the impacts concerning the membrane proteins, are already explored. The lack in the availability of phosphorus results in the release of citrate compounds which occur due to the improved movement of hydrogen ATPase & hydrogen passage inside the plasma layer in Lupin. Zinc & cadmium are transported through AtHMA4 which is the P-1B-ATPase of *Arabidopsis thaliana*. The activity of AtHMA4 is restricted to the cells, compassing the vessels of vascular tissues of roots. The development of the roots within the sight of poisonous groupings of zinc, cadmium & cobalt is improved by the ectopic overexpression of AtHMA4. An invalid freak is displaying a lower translocation of zinc & cadmium collection by the stem is shown by the overexpression lines of AtHMA4. The overwhelming metals & improved ingestion by roots could be activated by the emission of organic acids. The natural cadmium in mixture cadmium discharge from assimilated cadmium could be impacted through many low sub-atomic natural acids in soil. Also with the shaping of cadmium, LMWOA building, there would be an increment in the cadmium solvency. There was the availability of lower5 sub – an atomic mass of natural acids like acidic & succinate inside the root zone of cadmium – amassing genotype of wheat i.e. Kyle, yet it is not considered in Arcola genotype of wheat (Cieslinski et al.). There are some alterations in the potential of the surface of plasma membrane & enactment of hydrogen ion ATPase was seen due to aluminium induced exudation of maleate which produces aluminium resistance mechanisms. Be that as it may, differentiating end possibilities were accounted for exudation from a root in *Thlaspi caerulescens*. According to few analysts, they presumed no connection b/w hyperaccumulation of metals, as well as exudation of roots, will exist in *Thlaspi caerulescens.* Meanwhile, there is an expanding in the solvency of the held zinc in the soil of root zone of zinc accumulator *T. caerulescens* as well as alterations in conditions of soils in the root zone because of microorganisms of roots was announced by different reports. But according to our review, we came to the finding that there could be large excoriate by *Sedum alfredii* Hance of zinc and lead from polluted soil because of the exudation of zinc & cadmium hyperaccumulating plant species. There was an improvement in the hyperaccumulation & transportation of nickel through histidine – an amino acid found in *Allysm.* The exudation of root for molecular bases % biochemical portrayal of root exudation helps in the explanation of substantial metals activation in the root zone can be considered additionally. In oats, it has been demonstrated that there could be the discharge of phytosiderophores for expanding the assembly of iron, manganese, copper, and zinc in the soil under the deficiency of iron. There are broad & intensive studies for the emission as well as phytosiderophore associated sub – atomic bases. But still, in any case, there are no confirmations accessible yet whether the hyperaccumulators could discharge PCs to the root zone for expanding the take-up of metals. Under lower supply of iron & copper ions due to root reductases for increasing the take-up of iron, copper, manganese & magnesium. Although, in the rhizosphere, the portrayal of the hyperaccumulator associated rhizosphere reductase for an alteration of metal is not comprehended. The root zone has a large population of microorganisms which include mycorrhizal growths & certain bacterial species in the bulk of soil. These microorganisms colonizing the rhizosphere might essentially expand the bioavailability of different overwhelming particles of take-up of metals. The bioavailability of these metals can be altered due to the change in the oxidation & reduction reactions. Furthermore, by encouraging the expansion in the dissolvability of the zinc non – labile inside the soil, there would be the appearance of microorganisms for gathering of zinc in the stem of hyperaccumulator *T. caerulescens* which in turn upgrading its bioavailability in the plants. Also, there is stimulation for radiating natural compounds which can animate the bioavailability & encouraging the ingestion by roots for an assortment of metal particles including manganese & potentially cadmium with the help of microorganisms. The stimulation of uptake of plant supplements including metal particles & fungal cooperative affiliations could upgrade retention in the rhizosphere.

The Phenomenon of Metal Assimilation and Transportation

The flow through the plasmalemma of the metals is mediated by the transport of intracellular higher affinity locations and proteins. To calculate the proper means for developing genetically modified plants which could gather certain metals, there should be a proper understanding of processes for the transportation of metals in the plants. For the transportation of heavy metals, a few set of proteins are to be utilized in plants. The homeostasis of particles of metals is a result of CPx – type ATPases i.e. a strong heavy metal. The characteristic blockage related to macrophage protein group of proteins (Nramp) & cation diffusion facilitator (CDF) family proteins and zinc-iron permease (ZIP), etc. For the transfer of certain toxic metals like copper, lead, cadmium and zinc along with the essential metals through the cellular membranes in a vast scope of life forms, CPx – type heavy metals are recognized. For the withdrawal of a set of a charged substrate through the layers of cells, the transporters use energy in the form of ATPs and during the reaction cycle, a charged compound is created. Transporters of heavy metals utilize a common component of intramembranous cysteine-proline-cysteine. Cysteine-proline or cysteine-proline-serine (CPx) which are characterized as IB type and also referred to as CPx – ATPases. These are helpful in the transduction of heavy metals. In the category of higher plants (PAA1), Arabidopsis P-type ATPases are initially known as CPx – ATPases. Copper and cadmium transporters utilize most of the recognized CPx- type ATPases. In higher plants, the physiological role of ATPases is still not recognized. Due to the less common characteristics are there in Arabidopsis CPx – ATPases, then there would be possibilities that they have transportation of different substrates. They are assigned with the role of expelling the toxic metals through the cytoplasm by the efflux pump, also may be assisted for the heavy metal compartalization and sometimes available at various intracellular layers e.g. sequestration in ER, Golgi apparatus and vacuoles. Heavy metals which have intracellular are involved as transporters act as great condensers. Conceivably, this could happen at the transcriptional level (control on inception rates, mRNA steadiness, differential mRNA grafting) or at the post-translational level (focusing on, solidness), which has been seen in bacteria and yeast. Nramp is a novel group of related proteins that have been embroiled in the transport of divalent metal ions. The corresponding genes that code for these proteins are known as the Nramp genes (the qualities are stressed to recognize them from the proteins). In higher plants, three Nramp homologs have been distinguished in rice. Along these lines, two Arabidopsis genes indicating comparability to Nramps have likewise been recognized. Beginning outcomes recommend that Arabidopsis Nramp homologs encode functional metal transporters. Northern examination of the rice Nramp genes shows that OsNramp1 is fundamentally communicated in the roots and OsNramp2 is communicated in the leaves, whereas OsNramp3 in the two tissues of rice. This appropriate example of articulation could show that they are controlled distinctively and have a particular function in various tissues. CDF proteins have been embroiled in the transport of Zn, Co, and Cd in case of some bacteria and a few eukaryotes. 12 nucleotide sequences of A. thaliana genome are anticipated to encode individuals from this group of metal-ion transporters. They may all the more suitably be assigned as "cation efflux" transporters, because out of the cytoplasmic compartment, the proteins seem to work for the most part in cation efflux, regardless of whether over the plasma layer to the cell outside or across endomembranes into intracellular compartments, for example, the vacuole. A related Zn carrier (ZAT1) from Arabidopsis was accounted for by Van der Zaal, which has a place with another significant superfamily of metal-ion transporters known as ZIP gene family. Also, 15 individuals from this family have been recognized in the A. thaliana genome. Fe, Zn, Mn, and Disc are transported by Different individuals of ZIP, and a ZIP gene homolog, ZNT1, has been recognized in the hyperaccumulator *T. caerulescens*. Working with T. caerulescens from an alternate source population, Assuncao et al. have cloned two ZIP cDNAs (ZNT1 and ZNT2) and also found them as exceptionally communicated in root tissue. Lombi et al. have likewise cloned from *T. caerulescens* an ortholog of the *A. thaliana* iron transporter IRT1, which additionally has a place with the ZIP gene family. So to improve metal up-take following points can be considered:

- the quantity of up-take sites could be expanded,
- by adjusting the particularity of the up-take proteins,
- sequestration limit could be upgraded by expanding the rate of transport across organelles or the quantity of intracellular high-affinity binding sites.

Dissemination and Detoxification of Metals in Plants

The distributing of metals to the apoplast, such as trichone and cell walls, and ligand chelation to the metals is a general process for the detoxification of Heavy Metals in plants and the sequestration into the vacuole of the metal-ligand complex. In-plant cells of Ni and Cu / Cd hyperaccumulating plant types, cell walls can play an important role in metal detoxification. The walls of the apoplastic cell are spread out over 60–70% of Ni and/or Zn. However, it is not well understood the molecular foundations of metal detoxification by cell walls. Ligands complexation may be extracellular, e.g. through the Al tolerance process through organic acid effluxes such as maleate and root citrate. The peptide ligands like metallothioneins (MTs) and PCs are involved in intracellular complexation. MTs are abundant in cysteine polypeptides which have first been identified as Cd-binding peptides in mammalian tissues. Plants have identified multiple MT genes and also proteins. PCs are enzymatically synthesized while MTs are gene-encoded. A variety of plant species have identified PCs, such as monocots, dicots, gymnosperms, algae. PCs are a peptide family found initially in yeast. In recent years, molecular genetic analysis of yeast and arabidopsis have provided much of the current knowledge of the peptides. In cells and tissues exposed to several heavy metal ions like Cd, Ni, Cu, Zn, Ag, Hg, and Pb, and anions like arsenate and selenite PCs are produced rapidly. Just 3 amino acids, glutamine (Glu), cysteine (Cys) and the glycine (Gly) are used in PCs. These are structurally related and synthesised by GSH as tripeptide glutathione (GSH). The PCs form a structural family with increased -Glu-Cys dipeptide repetitions and followed by the Gly (-Glu-Cys)n-Gly Terminal where n is usually between 2–5, but can be as high as 11. GSH-deficient Arabidopsis mutants are PC-deficient and Cd sensitive. In cultivated cells of Silene Cucubalis, GSH-dependent PC synthase activity was established. Also, metal ions like Cd, Cu, Zn, Ag, Hg, Pb are found in an enzyme. In Arabidopsis, tomatoes, and pea related activities were found. Several studies have shown that the activity of PC synthase is regulated by metal ions at the activation level. It was found in *Vigna angularis* (Azuki beans), which are hypersensitive to Cd, that PC synthase activity plays an important role in Cd tolerance. The Azuki bean cell-suspension cultures did not synthesize Cd PCs. The cells lacked the action of the PC synthesis. External GSH was not able to improve or confer cell tolerance on PC's synthesis or to Cd that demonstrated the importance of PC synthesis in plant Cd-tolerance. Although the PC synthase activity was identified a decade ago, it has not occurred until recently that a particular gene was identified. First, two research groups isolated PC synthase genes almost simultaneously. Arabidopsis cDNA, called AtPCS1, described by Vatamaniuk et al.AtPCS1 protein expression induced an increase of the accumulation of Cd, suggesting a possible role for Cd chelation or sequestration. TaPCS1, which increases cd resistance in wild-type yeast, was described by Clemens et al. Like AtPCS1, TaPCS1 mediated Cd resistance had been linked to an increase in Cd buildup and was based on GSH. The GSH-dependent Cd tolerance mediated by AtPCS1 as well as TaPCS1 is found in the vacuoles, indicates a cytosolic position and media in vivo PC yeast biosynthesis. The overexpression of bacterial glutathione synthetase, a GSH synthesis-like enzyme, by Indian mouth-plants provides further evidence of the role of PCs in metal hyperaccumulation. Such transgenic plants have increased GSH and PC levels and improved Cd tolerance and control build-up. In transgenic yeast over expressions also improve the resistance of Cd and accumulation of plant PC synthetase. These studies show a major potential for increasing heavy metal accumulation by plants using manipulation of the GSH and PC concentrations. PC synthase activity regulation is probably the most important regulatory mechanism for PC biosynthesis pathway. Evidence showing the role of post-translation activation using plant cell cultures have been obtained. PC biosynthesis has been recorded in minutes after Cd exposure and is independent of de novo protein synthesis. PC synthase expression is independent of heavy metal exposure in *Silene cucubalis* cell cultures, tomatoes and arabidopsis. Such studies show that PC synthase is regulated by heavy metal enzyme activation. However, the induction of transcript levels in response to heavy metal exposure studies are contradictory. PC biosynthesis is likely to vary among different plant species at or after transcription or both. This shows that in various plant species the PC synthase activity is regulated differently. PC biosynthesis can also be controlled through glutathione biosynthesis. The increase in enzyme expression of GSH biosynthesis pathways resulted in the increase in PC biosynthesis and the resistance to Cd in transgenic plants of Indian mustard (*Brassica juncea*). Increment of the Cd exposure with increased G-glutamylcysteine Synthetase (G-ECS) transcript rates, which codes for the first enzyme of the pathway of PC-biosynthesis, Indian mustard plants reacted. Similarly, evidence supports g-ECS expression post-transcription regulation. Plants with metal-binders can increase metal tolerance and accumulation of proteins and peptides. These metal-binding peptides or proteins should be metal-specific preferentially to the point where sequestered metals such as Zn or Cu are only metals with toxic effect (e.g. Cd or Hg, Pb). *Littorina breviculos*, which are highly resistant to various heft metal concentrations and their metal-binding proteins (MP), are induced in presence of Cd and Zn, in a novel copper-binding protein (Cu-BP) isolated and characterized in its Asian periwinkle. The molecular weight is 11.38 kDa in the purified protein. It contains slightly less cys (eight residues) and a large volume of aromatic amino acids, tyr and phe, and is distinct from ordinary MT molluscs. The protein also includes His and Met, which are not present in the Cd-BP of L, close to MT. the patent. the patent. L's Cu-BP. Brevicula acts in the regulation of both Zn and Cu, which, under physiological conditions is an integral component of hemocyanin. This protein may be involved in the mechanism for detoxification of Cu's heavy burden.

Sequestration of Metals in Vacuoles

The main storage site for metals in yeast and plant cells is vacuole and there is some evidence that proves that in fission yeast & plants, PC-metal complexes are pumped into the vacuole. In the case of Ni resistance in yeast Schizo Saccharomyces cerevisiae, vacuolar accumulation of Ni is considered to be essential. This accumulation of Ni is done by the pH gradient that persists across the vacuolar membrane of yeast. Surprisingly, this kind of pH inclination subordinate Ni transport couldn't be seen in the root system of Ni delicate oat seedlings, and just a minor gathering of Ni could be distinguished in vacuoles secluded from leaves of Ni sensitive barley. In vacuole, the compartmentalization of metals is likewise a part of a resistant system of some metal hyperaccumulators. The Ni resistance of Ni hyperaccumulator Thlaspi goes intense is upgraded by compartmentalizing the majority of the intracellular leaf Ni into the vacuole. In T. goes intense, high-level articulation of a vacuolar metal-ion transporter TgMTP1 was proposed to represent the upgraded capacity to accumulate metal ions inside shoot vacuoles. Inside plant cells, PC-metal buildings limited by GSH or PCs are moved to the vacuole by an ABC-type carrier protein/transporter protein in the tonoplast. A similar sort of carrier/transporter is engaged with carrying GSH-conjugated anthocyanins to the vacuole. Likewise, anthocyanins can also tie metals, and have been proposed to play a role in metal sequestration. However other metal-restricting molecules that are associated with metal complexation in the vacuole are organic acids. Until this point in time, the best portrayed of the known vacuolar transporter and channel associated with metal resistance is YCF1 from Saccharomyces cerevisiae. YCF1 is a MgATP invigorated glutathione S-conjugate carrier responsible for vacuolar sequestration of organic compounds after their S-conjugation with glutathione, just as GSHmetal complexes. It catalyzes the transport of bis (glutathione) cadmium (Compact disc GS2) into vacuoles, just As-GS3 and Hg-GS2.

Physiology of Metal Hyperaccumulation in *Thlaspi Caerulescens*

It seems to include at the conservative estimate of four physiological occasions for the metal carrier constituent of hyperaccumulation in *T. caerulescens*. The metal that inflow over the root cell plasma layer is the initial restorative metal, the second diminished metal sequestration in the root vacuole, the tertiary expanded sliding into the xylem for transport to the shoots and, at long last, there is an animated metal deluge over the leaf cell plasma film and sequestration in the leaf vacuole in the fourth one. These are the summed up transmission steps in the model portrayed in Fig.3 (For transport in and over the root) that are bringing about metal hyper accumulation and Fig. 4 for transport in the leaf.

Fig. 3. Transport in and across the root

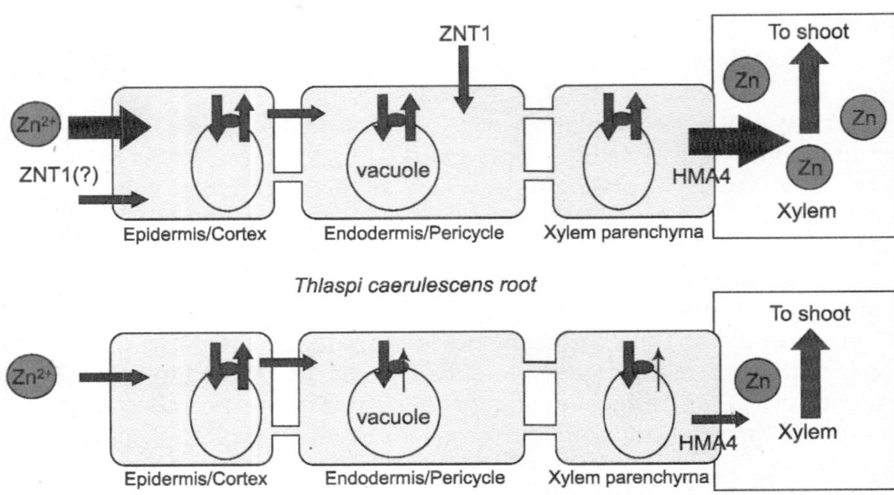

Fig. 5: Transport in the leaf

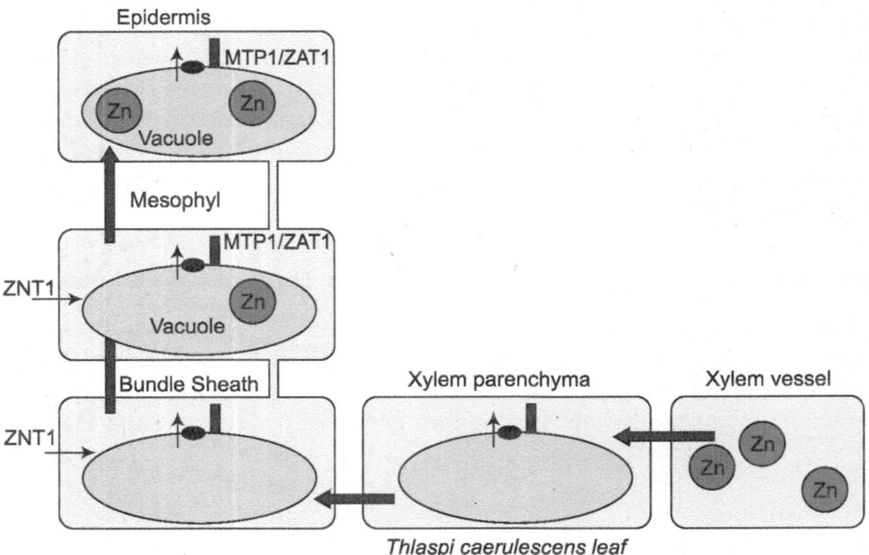

Based on a comprehensive concentration-dependent root Zn-uptake kinetic analysis for *T. caerulescens* collating with a closely related non-accumulator species, *T. arvense*, it was found that the harmony of the root Zn transporter for Zn was comparative between the two species (Km estimations of 6 mM for *T. caerulescens* and 8 mM for *T. arvense*); be that as it may, there is approx. six-folds higher Zn take-up (Vmax) in *T. caerulescens* was seen (Laset et al., 1996). This recommends that the Zn transporters are more efficient at transporting Zn in non-collector plants than in the underlying foundations of *T. caerulescens*; rather, apparently, in the root-cell plasma layer, Zn transporter thickness is a lot higher in *T. caerulescens* versus *T. arvense*. Root Zn take-up comes into view that it is controlled by plant Zn status in the two Thlaspi species. In *T. arvense,* like other non-(2007). In this investigation, it was discovered that the plants were more Cd forbearing and collected more Cd when the development of *T. caerulescens* was on high Zn levels. The opposite was likewise evident, Zn aggregated by the high Cd-developed plants are more than low Cd-developed plants. This invigorated shoot metal aggregation was related with upgraded root metal convergence, and xylem transport of these metals from the root to shoot. The creators estimated that as xylem stacking is a key advance in the hyperaccumulation procedure, the upgraded xylem stacking activated by presentation to high substantial metal levels for expanded periods may convert into improved overwhelming metal resilience, as the metals are all the more productively translocated to the shoots where exceptionally powerful metal resistance components work. At the point when these examinations are viewed as together, the guideline of articulation of micronutrient/substantial metal transporters by plant metal status is modified in an obscure manner in the hyperaccumulating Thlaspi species, bringing about the higher articulation of a few unique transporters along the metal vehicle pathway from the dirt to the shoot. When the Zn enters the root, the plant manages the Zn in one of two different ways: by putting away Zn in the root vacuole or by moving the Zn radially through the root to be stacked into the xylem for transport to the shoots. In the two *Thlaspi* species root Zn categorizaton through efflux examination was concentrated (Lasat et al., 1998). It was discovered that *T. arvense* put away approx. 2.5 occasions more Zn in the root vacuole contrasted and *T. caerulescens* and this vacuolar Zn was discharged from the vacuole twice as gradually in *T. arvense*. Longer-term considers demonstrated that as much as six-fold the amount of Zn was sequestered in the vacuoles of *T. arvense* contrasted and *T. caerulescens* over a 46-h timeframe (Lasat et al., 1998). These discoveries show that, alongside a lot higher paces of Zn passage into the root, the hyperaccumulator species likewise keep up the root Zn in a progressively versatile pool that is moved to the xylem considerably more promptly. The following key advance concerning the development of overwhelming metals from the dirt to the shoot is the stacking of Zn from xylem parenchyma into xylem vessels for translocation to the shoot. Xylem sap from *T. caerulescens* was found to contain approx. a 5-overlap higher Zn fixation contrasted and the xylem sap from *T. arvense* for plants developed on a similar Zn level. While the Zn fixation in the xylem expanded with expanded outer groupings of Zn, the proportion of xylem sap Zn focus between the two Thlaspi species remained the equivalent (Lasat et al., 1998). The significantly more noteworthy metal stacking into the xylem is a sign of metal hyperaccumulators and might be because of the movement of a P-type ATPase, HMA4, which will be talked about in more prominent detail in the sub-atomic investigations area later in this survey. These root-related vehicle steps are summed up in the model portrayed in Fig. 1. In the shoot, the capacity of Zn and Cd to elevated levels seems to require some co-appointment between various cell types. The most elevated convergences of Zn and Cd are found in accumulators; root Zn take-up increments as the plant changes from Zn adequacy to inadequacy. Thlaspi caerulescens, then again, keeps up its a lot more noteworthy root Zn deluge in both Zn-inadequate and Zn-adequate plants. Just when *T. caerulescens* is developed on high Zn focuses (supplement arrangement mind 50–1000 mM Zn) does the Vmax for Zn take-up decline, however, is as yet kept up at motion esteems higher than what is found in *T. arvense*. Another intriguing part of the connection between plant metal status and amassing in *T. caerulescens* originates from crafted by Papoyan et al. leaf epidermal cells, with focuses multiple times that of mesophyll cells (Kupper et al., 1999). This favoured stockpiling in leaf epidermal cells might be related with evasion of substantial metal harm to photosynthesis, as epidermal cells (aside from monitor cells) need chloroplasts. Although these metals are amassed to a lot more elevated levels in the epidermis, a noteworthy division of the all-out leaf metal gathering despite everything happens in the mesophyll, as a bigger portion of the leaf biomass is related with the mesophyll. It has been determined that 65–70% of the all-out Zn in the leaves is in the mesophyll (Ma et al., 2005). The vehicle ventures for metal development in the leaf, which is more ineffectively comprehended than root metal vehicle in *T. caerulescens*, are summed up in the model delineated in Fig. 2. It is assumed that a significant part of metal stockpiling includes metal chelation with natural ligands. Be that as it may, it is as yet not satisfactory which explicit ligands are included. Salt et al. (1999), utilizing X-beam ingestion, got results demonstrating that the type of Zn in *T. caerulescens* contrasted relying upon the area in the plant. It was proposed that a critical division of the Zn inside the roots is related with histidine; in the xylem sap Zn was, for the most part, found as the free hydrated Zn2þ particle with a little bit bound to natural acids, and in the shoots, Zn was most usually found related with natural acids, with littler portions found as the free particle and bound to histidine or the cell divider. It would be normal that metals, for example, Zn and Cd would be related to sulfur ligands, for example, those in cysteine, glutathione or phytochelatins. Kupper et al. (2004), in light of X-beam retention spectroscopy examination, proposed that an enormous segment of the Cd in the leaves of *T. caerulescens* will undoubtedly sulfur ligands. Then again, utilizing 113Cd-NMR, Ueno et al. (2005) introduced discoveries proposing a huge part of the leaf Cd was bound to natural acids, for the most part, malate. Malate is not an especially solid ligand for Cd, in this manner, Ueno et al. (2005) estimated that proficient vehicle of Cd into the vacuole, which as of now contains genuinely high malate focuses, would encourage the malate–Cd communication. While there were at that point a lot of malates in the vacuole, Cd treatment didn't encourage any increments in leaf malate levels. It would be normal that phytochelatins, that are accepted to assume a job in plant Cd resilience, would be engaged with Cd hyperaccumulation in *T. caerulescens*. Be that as it may, no distinctions were found in leaf or root phytochelatin levels in these two plant species in an investigation looking at phytochelatin levels in leaves of *T. caerulescens* and *T. arvense* when plants were presented to high Cd levels(Ebbs et al., 2002). An ongoing report taking a gander at the impact of hindrance of phytochelatin biosynthesis on Cd resistance demonstrated that the *T. caerulescens* ecotype, Prayon, showed an expanded affectability to Cd stress when a key protein in the phytochelatin manufactured pathway, g-glutamylcysteine synthetase, was hindered, recommending that phytochelatins may play in any event an underlying job in Cd resilience (Herna'dez-Allica et al., 2006). It is fascinating to take note of that, in a similar report, the Gange ecotype, which amasses impressively more shoot Cd than other *T. caerulescens* ecotypes (see conversation underneath), demonstrated no adjustment in Cd resistance when g-glutamylcysteine synthetase was hindered, recommending Gange may utilize an alternate Cd resilience system than other *T. caerulescens* ecotypes. The creators theorized that a low-atomic - weight thiol other than phytochelatins that have not yet been recognized may assume a job as a Cd ligand in the Gange ecotype. As referenced above, *T. caerulescens* ecotypes from the south of France, for example, Gange, gather considerably more leaf Cd than different ecotypes (with Prayon being the most generally contemplated 'other' ecotype), while the entirety of the ecotypes is around equivalent in their capacity to hyper accumulate Zn. An itemized dynamic examination of root Cd and Zn flood appeared there was an approx. 5-overlay bigger Vmax for root Cd take-up in Gange versus Prayon in present moment radiotracer contemplates, while the Km esteem for Cd take-up was not unique (Lombi et al., 2001). There is no distinction in either Vmax or Km when the examination of Zn convergence between the two ecotypes demonstrated (Lombi et al., 2001). These physiological examinations recommended there is a high-fondness Cd take-up transporter in the underlying foundations of the Gange ecotype that is absent in Prayon, which adds to the improved Cd hyperaccumulation in Gange. It isn't clear what the personality of this Cd transporter is. In any case, in a resulting study, Lombi et al. (2002) indicated that the inconvenience of iron (Fe) insufficiency on the two ecotypes explicitly initiated high-proclivity root Cd take-up in Gange. They likewise found that the root Fe take-up transporter IRT1 that has additionally been appeared to ship Cd (Connolly et al., 2002; Vert et al., 2002; Cohen et al., 2004), was instigated by Fe insufficiency in underlying foundations of Gange yet not Prayon. In the area on the atomic science of metal hyperaccumulation, these discoveries are portrayed in more detail.

Field and Soil Studies

In the course of recent years some fascinating field analysis has been performed that may demonstrate that, under specific conditions, despite its low biomass creation, it may be feasible to utilize *T. caerulescens* for the phytoremediation of certain metal-tainted soils. Certain factors, for example, season length, strategy for planting a seed, and soil pH have consequences for the Zn and Cd extraction capacities of *T. caerulescens* from the dirt (McGrath et al., 2006; Yanai et al., 2006). It was discovered that a soil pH of somewhere in the range of 5 and 6 is by all accounts ideal to encourage Cd extraction just as plant biomass creation (Yanai et al., 2006). Additionally leaving the plants over winter expanded the biomass creation just about 5-crease and expanded the measure of Zn taken up by 57 % versus that of a solitary 4-month developing period (McGrath et al., 2006). There was no distinction found in Cd gathering between the 4-and 14-month development conditions. As far as the planting of *T. caerulescens* plants, there was a .2-overlap increment in the Cd and Zn taken up from the dirt when the plants were planted as seedlings from a subplot versus that of planting seeds straightforwardly into the dirt (McGrath et al., 2006). These couple of studies demonstrates that for agronomically based experimentation on the experimental condition of metal hyperaccumulation and phytoremediation is required. Concerning metal hyperaccumulation from the dirt, it has been proposed in before examines that non-gatherers plants are not so efficient to dig the soil for metals whereas *T. caerulescens* might have the option to dig the dirt for metals more effectively. (Knight et al., 1997; McGrath et al., 1997). That is, it might be that *T. caerulescens* may approach an alternate, more firmly bound pool of soil metals than non-collecting plants, and it has been estimated this may include root natural corrosive exudation as a method for expanding metal accessibility in the rhizosphere. Nonetheless, it ought to be referenced that in hydroponic examinations, where all the Zn and Cd are accessible, *T. caerulescens* still gather substantially more Zn and Cd than *T. arvense*, even at low Zn/Cd focuses (Lasat et al., 1996; Pence et al., 2000). There have been a couple of studies proposing that rhizosphere metal accessibility assumes a job in metal hyperaccumulation. For instance, when *T. caerulescens* was developed in axenic versus non-sterile soil, the axenically developed plants amassed 25–42 % less Zn in the shoots, recommending that rhizosphere microorganisms may assume a job in metal accessibility for the hyperaccumulator (Whiting et al., 2001). Additionally, Ingwersen et al. (2006) found that *T. caerulescens* can build the accessibility of extractable Cd in the rhizosphere basically through the actively restricted desorption of Cd from soil particles. Notwithstanding, various examinations didn't find that *T. caerulescens* has any extraordinary properties concerning digging the dirt for substantial metals. Mallet et al. (2006) found the *T. caerulescens* didn't collect increasingly metal because of an expanded pool of metals in the dirt arrangement Zn and Cd than *T. arvense*, even at low Zn/Cd focuses (Lasat et al., 1996; Pence et al., 2000). There have been a couple of studies proposing that rhizosphere metal accessibility assumes to have an important role to play in metal hyperaccumulation. For instance, when T. caerulescens was developed in axenic versus non-sterile soil, the axenically developed plants amassed 25–42 % less Zn in the shoots, recommending that rhizosphere microorganisms may assume a job in metal accessibility for the hyperaccumulator (Whiting et al., 2001). Additionally, Ingwersen et al. (2006) found that *T. caerulescens* can build the accessibility of extractable Cd in the rhizosphere basically through the actively restricted desorption of Cd from soil particles. Notwithstanding, various examinations didn't find that *T. caerulescens* has any extraordinary properties concerning digging the dirt for substantial metals. Mallet et al. (2006) found the *T. caerulescens* didn't collect increasingly metal because of an expanded pool of metals in the dirt arrangement as has been recommended previously. They found that *T.caerulescens* got to similar soil metal pools as that of the non-gatherer, *Brassica napus,* under similar conditions. Concerning root exudation of natural acids, Zhao et al. (2001) found no noteworthy contrasts in natural corrosive exudation from the underlying foundations of *T. caerulescens* contrasted and *T. arvense*. In light of the proof that T. caerulescens has indistinguishable capacities from non-aggregators concerning expanding metal accessibility in the rhizosphere, it was then recommended that it may be conceivable to revise the dirt with materials, for example, EDTA or citrate to build metal dissolvability, and encourage better metal extraction by *T. caerulescens*. Nonetheless, McGrath et al. (2006) found that the expansion of specific revisions, for example, EDTA may diminish biomass creation of the hyperaccumulators, in this manner lessening their adequacy, perhaps by expanding the solvency of metals, Cu to levels that are and citrus extract were additionally taken a gander at under field conditions and appeared to have no impact on expanding metal substance or development of T. caerulescens (McGrath et al., 2006).

Genetic and Molecular Basis of Metal Hyper Accumulation

Longer than a century Metal hyperaccumulation have concerned researchers and also have a compelling occurrence. Hyper accumulators comprise a gathering of outstanding plant species and they have hereditarily acquired qualities of metals hyperaccumulation and resilience. Because of the advancement of a sub-atomic instrument, the comprehension of metal hyperaccumulation physiology has as of late improved (Verbruggen et al., 2009). Late research uncovered that Arabidopsis thaliana has eight qualities encoding individuals from the sort 1 B overwhelming metal shipping subfamily of the P-type ATPases. Three of these transporters, HMA2, HMA3 and HMA4, are firmly identified with one another and are generally comparative in grouping to the divalent overwhelming metal cation transporters of prokaryotes (Hussain et al., 2004). Quantitative mRNA in situ hybridization (QISH) in Thlaspi caerulescens shows that transporter quality articulation changes during cadmium (Cd)/zinc (Zn) hyper accumulations. For the first time, individuals from the ZIP quality family, a novel metal transporter family distinguished in quite a while are equipped for shipping an assortment of cations including Cd, Fe, Mn and Zn. By Kupper and Kochian (2010) the distinctive cell articulation designs for ZNT1 and ZNT5 were recorded. Both having a place with the ZIP group of progress metal transporters. ZNT1 may work in micronutrient nourishment while ZNT5 might be engaged with metal stockpiling related to hyperaccumulation. Cadmium instigated changes in cell articulation for ZNT1, ZNT5 and MTP1 could likewise be a piece of plants acclimatization to Cd harmfulness. The vital film protein Thlaspi goes intense metal resilience protein 1 (TgMTP1) has been proposed to assume a significant job in Zn hyperaccumulation. Significant levels of TgMTP1 at the vacuolar film in shoot tissue of the Zn hyperaccumulator T. goes intense has been assumed a significant job in both Zn resilience and upgraded Zn take-up and gathering, through the enactment of a fundamental Zn insufficiency reaction (Gustin et al., 2009). An antiporter is a necessary layer protein which is engaged with the auxiliary dynamic vehicle of at least two particles over a phospholipid film in inverse ways. An ongoing genomic examination gives an achievement in the procurement of zinc hypertolerance and hyperaccumulation characters using the contribution of Zn2+/H+ antiporter. Metal Tolerance Protein 1 (MTP1) is one of the qualities present in *A. halleri* which encodes a Zn2+/H+ antiporter associated with cytoplasmic zinc detoxification and built up a zinc resilience trademark (Shahzad et al., 2010). Plant tissues societies, for example, callus, cell suspensions and shaggy roots are broadly utilized in phytoremediation inquire about as model plant frameworks. Numerous examinations have exhibited that plant tissue societies are an incredibly significant implement in phytoremediation explore. The outcomes got from tissue societies can be utilized to foresee the reactions of plants to ecological contaminants and to improve the structure and in this manner lessen the expense of resulting ordinary entire plant tests (Doran, 2009).

Antioxidants Defense System in Metal Hyper tolerance

By the catalyst phytochelatin synthase oligomers of glutathione created and phytochelatins are oligomers of glutathione. They are determined in plants, growths, nematodes and all gatherings of green growth as well as cyanobacteria. Phytochelatins behave as chelators and are significant for substantial metal detoxification (Ha et al., 1999). They are contracted PC2 through PC11. For searching for harmful free radicals to shield themselves from the oxidant stress including that brought about by overwhelming metals plants have a few antioxidative resistance frameworks. The antioxidative guard framework falls into two general classes:

- low atomic weight oxidation inhibitor, which comprises of lipid-dissolvable layer related cell reinforcements (e.g., α-tocopherol and β-carotene) and water-solvent reductants (e.g., glutathione and ascorbate) and
- (antioxidative compounds: Superoxide Dismutase (SOD) Ascorbate Peroxidase (APX), Catalase (CAT) and Glutathione Reductase (GR). Glutathione (GSH), sulfur-containing tripeptide, is viewed as the most significant oxidation inhibitor associated with cell safeguard against toxicants (Scott et al., 1993) and work straightforwardly as a free extreme forager.

Conclusion

In recent years, metal hyperaccumulators have gotten a great consideration lately, because of the capability of utilizing these plants for phytoremediation of metal debased soils. To become effectual and cheaper on a business scale, there are some obstructions in this like mostly hyperaccumulating plants are not adjustable to severe environment and also have little biomass. A good insight has been needed to get control of these constraints by grasping the mechanisms of metal hyperaccumulation in plants. There is betterment in the physiological mechanisms of hyperaccumulation in plants because in the previous years, almost experimenters concentrating on this to achieve more developments in this; though, our conception of metal procurement and homeostasis in higher plants have been no doubt changed because of good insight of the range of atomic/cell components. The full array of genes that are possibly associated with overwhelming metal homeostasis and aggregation will be recognized with the fulfilment of the Arabidopsis genome venture, after some time followed by genome successions for different plants. By expanding plant yield and metal take-up by building normal plants with hyperaccumulating qualities the issue of low biomass Phyto remediators can be overwhelmed. Presuming that the non-local transgenic plants are utilized for phytoremediation, appropriate control of their spread must be embraced to keep away from the presentation of new weed species. Significant focuses of phytoremediation innovation are probably going to be bigger defiled regions, for example, rural grounds dirtied by mechanical or mining exercises. Extended research programs on the essential and applied procedures and issues influencing every individual class of toxins are required before critical advancement is to be made. Organization of phytoremediation under various conditions and contaminants require an assessment of field execution, which is confounded by the trouble to describe the mass equalization of metal contaminants just as the multifaceted nature of communications that happen between soil, metal, and plants. A multidisciplinary explore exertion that coordinates efforts by plant researcher, soil scientific experts, microbiologists, and natural architects is basic for more noteworthy achievement of phytoremediation as a feasible soil cleanup procedure.

References

Chand, J., & Kumar, P. (2020). Biochemical shift of mustard grown under cadmium contaminated soil. *Journal of Pharmacognosy and Phytochemistry, 9*(3), 178-183.

Chand, J., & Kumar, P. (2020). Yield attribute shift of mustard grown under cadmium contaminated soil. *Plant Archives, 20*(2), 3518-3523.

Chand, J., & Kumar, P. (February 2021). Retrieved from http;//www.ijrar.org/IJRARJFM1204.pdf. 'A Review on cost of production and cost Benefit Analysis in Rice Crop', IJRAR. *International Journal of Research and Analytical Reviews(IJRAR),* E-ISSN 2348-1269, P-ISSN 2349-5138, 8, Issue 1, Page No pp.672-679.

Chand, J., Kumar, P., Upadhyay, H. (March 2021). Retrieved from http;//www.jetir.org/papers/JETIR2103212.pdf.Geographical Distribution of Various Crops Like Cereals, Legumes, Oilseed, Vegetables, Fodder and Forages, Commercial Crop, Condiments and Species, Medical and Aromatic Plant: An Overview. *Int. J. Emerg. Technol. Innov. Res,* ISSN: 2349-5162,*8*(3), Page No.1679-1725.

Chaney, R. L., Malik, M., Li, Y. M., Brown, S. L., Brewer, E. P., Angle, J. S., & Baker, A. J. (1997). Phytoremediation of soil metals. *Current opinion in Biotechnology, 8*(3), 279-284.

ChitraMani, P. K. (2020). Evaluation of antimony induced biochemical shift in mustard. *Plant Archives, 20*(2), 3493-3498.

Ebbs, S. D., Zambrano, M. C., Spiller, S. M., & Newville, M. (2009). Cadmium sorption, influx, and efflux at the mesophyll layer of leaves from ecotypes of the Zn/Cd hyperaccumulator Thlaspi caerulescens. *New Phytologist, 181*(3), 626-636.

Ebbs, S., Lau, I., Ahner, B., & Kochian, L. (2002). Phytochelatin synthesis is not responsible for Cd tolerance in the Zn/Cd hyperaccumulator Thlaspi caerulescens (J. & C. Presl). *Planta, 214*(4), 635-640.

Faucon, M. P., Colinet, G., Mahy, G., Luhembwe, M. N., Verbruggen, N., & Meerts, P. (2009). Soil influence on Cu and Co uptake and plant size in the cuprophytes Crepidorhopalon perennis and C. tenuis (Scrophulariaceae) in SC Africa. *Plant and Soil, 317*(1), 201-212.

GuimarŃes, M. D. A., Gustin, J. L., & Salt, D. E. (2009). Reciprocal grafting separates the roles of the root and shoot in zinc hyperaccumulation in Thlaspi caerulescens. *New Phytologist, 184*(2), 323-329.

Ha, S. B., Smith, A. P., Howden, R., Dietrich, W. M., Bugg, S., O'Connell, M. J., ... & Cobbett, C. S. (1999). Phytochelatin synthase genes from Arabidopsis and the yeast Schizosaccharomyces pombe. *The Plant Cell, 11*(6), 1153-1163.

Hernández-Allica, J., Becerril, J. M., Zárate, O., & Garbisu, C. (2006). Assessment of the efficiency of a metal phytoextraction process with biological indicators of soil health. *Plant and Soil, 281*(1), 147-158.

Hernandez-Allica, J., Garbisu, C., Becerril, J. M., Barrutia, O., Garcia-Plazaola, J. I., Zhao, F. J., & McGrath, S. P. (2006). Synthesis of low molecular weight thiols in response to Cd exposure in Thlaspi caerulescens. *Plant, Cell & Environment, 29*(7), 1422-1429.

Jaffré, T., Brooks, R. R., Lee, J., & Reeves, R. D. (1976). Sebertia acuminata: a hyperaccumulator of nickel from New Caledonia. *Science, 193*(4253), 579-580.

Kerkeb, L., Mukherjee, I., Chatterjee, I., Lahner, B., Salt, D. E., & Connolly, E. L. (2008). Iron-induced turnover of the Arabidopsis IRON-REGULATED TRANSPORTER1 metal transporter requires lysine residues. *Plant Physiology, 146*(4), 1964-1973.

Knight, B., Zhao, F. J., McGrath, S. P., & Shen, Z. G. (1997). Zinc and cadmium uptake by the hyperaccumulator Thlaspi caerulescens in contaminated soils and its effects on the concentration and chemical speciation of metals in soil solution. *Plant and soil, 197*(1), 71-78.

Komatsu, M., Waguri, S., Ueno, T., Iwata, J., Murata, S., Tanida, I., ... & Chiba, T. (2005). Impairment of starvation-induced and constitutive autophagy in Atg7-deficient mice. *Journal of Cell Biology, 169*(3), 425-434.

Koren, Š., Arčon, I., Kump, P., Nečemer, M., & Vogel-Mikuš, K. (2013). Influence of CdCl 2 and CdSO 4 supplementation on Cd distribution and ligand environment in leaves of the Cd hyperaccumulator Noccaea (Thlaspi) praecox. *Plant and Soil, 370*(1), 125-148.

Küpper, H., & Kochian, L. V. (2010). Transcriptional regulation of metal transport genes and mineral nutrition during acclimatization to cadmium and zinc in the Cd/Zn hyperaccumulator, Thlaspi caerulescens (Ganges population). *New Phytologist, 185*(1), 114-129.

Küpper, H., Mijovilovich, A., Meyer-Klaucke, W., & Kroneck, P. M. (2004). Tissue- and age-dependent differences in the complexation of cadmium and zinc in the cadmium/zinc hyperaccumulator Thlaspi caerulescens (Ganges ecotype) revealed by X-ray absorption spectroscopy. *Plant physiology, 134*(2), 748-757.

Küpper, H., Zhao, F. J., & McGrath, S. P. (1999). Cellular compartmentation of zinc in leaves of the hyperaccumulator Thlaspi caerulescens. *Plant physiology, 119*(1), 305-312.

Lasat, M. M., Baker, A. J., & Kochian, L. V. (1998). Altered Zn compartmentation in the root symplasm and stimulated Zn absorption into the leaf as mechanisms involved in Zn hyperaccumulation in Thlaspi caerulescens. *Plant physiology, 118*(3), 875-883.

Lasat, M. M., Fuhrmann, M., Ebbs, S. D., Cornish, J. E., & Kochian, L. V. (1998). *Phytoremediation of a radiocesium-contaminated soil: Evaluation of cesium-137 bioaccumulation in the shoots of three plant species* (Vol. 27, No. 1, pp. 165-169). American Society of Agronomy, Crop Science Society of America, and Soil Science Society of America.

Lasat, M. M., Pence, N. S., Garvin, D. F., Ebbs, S. D., & Kochian, L. V. (2000). Molecular physiology of zinc transport in the Zn hyperaccumulator Thlaspi caerulescens. *Journal of Experimental Botany, 51*(342), 71-79.

Lee, J., Reeves, R. D., Brooks, R. R., & Jaffré, T. (1977). Isolation and identification of a citrato-complex of nickel from nickel-accumulating plants. *Phytochemistry, 16*(10), 1503-1505.

Lombi, E., Zhao, F. J., Dunham, S. J., & McGrath, S. P. (2001). Phytoremediation of heavy metal–contaminated soils: Natural hyperaccumulation versus chemically enhanced phytoextraction. *Journal of Environmental Quality, 30*(6), 1919-1926.

Ma, Z., & Michailides, T. J. (2005). Advances in understanding molecular mechanisms of fungicide resistance and molecular detection of resistant genotypes in phytopathogenic fungi. *Crop Protection, 24*(10), 853-863.

Mani, C. (2019). Impacts Of Foliar Feeding Of Zinc And Polyamine On The Productivity Of Wheat. *International Journal of Emerging Technologies and Innovative Research,* 6 **(2)**, 354-372.

Maxted, A. P., Black, C. R., West, H. M., Crout, N. M. J., McGrath, S. P., & Young, S. D. (2007). Phytoextraction of cadmium and zinc from arable soils amended with sewage sludge using Thlaspi caerulescens: development of a predictive model. *Environmental Pollution, 150*(3), 363-372.

Nakamura, S., Senoh, M., Nagahama, S. I., Iwasa, N., Yamada, T., Matsushita, T., ... & Sugimoto, Y. (1996). InGaN-based multi-quantum-well-structure laser diodes. *Japanese Journal of Applied Physics, 35*(1B), L74.

Poll, C., Ingwersen, J., Stemmer, M., Gerzabek, M. H., & Kandeler, E. (2006). Mechanisms of solute transport affect small-scale abundance and function of soil microorganisms in the detritusphere. *European Journal of Soil Science, 57*(4), 583-595.

Reuse, F. A., Khanna, S. N., & Bernel, S. (1995). Electronic structure and magnetic behavior of Ni 13 clusters. *Physical Review B, 52*(16), R11650.

Salt, D. E., Prince, R. C., Baker, A. J., Raskin, I., & Pickering, I. J. (1999). Zinc ligands in the metal hyperaccumulator Thlaspi caerulescens as determined using X-ray absorption spectroscopy. *Environmental Science & Technology, 33*(5), 713-717.

Verbruggen, N., Hermans, C., & Schat, H. (2009). Molecular mechanisms of metal hyperaccumulation in plants. *New phytologist, 181*(4), 759-776.

Vert, G., Grotz, N., Dédaldéchamp, F., Gaymard, F., Guerinot, M. L., Briat, J. F., & Curie, C. (2002). IRT1, an Arabidopsis transporter essential for iron uptake from the soil and for plant growth. *The Plant Cell, 14*(6), 1223-1233.

White, P. J., & Broadley, M. R. (2011). Physiological limits to zinc biofortification of edible crops. *Frontiers in plant science, 2*, 80.

Whiting, S. N., de Souza, M. P., & Terry, N. (2001). Rhizosphere bacteria mobilize Zn for hyperaccumulation by Thlaspi caerulescens. *Environmental science & technology, 35*(15), 3144-3150.

Xue, S. G., Chen, Y. X., Reeves, R. D., Baker, A. J., Lin, Q., & Fernando, D. R. (2004). Manganese uptake and accumulation by the hyperaccumulator plant Phytolacca acinosa Roxb.(Phytolaccaceae). *Environmental Pollution, 131*(3), 393-399.

Yaman, & Kumar, P. (2019). Effect Of Azospirillum And Foliar Application Of Iron (Fe) On Growth And Yield Attributes Of Barley. *International Journal of Emerging Technologies and Innovative Research,* 6 **(2)**, 373-393.

Yaman, & Kumar, P. (2019). Trichoderma Mediated Mitigation of Lead Toxicity in Mustard with Special Reference to its Yield and Attributes. *International Journal of Emerging Technologies and Innovative Research,* 6 **(2)**, 311-324.

Yaman, & Kumar, P. (2019). Trichoderma Mediated Mitigation Of Lead Toxicity In Mustard. *International Journal of Emerging Technologies and Innovative Research,* 6 **(2)**, 343-353.

Yanai, J., Zhao, F. J., McGrath, S. P., & Kosaki, T. (2006). Effect of soil characteristics on Cd uptake by the hyperaccumulator Thlaspi caerulescens. *Environmental Pollution, 139*(1), 167-175.

Zhao, F. J., Hamon, R. E., Lombi, E., McLaughlin, M. J., & McGrath, S. P. (2002). Characteristics of cadmium uptake in two contrasting ecotypes of the hyperaccumulator Thlaspi caerulescens. *Journal of Experimental Botany, 53*(368), 535-543.